Rare Earth
A tribute to the late Mr. Rare Earth, Professor Karl Gschneidner

Edited by

Sooraj H. Nandyala

School of Metallurgy and Materials, College of Engineering and Physical Sciences, University of Birmingham, Birmingham B15 2TT, UK

Published by **Materials Research Forum LLC**
Millersville, PA 17551, USA

Published as part of the book series
Materials Research Foundations
Volume 164 (2024)
ISSN 2471-8890 (Print)
ISSN 2471-8904 (Online)

Print ISBN 978-1-64490-304-9
eBook ISBN 978-1-64490-305-6

Distributed worldwide by

Materials Research Forum LLC
105 Springdale Lane
Millersville, PA 17551
USA
https://www.mrforum.com

Manufactured in the United States of America
10 9 8 7 6 5 4 3 2 1

Dedicated to

Late Professor S.V.J. Lakshman D.Sc.,F.N.A.Sc.
(23 -9-1933 to 21-5-1998)
**Formerly Professor of Physics and Vice-Chancellor of
S.V. University, Tirupati and A. N. University, Guntur, India**

Table of Contents

Preface

One of the inspirational men of my research (and also for the world researchers) was Prof. Karl Gschneidner. He died on April 27, 2016, (85Yrs) and he was a Distinguished Professor of Materials Science and Engineering at the Iowa State University, a Senior Metallurgist at the Ames Laboratory, and the Chief Scientist of the Critical Materials Institute. He was a member of the National Academy of Engineering and earned a lengthy list of awards for his research. He was recognised as "Mr. Rare Earth". Therefore, in the memory of his inspiration, I edited "A monograph on "Rare Earth".

The monograph is based on chapters provided by many international researchers and the editor wishes to thank the contributors of this book for

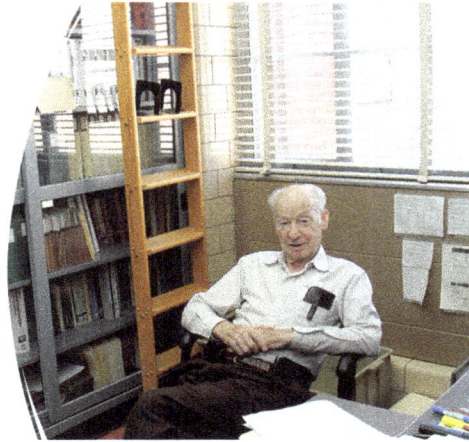

Credit: AMES Labratory/Iowa State University

their help to build a source of knowledge on Rare Earth materials. I hope this book will have a very useful and good impact on world researchers.

Sooraj H Nandyala
University of Birmingham, United Kingdom

Contributors

Adam J. Schwartz Ames National Laboratory of US DOE, Iowa State University, 2408 Pammel Dr. Ames, IA 50011-3020 USA

Aachal A. Sharma Luminescent Materials and Devices Group, Department of Physics, National Institute of Technology, Warangal 506004, Telangana, India

Aaron Akah R & D Center, Saudi Aramco, Dhahran 31311, Saudi Arabia

Anthapu Pranav Biomedical Engineering, New Jersey Institute of Technology, Newark, NJ, USA

Brian M. Walsh NASA Langley Research Center, Hampton, VA 23681, USA

Chaogang Lou School of Electronic Science and Engineering, Southeast University, P. R. China

C.S. Kamal Luminescent Materials and Devices Group, Department of Physics, National Institute of Technology, Warangal 506004, Telangana, India

D. Haranath Luminescent Materials and Devices Group, Department of Physics, National Institute of Technology, Warangal 506004, Telangana, India

D. Ranjith Kumar Department of Nanoscience and Technology, Bharathiar University, Coimbatore-641 046, Tamil Nadu, India

Farzin Amzajerdian NASA Langley Research Center, Hampton, VA 23681, USA

Jihong Geng AdValue Photonics, Inc., Tucson, AZ 85706, USA

Karima Ouannes Faculté des Sciences et de la Technologie, Université de Biskra, BP 145, Biskra, Algeria

M. Reza Dousti Unidade Acadêmica do cabo de Santo Agostinho, Universidade Federal Rural de Pernambuco, Brasil

Payal P. Pradhan Luminescent Materials and Devices Group, Department of Physics, National Institute of Technology, Warangal 506004, Telangana, India

R. Ramakrishna Reddy Department of Physics, Sri Krishnadevaraya University, Anantapur - 515 001, Andhra Pradesh, India

R.T. Rajendra Kumar Department of Nanoscience and Technology, Bharathiar University, Coimbatore-641 046, Tamil Nadu, India

Shibin Jiang AdValue Photonics, Inc., Tucson, AZ 85706, USA

Yuvaraj Haldorai Department of Nanoscience and Technology, Bharathiar University, Coimbatore-641 046, Tamil Nadu, India

Yaroslav Mudryk Ames National Laboratory of US DOE, Iowa State University, 2416 Pammel Dr, Ames, IA 50011-2416 USA

About Mr. Rare Earth
– Professor Karl. A. Gschneidner, Jr.

Karl Albert Gschneidner, Jr. (1930 – 2016) is an American scientist famous for his research work on rare earth alloys and compounds and for his societal impact on the rare earth community, which transcended the laboratory spaces and academic journals, and earned him a worldwide recognition as "Mr. Rare Earth". Born in a family of German immigrants near Detroit, Michigan, Karl graduated from the University of Detroit with a Bachelor of Science Degree in Chemistry, minoring in Physics and Mathematics. His next step was a graduate program at Iowa State University in Ames, Iowa, under the supervision of Frank H. Spedding and Adrian H. Daane, where he did his thesis work on the rare earth-carbon systems using high-purity rare earth metals produced by Ames Laboratory; Karl obtained his Ph.D. in 1957. This was his first stay at the Ames Laboratory (now the Ames National Laboratory of US Department of Energy), which brimmed with excellence and expertise in lanthanide and actinide processing following the successful conclusion of the Manhattan Project, in which Ames played a small but critical role. This environment infected Karl with a strong passion for the rare earth science, which he would carry through his life until the very end.

In the year of completing his Ph.D., Karl married the love of his life, Melba (they had four children) and they drove their car together to Los Alamos Scientific Laboratory (now Los Alamos National Laboratory), where Karl worked as a scientific staff member from 1957 to 1963. Being interested in a faculty position, he first took a short leave from his work at Los Alamos to serve as an Assistant Professor in the Physics Department at the University of Illinois at Urbana in 1962. When a position opened in the Metallurgy Department (now Department of Materials Science and Engineering) at Iowa State University, Karl returned to Ames, becoming both a Professor at the University and a group leader in the Ames Laboratory, where he remained until his passing. Even though Karl formally stepped down from teaching students later in his career, he never retired from research and continued working on the science of rare earth even at the age of 86.

Karl A. Gschneidner, Jr. is credited with many achievements, for example, the construction of a generalized phase diagram capable of predicting phase relationships in any intra-rare-earth alloy system (excluding divalent lanthanides, Eu and Yb). He, together with Vitalij K. Pecharsky, discovered a giant magnetocaloric effect near room temperature in a rare earth compound, $Gd_5Si_2Ge_2$, starting a new research field of room temperature caloric cooling, which is more efficient and environmentally friendly than a conventional vapor-compression refrigeration approach. Even more visionary, perhaps, was his work on rare earth informatics, when in 1966 he founded and maintained for the next ~30 years the highly respected Rare Earth Information Center (RIC), a computerized database of rare earth scientific literature containing over 100,000 entries – nearly every published article related to rare earth. The articles were available per request to industry and academia researchers. Additionally, Karl edited an RIC News newsletter devoted to major events of rare earth science and industry, which was distributed globally. He is also known a founding editor of the influential series Handbook on the Physics and Chemistry of Rare Earths, which regularly published the reviews dedicated to many areas of rare earth research.

In his later years, Karl used all his energy and influence to promote science. His famous speech at the hearings organized by the Investigations and Oversight Subcommittee of the US House Committee on Science and Technology brought the attention of US government officials to the insufficient - in his words "virtually zero" - funding for rare-earth research in the United States. As a result, a Critical Materials Institute was established in Ames to address problems with critical materials, on which Karl served as a Chief Scientist.

Over the years, Karl received multiple awards and was elected as a fellow of many professional societies, including the U.S. National Academy of Engineering. His publication record includes over 550 manuscripts in peer-reviewed scientific journals and 16 U.S. Patents. His main contribution to science, however, is in his constant leadership and mentorship that guided and touched many generations of students, scholars, and engineers. There were, are, and will be many distinguished scientists who advance the science of rare earth to new heights. But there will be only one Mr. Rare Earth – Professor Karl. A. Gschneidner, Jr.

Yaroslav Mudryk and Adam Schwartz
Ames National Laboratory, USA

Rare Earth - A tribute to the late Mr. Rare Earth, Professor Karl Gschneidner   Materials Research Forum LLC
Materials Research Foundations 164 (2024) 1-66   https://doi.org/10.21741/9781644903056-1

Chapter 1

Rare Earth Glass Spectroscopy and Fiber Lasers

Brian M. Walsh*,1, Karima Ouannes[2], Jihong Geng[3], Farzin Amzajerdian[1], Shibin Jiang[3]

[1]NASA Langley Research Center, Hampton, VA 23681, USA

[2]Faculté des Sciences et de la Technologie, Université de Biskra, BP 145, Biskra, Algeria

[3]AdValue Photonics, Inc., Tucson, AZ 85706, USA

* brian.m.walsh@nasa.gov / brianmwalshphd@gmail.com

Abstract

The ability to make new materials is often key to major progress in fundamental physics and numerous applications. This is the focus of our research. Therefore, we have synthesized a novel glass compositions doped with erbium. Glasses based on antimony oxide (Sb_2O_3) make one of the major classes of heavy metal oxide glasses, which have specific properties such as low phonon energy, a high non-linearity associated with a high refractive index, good mechanical properties, and better stability than that fluoride and tellurite glasses. They present the advantage to be easy to elaborate at a low price compared to single crystals. It is also possible to modulate the optical properties by moderating their chemical composition. We have demonstrated favorable spectroscopic properties in these Er antimonate glasses which compared with Er silica glass. We will present the data, and these glasses are able to accommodate other rare earth ions offering thereby a potential for developing new laser materials.

Keywords

Antimony Oxide Glass, Er Silica Glass, Er Antimonate Glass, Spectroscopic Properties, Fiber Lasers

Contents

1.   Introduction

The rare earths are comprised of the elements Y, Sc, La, Ce, Pr, Nd, Pm, Sm, Eu, Gd, Tb, Dy, Ho, Er, Tm, Yb and Lu. The last 15 make up the lanthanide series. In their trivalent state, lanthanide ions can be doped into crystals and glasses to produce solid-state laser materials, from which various laser devices can be realized. There are a variety of differences between glasses and crystals that impact their use as host materials. Crystalline materials are expensive to grow and are limited in the size they can be produced. Crystals are characterized by sharp, well-defined laser transitions with typical bandwidths of several nanometers. Glasses, on the other hand, exhibit broad, smeared out transitions with typical bandwidths of tens of nanometers. This difference results from inhomogeneous broadening in glasses where optically active ions see many different environments due to the amorphous structure. In crystals the optically active ions see more nearly uniform environments, and although they also exhibit inhomogeneous broadening, it is much weaker than in glasses.

Lanthanide doped crystals and glasses differ in their physical properties as well, which impacts the manufacturing. Glasses can be produced much more inexpensively and offer more flexibility in the size and shape. They can be drawn into fibers that are microns in diameter and meters of length, or they can be made into bulk rods that are centimeters in diameter and meters long. Glasses also have larger flexibility in their physical properties through selection of the base material. The refraction of index, for instance, can be varied from 1.5 to 2.0 and stress optic coefficients can be minimized for more thermally stable laser cavities.

There are often tradeoffs in the use of glasses or crystals for laser applications. Glasses are more useful for high energy per pulse applications because they have smaller emission cross sections and can be manufactured in large sizes, possess flexibility in their physical parameters and have large bandwidths. These characteristics are advantageous for Q-switching, amplifier applications, wavelength tuning, and production of ultra-short pulses. However, due to the strong inhomogeneous broadening in glasses, they often exhibit smaller absorption and emission cross sections. Crystals are more useful for CW and high repetition rate lasers due to their narrow emission and higher thermal conductivity. In addition, they exhibit higher cross sections than glasses, which offers performance tradeoffs for use in pulsed applications as well.

Section 2 discusses some of the fundamental aspects of ions in solids, with emphasis on materials in the glassy state. Section 3 discusses the spectroscopy of rare earth doped glasses with emphasis on Er-doped systems, specifically Er-doped silicon dioxide (SiO_2) based glass (Er:silica) and Er-doped antimony trioxide (Sb_2O_3) based glasses (Er:antimonate). Section 4 discusses rare earth doped fiber lasers with a survey of Nd, Tm, Yb systems and Er systems. Section 5 discusses the fundamentals of fiber lasers with emphasis on the advantages they offer over bulk solid-state lasers. Section 6 discusses Er-doped fiber lasers. Section 7 discusses some applications of rare earth doped glass and fiber laser devices in materials processing, medicine and remote sensing.

2.   Ions in solids

2.1   Terminology of the glassy state

Because of the different nature of the glassy state and the crystalline state, the terminology used requires some clarification [1]. All glasses are amorphous, but not all amorphous substances are glasses. Plastics, for instance, are amorphous solids, but are not glasses. The term amorphous simply means lacking any distinct crystalline structure. The glassy state is structurally analogous to a liquid state, however, the molecules that compose a glass form in a fixed but disordered arrangement, in contrast to a liquid where the molecules do not remain fixed. The important distinction is a kinetic one not a thermodynamic one, i.e., the glassy state does not have sufficient kinetic energy to overcome the potential energy barriers required for movement of the molecules past one another. Glasses and super cooled liquids are metastable phases, unlike a true thermodynamic phase in the case of a *crystalline solid*. In fact, glasses can undergo spontaneous crystallization, a process known as *devitrification*. The term *vitreous* is also used to classify glasses and is generally used synonymously with *amorphous solid*, although the later terminology is more precise. An important feature to remember about the glassy state is that it does not have a unique melting point like a crystal, but instead has a range of temperatures over which viscosity changes.

The equivalent of a crystalline solid lattice is the network or matrix in a glass. A network may be composed of single atoms, but more generally it is composed of molecular units. The arrangement of these molecular units in a glass is analogous to those in a crystal, but in the former the bond lengths and angles can vary randomly. In a glass the constituents that compose it are either *network formers* (NWF) or *network modifiers* (NWM) [2]. In silica glass, SiO_2 is the primary network former. Some other non-metal oxides are often included as additional network formers to replace some of the Si, such as Al_2O_3 (aluminosilicates) and B_2O_3 (borosilicates). Similarly, in germanate glass GeO_2 is the primary network former, with Al_2O_3 perhaps added as an additional network former. The addition of extra network formers such as Al_2O_3 in oxide glasses increases the chemical durability and suppresses the crystallization rate as well as increasing the viscosity of the glass at all temperatures. The discussion now turns to the network modifiers. As their name suggests, they serve to modify the random network by disrupting the connectivity of bonds. Common network modifiers in silica glass are alkali metal oxides such as Na_2O, Li_2O and alkaline earths such as CaO, BaO. The role of the network modifiers is to introduce some ionic bonding to the structure and interrupting the largely covalent network, giving more flexibility to the structure. Pure SiO_2 (fused silica) or SiO_2 with Al_2O_3 added has a very high softening temperature > 1400 °C, making the material more expensive and difficult to manufacture. Modifiers can aid in reducing the softening temperature. Glasses based on antimony oxide (Sb_2O_3) are one of the major classes of heavy metal oxide glasses. One advantage is that they are easy to produce at a low price compared to silica glass. Essentially, the network modifiers force the rest of the network to form around them. A two-dimensional representation of silica glass is shown in Fig. 1. It is evident that glasses may have local short-range order, but no long-range periodic order. Large ions such as the lanthanides and actinides acting as optically active centers are, in general, too large to enter as network formers, and must enter as modifiers. In oxide glasses particularly, La_2O_3 is often added as a network modifier in order to incorporate rare earth ions as dopants to make the material optically active as a laser material. For example, to dope the glass with Er, La_2O_3 would be replaced substitutionally in small quantities with Er_2O_3.

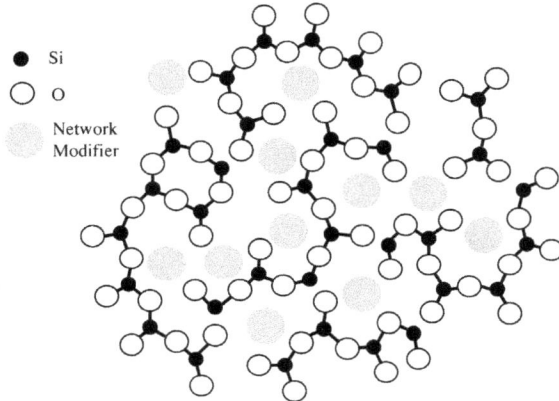

Figure 1. Two-dimensional representation of silica glass.

In taking the time here to define the glassy state, a point of departure to discuss optically active impurity centers can be approached with greater understanding. There is an analogous situation with regards to the crystalline state and the glassy state. The main difference lies in the fact that the spectroscopic properties of optically active ions in the glassy state will be affected by the disordered nature and randomness of the host. The host material is considered as a collection of charges that affect the symmetry of the local environment in which the optically active ion resides. This is true for both crystals and glasses. Some assumptions for crystals can be applied to glasses in this regard, keeping in mind that the glassy state is amorphous with no long-range order as occurs in crystals.

2.2   Impurity centers

An ion in a solid can be considered as an impurity embedded in a solid host material, usually in small quantities. These impurities form optically active centers that exhibit luminescence when pumped by an appropriate excitation source. When speaking of solids in general, a glass or crystal is implied. A glass is amorphous over a long range, but it may contain local order. A crystal has definite long-range order in a lattice structure. *Ceramics* have gained a lot of attention in recent years and represent a class of materials intermediate between glasses and crystals [3]. Glasses and crystals are insulators, distinguishing them from semiconductors, and have band gaps greater than 5 eV, which corresponds to a wavelength of photons in the deep UV. The electric field produced by the host material plays a fundamental role in determining the nature of the observed spectra of the optically active impurity ions. This is known as *Ligand Field Theory* in general and *Crystal Field Theory* in the case of an ordered periodic lattice. While glasses are amorphous and have no long-range periodic order like crystals, they may possess local order over short ranges. However, each area of local order surrounding the optically active ions is slightly different, producing slightly different line centers. The resulting effect of which is *inhomogeneous broadening* of the spectral lines from the combined aggregate of many rare earth ions with different surroundings. While inhomogeneous broadening exists in crystals, it is more pronounced in glasses, resulting in broad, undefined features in contrast to crystals, having very sharp, well-defined features.

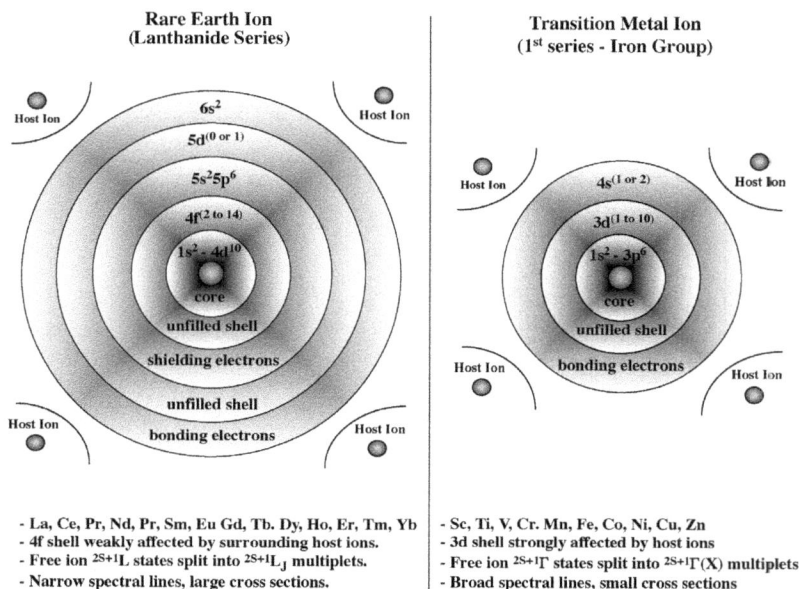

Rare Earth Ion (Lanthanide Series)

Host Ion — Host Ion
$6s^2$
$5d^{(0 \text{ or } 1)}$
$5s^2 5p^6$
$4f^{(2 \text{ to } 14)}$
$1s^2 - 4d^{10}$
core
unfilled shell
shielding electrons
unfilled shell
Host Ion — Host Ion
bonding electrons

Transition Metal Ion (1st series - Iron Group)

Host Ion — Host Ion
$4s^{(1 \text{ or } 2)}$
$3d^{(1 \text{ to } 10)}$
$1s^2 - 3p^6$
core
unfilled shell
Host Ion — bonding electrons — Host Ion

- La, Ce, Pr, Nd, Pr, Sm, Eu Gd, Tb. Dy, Ho, Er, Tm, Yb
- 4f shell weakly affected by surrounding host ions.
- Free ion ^{2S+1}L states split into $^{2S+1}L_J$ multiplets.
- Narrow spectral lines, large cross sections.

- Sc, Ti, V, Cr. Mn, Fe, Co, Ni, Cu, Zn
- 3d shell strongly affected by host ions
- Free ion $^{2S+1}\Gamma$ states split into $^{2S+1}\Gamma(X)$ multiplets
- Broad spectral lines, small cross sections

Figure 2. Rare earth and transition metal ion atomic structure.

The impurity ions, usually called dopant ions, are the optically active centers. The host is generally transparent. The impurities, dopant ions, are transition metal or lanthanide series ions. The latter are characterized by unfilled shells in the interior of the ion. The atomic structure of rare earth ions in the lanthanide series and transition metal ions of the iron group are shown in Fig. 2. These representations are not drawn to scale and are shown simply to give an overall visual representation of their structure. All lanthanide ions are characterized by a Xe core, an unfilled $4f$ shell, and filled $5s$ and $5p$ shells that screen the $4f$ shell from outside perturbing influences. This screening effect protects the optically active electrons to a great extent from the influence of the crystal field, giving the lanthanides their characteristic sharp and well-defined spectral features. In other words, they are very similar to free ion spectra. This contrasts to transition metals where the unfilled $3d$ shell is not as well screened due to only a single outer shell. The 4s electrons participate in the bonding process rather than forming filled shielding shells. The transition metal ions are characterized by broad, undefined features, although some sharp lines are observed. The R_1 and R_2 lines of $Cr:Al_2O_3$, ruby, for example, are sharp lines. Transition metals are, therefore, strongly coupled to the lattice and susceptible to the vibrational motions of the host lattice ions. These transitions occur within the bandgap of the material and have a range out to about 5 eV. In terms of wavelength, these transitions are observed from about 0.2 to 5.0 μm.

As indicated in Fig. 2 lanthanide series ions are somewhat larger than transition metal ions. Due to the large size of lanthanide ions, they cannot generally enter in the glassy state as network formers and must be included as network modifiers. Under these circumstances the 4f ions and their larger cousins, the 5f actinides, do not enter the glass host substitutionally the way they do in crystals, but instead must fit in the spaces provided in the network. The 3d transition metals, being intermediate in size, can enter the glassy state as either network formers or modifiers. At low concentrations they readily fill up the spaces as modifiers and then begin replacing formers at higher concentrations. When impurity ions begin to replace network formers, a glass material can undergo significant changes leading to crystallization.

2.3 Effect of the glassy network

To a first approximation, impurity centers in a solid can be considered as isolated centers and the host network or lattice simply provides the charges that serve to alter the environment of the center. This approximation neglects intercenter interactions and any dynamical influence of the network or lattice. Nevertheless, these charges play an important role in influencing the features of optical spectra and play an obvious role in the *Stark effect* regarding the splitting of energy levels of ions in solids. Even order terms of the crystal field influence energy level splitting, while the odd order terms mix states of opposite parity into the 4f wavefunctions to make intra-4f transitions possible. In a glass the distribution of nearest neighbors quite closely approximates the situation observed in a crystalline solid. In essence, the short-range order in glasses can be treated as locally crystalline. In the discussion that follows the principles of crystal field theory can be applied to glasses, keeping in mind that the long-range disorder affects the nature of the observed spectra. The nomenclature of 'crystal field' will be retained in the discussion that follows to refer to the effect of the network or lattice charges on the impurity ions in glasses or crystals.

The crystal field is independent of the optically active dopant ion, with symmetry determined by chemical composition of the host. In ionic solids, the dopant ions feel the influence of electrons, belonging to the crystal or glass ions, as a repulsion, and vice-versa of the nuclei as an attraction. The accumulation of these influences can be considered as a net electric field from the host ions. The crystal field plays a fundamental role in making many laser transitions between 4f energy levels possible. The energy level structure for rare earth Er^{3+} ions is shown in Fig. 3.

An obvious question is: what is the origin of these energy levels and what role does the crystal field play? To begin to answer this question, consider the free ion Hamiltonian

$$H_F = -\frac{\hbar^2}{2m}\sum_{i=1}^{N}\nabla_i^2 - \sum_{i=1}^{N}\frac{Ze^2}{r_i} + \sum_{i<j}^{N}\frac{Ze^2}{r_{ij}} + \sum_{i=1}^{N}\xi(r_i)(\mathbf{s}_i \cdot \mathbf{l}_i) \tag{1}$$

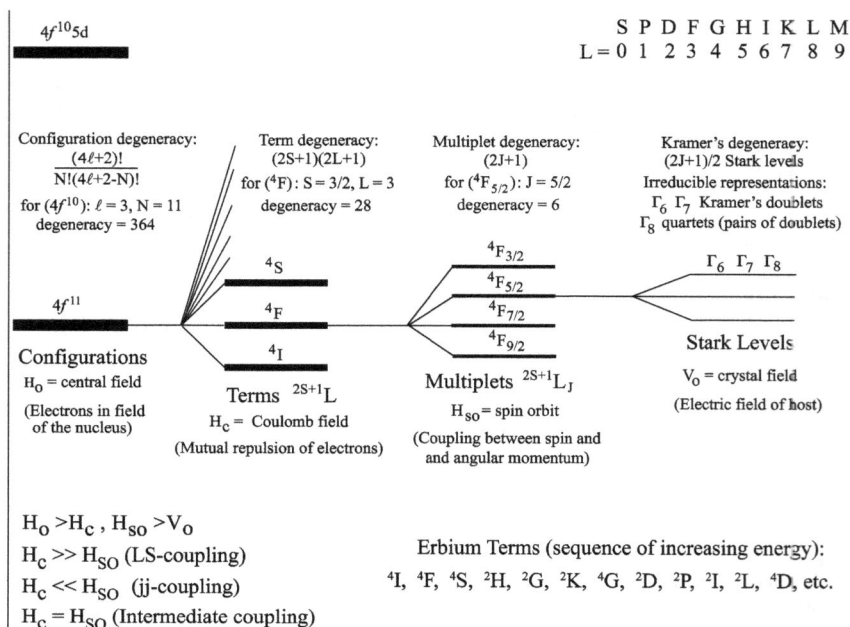

Figure 3. Energy level splitting for Er^{3+} ions.

The first term is the sum of the kinetic energies of all the electrons of a 4f ion, the second term, H_o, is the sum of the potential energy of all the electrons in the field of the nucleus. The third term, H_c, is the repulsive Coulomb potential of the interactions between pairs of electrons, and the last term, H_{SO}, is the spin-orbit interaction, describing coupling between the spin angular momentum and the orbital angular momentum. In terms of the *central-field approximation*, each electron can be considered as moving independently in the field of the nucleus and the spherically averaged potential of all the other electrons. The Coulomb interaction produces different SL terms with different energies, but independent of the total angular momentum J of the electrons. The spin-orbit interaction allows coupling between states of different SL (the spin and angular momentum of the electrons) and thus is dependent on J (the total angular momentum). In the language of quantum mechanics, the spin orbit operator does not commute with **LS** and S^2, but it does commute with J^2 and J_z. In simple terms this means that the Coulomb interaction removes degeneracy in S and L, while the spin orbit interaction removes degeneracy in J. The M_J degeneracy remains and is removed by the crystal field, V_{CF}.

The answer to the question posed earlier is that the energy levels arise from a combination of the Coulomb, spin-orbit, and crystal field interactions. The electrostatic interaction leads to ^{2S+1}L splitting on the order of $\sim 10^4$ cm^{-1}. The spin-orbit interaction splits the levels further into $^{2S+1}L_J$, separating the J states by $\sim 10^3$ cm^{-1}. Finally, the crystal field removes or partially reduces the degeneracy in J yielding an energy level separation on the order of $\sim 10^2$ cm^{-1}. The extent to which the Stark split sublevels spread is dependent on the strength of the crystal field. The larger the crystal field, the larger will be the spread of the J sublevels. So, in the free atom there exists spherical symmetry and each level has 2J+1 degeneracy. When the ion is placed in a network or lattice environment the spherical symmetry is destroyed and each level splits under the influence of the crystal field. In fact, the spherical symmetry is reduced to the point symmetry at the ion site. The degree to which the 2J+1 degeneracy is removed will depend on the point symmetry surrounding the ion. This aspect will become clear shortly. The perturbed free ion Hamiltonian is

$$H = H_F + V_{CF} \tag{2}$$

where V_{CF}, the perturbation Hamiltonian, is due to the potential provided by the network or lattice environment around the ion. Because the eigenfunctions of the free ion Hamiltonian possess complete spherical symmetry and are expressible in terms of spherical harmonics, it is natural to expand V_{CF} in terms of spherical harmonics

$$V_{CF} = \sum_{kq} A_{kq} \sum_i r_i^k Y_{kq}(\vartheta_i, \varphi_i) \tag{3}$$

where the summation over i involves all electrons, with position r_i, of the ion of interest. The A_{kq} are structural parameters in the static crystal field expansion. They depend only on the host and can be calculated in a point charge lattice sum using crystallographic data and charges of the host lattice. The point charge model assumes that the charges of the host are all point charges. The A_{kq} are then given by

$$A_{kq} = -q_e \sum_i \frac{Z_i Y_{kq}(\vartheta_i, \varphi_i)}{R_i^{k+1}} \tag{4}$$

where q_e is the electronic charge, Z_i is the size of the charge at position R_i corresponding to the surrounding atoms composing the crystal. In the calculation of matrix elements of the crystal field operator, $\langle \alpha | V_{CF} | \beta \rangle$, there results matrix elements of the form $\langle r^k \rangle = \langle nl | r^k | nl \rangle$, that represent the average value of r^k. This leads to terms of the form [4]

$$B_{kq} = A_{kq} \langle r^k \rangle \tag{5}$$

These terms enter prominently in the calculation of energy levels. They are generally determined empirically from experimental data, e.g., by fitting the measured energy levels to the theory by a

Rare Earth - A tribute to the late Mr. Rare Earth, Professor Karl Gschneidner   Materials Research Forum LLC
Materials Research Foundations 164 (2024) 1-66   https://doi.org/10.21741/9781644903056-1

least squares iterative fitting procedure. Once the point symmetry and the appropriate form of the crystal field are known, it is possible to construct the crystal field energy matrix. This matrix is then diagonalized using an estimated set of B_{kq} starting parameters. The set of theoretical energy levels are then compared to the set of experimental levels and, by an iterative fitting procedure, the B_{kq} parameters are adjusted to obtain the best overall fit to experiment. In principle this procedure can be done for impurity ions in glassy materials, but due to inhomogeneous broadening, the energy levels are difficult to determine.

2.4   Spectra of lanthanide ions in glasses

The concept of parity must be introduced here for further discussion. Parity refers to the invariance of systems under spatial reflection. In the context discussed here it refers to the wavefunctions being even (+1 parity) or odd (-1 parity) under spatial reflection. The wavefunctions for some 1s to 6f states are pictured in Fig. 4. This figure provides a nice pictorial representation of the parity of the wavefunctions for various orbitals. For instance, f-orbitals clearly have odd parity since there is a change in sign on reflection about the origin. On the other hand, d-orbitals have even parity as the sign is preserved.

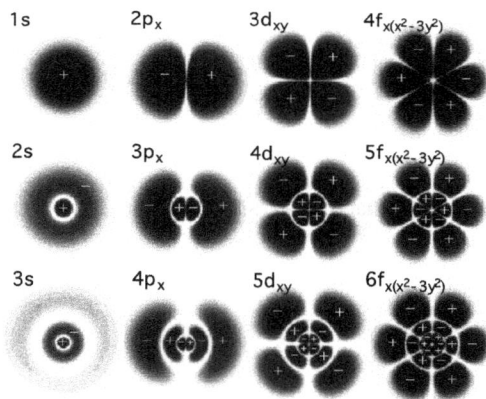

Figure 4. Wavefunctions of some s, p, d and f orbitals.

If the initial and final states (wavefunctions) have the same parity, then k in Eq. 3 must be an even number. If the initial and final states have opposite parity, then k must be odd. Otherwise, the matrix elements of V_{CF} are zero. This is a statement of *Laporte's selection rule*, which says that states with even parity can be connected by electric dipole (ED) transitions only with states of odd parity, and odd states only with even ones. Another way of saying this is that the algebraic sum of the angular momenta of the electrons in the initial and final state must change by an odd integer. So, if a matrix element of an operator of rank k connects angular momenta l and l' then the triangle condition, $l + l' \geq k \geq |l - l'|$, must hold. For 4$f$ electrons, $l = l' = 3$, and k must be even for transitions within the f^n configuration, and must have values k = 0, 2, 4, 6. However, if states of the f^n configuration are coupled to states of opposite parity in higher lying configurations, such as $4f^{(n-1)}5d$, then $l = 3$ and $l' = 2$. In this case k is odd and is limited to values k = 1, 3, 5. These odd-order terms play a key role in the *Judd-Ofelt theory* for forced electric dipole transitions in lanthanide and actinide ions in solids. The values for k and q are also limited by the point symmetry of the ion. That is, the number of nonzero terms is dependent on the point symmetry. This arises from the fact that the Hamiltonian must be invariant under operations of the point symmetry group. Thus, the crystal field must also exhibit the same symmetry as the point symmetry of the ion, since it is part of the total Hamiltonian. Equating the crystal field expansion with the expansion that has been transformed through operations of the point symmetry group gives the allowed crystal field parameters for a particular ionic point symmetry. Thus, the spherical symmetry of ions in solids is reduced to the point symmetry at the site of the ion. It is noted that terms with k = 0 and q = 0 are spherically symmetric and affect all energy levels in the same way, resulting in only a uniform shift of all levels in the configuration. The cases where q = 1 and q = 5 occur only when there is no symmetry. In general, values of q are restricted to q ≤ k, but the point symmetry introduces further restrictions and determines allowed values for q.

Lanthanide ions are characterized by a shielded $4f$-shell where atomic like transitions take place. The $4f$ states all have the same parity, that is $(-1)^{\Sigma l_i}$, where $l = 3$ for lanthanides. The question then arises, how do these transitions occur when they are forbidden by the Laporte selection rule? The answer, proposed by Van Vleck in 1937 [5], is that the $4f$ states have opposite parity mixed in from higher lying configurations, e.g., from shells above the $4f$, such as the $5d$. The d electrons have $l = 2$ and have opposite parity to f electrons. How does this *parity mixing* occur? The answer considers a distortion of the electronic motion by crystalline fields in solids, so that the selection rules for free atoms no longer apply. A typical wavefunction for an ion in a crystal can be expressed as a linear combination of states in the free ion, which are composed of sums of both even and odd parity wavefunctions, forming a complete orthonormal set of basis-functions. This allows for mixed parity states via the odd-order terms of the crystal field. As a result, electric dipole transitions can be forced through parity mixing resulting from the perturbation caused by the odd-order terms of the crystal field. Nature has found a way around Laporte's rule. Since these forced ED transitions come about as a result of a perturbation, they are orders of magnitude smaller than in free ions, but strong enough to produce a plethora of f–f rare earth transitions from which a multitude of laser transitions can be realized. This represents a remarkable circumstance of opportunity for development of rare earth lasers. The theory that predicts the intensities of such f–f rare earth transitions is the Judd-Ofelt theory, mentioned earlier. This topic has been covered extensively in the literature [6-8], and an overview of the Judd-Ofelt theory will be given and discussed further in section 3.

2.5   Optical transitions in lanthanide ions

Energy exchanges between rare earth ions and electromagnetic radiation play a fundamental role in all the processes that take place in laser materials. These interactions generate optical transitions that can be either radiative, with emission of photons, or non-radiative, without emission of photons but with emission of *phonons* or localized vibrations. Various interaction mechanisms of radiation with lanthanides is presented in this section, which aides in interpretation of experimental results for Er:glass materials.

2.5.1 Absorption, spontaneous emission, and stimulated emission

Consider the radiative transitions, interactions between the electrons of lanthanide ions (Ln^{3+}) and photons. The absorption of light by the electrons of Ln^{3+} ions induce transitions between two energy levels. The absorbed photons are those which have an energy corresponding exactly to the energy difference between these two levels, which results in the promotion of the Ln^{3-} ion to an excited state. However, an electron cannot remain indefinitely in an excited state, and subsequently returns to initial state with the emission of photons, with energies of the photons corresponding to a difference in the energies between the initial and final energy states related to the transitions. The return to the ground state can be done in various ways, which will be described later.

In the case of radiative transitions, there are *spontaneous emission* and *stimulated emission*. Spontaneous emission is a radiative process in which an excited electron decays in a certain lifetime and a photon is emitted. In contrast, for the case of stimulated emission, the incident light induces a radiative transition of an excited electron. The emitted light due to the stimulated emission has the same wavelength, phase, and direction as the incident light. Therefore, the light generated by the stimulated emission is highly monochromatic, coherent, and directional. In the stimulated emission, one incident photon generates two photons; one is the incident photon itself, and the other is an emitted photon due to the stimulated emission. As a result, the incident light is amplified by the stimulated emission.

2.5.2 Excited state absorption (ESA)

Excited State Absorption (ESA) is an *upconversion* process. This process is more likely as the lifetime of the excited level is long. The photon can be a pump photon (excitation) or a signal (emission). In the latter case, the ESA is a source of signal degradation. The positive effects of the ESA are the depopulation of the terminal level in the case of a radiative transition and the excitation of energy levels higher than that of the pump. This process allows for upconversion lasers. Fig. 5 schematically shows ESA principle.

The process can occur in a more complex way, for example, by excitation in a higher level followed by multi-phonon relaxation. This is the case of the ESA in Er^{3+} ions under infrared pumping ($\lambda = 800$ nm) which leads to an intense green emission.

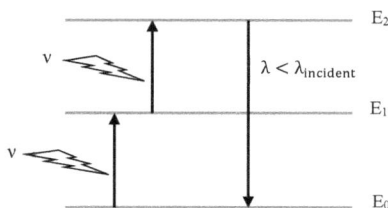

Figure 5. Excited State Absorption principle (ESA).

2.5.3 Multiphonon relaxation

For a low Ln^{3+} concentrations, or doping amount, the energy transfers between neighboring ions are weak, the probability of non-radiative deexcitation, W_{NR}, is thus limited to the probability of *multiphonon* relaxation, W_{MP}. In this case, the measured lifetime, τ_{meas}, of the emission level is defined as follows

$$\frac{1}{\tau_{meas}} = W_R + W_{MP} = W_{tot} \tag{6}$$

where, W_R = is the radiative decay rate,

W_{MP} = is the non-radiative decay rate due to the multiphonon relaxation.

When lanthanide ions in excited levels, they can relax to lower energy levels by directly transferring their surplus energy to the host network by simultaneous creation of multiple phonons. The probability of multiphonon relaxation between two energy levels is an exponential law as a function of the number of phonons required to pass the gap energy, commonly called *gap law* [9,10]:

$$W_{MP}(T, \Delta E) = Ce^{-\alpha\Delta E}\left[\frac{e^{\hbar\omega/kT}}{e^{\hbar\omega/kT} - 1}\right]^p \tag{7}$$

where, C (s^{-1}) and α (cm) are two positive constants characteristic of the host matrix and independent of the considered lanthanide ion as well as the electronic levels involved, and p is the number of phonons consumed during multiphonon relaxation. In practice, these empirical constants are determined experimentally for a given matrix by comparing the radiative lifetimes with the experimental lifetimes as a function of the difference between the energy levels. In oxide glasses, phonons have an energy of between 1100 cm^{-1} and 1400 cm^{-1}. However, in some heavy-metal oxide glasses, the phonon energy is still low and comparable to those of fluoride glasses [11].

In general, to avoid the fluorescence extinction of transitions covering the middle infrared ($\Delta E < 2500$ cm^{-1}), it is imperative to use glass hosts with low cutoff frequency (less than 400 cm^{-1}). Thus, the multiphonon relaxation process will no longer compete with the radiative processes.

2.5.4 Energy transfer

The energy transfer (ET) process due to ion-ion interaction becomes dominant for higher lanthanide ions concentration (generally greater than 10000 ppm), ET process are no longer negligible and intervene in the measurement of fluorescence decay. The total decay rate which is equal to a reciprocal of fluorescence lifetime τ_{meas}, is given by

$$\frac{1}{\tau_{meas}} = W_R + W_{MP} + W_{ET} \tag{7}$$

These process of interaction between neighboring ions are a function of the inter-ionic distance which decreases with the increase of the concentration. During the ET process, a donor ion will give up its energy to an acceptor atom without radiative emission. There are different types of interaction between ions leading to the depopulation mechanisms of an excited level:

ENERGY MIGRATION (EM): this process is most effective if the concentration of lanthanide ions is large, resulting in distances between RE ions that are short. The energy can, while migrating, encounter traps in the material. These traps can be impurities (OH-, ...) or a network defect. There is then fluorescence extinction by non-radiative trapping. This occurs when the samples are highly doped, (too concentrated) or contain impurities.

CROSS-RELAXATION (CR): the energy transfer by cross relaxation, when at the end of the transfer, the electrons of the donor ion are not on the fundamental level but on a level of energy between the initial excited state and the fundamental state. This process is involved in the case of co-doped waveguides. When this transfer takes place between two ions of the same nature, we speak of *self-quenching* [12].

UPCONVERSION: in the Er^{3+} doped or Er^{3+}/Yb^{3+} co-doped compounds, these up-conversion energy transfer processes are numerous. They result in the following,

- An addition of photons by energy transfer, called APET. Two ions are in an excited state of energy hv. One of them deexcites and gives its energy to the second ion which passes into an excited state of energy 2hv.
- Excited state absorption (ESA). An ion in the excited state of energy hv absorbs a second photon and is carried into an excited state of energy 2hv. This second process does not involve energy migration. It is in competition with APET and is not always easy to distinguish from the latter.

For an amplification at 1.5 μm by Er^{3+} ions, these upconversion processes should be avoided. Let us now summarized all the possible optical transitions resulting from the radiation-matter interaction are classified as follows,

Rare Earth - A tribute to the late Mr. Rare Earth, Professor Karl Gschneidner   Materials Research Forum LLC
Materials Research Foundations 164 (2024) 1-66   https://doi.org/10.21741/9781644903056-1

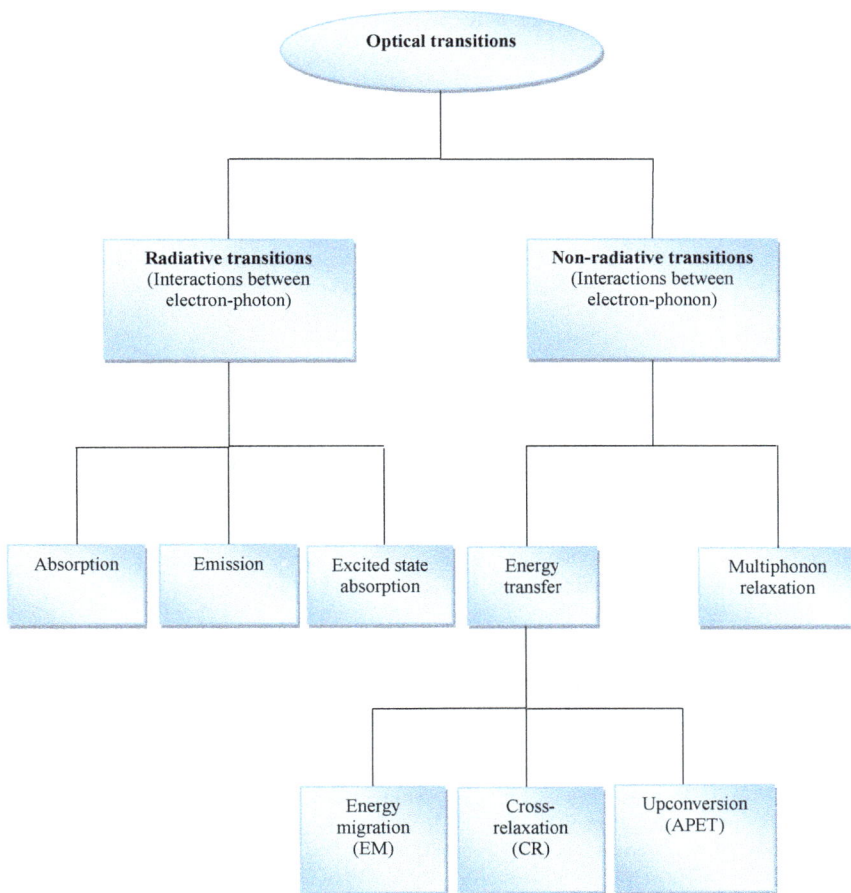

Figure 6. Classification of optical transitions in lanthanides ions.

Rare Earth - A tribute to the late Mr. Rare Earth, Professor Karl Gschneidner   Materials Research Forum LLC
Materials Research Foundations 164 (2024) 1-66   https://doi.org/10.21741/9781644903056-1

3. Rare Earth glass spectroscopy

3.1 Erbium:glass – Silica, Antimonate

Glasses doped with rare earth ions were investigated for their optical properties as potential laser materials shortly after the demonstration of the first laser in 1960. Laser oscillation in Erbium (Er) glass was first demonstrated in 1965 [13, 14]. Snitzer and Woodcock utilized Yb^{3+} to Er^{3+} energy transfer to produce laser operation at 1.54 µm in a silica-based glass. The glass they used contained 15 wt% Yb_2O_3 and 0.25 wt% Er_2O_3 dissolved in the base composition: 75 wt% SiO_2, 8 wt% Na_2O, 12 wt% K_2O, and 5 wt% BaO. In this composition with a density of 2.85 g/cm^3, the Er concentration was given to be 0.23×10^{20} ions/cm^3. This Yb:Er:glass laser was a room temperature, Yb-pumped laser behaving almost like a *three-level laser*. Researchers at the U.S. Naval Research Laboratory used a singly doped Er glass, lithium-magnesium aluminosilicate, to produce laser oscillation at 1.55 µm. The Er concentration was given as 3×10^{20} ions/cm^3. This Er:glass laser was a 77K, flashlamp pumped laser, behaving almost like a *four-level laser*. The types of lasers, such as three level and four level, will be discussed more in section 4.

Table 1. Wavelength spectral ranges.

Division name	Abbreviation	Wavelength (µm)
Ultraviolet	UV	0.01 – 0.40
Visible	VIS	0.40 – 0.75
Near infrared	NIR	0.75 – 1.4
Short wavelength infrared	SWIR	1.4 – 3.0
Mid wavelength infrared	MWIR	3 – 8
Long wavelength infrared	LWIR	8 – 15
Far infrared	FIR	15 – 1000

There are two significant Er laser transitions of interest in the infrared (IR) region, one in the short wavelength infrared (SWIR), operating on the $^4I_{13/2} \rightarrow {}^4I_{15/2}$ at ~ 1.6 µm, and another near the mid wavelength infrared (MWIR) operating on the $^4I_{11/2} \rightarrow {}^4I_{13/2}$ at ~ 2.9 µm. The transition at ~ 1.6 µm is suitable for fiber laser applications with optimal transmission (> 90%) through most glass materials, while the later transition ~ 2.9 µm is on the tailing end of declining transmission (< 50%) for most glass materials. The laser transition at ~ 1.6 µm is, therefore, more suitable for fiber lasers than the ~ 2.9 µm transition. Laser wavelengths beyond ~ 1.4 µm are characterized as eye safe since the laser light is predominantly absorbed in the cornea and lens, limiting exposure to the retina which is most susceptible to damage by a laser. Eye safe wavelengths have a maximum permissible exposure of around 3 to 4 orders of magnitude larger than is allowed for the visible and near infrared (NIR) wavelengths. This is important for applications where eye safety is an issue, such as in the field of remote sensing. The various wavelength spectra ranges are defined in table 1 for reference.

In order to understand the operation of Er based lasers in general, an understanding of the underlying spectroscopy of Er-doped materials is necessary, with the focus here on Er:glass materials. Among the many glass systems available today, this section discusses Er^{3+} doped silica

(SiO_2) based, and antimony based (Sb_2O_3) oxide glasses (*antimonate*, antimonite). Each of these Er-doped glass systems exhibit broad absorption and emission features due to the amorphous glass host, but they possess very different phonon energies due to the different composition of the glass network. Interestingly, phonons are not an accurate term when discussing the glassy state since the term phonon generally applies to collective vibrational modes of an ordered lattice. The amorphous structure implies that there is no ordered lattice. Vibrational excitation is a more nearly correct term. Clustrons have been suggested as a possible name to distinguish them from phonons [15]. Nevertheless, the use of the term phonon will be retained with this understanding in mind. The maximum phonon energy in silica and antimony-based glasses and are around 1100 and 700 cm^{-1}, respectively [16, 17]. The wide difference in the phonon energies of these glassy materials greatly influences the multiphonon relaxation rates of all the Er manifolds and affects both the manifold-to-manifold decay times and energy transfer processes.

Phonons play an important role not only in their effect on the spectroscopic dynamics, but also on the heat load they introduce for lasers, which cause many problems. Multiphonon decay depends on the number of phonons required to bridge a energy gap to the next lower lying manifold. Because the number of available phonons depends on Bose-Einstein statistics, a factor of two in the phonon energy, E_P, can make an order of magnitude difference in the Bose-Einstein occupation number, $[\exp (E_p / kT)]^{-1}$. This factor profoundly affects the decay dynamics of lanthanide ions in solids. The higher the phonon energy, the fewer the number of phonons required to bridge the gap between two manifolds, and thus, the higher the probability of the nonradiative quenching process.

The energy level structure of Er doped systems up to 2.0×10^4 cm^{-1} is shown in Fig. 7 for reference. Also shown in Fig. 7 are the various processes of nonradiative multiphonon decay (nr) and upconversion energy transfer (1 & 2). The nonradiative multiphonon decay was just discussed, and the energy transfer processes will be discussed later in section 3.3 on the dynamics of Er-doped systems. Referring to this figure, the $^4I_{15/2}$ is the ground state, and excited state manifolds of higher energy are the $^4I_{13/2}$, $^4I_{11/2}$, $^4I_{9/2}$, etc. The absorption cross sections of Er:silica, and [Karima glass] are shown in Fig. 8 and 9. The Er absorption from the $^4I_{15/2}$ manifold to the $^4G_{11/2}$, $^4F_{7/2}$, $^2H_{11/2}$, $^4F_{9/2}$, $^4I_{9/2}$, $^4I_{11/2}$, and $^4I_{13/2}$ manifolds are clearly identifiable. There are some others between $0.40 - 0.47$ μm including $^2H_{9/2}$, $^4F_{3/2}$ and $^4F_{5/2}$ not labeled in the figure due to lack of space for the labels. These spectra all exhibit the broad spectra typical of amorphous rare earth doped glass.

Rare Earth - A tribute to the late Mr. Rare Earth, Professor Karl Gschneidner   Materials Research Forum LLC
Materials Research Foundations 164 (2024) 1-66   https://doi.org/10.21741/9781644903056-1

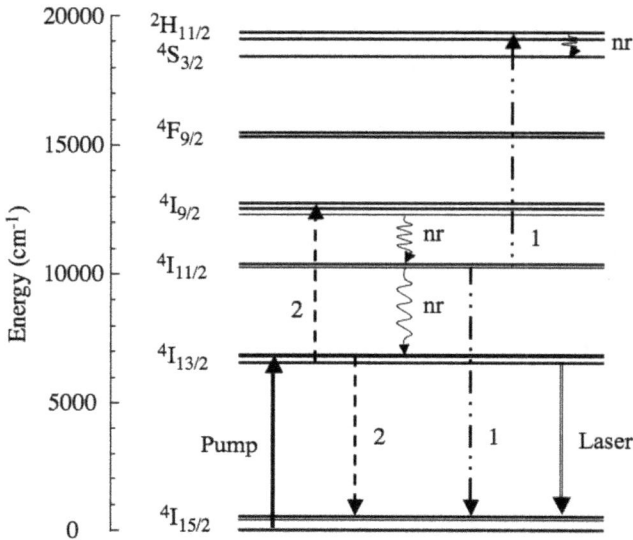

Figure 7. Energy level diagram of Er with schematic of energy transfer processes.

There are some differences in the relative strength of the cross sections among the various manifolds for each of these materials. For the purposes of laser diode pumping in the near infrared, the $^4I_{9/2}$ and $^4I_{11/2}$ manifolds of particular interest. Er:silica exhibits absorption with a peak cross section of 0.685×10^{-21} cm^2 at 799.5 nm, and 1.296×10^{-21} cm^2 at 979.5 nm. Er:antimonate has a peak cross section of 1.366×10^{-21} cm^2 at 800.0 nm, and 3.561×10^{-21} cm^2 at 976.0 nm. Since the absorption of the $^4I_{9/2}$ and $^4I_{11/2}$ manifolds is rather low, in-band pumping of the $^4I_{13/2}$ manifold can alternatively be employed. Er:silica exhibits absorption with a peak cross section of 2.219×10^{-21} cm^2 at 1485 nm. Er:antimonate glass has a peak cross section of 6.415×10^{-21} cm^2 at 1496 nm. Due to the importance of the various wavelengths of manifold absorption in Er-doped materials for laser diode pumping, an expanded view of absorption in the 0.75 to 1.05 μm region is shown in Fig. 10, and the 1.4 to 1.7 μm region in Fig. 11.

As with all laser materials, a consideration of the absorption properties alone does not provide a sufficient basis on which to assess the relative merits of one laser material over another. Thermal and mechanical properties offer various tradeoffs as well, depending on the application. In addition, the intrinsic losses of the material should also be taken into consideration. For lower absorption cross sections, glass materials with very low intrinsic loss are very important for fiber applications where long lengths are used.

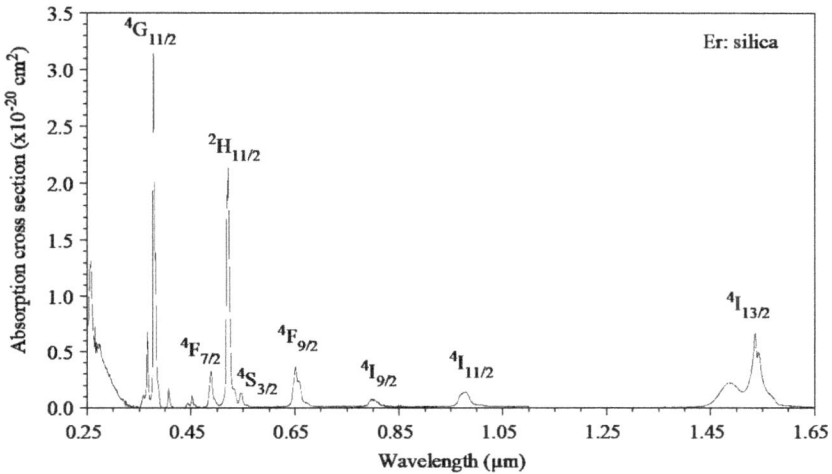

Figure 8. Absorption cross section of Er:silica glass.

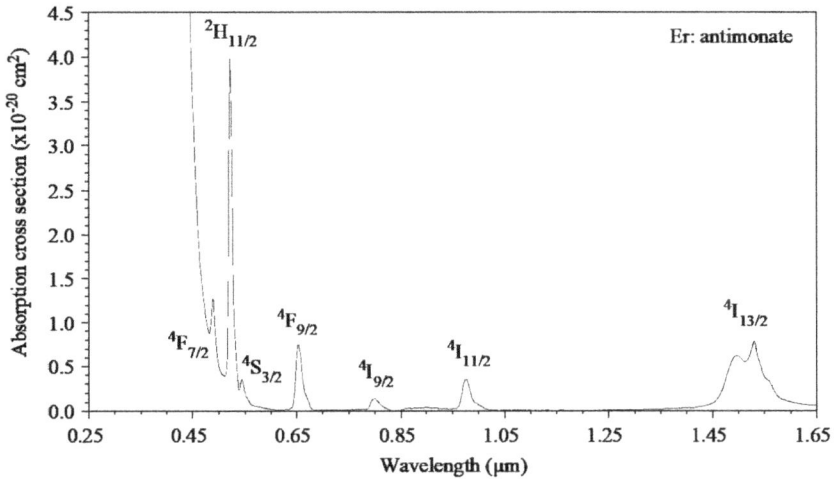

Figure 9. Absorption cross section of Er:antimonate glass.

19

Rare Earth - A tribute to the late Mr. Rare Earth, Professor Karl Gschneidner   Materials Research Forum LLC
Materials Research Foundations 164 (2024) 1-66   https://doi.org/10.21741/9781644903056-1

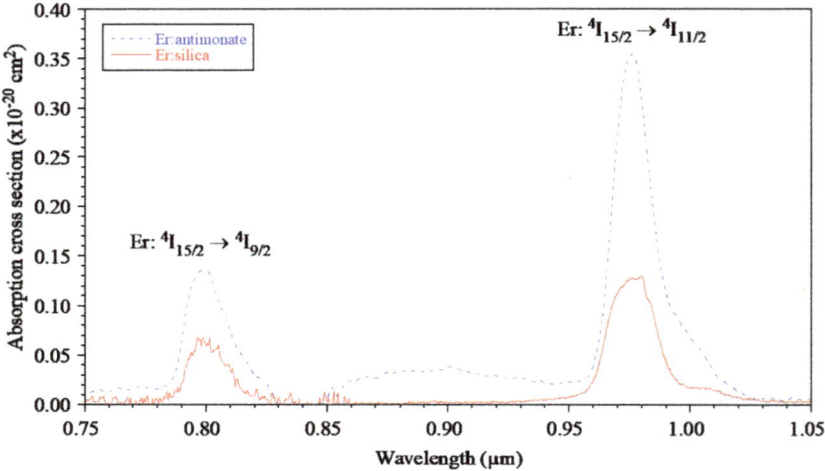

Figure 10. Absorption cross sections of the $^4I_{9/2}$ and $^4I_{11/2}$ manifolds in Er:glass.

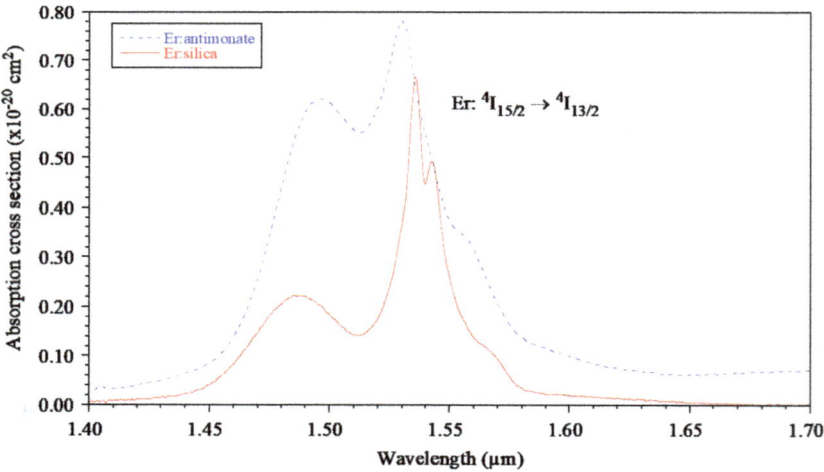

Figure 11. Absorption cross sections of the $^4I_{13/2}$ manifold in Er:glass.

Rare Earth - A tribute to the late Mr. Rare Earth, Professor Karl Gschneidner   Materials Research Forum LLC
Materials Research Foundations 164 (2024) 1-66   https://doi.org/10.21741/9781644903056-1

Another factor to consider in the assessment of laser materials is the emission cross section. This is an important factor in determining the laser gain. For 1.5 µm lasers the most important emission is the Er $^4I_{13/2} \rightarrow {}^4I_{15/2}$ transition. A comparison of the emission cross section in Er:silica and Er:antimonate glass is shown in Fig. 12. As can be seen, these are rather low gain materials, and typical of Er-doped materials in general. This has typically been an impediment to their usefulness for applications requiring high energy per pulse. Er-doped fiber lasers have traditionally been the preferred type of laser over Er-doped bulk lasers, and the fiber laser aspects will be discussed in section 5.

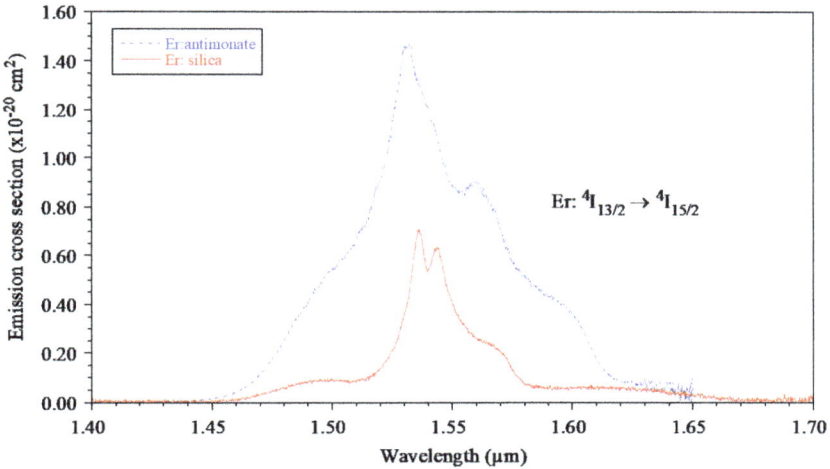

Figure 12. Emission cross sections of the $^4I_{13/2}$ manifold in Er:glass.

3.2   Judd-Ofelt analysis

The Judd-Ofelt theory [6-8] allows for the calculation of manifold-to-manifold transition probabilities, from which the *radiative lifetimes*, τ (inverse of the total transition probability), and *branching ratios*, β (fraction of the total photon flux), of emission can be determined. Simply stated, the Judd-Ofelt theory describes the intensities of lanthanide and actinide transitions in solids and solutions. The utility of the Judd-Ofelt theory is that it provides a theoretical expression for the line strength

$$S_{ED}(J;J') = \sum_{\lambda=2,4,6} \Omega_\lambda \left| \left\langle f^n[SL]J \| U^{(\lambda)} \| f^n[S'L']J' \right\rangle \right|^2 \tag{9}$$

where Ω_λ are the *Judd-Ofelt* parameters. The terms in brackets are doubly reduced matrix elements for intermediate coupling. In practice, this equation is used to determine a set of phenomenological parameters, Ω_λ (λ = 2, 4, 6), by fitting the experimental absorption or emission *linestrength* measurements, in a least squares difference sum, with the Judd-Ofelt expression, $S_{ED}(J;J')$. This is most efficiently done in the following way. First, the measured linestrength is written as a 1xN column matrix, S_j^m, and Eq. 9 for the theoretical linestrength. S_j^t is also written in component matrix form as,

$$S_j^t = \sum_{i=1}^{3} M_{ij}\Omega_i \tag{10}$$

where M_{ij} are components of a Nx3 matrix for the square matrix elements of $U^{(2)}$, $U^{(4)}$ and $U^{(6)}$. The Ω_i are components of a 1x3 matrix for the Judd-Ofelt parameters Ω_2, Ω_4 and Ω_6. Note that N represents the number of transitions to fit, which depends on the number of absorption manifolds measured. Obviously, since there are only three Judd-Ofelt parameters, N>3, and the Judd-Ofelt theory cannot be applied unless at least four measurements are available. Obviously, the more measurements that can be made the larger will be the sample size for the linear regression used to fit the Judd-Ofelt theory.

As is standard in linear regression models using a least squares approach, the sum of the squared difference is formed,

$$\sigma^2 = \sum_{j=1}^{N}\left(S_j^m - \sum_{i=1}^{3}M_{ij}\Omega_i\right)^2 \tag{11}$$

and subsequently minimized by taking the derivative with respect to Ω and setting the result equal to zero,

$$\frac{\partial\left(\sigma^2\right)}{\partial\Omega_k} = -2\sum_{j=1}^{N}M_{jk}\left(S_j^m - \sum_{i=1}^{3}M_{ij}\Omega_i\right) = 0 \tag{12}$$

and the solution to this equation is

$$\sum_{j=1}^{N}M_{jk}S_j^m = \sum_{j=1}^{N}M_{ij}\sum_{i=1}^{3}M_{ji}\Omega_i \tag{13}$$

Noting that $M_{jk} = M_{kj}^\dagger$ allows us to write

Rare Earth - A tribute to the late Mr. Rare Earth, Professor Karl Gschneidner   Materials Research Forum LLC
Materials Research Foundations 164 (2024) 1-66   https://doi.org/10.21741/9781644903056-1

$$A_k = \sum_{j=1}^{N} M_{kj}^{\dagger} S_j^m \tag{14}$$

$$B_{ki} = \sum_{j=1}^{N} M_{kj}^{\dagger} M_{ji} \tag{15}$$

which are the component expressions for $\mathbf{A} = \mathbf{M}^{\dagger}\mathbf{S}$ and $\mathbf{B} = \mathbf{M}^{\dagger}\mathbf{M}$, respectively, where $\mathbf{M}^{\dagger}$ is a 3xN matrix, $\mathbf{M}$ is a Nx3 matrix and $\mathbf{S}$ is a Nx1 matrix. We can then write

$$A_k = \sum_{i=1}^{3} B_{ki} \Omega_i \tag{16}$$

which is the component expression for $\mathbf{A} = \mathbf{B}\Omega$. It follows that $\mathbf{B}^{-1}\mathbf{A} = \Omega$, but $\mathbf{B} = \mathbf{M}^{\dagger}\mathbf{M}$ and $\mathbf{A} = \mathbf{M}^{\dagger}\mathbf{S}$, so the result is

$$\Omega = (\mathbf{M}^{\dagger}\mathbf{M})^{-1} \, \mathbf{M}^{\dagger}\mathbf{S} \tag{17}$$

which is the set of Judd-Ofelt parameters that minimizes the sum of the squared difference of measured and theoretical linestrengths. Due to the large number of calculations to be made, matrices are suitable for computer-based calculations. The sum of square errors SSE (or residual sum of squares), and mean square error MSE (or residual-mean-square) are written as

$$SSE = \sum_{j=1}^{N} \left(S_j^m - S_j^t \right)^2 \tag{18}$$

$$MSE = \frac{SSE}{N-p} \tag{19}$$

where N is the number of observations and p is the number of parameters to fit. In this case there are p = 3 for the three Judd-Ofelt parameters. The root-mean-square error, δ, is then

$$RMSE = \sqrt{MSE} \tag{20}$$

This is most often used as a judge of how good the fit is. The variance-covariance matrix is given by (MSE) x $(\mathbf{M}^{\dagger}\mathbf{M})^{-1}$, which contains information on the error associated with the fitted parameters. Taking the square root of the diagonal elements of the variance-covariance matrix gives the errors associated with the Judd-Ofelt fitting parameters. That is

$$\Delta\Omega_i = \pm RMSE\sqrt{B_{ii}} \tag{21}$$

where B_{ii} are the diagonal elements of the matrix $\mathbf{B}^{-1} = (\mathbf{M}^{\dagger}\mathbf{M})^{-1}$. These errors are seldom reported in the literature. There is really no reason not to report them since they come right out of the analysis. Hopefully this article serves to illustrate that it is a simple matter to find the error associated with Judd-Ofelt parameters, and that this will become standard procedure when presenting a Judd-Ofelt analysis.

A Judd-Ofelt analysis relies on accurate absorption measurements, specifically the integrated absorption cross section over the wavelength range of respective manifolds. From the integrated absorption cross section, the line strength, S_m, can be found

$$S_m = \frac{3ch(2J+1)}{8\pi^3 e^2 \bar{\lambda}} n \left(\frac{3}{n^2+2}\right)^2 \int_{manifold} \sigma(\lambda)d\lambda \tag{22}$$

where J is the total angular momentum of the lower state, found from the $^{2S+1}L_J$ designation. $\sigma(\lambda)$ is the absorption cross section as a function of wavelength. The mean wavelength, $\bar{\lambda}$, is found by the first moment of the absorption spectral data,

$$\bar{\lambda} = \frac{\sum \lambda\sigma(\lambda)}{\sum \sigma(\lambda)} \tag{23}$$

and the other symbols have their usual meaning. A Judd-Ofelt analysis minimizes the square of the difference between S_m and S_{ED}, with Ω_t as adjustable parameters. Once the Judd-Ofelt parameters are obtained and S_{ED} determined, then the electric dipole transition probability, A_{ED}, for any excited state transition can be calculated from

$$A_{ED} = \frac{64\pi^4 e^2}{3h(2J'+1)\bar{\lambda}^3}\left[n\left(\frac{n^2+2}{3}\right)^2 S_{ED} + n^3 S_{MD}\right] \tag{24}$$

where J' is the total angular momentum of the upper state. Notice that this equation contains a term S_{MD}, the magnetic dipole (MD) line strength. MD transition probabilities, A_{MD}, can be

Materials Research Foundations 164 (2024) 1-66
https://doi.org/10.21741/9781644903056-1

calculated separately [8,18]. While MD transitions are normally orders of magnitude smaller than ED transitions, because ED transitions for lanthanides in solids occur as a result of a perturbation, some MD transitions make significant contributions. The Judd-Ofelt parameters for Er:silica and Er: antimonate are shown in table 2. The results of the analysis are shown in tables 3 and 4. Eight absorption measurements were used in the fit to obtain the Judd-Ofelt parameters. The manifolds used in the fit include: $^2G_{9/2}$, $^4F_{3/2}$ + $^4F_{5/2}$, $^4F_{7/2}$, $^2H_{11/2}$, $^4S_{3/2}$, $^4F_{9/2}$, $^4I_{9/2}$ and $^4I_{11/2}$, with the $^4I_{13/2}$ excluded due to its strong magnetic dipole contribution.

Table 2. Comparison of Judd-Ofelt parameters in Er:glass.

Glass material	Ω_2 (x10^{-20} cm^2)	Ω_4 (x10^{-20} cm^2)	Ω_6 (x10^{-20} cm^2)	d*
Er:silica	3.985 ± 0.172	1.373 ± 0.214	0.793 ± 0.121	0.10
Er:antimonate	5.926 ± 0.051	1.597 ± 0.059	1.853 ± 0.020	0.03

* δ is the root-mean-square error (RSME) of the fit as given in Eq. 20

The Judd-Ofelt parameters given here compare in a favourable way to those found in the literature for other glass compositions [19,20]. Discrepancies can arise for a variety of reasons because the Judd-Ofelt parameters may be sensitive to the accuracy of the absorption measurements as well as the transitions used in the fit.

The Judd-Ofelt theory [6-8] is often used to theoretically describe the intensities of rare-earth ions in solids. It is a phenomenological theory, meaning that it relates empirical observations in a way that is consistent with the fundamental theory, but that is not derived directly from the theory. In essence it uses experimentally measured quantities to determine phenomenological parameters that can then be used for theoretical predictions in the context of the theory. With this understanding, the theory is susceptible to measurement error as well as errors associated with the approximations inherent in the theory.

Table 3. Judd-Ofelt analysis of Er:silica glass

TRANSITION	λ (nm)	S_{ED} (x10^{-20}cm^2)	A_{ED} (s^{-1})	A_{MD} (s^{-1})	β	τ (ms)
$^4F_{3/2} \rightarrow {}^4F_{5/2}$	28736	0.2942	0.008	0.007	0.0000	
$^4F_{3/2} \rightarrow {}^4F_{7/2}$	5179	0.0915	0.444		0.0003	
$^4F_{3/2} \rightarrow {}^2H_{11/2}$	3029	0.0030	0.073		0.0000	
$^4F_{3/2} \rightarrow {}^4S_{3/2}$	2435	0.1036	4.844	6.215	0.0063	
$^4F_{3/2} \rightarrow {}^4F_{9/2}$	1402	0.0527	12.972		0.0074	
$^4F_{3/2} \rightarrow {}^4I_{9/2}$	1011	0.3599	238.039		0.1350	
$^4F_{3/2} \rightarrow {}^4I_{11/2}$	822	0.5127	636.248		0.3609	
$^4F_{3/2} \rightarrow {}^4I_{13/2}$	631	0.0274	76.455		0.0434	
$^4F_{3/2} \rightarrow {}^4I_{15/2}$	454	0.1009	787.511		0.4467	0.5673
$^4F_{5/2} \rightarrow {}^4F_{7/2}$	6317	0.4541	0.809	0.744	0.0008	
$^4F_{5/2} \rightarrow {}^2H_{11/2}$	3385	0.2251	2.608		0.0013	
$^4F_{5/2} \rightarrow {}^4S_{3/2}$	2660	0.0381	0.910	0.637	0.0008	
$^4F_{5/2} \rightarrow {}^4F_{9/2}$	1474	0.6166	87.003		0.0424	

TRANSITION	λ (nm)	S_{ED}	A_{ED}	A_{MD}	β	τ(ms)
$^4F_{5/2} \rightarrow {}^4I_{9/2}$	1048	0.2028	80.231		0.0391	
$^4F_{5/2} \rightarrow {}^4I_{11/2}$	846	0.1366	103.416		0.0504	
$^4F_{5/2} \rightarrow {}^4I_{13/2}$	645	0.5169	898.728		0.4382	
$^4F_{5/2} \rightarrow {}^4I_{15/2}$	461	0.1770	875.811		0.4270	0.488
$^4F_{7/2} \rightarrow {}^2H_{11/2}$	7294	0.8293	0.719		0.0003	
$^4F_{7/2} \rightarrow {}^4S_{3/2}$	4596	0.0082	0.029		0.0000	
$^4F_{7/2} \rightarrow {}^4F_{9/2}$	1923	0.1073	5.102	12.475	0.0063	
$^4F_{7/2} \rightarrow {}^4I_{9/2}$	1257	0.5350	91.713	13.259	0.0378	
$^4F_{7/2} \rightarrow {}^4I_{11/2}$	977	0.4978	182.609		0.0657	
$^4F_{7/2} \rightarrow {}^4I_{13/2}$	718	0.4630	433.654		0.1560	
$^4F_{7/2} \rightarrow {}^4I_{15/2}$	498	0.6986	2039.753		0.7339	0.360
$^2H_{11/2} \rightarrow {}^4S_{3/2}$	12422	0.2811	0.033		0.0000	
$^2H_{11/2} \rightarrow {}^4F_{9/2}$	2611	1.4786	18.681	0.154	0.0033	
$^2H_{11/2} \rightarrow {}^4I_{9/2}$	1518	1.1449	73.979	0.680	0.0130	
$^2H_{11/2} \rightarrow {}^4I_{11/2}$	1128	0.3615	57.185	8.899	0.0115	
$^2H_{11/2} \rightarrow {}^4I_{13/2}$	797	0.2176	98.954	71.100	0.0295	
$^2H_{11/2} \rightarrow {}^4I_{15/2}$	534	3.4790	5426.424		0.9427	0.174
$^4S_{3/2} \rightarrow {}^4F_{9/2}$	3306	0.0213	0.398		0.0004	
$^4S_{3/2} \rightarrow {}^4I_{9/2}$	1729	0.3099	40.570		0.0374	
$^4S_{3/2} \rightarrow {}^4I_{11/2}$	1241	0.0644	22.924		0.0211	
$^4S_{3/2} \rightarrow {}^4I_{13/2}$	851	0.2746	306.140		0.2820	
$^4S_{3/2} \rightarrow {}^4I_{15/2}$	558	0.1754	715.567		0.6591	0.921
$^4F_{9/2} \rightarrow {}^4I_{9/2}$	3625	0.5400	3.057	2.758	0.0052	
$^4F_{9/2} \rightarrow {}^4I_{11/2}$	1987	1.3139	45.305	7.127	0.0465	
$^4F_{9/2} \rightarrow {}^4I_{13/2}$	1147	0.3074	55.527		0.0492	
$^4F_{9/2} \rightarrow {}^4I_{15/2}$	672	1.1014	1014.586		0.8992	0.886
$^4I_{9/2} \rightarrow {}^4I_{11/2}$	4396	0.2053	0.651	1.377	0.0130	
$^4I_{9/2} \rightarrow {}^4I_{13/2}$	1678	0.5842	33.500		0.2141	
$^4I_{9/2} \rightarrow {}^4I_{15/2}$	824	0.2458	120.919		0.7729	6.392
$^4I_{11/2} \rightarrow {}^4I_{13/2}$	2713	1.2281	13.828	9.679	0.2018	
$^4I_{11/2} \rightarrow {}^4I_{15/2}$	1015	0.4262	92.976		0.7982	8.585
$^4I_{13/2} \rightarrow {}^4I_{15/2}$	1621	1.3743	62.446	35.474	1.0000	10.213

Table 4. Judd-Ofelt analysis of Er:antimonate glass

TRANSITION	λ (nm)	S_{ED} ($\times 10^{-20}$ cm^2)	A_{ED} (s^{-1})	A_{MD} (s^{-1})	β	τ(ms)
$^4F_{3/2} \rightarrow {}^4F_{5/2}$	28736	0.4220	0.036	0.019	0.0000	
$^4F_{3/2} \rightarrow {}^4F_{7/2}$	5179	0.1101	0.872	0.000	0.0001	
$^4F_{3/2} \rightarrow {}^2H_{11/2}$	3029	0.0062	0.274	0.000	0.0000	
$^4F_{3/2} \rightarrow {}^4S_{3/2}$	2476	0.1541	12.598	10.083	0.0030	
$^4F_{3/2} \rightarrow {}^4F_{9/2}$	1398	0.1166	54.607	0.000	0.0073	
$^4F_{3/2} \rightarrow {}^4I_{9/2}$	1008	0.4706	599.924	0.000	0.0798	
$^4F_{3/2} \rightarrow {}^4I_{11/2}$	821	1.0487	2527.202	0.000	0.3362	

Transition						
$^4F_{3/2} \rightarrow {}^4I_{13/2}$	631	0.0640	352.556	0.000	0.0469	
$^4F_{3/2} \rightarrow {}^4I_{15/2}$	450	0.2357	3959.650	0.000	0.1330	0.5267
$^4F_{5/2} \rightarrow {}^4F_{7/2}$	6317	0.7213	1.878	1.045	0.0004	
$^4F_{5/2} \rightarrow {}^2H_{11/2}$	3385	0.4317	8.947	0.000	0.0011	
$^4F_{5/2} \rightarrow {}^4S_{3/2}$	2709	0.0549	2.269	1.024	0.0004	
$^4F_{5/2} \rightarrow {}^4F_{9/2}$	1469	1.0503	281.508	0.000	0.0338	
$^4F_{5/2} \rightarrow {}^4I_{9/2}$	1045	0.3447	262.432	0.000	0.0315	
$^4F_{5/2} \rightarrow {}^4I_{11/2}$	845	0.1614	236.990	0.000	0.0284	
$^4F_{5/2} \rightarrow {}^4I_{13/2}$	645	0.9204	3150.703	0.000	0.3780	
$^4F_{5/2} \rightarrow {}^4I_{15/2}$	457	0.4136	4389.187	0.000	0.5265	0.1200
$^4F_{7/2} \rightarrow {}^2H_{11/2}$	7294	1.4970	1.605	0.000	0.0002	
$^4F_{7/2} \rightarrow {}^4S_{3/2}$	4742	0.0097	0.051	0.000	0.0000	
$^4F_{7/2} \rightarrow {}^4F_{9/2}$	1914	0.1545	13.858	21.777	0.0034	
$^4F_{7/2} \rightarrow {}^4I_{9/2}$	1251	1.0413	341.631	23.409	0.0347	
$^4F_{7/2} \rightarrow {}^4I_{11/2}$	975	0.7246	511.474	0.000	0.0486	
$^4F_{7/2} \rightarrow {}^4I_{13/2}$	719	0.5388	983.056	0.000	0.0935	
$^4F_{7/2} \rightarrow {}^4I_{15/2}$	493	1.3956	8621.185	0.000	0.8197	0.0951
$^2H_{11/2} \rightarrow {}^4S_{3/2}$	13550	0.3364	0.158	0.000	0.0000	
$^2H_{11/2} \rightarrow {}^4F_{9/2}$	2595	2.1904	51.618	0.267	0.0030	
$^2H_{11/2} \rightarrow {}^4I_{9/2}$	1511	1.8657	229.663	1.197	0.0133	
$^2H_{11/2} \rightarrow {}^4I_{11/2}$	1126	0.5011	151.658	15.692	0.0096	
$^2H_{11/2} \rightarrow {}^4I_{13/2}$	797	0.3320	291.651	126.049	0.0240	
$^2H_{11/2} \rightarrow {}^4I_{15/2}$	529	5.0527	16500.849	0.000	0.9500	0.0576
$^4S_{3/2} \rightarrow {}^4F_{9/2}$	3210	0.0494	1.810	0.000	0.0003	
$^4S_{3/2} \rightarrow {}^4I_{9/2}$	1700	0.5970	153.764	0.000	0.0297	
$^4S_{3/2} \rightarrow {}^4I_{11/2}$	1227	0.1437	100.014	0.000	0.0193	
$^4S_{3/2} \rightarrow {}^4I_{13/2}$	847	0.6417	1400.010	0.000	0.2703	
$^4S_{3/2} \rightarrow {}^4I_{15/2}$	550	0.4098	3523.454	0.000	0.6803	0.1931
$^4F_{9/2} \rightarrow {}^4I_{9/2}$	3614	0.8193	8.303	4.607	0.0036	
$^4F_{9/2} \rightarrow {}^4I_{11/2}$	1987	2.8136	180.069	12.246	0.0539	
$^4F_{9/2} \rightarrow {}^4I_{13/2}$	1151	0.4370	148.232	0.000	0.0416	
$^4F_{9/2} \rightarrow {}^4I_{15/2}$	664	1.7109	3213.668	0.000	0.9009	0.2803
$^4I_{9/2} \rightarrow {}^4I_{11/2}$	4415	0.3610	1.935	2.184	0.0251	
$^4I_{9/2} \rightarrow {}^4I_{13/2}$	1689	1.3464	141.597	0.000	0.8637	
$^4I_{9/2} \rightarrow {}^4I_{15/2}$	813	0.0184	18.217	0.000	0.1111	6.1001
$^4I_{11/2} \rightarrow {}^4I_{13/2}$	2735	2.4820	49.817	16.023	0.1426	
$^4I_{11/2} \rightarrow {}^4I_{15/2}$	997	0.9000	396.015	0.000	0.8574	2.1652
$^4I_{13/2} \rightarrow {}^4I_{15/2}$	1568	2.9557	278.386	67.948	1.0000	2.8874

It should be noted that caution must be exercised in comparing intensity parameters found in the literature because some older articles use the τ_λ parameter, instead of the Ω_λ parameter commonly used today. These forms of intensity parameters are related by

$$\Omega_\lambda = \frac{3h}{8\pi^2 mc} n \left(\frac{3}{n^2 + 2} \right)^2 \tau_\lambda \tag{25}$$

where n is the index of refraction and $3h/8\pi^2 mc = 9.2253 \times 10^{-12}$ cm. Many of the articles by Reisfeld in the 1970's used the τ_λ parameter and this should be converted to Ω_λ parameters for proper comparison with modern articles.

3.3   Dynamics of Erbium-doped materials

Laser diodes operating ~1.4 μm can selectively pump Er-doped materials directly by raising Er atoms from the $^4I_{15/2}$ manifold to the $^4I_{13/2}$ manifold. For ~ 1.6 μm operation on $^4I_{13/2} \to {}^4I_{15/2}$ transitions, the $^4I_{13/2}$ is the upper laser manifold. Energy transfer processes can take place which can affect the laser operation. Referring to Fig.7, process 2 is an energy transfer *upconversion* that takes 2 Er atoms from the upper laser manifold, $^4I_{13/2}$, demoting 1 Er atom back in the ground manifold and promoting 1 Er atom up to the $^4I_{9/2}$ manifold. In this case, upconversion is deleterious. To minimize upconversion in this case, low Er concentrations are needed. However, reducing the Er concentration affects the absorption efficiency. Conversely, for operation on the at $^4I_{11/2} \to {}^4I_{13/2}$ transition at ~ 2.9 μm, an Er atom is raised to the $^4I_{9/2}$. From there, the Er atom most probably relaxes by a nonradiative transition to the $^4I_{11/2}$ manifold. Therefore, upconversion increases the population inversion by 3, taking 2 Er atoms out of the lower laser level, $^4I_{13/2}$, and adding 1 Er atom to the upper laser manifold, $^4I_{11/2}$. In this case upconversion is beneficial. Therefore, to encourage upconversion in this situation, high Er concentrations are needed. Consequently, it is useful to measure the upconversion energy transfer parameters as a function of the Er concentration. There is another upconversion process, which is denoted by process 1 in Fig. 7, which contributes to observable green fluorescence. This process is considered minimal and safe to neglect. A model for the population dynamics in the upper laser level for the $^4I_{13/2}$ was developed previously and used to measure upconversion parameters of process 2 in Er:YAG materials [21], but it is applicable to Er:glass materials as well.

Although it is not always considered, all energy transfer processes have a forward and reverse process. The reverse process of the upconversion is a self-quenching or cross-relaxation process. In general, however, the reverse processes are smaller than the forward processes, and usually neglected. There are situations where the reverse processes can be important. In the cases being considered here, the weaker reverse processes coupled with nonradiative multiphonon decay makes them quite negligible. For example, in process 2, the efficient nonradiative multiphonon decay from the $^4I_{9/2}$ to $^4I_{11/2}$ would make a reverse process taking 1 Er atom from the $^4I_{9/2}$ to the $^4I_{13/2}$ and 1 Er atom from the $^4I_{15/2}$ to the $^4I_{13/2}$, which would be very inefficient. The same reasoning applies to the reverse of process 1.

The Er $^4I_{13/2}$ lifetimes, when directly pumped, show some slight non-exponential behavior at early times, which increases with increasing concentration, and some green light is also observed by the naked eye at higher concentration. These are some signature indications of active upconversion processes. The decay profiles for various Er concentrations in silica and antimonate glass on a log scale normalized intensity are shown in Fig. 13, and the exponential lifetimes given in table 5.

Rare Earth - A tribute to the late Mr. Rare Earth, Professor Karl Gschneidner   Materials Research Forum LLC
Materials Research Foundations 164 (2024) 1-66   https://doi.org/10.21741/9781644903056-1

Figure 13. Decay of the $^4I_{13/2} \rightarrow {}^4I_{15/2}$ transition at ~ 1.6 μm in Er:glass.

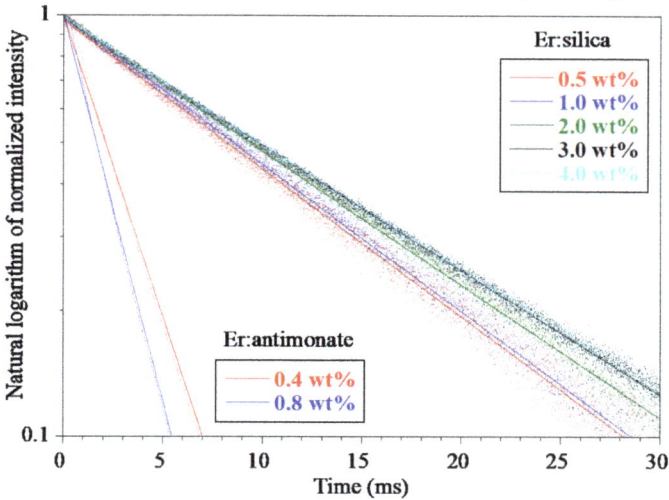

All the measured lifetimes in silica glass are longer than those predicted by the Judd-Ofelt theory. This situation may be explained by radiative trapping which tends to lengthen lifetime with concentration. For the antimonate glass, the agreement with Judd-Ofelt theory is reasonably good. There is, however, an observed shortening of the lifetime with concentration that is indicative of energy transfer processes.

Table 5. Measured and Judd-Ofelt lifetimes in Er:silica and Er:antimonate

Glass material	Er_2O_3 wt%	Er density $(\times 10^{20}\ cm^{-3})$	$\tau_{meas}\ ^4I_{13/2}$ (ms)	$\tau_{JO}\ ^4I_{13/2}$ (ms)	$\Delta\tau$ (ms)
silica	0.50	0.498	12.58	10.21	2.37
silica	1.00	0.995	12.38	10.21	2.17
silica	2.00	1.990	13.71	10.21	3.50
silica	3.00	2.985	14.73	10.21	4.52
silica	4.00	3.075	14.81	10.21	4.60
antimonate	0.40	0.636	2.75	2.89	0.14
antimonate	0.80	1.232	2.12	2.89	0.77

4.   Rare earth fiber lasers

4.1   Quasi-four level lasers

The first Er doped glass laser was demonstrated in 1965 using Yb:Er co-doped silica glass [13]. This flashlamp pumped laser produced optical pulses at 1.55 μm in LiMgAlSiO₃ glass singly doped with Er, making Er the third rare earth ion to exhibit optical oscillations in a glass matrix, after Nd and Yb, and followed by and Ho and Tm as the fourth and fifth [22].

Er lasers operating on the $^4I_{13/2} \rightarrow {}^4I_{15/2}$ transition are *quasi-four level* lasers. This terminology can be best understood by an analysis of the laser gain. One can start the analysis by considering the small signal gain coefficient, g_0. That is,

$$g_0 = \sigma n_2 - \sigma n_1 \tag{26}$$

where σ is the emission cross section and n_2 and n_1 are the population densities of the upper and lower *laser levels*. However, because the time for thermalization of the *laser manifold* is very fast, all the atoms in all levels of the upper and lower manifolds can participate. Under this approximation, the small signal gain coefficient becomes

$$g_0 = \sigma f_2 N_2 - \sigma f_1 N_2 \tag{27}$$

where f_2 and f_1 are thermal occupation factors of Boltzmann statistics for the levels and N_2 and N_1 are the population densities of the upper and lower laser manifolds. However, what is usually reported in the literature is the effective stimulated emission cross section, σ_e,

$$\sigma_e = f_2 \sigma \tag{28}$$

Factoring the small signal gain coefficient equation and substituting yields

$$g_0 = \sigma_e \left(N_2 - \frac{f_1}{f_2} N_1 \right) \tag{29}$$

For many lasers like the Ho, Tm and Er laser, and certainly for the Yb laser, all the atoms reside either in the upper or lower laser manifold. Thus,

$$N_1 + N_2 \approx C_A N_s \tag{30}$$

Rare Earth - A tribute to the late Mr. Rare Earth, Professor Karl Gschneidner   Materials Research Forum LLC
Materials Research Foundations 164 (2024) 1-66   https://doi.org/10.21741/9781644903056-1

where $C_A N_s$ represents the product of the concentration of the active atoms and the number density of the sites where active atoms could reside. Using this to eliminate N_1 yields

$$g_0 = \sigma_e \left[\left(1 + \frac{f_1}{f_2} \right) N_2 - \frac{f_1}{f_2} C_A N_s \right] \tag{31}$$

Often the factor of γ is used,

$$\gamma = 1 + \frac{f_1}{f_2} \tag{32}$$

With this notation, the small signal gain coefficient becomes

$$g_0 = \sigma_e \left[\gamma N_2 - (\gamma - 1) C_A N_s \right] \tag{33}$$

The factor γ is 1 for a true *four-level laser* and 2 for a true *three-level laser*. Therefore, a quasi-four level laser has a value of γ that is closer to 1.0 than 2.0. To see why this is so, consider Fig. 14. This figure shows the energy level schematic for three and four level lasers. As can be seen from this figure, the lower laser level in a three-level laser is the ground state and, as such, has some thermal population. In a four-level laser the lower laser level is some excited state and has no thermal population. Since $\gamma = 1 + f_{lower}/f_{upper}$ then γ must be 1 when $f_{lower} = 0$ for a four-level laser and 2 when $f_{lower} = f_{upper}$ for a three-level laser. Although a quasi-four level laser resembles the three-level laser in structure, it is more precisely quasi-four-level because f_{lower} is close to zero. In other words, it looks like a three-level laser but behaves more nearly like a four-level laser, hence the name quasi-four level.

A few definitions are in order, and the various types of solid-state lasers are shown in Fig. 14 for reference. When an atom in a 4 level laser transitions from the upper laser manifold to the lower laser manifold the population inversion decreases by 1 quantum. There is 1 less atom in the upper laser manifold and the added population in the lower laser manifold rapidly decays. Nd:YAG operating at 1.06 µm is a good example of a 4 level laser. If the lower laser level does not decay rapidly, it is called a terminated 4 level laser. Dy:YLF operating at 4.3 µm is an example of a terminated 4-level laser. When an atom in a 3-level laser transitions from the upper laser manifold to the lower laser manifold, there is 1 less atom in the upper laser manifold and 1 more atom in the lower laser manifold. This decreases the population inversion by 2. The ruby laser ($Cr:Al_2O_3$) operating at ~ 0.7 µm is an example of a 3-level laser. Lasers that are between 3 and 4-level lasers are characterized by a factor describing how the population inversion changes when 1 atom transitions. If this factor is less than 1.5, it is quasi-four-level laser. Quasi 4-level lasers have a high threshold because the lower laser level thermal population must be overcome to achieve a positive gain. Some examples of quasi-four level laser systems besides the Er $^4F_{13/2} \rightarrow {}^4I_{15/2}$, include the Tm $^3F_4 \rightarrow {}^3H_6$, Nd $^4F_{3/2} \rightarrow {}^4I_{9/2}$, Yb $^2F_{5/2} \rightarrow {}^2F_{7/2}$ and Ho $^5I_7 \rightarrow {}^5I_8$.

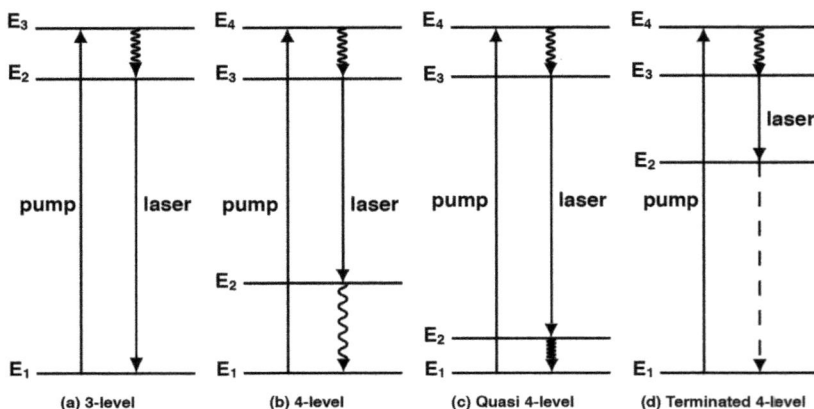

Figure 14. Types of Solid-State Lasers: (a) 3-level; (b) 4-level;
(c) Quasi 4-level; (d) Terminated 4-level

These types of lasers present manageable challenges due to the small thermal occupation factors residing in the lower laser level. Due to the lack of sharply defined energy levels in a glass, level populations become more difficult to quantify. It is, therefore, difficult to theoretically assess the situation in glasses. Nevertheless, as was discussed in the last section, the emission of phonons adds to the heat load, which adversely affects quasi-four-level lasers by increasing the lower laser level population and, therefore, increasing the threshold of the laser. There are often trade-offs in considering a glass for a particular application. High phonon glasses have the disadvantage of having to bear a higher heat load, but are more mechanically durable, whereas low phonon glasses have a smaller heat load, but are more fragile. The effect of the phonon energy regarding energy transfer processes is important as well.

4.2   Fiber lasers – ND, ER, YB AND TM

Nd, Er, Yb, and Tm are among the more common fiber lasers. A diagram showing the energy levels of these atoms, often referred to as a *Dieke diagram*, appears in Fig. 15. The early research on fiber laser oscillators utilized flash lamps, but the advent of diode lasers as pump sources advanced their efficiency greatly and moved them to the position of prominence they play today in the forefront of laser glass research. Fiber lasers are usually laser diode pumped, although they are occasionally pumped by other solid-state lasers. In principle, laser diodes could be fabricated at a multitude of wavelengths. In practice, high power laser diodes are available commercially around 0.80 and 0.95 μm. Nd, Er, and Tm can be pumped using laser diodes around 0.80 μm. Er and Yb can be pumped using laser diodes around 0.95 μm. Er fiber amplifiers are widely employed

Rare Earth - A tribute to the late Mr. Rare Earth, Professor Karl Gschneidner   Materials Research Forum LLC
Materials Research Foundations 164 (2024) 1-66   https://doi.org/10.21741/9781644903056-1

in the communications industry. Er doped fiber amplifiers are a field too large to be included here. Instead, fiber laser oscillators will be discussed.

Laser oscillator performance can be characterized with a *threshold* and *slope efficiency*. Above threshold, the laser output power increases nearly linearly with pump power at a rate characterized by the slope efficiency. Threshold and slope efficiency depend on the length of the fiber. There are two main contributions to the laser threshold: (1) overcoming the population density in the lower laser level and (2) overcoming the losses, including the output coupling. The first contribution is applicable to quasi-four level lasers, while the second contribution depends on the sum of the losses including the output coupler.

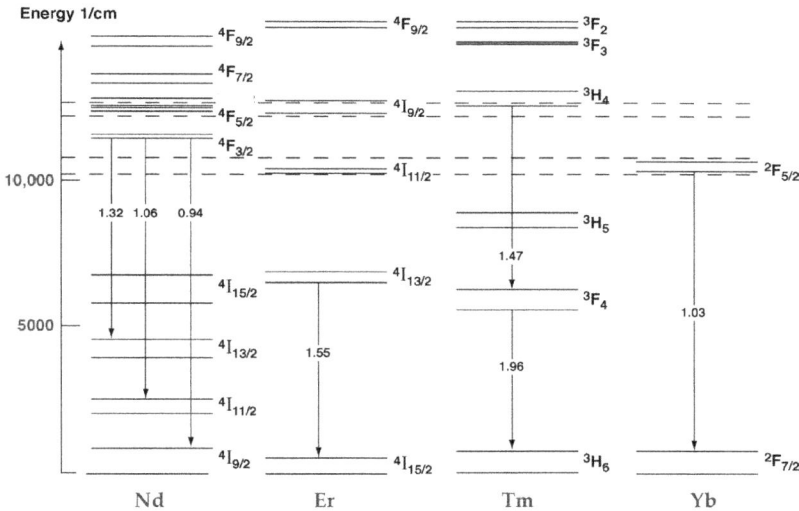

Figure 15. Laser transitions in lanthanide series atoms. The range of traditionalNIR laser diode wavelengths lie between the dashed lines.

Threshold and slope efficiency depend on the *launch efficiency*, defined as the fraction of the incident pump power that begins to propagate in the core and inner cladding of the fiber. For pump radiation concentrated into a fiber, the fraction falling within the inner cladding and numerical aperture can be launched. The launch efficiency is determined using the *cut back method*, where the fraction of incident pump power transmitted through the fiber is measured as a function of length. A curve fit of log of transmission vs. fiber length, Fig. 16, gives launch efficiency (intercept) and effective absorption coefficient (slope). Studies in the literature quote threshold and slope efficiency in terms of incident pump power and sometimes launched pump power, without quoting launch efficiency.

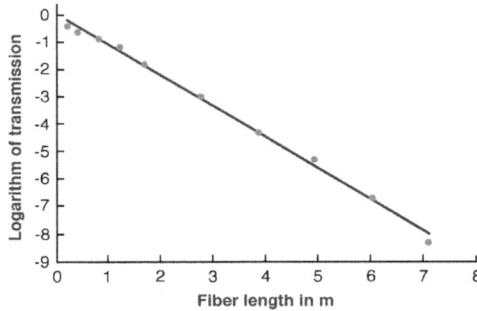

Figure 16. Natural logarithm versus fiber length in cut-back method for determination of launch efficiency (0.95) and effective absorption (1.09 m^{-1}).

Threshold and slope efficiency are dependent on the ratio of the pump wavelength (λ_P) and the laser wavelength, λ_L, and the quantum efficiency, η_Q. For each pump photon absorbed, the fiber laser gains energy (hc / λ_P) and for each laser photon emitted the fiber laser contributes energy (hc / λ_L) where h is Plank's constant, and c is the speed of light. The quantum efficiency is the ratio of the number of atoms reaching the upper laser manifold to the number of photons absorbed. Therefore, slope efficiency is proportional to ($\lambda_P \eta_Q / \lambda_L$). The quantum efficiency plays a large role in lasers for which energy transfer occurs.

4.2.1   Neodymium fiber lasers

Neodymium (Nd) operating on the $^4F_{3/2} \rightarrow {}^4I_{11/2}$ transition around 1.06 μm is a four-level laser. A nominal 0.80 μm pump photon can elevate a Nd atom from the $^4I_{9/2}$ manifold, the ground manifold, to the $^4F_{5/2}$ manifold. From there the excited Nd atom most likely relaxes to the $^4F_{3/2}$ manifold, the upper laser manifold. Because the $^4F_{3/2} \rightarrow {}^4I_{11/2}$ transition is a four-level laser and the gain is relatively high, the threshold is very low. Silica Nd:fibers are often co-doped with Al_2O_3 or P_2O_5 to increase the solubility of Nd. A Nd:fiber laser was demonstrated early in the development of fiber lasers. A Nd:fiber with a 12 μm core had a negligible threshold and a slope efficiency of 0.26, launched optical to optical. A laser output power of 9.2 W was produced [23].

A Nd:silica fiber laser with better efficiency was demonstrated more recently. Al_2O_3 was co-doped in a 10:1 ratio to improve solubility of the Nd. The Nd concentration was 6000 ppm. The core diameter was 5.56 μm and a numerical aperture of 0.14. The pump cladding was 250 μm. A high power, 0.805 μm laser diode pumped the Nd:silica fiber laser. Lasing was obtained at both 1.06 and 1.09 μm. The threshold was essentially negligible, and the slope efficiency was 0.39, launched optical-to-optical [24].

4.2.2   Erbium fiber lasers

Erbium (Er) operating on the $^4I_{13/2} \rightarrow {}^4I_{15/2}$ transition around 1.55 μm is a quasi-four level laser. A quasi-four level laser has a nontrivial population density in the lower laser level. Because of this, the pumping must be sufficiently intense to overcome the thermal population. Er lasers can be pumped to the $^4I_{9/2}$ manifold using laser diodes around 0.80 or to the $^4I_{11/2}$ manifold using laser diodes around 0.95 μm. In the former case, the Er laser should be designed to promote self-quenching of the $^4I_{9/2}$ manifold. In this self-quenching process, an Er atom in the $^4I_{9/2}$ manifold and an Er atom in the $^4I_{15/2}$ manifold interact to produce 2 Er atoms in the $^4I_{13/2}$ manifold, the upper laser manifold. In the latter case, the Er laser should be designed to promote nonradiative quenching of the $^4I_{11/2}$. In addition, Yb can be co-doped with Er to increase absorption of the pump. In this case, the pump radiation is absorbed by the $Yb^2F_{5/2}$ manifold which then transfers the energy to the Er $^4I_{11/2}$ manifold. Co-doping with Yb has the advantage of having good absorption by a low Er population density in the lower laser level.

Efficient laser operation was obtained with an Er:Yb:silica fiber laser. The phospho-silicate core had a diameter of 30 μm and 0.22 numerical aperture. A 400 μm inner cladding with a numerical aperture of ≈ 0.4 had a flat ground on its lateral surface to promote efficient pumping. The fiber laser was pumped from both ends using 0.975 μm diode stacks. When the Er laser operated at 1.565 μm, threshold was essentially nil and the slope efficiency was 0.41, launched optical to optical. 188 W of continuous power could be produced [25].

4.2.3 Ytterbium fiber lasers

Ytterbium (Yb) has several desirable spectroscopic properties when compared with Nd. Yb lasers operate around 1.03 μm on the $^2F_{5/2} \rightarrow {}^2F_{7/2}$ transition. Nd usually operates around 1.06 μm, a similar wavelength as Yb. The 1.06 μm Nd transition is a four-level laser whereas the 1.03 μm Yb transition is a quasi-four level laser. The quasi-four level nature militates against Yb. On the other hand, Yb has a longer lifetime than Nd, by a factor of roughly 2, a wider spectral bandwidth, as well as the possibility of higher Yb concentrations. Because Yb has only 2 manifolds, high Yb concentrations do not promote deleterious energy transfer processes such as self-quenching or up conversion. In addition, it can have a larger slope efficiency because the ratio of pump to laser wavelength is ≈ 0.76 for Nd and ≈ 0.92 for Yb. The possibility of a high slope efficiency for Yb has the additional advantage that a higher slope efficiency implies less heat deposition. Absorbed energy that does not appear in the form of photons, either spontaneously emitted or stimulated, appears as heat. For the slope efficiencies that are quoted above, the heat load for Yb could be 1/3 that for Nd.

Yb:fiber lasers can produce high power. A Yb:silica laser produced 1.36 kW by using a large core fiber [26]. The threshold was 40 W, and the slope efficiency was 0.83, launched optical to optical.

4.2.4 Thulium fiber lasers

Thulium (Tm) fiber lasers offer an attractive alternative to more conventional Tm-doped bulk lasers because Tm-doped fibers can provide high gain, despite the low emission cross section, by providing long lengths. In fact, one of the first applications of fiber lasers was as power amplifiers. Another attractive feature of Tm-doped fibers is that they can be pumped with commercially available diode lasers since they possess strong absorption features around 780-800 nm, allowing for incorporation of the laser gain material and the pump delivery system in a double clad fiber.

Tm operating on the $^3F_4 \rightarrow {}^3H_6$ transition around 1.8 μm is a quasi-four level laser. It is pumped with nominal 0.80 μm laser diodes. Although the ratio of pump wavelength to laser wavelength does not favor high efficiency, the quantum efficiency can approach 2.0, whereby a Tm ion in the

$^3H_4$ manifold interacts with another Tm ion in the $^3H_6$ ground manifold, to produce 2 excited ions in the $^3F_4$ manifold. A high quantum efficiency is possible when cross-relaxation of the $^3H_4$ manifold, the pump manifold, is dominant. To obtain a quantum efficiency approaching 2.0, the cross-relaxation process must dominate all other possible decay processes of the $^3H_4$ manifold. A possible competitive process is nonradiative decay. If a Tm atom in the $^3H_4$ manifold nonradiatively decays, it populates the Tm $^3H_5$ manifold which then nonradiatively decays to the $^3F_4$ manifold, the upper laser manifold. Thus, the quantum efficiency varies approximately from 1.0 to 2.0. Nonradiative processes are highly dependent on the maximum phonon energies associated with a particular laser material.

Diode laser pumped operation of tunable Tm:silica fiber lasers operating from 1.85 to 2.1 μm on the Tm $^3F_4 \rightarrow ^3H_6$ transition been demonstrated in continuous operation [27-32]. Tunable operation of Ti:Al$_2$O$_3$ pumped Tm:silica around 1.8 μm has been demonstrated in pulsed operation [33]. Tunable operation of diode pumped Q-switched Tm:silica around 1.8 μm has also been demonstrated in pulsed operation [34]. Diode pumped Tm:ZBLAN operating at ~1.8 μm is much less studied [35]. Many studies focus on continuous operation with various pumping schemes [33]. Walsh, et. al, has demonstrated diode laser pumped tunable laser operation of Tm:silica, Tm:ZBLAN, and Tm:germanate fibers in pulsed mode around 1.9 μm [36, 37]. This is not intended to be a complete survey of the literature. The references provided here give the reader a basis to explore this topic further.

5. Fiber lasers

5.1 Fiber fabrication

There are a variety of different techniques available for the fabrication of glass fibers. A popular method is based on *chemical vapor deposition* (CVD). This technique relies on the vaporization of the chemical components comprising the glass and reacting them with oxygen. This process is illustrated in Fig. 17 for silica-based fibers.

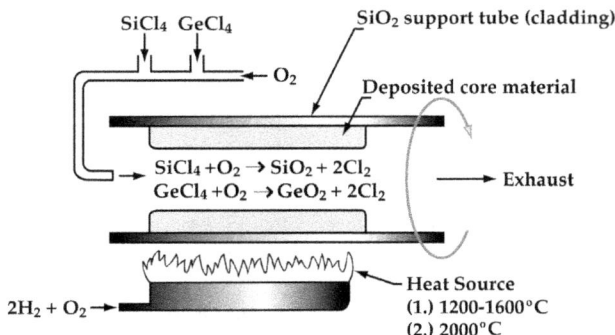

Figure 17. Fabrication of glass fibers by chemical vapor deposition (CVD).

After a sufficient amount of core material is deposited, an oxy-hydrogen torch is used to raise the temperatures to approximately 1200–1600 °C, until it sinters into a clear, bubble free glass. At this stage the vapor flow is stopped, and the temperature is raised to ~ 2000 °C. In this way the deposited core material is trapped in the support cylinder, which forms the cladding, and the materials soften and collapses into a solid rod that forms a *preform*. This perform can then be placed in a vertical drawing tower like the one pictured in Fig. 18. The preform is heated to 2000 °C until it softens and allows for molten flow of the glass. This molten material can now be drawn into a fiber by controlling the temperature and the pulling force. The uniformity of the glass fiber and the final dimensions are determined by manipulation of the drawn molten material. Immediately after being pulled, the fiber can be coated with a plastic sheath to protect the fiber from water vapor intrusion and mechanical damage. The choice of the plastic sheath material is also of importance to prevent degradation of the fiber over time. Once coated, a drawing tractor and tensile strength motor puts the fiber on a winding drum.

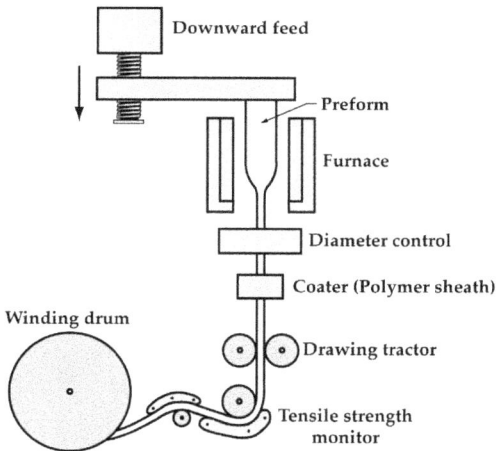

Figure 18. Vertical fiber drawing tower.

An advantageous outcome of the drawing process is that the refractive index profile of the fiber is a very accurate reproduction of that of the preform. Therefore, it is possible to ascertain some fundamental properties of the fiber from the perform itself, avoiding additional expense in the drawing procedure if the perform is flawed. The final fiber material is composed of a plastic sheath covering the cladding, made of pure silica, outside the core, which is a mixture of silica (SiO_2) and germania (GeO_2) and some rare earth optically active ions. The addition of germania to the reactants increases the index of refraction of the glass core above that of the cladding in order to form a step index to allow the fiber to act as a waveguide by total internal reflection.

5.2   Advantages of fiber lasers

Fiber lasers have several inherent advantages compared with conventional bulk solid-state lasers. Among these potential inherent advantages of a well-designed fiber laser are: good mode quality, high gain, good reliability, high efficiency, high average power, and continuous tuning capability. Some advantages are inherent to fiber lasers while others are dependent on good design. Each of the advantages will be discussed below with the basic physics necessary to understand them.

5.2.1   Fiber MODE quality

Fiber lasers can display excellent beam quality because they are guided modes. Typically, a fiber is a thin cylinder of glass composed of a small cylindrical core surrounded by a glass *cladding*. The cladding has a slightly lower index of refraction than the core, thus providing for *total internal reflection*. Any radiation that is initially within the core and is propagating essentially along the direction of the fiber will remain in the core because of total internal reflection, see Fig. 19.

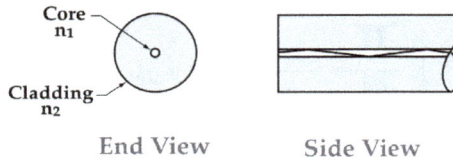

Figure 19. Typical optical fiber.

Fig. 20 illustrates the conditions for total internal reflection in fiber lasers. Optical rays with an angle $\theta > \theta_c$ at the core cladding boundary will exhibit total internal reflection. The critical angle for this to occur is given by

$$\theta_c = \sin^{-1}\left(n_2 / n_1\right) \tag{34}$$

where n_1 and n_2 are the index of refraction of the core and cladding, respectively.

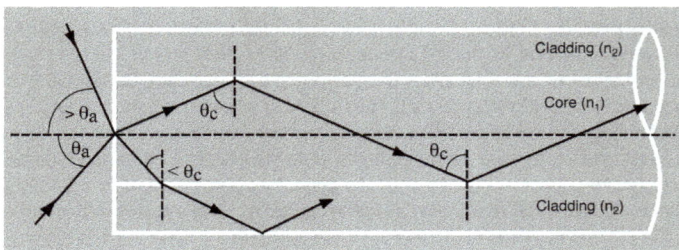

Figure 20. Total internal reflection conditions in fiber laser materials.

Rare Earth - A tribute to the late Mr. Rare Earth, Professor Karl Gschneidner   Materials Research Forum LLC
Materials Research Foundations 164 (2024) 1-66   https://doi.org/10.21741/9781644903056-1

If the initial angle, θ_a, becomes too great, then the critical angle will not be exceeded and will not be reflected at the core-cladding interface. The sine of this critical angle, θ_{ac}, is known as the *numerical aperture*, abbreviated as N.A. The numerical aperture is, therefore, a measure of the ability of the fiber to capture light. To enjoy total internal reflection at the interface between the core and the cladding, the incident angle must be within the numerical aperture. The numerical aperture is defined as

$$N.A. = \sin(\theta_{ac}) = \left(n_1^2 - n_2^2\right)^{1/2} \tag{35}$$

The guided mode fills the core and encroaches into the cladding. Properties of the guided mode are described with the help of the parameter V, given by

$$V = \frac{2\pi a_c}{\lambda}\left(n_1^2 - n_2^2\right)^{1/2} \tag{36}$$

where a_c is the radius of the core and λ is the wavelength. If V is less than 2.405, the fiber will only transmit the lowest order fiber mode. The modes are usually designated as LP$_{lm}$, where l and m are integers describing the radial and azimuthal characteristics of the guided mode. In the core, the intensity profile is described by a *Bessel function* of order l, $J_l(k_T\rho)$, where ρ is the radial coordinate. In the cladding, the intensity profile is described by a modified Bessel function $K_l(\gamma\rho)$. If γ is the propagation constant and k_0 is $2\pi/\lambda$,

$$\text{core:} \qquad k_T^2 = n_1^2 k_0^2 - \beta^2 \tag{37}$$

$$\text{inner cladding:} \qquad \gamma^2 = \beta^2 - n_2^2 k_0^2 \tag{38}$$

Graphing the profile of the lowest order of the Bessel and modified Bessel functions in their respective regions, the composite mode resembles a *Gaussian beam* profile and is shown in Fig. 21. The Gaussian beam radius, w, can be approximated by

$$w = a_c\left(0.65 + 1.619/V^{1.3} + 2.879/V^6\right) \tag{39}$$

Because the guided and the free space beam profiles closely resemble each other, coupling efficiency between these beam profiles can be very high.

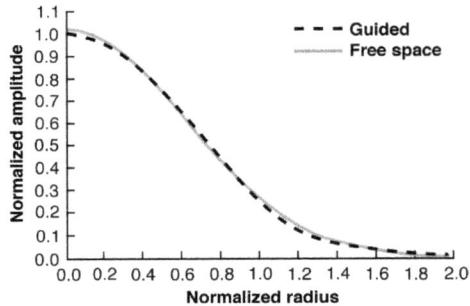

Figure 21. Guided and free space beam profiles.

5.2.2 Fiber gain

Fiber lasers can have long lengths and thus a very high gain. Gain is exponentially dependent on the gain coefficient, g_0, and length, l, product; that is $g_0\,l$. The gain coefficient is linearly related to the product of the emission *cross section*, σ_e, and the population density of the excited state, N_2; that is $\sigma_e N_2$. In turn, N_2 increases with the level of the pump per unit volume. Given a finite pump capability, a finite volume of laser material can be excited to a sufficiently high level. In a fiber laser, the excited state population density is concentrated in the core of the fiber. Because the fiber core has a small cross-sectional area, long lengths can be excited and still maintain a finite pump volume. Although the emission cross section may be low, a long length can compensate. In addition, without the deleterious effects of diffraction, a guided mode can enjoy the entire pumped volume. Therefore, even laser materials with low emission cross sections can produce high gain fiber lasers.

5.2.3 Fiber reliability

Fiber lasers can enjoy virtual misalignment immunity. Reliability demands that the laser is resistant to misalignment and component failure. Bulk solid-state lasers consist of an excited volume of laser material between 2 mirrors. Lasing requires that the mirrors be very nearly parallel. Depending on the design of the laser resonator, the mirrors may have to be parallel to within a small fraction of a milliradian. Temperature fluctuations and mechanical vibrations can often cause mirror misalignment. On the other hand, fiber lasers can employ cleaved fiber ends and fiber gratings for mirrors. Fiber gratings can be fused directly on to a length of doped fiber serving as the active material. With this technology, the fiber laser becomes a single piece, complete with mirrors, of fiber and thus impervious to misalignment.

Fiber laser reliability benefits by being pumped with laser diodes. Bulk solid-state lasers were traditionally pumped with flash lamps or arc lamps. Typical lifetimes for these components are 10^8 shots and 200 hours, respectively. As the price of laser diodes decrease, more and more bulk lasers are changing to laser diode pumping. A reason for the change was to increase laser lifetime and reliability. However, another reason for the change was to increase laser efficiency. On the other hand, fiber lasers became practical because of laser diode pumping. Laser diode lifetimes are roughly 10^9 shots and 5,000 to 50,000 hours.

5.2.4 Fiber laser efficiency

A prime reason for fiber laser efficiency is the efficiency of the pump source, that is laser diodes. Fiber lasers are essentially all end pumped. The core of a fiber laser is usually too small to support side pumping. Efficient end pumping requires a bright pump source to deliver sufficient pump power to the small cross-sectional area and within the numerical aperture of the fiber laser. High brightness and high efficiency laser diodes have sufficient capability for this application. Laser diode efficiency, electrical to optical, is commonly in the range of 0.4 to 0.5. However, laser diode efficiencies have recently exceeded 0.7.

High power laser diode bars can be efficiently employed by designing a fiber with an inner cladding. A single laser diode emitter has dimensions on the order of 1.0 by 100 μm. Even focusing a single laser diode emitter onto the core of a fiber is challenging. However, a single laser diode emitter is limited to pump powers on the order of 5 W. For applications requiring higher pump powers, several emitting areas, typically ~ 20, are fabricated on a 10 mm long laser diode bar. To couple all the pump power from a laser diode bar into a fiber, the fiber can be fabricated with an inner cladding. Typically, the core is small, roughly 10 μm, and is surrounded by a much larger inner cladding, often >100 μm. In turn, the inner cladding is surrounded be an outer cladding. The numerical aperture between the core and inner cladding is usually small. However, the numerical aperture between the inner and outer cladding is usually large. With a large inner cladding area and numerical aperture, the pump power from a laser diode bar can be imaged onto the inner cladding. Any pump radiation confined to the inner cladding will tend to travel through the core, eventually, where it can be absorbed. Although the average absorption is small, the fiber laser can be long, promoting high efficiency.

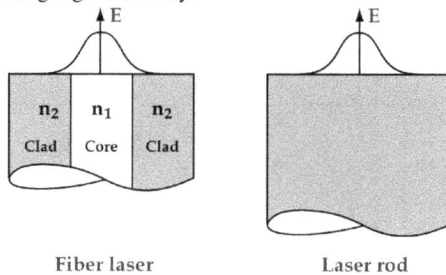

Fiber laser   Laser rod

Figure 22. Beam profiles in a fiber laser and a laser rod.

The guided mode design provides for high extraction efficiency. In a side pumped laser, gain tends to be distributed, more of less uniformly, throughout the active region. The gain near the center of the laser rod can be extracted with ease. However, the gain near the lateral surfaces of the laser rod is difficult to extract without diffraction effects, Fig. 22. On the other hand, in the fiber the extracting beam fully illuminates the core and a portion of the inner cladding as well. Because the entire core is illuminated with significant intensity, gain can be extracted from the entire core thereby improving laser efficiency.

5.2.5 High average power capability

Fiber lasers can generate high average powers, in part, because of their high surface to volume ratio. For a long cylinder, the surface to volume ratio is $\approx 2/a_c$, where a_c is the radius of the cylinder. Compared to a typical laser rod, a fiber laser can have a larger surface to volume ratio by a factor of 50 or more. Because cooling almost invariably occurs on the surface, a large surface to volume ratio promotes efficient cooling. Efficient cooling produces lower operating temperatures. Lower operating temperatures promote efficient operation, especially for *quasi-four-level* lasers. Quasi-four level lasers were discussed in section 4.

The guided wave nature of fiber lasers is also beneficial for high power applications because it offsets the thermal lensing effects. In cylindrical laser rods that are uniformly heated and cooled on the surface, a radial temperature gradient is established. Because the index of refraction is dependent on the temperature, a radial variation of the refractive index results. This radial variation acts like a lens. Under heavy pumping, the focal length of the lens can be very short making the laser resonator unstable. Without compensation, the mode quality will suffer. With compensation, the efficiency is apt to suffer. Other laser designs, such as the zigzag slab, were conceived to mitigate these effects. These designs have met with some success. On the other hand, fiber lasers have demonstrated a degree of immunity to thermal lensing. Essentially, the guided wave nature can dominate the thermal lensing.

5.2.6 Tuning of fiber lasers

The active atoms used in fiber lasers are lanthanide series atoms. When placed in a crystalline environment, the crystal field splits the free atom energy levels into sets of energy levels, referred to as a manifold. Transitions between a pair of energy levels of different manifolds are usually narrow compared with the splitting of the energy levels. Therefore, the spectra of lanthanide series atoms in crystals appear as a series of relatively narrow lines. In glasses however, the regular structure of crystals is no longer applicable. The disordered nature of a glass smears the energy levels out. In this case, spectra of lanthanide series atoms in glasses appear as a continuous function.

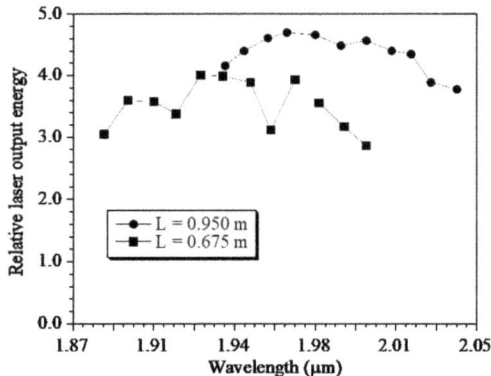

Figure 23. Tm:germanate tuning range depends on the length of the fiber.

Materials Research Foundations 164 (2024) 1-66   https://doi.org/10.21741/9781644903056-1

Fiber lasers display more tuning capability than their crystalline counterparts because the disordered nature of a glass smears the energy levels out. In this case, spectra of lanthanide series atoms in glasses appear as a continuous function, thereby affording them more continuous tuning capability. A typical tuning curve appears in Fig. 23. This figure shows the tuning range in a Tm:germanate fiber and the dependence on fiber length. Shorter fiber lengths are seen to shift the tuning range to shorter wavelengths in this material.

6.   Fiber lasers

6.1   Er-doped silicate glass fiber lasers

Er-doped glass fiber is the most extensively studied optical gain fiber doped with rare-earth-ions. As its invention in the late 1980s, Er-doped glass fiber was immediately considered the best gain fiber candidate for optical fiber amplifier application in new-generation tele-communication systems. Since then, it has revolutionized information communication technology, and nowadays has been widely used in fiber communication systems worldwide, changing the life of human beings in many aspects. Er-doped glass fiber technology has also dramatically promoted/accelerated the research of itself and other rare-earth-ions (Yb, Tm, Ho) glass fibers for industrial applications, including revolutionizing fiber laser technology.

With the help of remarkable technological progress and huge commercial success in optical fiber communication industry, fiber laser technology has emerged as a cutting-edge laser technology, experiencing a remarkable progress in the past two decades. Due to their superior performances [38], in terms of high average power, high beam quality, high efficiency, and high reliability, fiber lasers (including Yb-doped fiber lasers at 1 μm, Er-doped fiber lasers at 1.55 μm, Tm/Ho-doped fiber lasers at 2 μm) have been replacing conventional solid-state and gas lasers in many applications. Er-doped fiber lasers, operating at eye-safe wavelength near 1.55 μm, are very useful for a variety of military and civil applications, such as remote sensing, material processing, and free space optical communications due to their unique features of retina-safe qualities and low attenuation in the atmosphere. To date, the highest continuous-wave laser power at 1.55 μm was limited at ~300 W [39], which was obtained from an erbium/ytterbium (Er/Yb) co-doped fiber laser pumped by laser diodes at 9xx nm. Compared to the well-established multi-kilowatt Yb doped fiber lasers near 1 μm [40], power scaling of an eye-safe fiber laser at 1.55 μm is still quite limited. It is even worse than power scalability of other kinds of popular eye-safe fiber lasers at longer wavelengths, i.e., Tm- or Ho-doped fiber lasers near 2 μm. The eye-safe wavelength in the 1.55 μm band is still preferable for some applications in the atmosphere, because the 2 μm band is overlapped with some strong absorption features of water in the atmosphere.

With the current fiber laser design for the 1.55 μm band, i.e., erbium/ytterbium (Er/Yb) co-doped gain fiber pumped by well-established 9xx nm laser diodes, the power scaling limitation is fundamental. Because of a large quantum defect from 9xx nm pumping wavelength to 1.55 μm laser wavelength, the highest slope efficiency of an Er/Yb-doped fiber laser is about 30% or less, much lower than that of an Yb-doped fiber laser (> 80%). As a result, a lot more heat is generated in an Er/Yb-doped fiber laser than in its 1 μm counterpart at a same power level. Heat issues are always one of the most important limits for laser power scaling at high power level. More importantly, those Yb ions co-doped in Er/Yb-doped fibers do not allow for very high-power laser operation at 1.55 μm, because of Yb-ion emission or the onset of parasitic lasing at 1 μm in Er/Yb-doped fiber lasers as the pump power is high.

In this section, we present our development of high-efficiency and high-energy Er-doped fiber lasers by using heavily Er-doped silicate glass fibers – a multi-component non-silica glass fiber. The first sub-section describes resonantly-pumped continuous-wave fiber lasers at 1.55 μm by using Yb-free Er-doped silicate glass fibers. The second sub-section presents high energy pulse lasers, and the third sub-section details particularly on a high-energy narrow-linewidth all-fiber laser at a long wavelength (1572 nm), by using heavily Er/Yb co-doped silicate glass fibers.

6.2   Resonantly-pumped, continuous-wave, ER fiber lasers

In high power continuous-wave 1.55μm fiber lasers, two engineering issues are typically associated with the most commonly used Er/Yb co-doped silica glass fibers. One issue is their poor laser efficiency, as compared to their counterparts at 1μm or even at 2μm spectral regions. The poor laser efficiency results in difficult thermal management in the lasers, which limit their capability of power scaling. The other issue is amplified spontaneous emission of the co-doped Yb ions or even the onset of parasitic lasing at 1μm in high-power Er/Yb-doped fiber lasers, which again limits their capability of power scaling.

A new design for high-power 1.55 μm fiber lasers was proposed to eliminate these two issues, which is a Yb-free Er-doped fiber laser pumped resonantly (so it is called as resonant pumping, or in-band pumping, or tandem pumping) [41]. This new design has attracted great interest in recent years with some successful experimental demonstrations of ~85% slope efficiency (with respect to absorbed pump power) in a core-pumped Yb-free single mode Er fiber laser [42], and 69% slope efficiency in a cladding-pumped laser [43], respectively. With such a high slope efficiency, Er-doped fiber lasers should have the potential for achieving a similar power scalability as multi-kilowatts Yb-doped fiber systems.

One of the critical challenging issues in realizing high-power resonantly-pumped Yb-free Er-doped fiber laser is to develop heavily Er-doped glass fiber with high pump absorption and high quantum conversion efficiency. The most commonly used Er-doped fiber is made of silica glass. However, the doping concentration of rare-earth ions in silica glass is limited due to its intrinsic glass network structure. Silica glass with a high Er-doped concentration will form the so-called Er ion-clusters and easily produce detrimentally cooperative up-conversion, resulting in a lower quantum conversion efficiency. Various approaches are developed to increase the doping concentration, including co-doping with Al_2O_3, B_2O_3, and P_2O_5 and using nanoparticles. The highest doping level for erbium ions without the so-called ion-cluster is less than 5,000 ppm.

In those above-mentioned demonstrations with high-efficiency Er-doped fiber lasers, however, very long pieces of Yb-free Er-doped fiber must be used (12 m for core pumping configuration [42] and 15m for cladding pumping configuration [43]), due to their low doping concentration (thereby low pumping absorption) in those Er-doped silica glass fibers used in the experiments. The use of such a long piece of gain fiber could raise another concern about its power scalability, i.e., optical nonlinear effects in the gain fiber (such as Brillouin or Raman scattering), especially for laser operation with narrow-band or pulsed output. Therefore, it is necessary to develop new glass fibers that enable high doping concentration of Er ions in order to achieve high-power resonantly-pumped Er-doped fiber lasers by using a relatively short piece of gain fiber.

Rare Earth - A tribute to the late Mr. Rare Earth, Professor Karl Gschneidner   Materials Research Forum LLC
Materials Research Foundations 164 (2024) 1-66   https://doi.org/10.21741/9781644903056-1

It is well known that multi-component glasses exhibit excellent solubility for rare-earth oxides. Common multi-component glasses include phosphate glasses, germanate glasses, borate glasses tellurite glasses, and silicate glasses with P_2O_5, GeO_2, B_2O_3, TeO_2, and SiO_2, as glass network formers, respectively. Phosphate glass is the best-known one of the multi-component glasses for high Er concentrations. One of the inherent drawbacks of phosphate glass fiber is its high propagation loss of ~ 0.2 dB/cm, resulting in low slope efficiency. The high propagation loss of phosphate glass fiber is mainly caused by the high volatility of P_2O_5. The vaporized P_2O_5 reacts with surrounding materials and the impurity will deposit back to the glass. During fiber drawing process, the surface vaporization of P_2O_5 produces additional scattering loss at the interface between the core and the cladding of the fiber. The typical propagation loss of erbium doped phosphate glass fiber is around 0.2 dB/cm while the silicate glass fiber is 0.007 dB/cm. The high propagation loss of phosphate glass fiber reduces the slope efficiency significantly. Although a short length (tens of centimeters) of phosphate glass fiber can provide sufficient high gain for power scaling in a fiber amplifier, due to its high gain per unit length (several dB/cm), heat dissipation becomes a big issue while power scaling at a very high level (> 100W) in such a short gain fiber, not to mention its low glass transition temperature.

Fortunately, silicate glass, multi-component glass with SiO_2 as the glass network former, also exhibits excellent solubility for rare-earth oxides. The glass network modifiers, such as sodium ions, potassium ions, barium ions, and calcium ions, break the well-defined glass network of silica, which produces many sites for rare-earth ions, thereby enabling high Er concentrations in silicate glass compositions. A doping concentration of 65 wt% rare-earth has been demonstrated at AdValue Photonics for a Tb_2O_3-doped all-fiber isolator project [44, 45]. Silicate glasses typically exhibit much stronger mechanical strength than phosphate because of the following two reasons. First, because of the pentavalency of phosphorus, only three out of four corners of the tetrahedron of PO_4, the principal building unit of phosphate glass, are connected, resulting in relatively weak glass structure network. Second, the bond strength of Si-O is stronger than that of P-O. Logically the mechanical strength of silicate glasses typically is much stronger than phosphate glasses.

In addition, silicate glasses are more compatible with silica glass. Although silica and silicate glass are two different types of glasses according to glass science definition, they have the same glass network former, which is Si-O tetrahedron. Silica glass contains near 100% SiO_2, silicate glass has approximately 70% SiO_2. Therefore, silicate glass is compatible with silica glass, which enables the low fusion splicing loss between silica fiber and silicate glass fiber. The fusion splicing loss between silicate glass fiber and silica fiber is lower than 0.1 dB. Considering most commercial passive fiber components on the market are pigtailed with silica glass fiber, good glass compatibility with silica glass fiber is another important benefit from the use of Er-doped silicate glass fiber, as compared with other multi-component glass fibers.

Yb-free Er-doped silicate glass fibers can be fabricated by using a standard rod-in-tube fiber drawing technique. Since it is easy to design and engineer the compositions of silicate glasses for different refractive index, it can readily fabricate all-glass double cladding (DC). A heavily Er-doped (1 ~ 2 wt%) silicate glass with a high refractive index (n ~ 1.6) can be used to form core glass, and the other two un-doped silicate glasses with low refractive index (n ~ 1.5) can be used as inner cladding and outer cladding glasses. The core glass rod was first drilled from the Er-doped silicate glass and its barrel was then high-surface-quality polished. Two cladding glass tubes were done similarly. Fig. 24 shows various fiber designs for rare-earth-ion doped silicate glass fibers fabricated by the rod-in-tube technique.

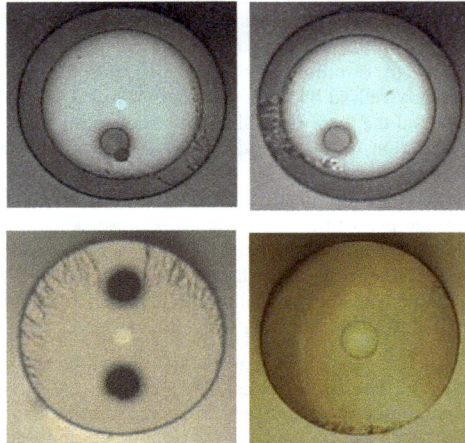

Figure 24. Rare-earth-ion doped silicate glass fibers. (a) double-cladding single-mode fiber. (b) double-cladding LMA fiber. (c) single-cladding polarization-maintaining fiber. (d) single-cladding LMA fiber.

As aforementioned, resonant pumping of an Er-doped fiber laser can limit high thermal load and result in high laser efficiency, leading to a great potential of power scaling. Resonant pumping wavelength of an Er-doped fiber laser can be either 1480 nm with a Raman fiber laser, or simply 1535 nm with another Er-doped fiber laser. For simplicity here, we only focus on resonant pumping at 1535 nm in the subsequent discussions. All measurements are then carried out with continuous wave (CW) operating mode. By using heavily Er-doped (Yb-free) silicate glass fibers, high laser efficiency can be achieved in a resonantly-pumped Er-doped fiber amplifier in both a core-pumping and clad-pumping configurations.

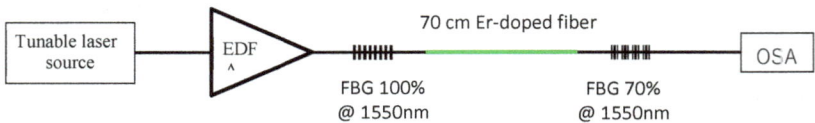

Figure 25. A schematic setup of a core-pumped Er-doped fiber laser oscillator.

46

Rare Earth - A tribute to the late Mr. Rare Earth, Professor Karl Gschneidner   Materials Research Forum LLC
Materials Research Foundations 164 (2024) 1-66   https://doi.org/10.21741/9781644903056-1

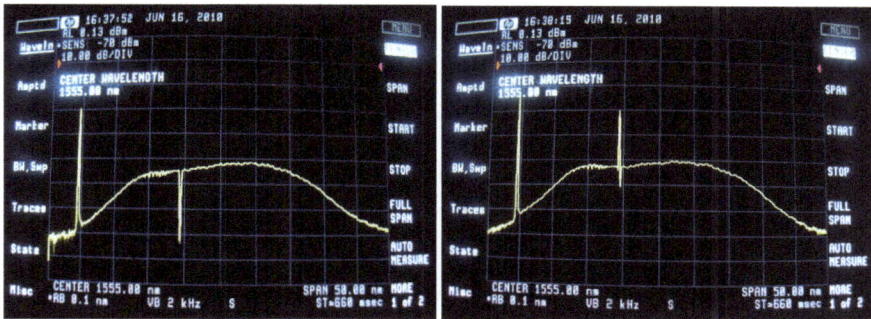

Figure 26. Optical spectra of the resonantly-pumped Er-doped fiber laser oscillator.
Left: below lasing threshold. Right: above lasing threshold.

Fig. 25 shows a schematic setup of a core-pumped Er-doped fiber laser oscillator, which was pumped at 1535 nm. The pump source was a tunable laser source with a 1 W-level fiber power booster (i.e., EDFA). The resonantly-pumped Er-doped fiber laser cavity was formed with a pair of fiber Bragg gratings at 1550 nm, spliced with 70 cm Yb-free Er-doped silicate glass fiber with 1 wt% Er doping concentration. The output coupler grating had a 70% reflectivity. The laser output at 1550 nm can be separated and monitored by using an optical spectrum analyzer (OSA). Fig. 26 shows the measured optical spectra of the resonantly-pumped Er-doped fiber laser oscillator, below lasing threshold and above lasing threshold, respectively. Lasing threshold of the resonantly-pumped Er-doped fiber laser oscillator at 1550 nm was about 100 mW for a pumping wavelength at 1535 nm.

The single-mode Yb-free Er-doped silicate glass fiber with 1 wt% Er doping concentration was also tested in a core-pumped fiber master oscillator power amplifier (MOPA) configuration [46]. A 500-mW single-mode fiber laser at 1560 nm was used as a seed laser, injecting into the core-pumped fiber amplifier. The pump laser was a single-mode Er-doped fiber laser at 1535 nm with a maximum power at about 8 W, which was combined with the seed laser by using a home-made 1535 nm/1560 nm wavelength division multiplexer (WDM). After the WDM, the available pumping power was about 6 W. A bulky diffraction grating was used to separate the pump and seed laser at the single-mode fiber output exit.

Fig. 27 shows the 1560 nm laser output power and amplification gain from a core-pumped fiber master oscillator power amplifier using 1-m long piece of single-mode Yb-free Er-doped silicate glass fiber. The laser output power increases linearly with increasing pump power with a slope efficiency of about 75%. A maximum 4 W output power was obtained in the experiment, limited by the available pump power at 1535 nm.

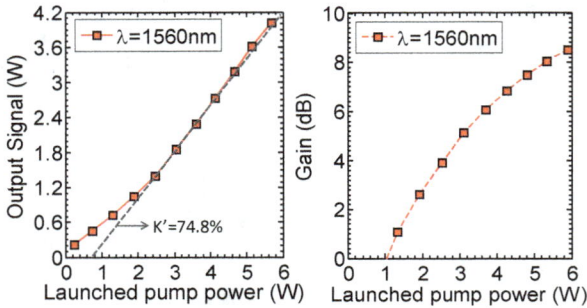

Figure 27. Laser output power and amplification gain at 1560 nm from a core-pumped fiber master oscillator power amplifier using 1-m long piece of single-mode Yb-free Er-doped silicate glass fiber. A maximum 4 W output power was obtained in the experiment, limited by the available pump power at 1535 nm. Copied from [46].

Double-cladding single-mode Yb-free Er-doped silicate glass fiber has been used in a resonantly cladding-pumped fiber amplifier [46], with an experimental setup shown in Fig. 28. Six identical fiber lasers at 1535 nm were used as the pump sources in the resonantly cladding-pumped fiber amplifier. Although these six pump sources had single-mode output beams, they were combined by using a commercial (6+1) x 1 pump combiner. After pump combining, the available maximum pump power was about 50 W at 1535 nm. The resonantly cladding-pumped fiber amplifier was seeded with 1W single-mode signal laser at 1560 nm. The double-cladding Yb-free Er-doped silicate glass fiber had a length of 2 m.

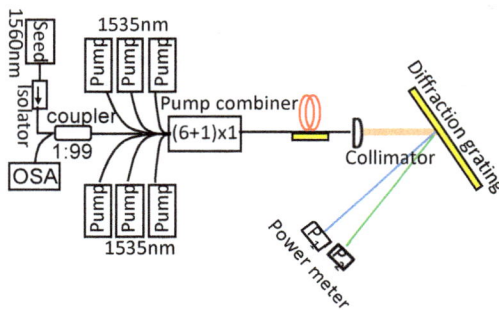

Figure 28. Experimental setup for a resonantly cladding-pumped fiber amplifier with 2 m long Yb-free Er-doped silicate glass fiber with 50 W of pump power at 1535 nm available for the fiber amplifier. Copied from [46].

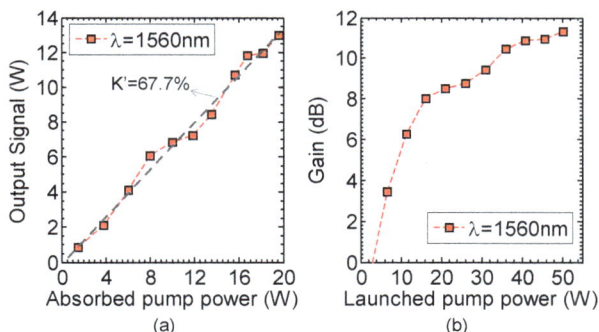

Figure 29. Laser output power and amplification gain from a resonantly clad-pumped Er-doped fiber amplifier with 2 m long Er-doped (Yb-free) silicate glass fiber, seeding with a 1 W laser at 1560 nm. The maximum available pump power was 50 W at 1535 nm. Copied from [46].

Fig. 29 shows laser output power from the resonantly cladding-pumped Er-doped fiber amplifier. Output power at the signal wavelength of 1560 nm increases linearly with pump power and has a slope efficiency of ~68% with respect to absorbed pump power. The output power was about ~ 13 W at 1560 nm when the maximum pump power of 50 W was launched, of which 40% pump power was absorbed by the 2 m gain fiber. As compared with the above-mentioned core-pumping configuration, the pump absorption in the cladding-pumping configuration was poorer, due to a small ratio of the doped area in the fiber core to pump beam area in fiber cladding (9 μm core diameter versus 140 μm cladding diameter).

6.2   High peak power, short pulse operation, Er fiber lasers

Over the last two decades, high-power continuous-wave fiber lasers have gained a tremendous commercial success, because they offer superior performances over other conventional lasers, in terms of laser efficiency, average output power, compactness, and robustness. At the same time, fiber-based pulsed lasers have also received considerable attention, because of their potentials for many applications, such as LIDAR, remote sensing, nonlinear frequency conversion, and various material processing. Although fiber-based pulsed lasers have already been widely used in a variety of applications, they don't gain the same success as continuous-wave fiber lasers because of their limited pulse energy and peak power as compared with their crystal-based solid-state counterparts. The major limitations to high pulse energy and high peak power in fiber lasers/amplifiers are various nonlinear optical effects, occurring over a long interaction length in optical fiber within its small fiber core, which can result in substantial degradations in laser pulse performance, including a limited maximum extractable output power, nonlinear wavelength shift, unwanted spectral broadening, and fiber damage.

For Er-doped pulsed fiber lasers using commercial silica-based gain fibers, currently the maximum achievable pulse energy and peak power are typically limited to a little more than 100 μJ and a few kW, respectively. However, if using heavily doped multi-component glass fiber as the gain fiber, the maximum achievable pulse energy and peak power can easily be increased by one order of magnitude. We have demonstrated > 1 mJ pulse energy corresponding to > 100 kW peak power with near diffraction limited beam quality ($M^2 < 1.2$) from all-fiber Er-doped silicate glass lasers.

For some military applications, such as 3D Lidar and target illuminating and tracking, high-peak-power short-pulse eye-safe lasers, capable of transmitting through the atmosphere with minimum absorption, are required. Such lasers in the 1.55 μm spectral band are particularly desirable in applications for target detection and identification, because it can take the advantage of commercial availability of highly sensitive, fast-response InGaAs-based imaging sensors.

Heavily Er-doped silicate glass fibers offer high pump absorption and high laser gain with a short length of gain fiber, which is greatly beneficial for mitigating detrimental nonlinear optical effects, allowing for generation of high pulse energy and peak power from fiber-based lasers. In general, for pulsed fiber lasers, there are two main concerns for their high-power operation. One is optical damage limit and the other is nonlinear effect limit. The onset of the nonlinear effect limit usually comes earlier than optical damage limit, especially for spectrally narrow-band laser operation. Different technical strategies may have to be used to overcome these two different limits. To test the capability of generating high pulse energy and high peak power with heavily Er-doped silicate glass fibers, two different experiments are designed to decouple these two issues, thereby investigating what the output power limits are when they are used as the gain fibers for high-power short-pulse lasers with broad or narrow spectral bandwidth. In the two different experiments, different seed laser sources are used, either a broadband laser diode (Fabry-Perot cavity) or a narrow-band DFB laser diode.

With a broadband Fabry-Perot laser diode (with ~ 5 nm spectral bandwidth centered at 1545 nm) used as the seeder, stimulated Brillouin scattering (SBS) effect is not a major concern in high-power short-pulse fiber lasers. Power/energy scalability of heavily Er-doped large-core silicate glass fibers have been tested when they are used as gain fibers in last-stage power amplifiers pumped resonantly at 1480 nm or regularly at 976 nm. In both pumping configurations, the seed laser pulses have 5 ns pulse duration generated from a direct-modulated broadband Fabry-Perot diode at 10 kHz repetition rate.

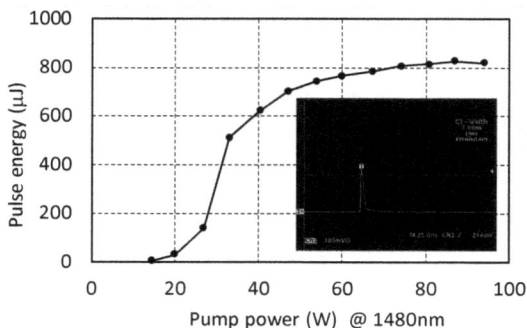

Figure 30. Laser pulse energy vs pump power from a single-mode 100 W Raman fiber laser (IPG Photonics) at 1480 nm. The gain fiber was Yb-free Er-doped silicate glass fiber with 80 cm length and 50 μm core diameter.

Rare Earth - A tribute to the late Mr. Rare Earth, Professor Karl Gschneidner   Materials Research Forum LLC
Materials Research Foundations 164 (2024) 1-66   https://doi.org/10.21741/9781644903056-1

Fig. 30 shows laser pulse energy generated from a broadband fiber laser system, including a last-stage fiber amplifier constructed with 80 cm long, 50 μm core diameter of Yb-free Er-doped (1.2 wt%) silicate glass fiber, core-pumped by commercial single-mode 100 W Raman fiber laser (IPG Photonics) at 1480 nm. The maximum pulse energy reached > 800 μJ with < 4 ns pulse duration, corresponding to 200 kW peak power. The laser pulse energy increases rapidly at low pump level, and gradually saturated at high pump level. The saturation of pulse energy extraction from the last-stage fiber amplifier is attributed to pump bleaching at high pump level. Higher pulse energy is expected if slightly increasing fiber gain length.

A similar test has been done with short-pulse laser amplification using Er-Yb-codoped silicate glass fiber as the gain medium in the last-stage fiber amplifier. Fig. 31 shows laser pulse energy generated from a last-stage fiber amplifier constructed with 30 cm long, 45 μm core diameter of Er-Yb-doped (1 wt% Er and 5 wt% Yb) silicate glass fiber, clad-pumped by pump diodes at 976 nm. When pump power was > 120 W, a maximum pulse energy > 927 μJ with < 4 ns pulse duration was obtained, corresponding to > 200 kW peak power. The laser pulse energy is gradually saturated at high pump power, similarly because of pump bleaching.

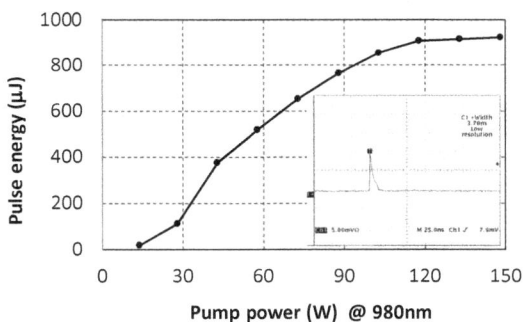

Figure 31. Laser pulse energy as a function of pump power from multimode pump diodes at 976nm. The gain fiber was 30cm-long 45mm-core diameter of Er-Yb-doped (1wt% Er and 5wt% Yb) silicate glass fiber.

It is noted that optical damage is not seen with > 200 kW peak power for nanosecond laser pulses at the surface of the heavily Er-doped silicate glass fibers, without using an endcap. Considering the 50 μm core diameter of the Yb-free Er-doped silicate glass fiber in 1480 nm core-pumping experiment and the 45 μm core of the Er-Yb-codoped silicate glass fiber in 976 nm clad-pumping experiment, the < 4 ns laser pulses with >200 kW peak power has a calculated laser intensity density of 10 GW/cm^2 at the end surface of the gain fibers, indicating their capability of high laser intensity handling. With such a high peak power and high intensity in the gain fibers, the laser pulses inevitably experience spectral broadening. Fig. 32 shows laser spectra of an input laser pulse (with < 10 μJ pulse energy) and output laser pulse (with >100 μJ pulse energy) after a fiber amplifier with ~ 40 cm long, 20 μm-core, Er-Yb-codoped silicate glass fiber and 1 m long passive delivery fiber (20 μm core diameter).

Rare Earth - A tribute to the late Mr. Rare Earth, Professor Karl Gschneidner   Materials Research Forum LLC
Materials Research Foundations 164 (2024) 1-66   https://doi.org/10.21741/9781644903056-1

When a directly-modulated DFB laser is used as the seed laser, high-peak-power narrow-bandwidth laser pulses can be generated from an all-fiber laser system with a short piece of heavily Er-Yb codoped silicate glass fibers. Fig. 33 shows optical spectrum of such an all-fiber narrow-bandwidth laser with > 350 µJ energy of ~ 10 ns laser pulses at 1545 nm, which were generated directly from large-mode-area Er-Yb-doped (1 wt% Er and 5 wt% Yb) silicate glass fiber with a length of 30 cm and a core diameter of 45 µm. Symmetrical optical side-bands are generated in the output laser spectrum, indicating the characteristic spectral feature of a specific nonlinear effect, i.e., four-wave mixing or modulation instability. A high-resolution spectral measurement, using a scanning Fabry-Perot interferometer, shows that the spectral bandwidth of the ~10 ns laser pulses was about several GHz, mainly caused by laser frequency chirping during direct diode modulation.

Transform-limited linewidth pulsed fiber lasers with a relatively long pulse width (several 100's of nanosecond) and high pulse energy are very useful for coherent lidar applications. This kind of high-energy, narrow-linewidth pulsed lasers can be constructed in an all-fiber format by taking advantage of heavily Er-doped silicate glass fibers, in terms of high pump absorption and high laser gain with a short length of gain fiber, as demonstrated above. Particularly, heavily Er-doped silicate glass fibers offer superior performance over standard Er-doped silica-based fibers when operating in a long wavelength with high-energy laser pulses. The next subsection will discuss our development of a high-energy transform-limited-linewidth pulsed fiber laser operating at 1572 nm.

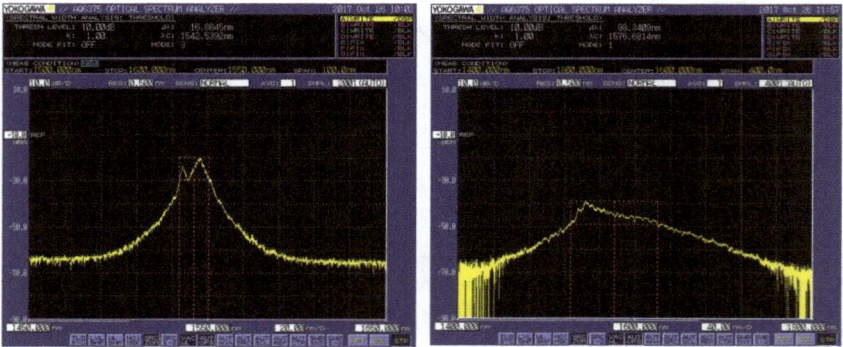

Figure 32. Nanosecond laser pulses experience spectral broadening in a high-peak-power Er-doped fiber amplifier. Left graph: input laser pulse with10 dB bandwidth of ~17 nm. Right graph: output laser pulse with 10 dB bandwidth of ~ 98 nm.

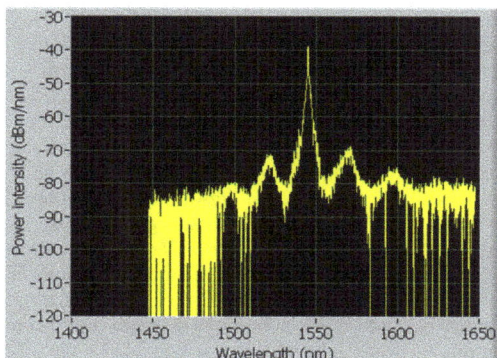

Figure 33. Optical spectrum of an all-fiber narrow-bandwidth laser with > 350 μJ energy of ~10 ns laser pulses at 1545 nm.

6.3   High-energy, narrow linewidth, pulsed Er fiber lasers

There has been a significant need of developing a high energy single frequency optical pulse transmitter for atmospheric CO_2 absorption column measurements from space, which was motivated by a NASA on-going research project - the Active Sensing of CO_2 Emissions over Nights, Days, and Seasons (ASCENDS) mission [47]. In this mission, it requires transform-limited narrow-linewidth optical pulses at 1572.335 nm with a high beam quality and high pulse energy (multiple mJ) at a repetition rate of several kHz, and a pulse width of several 100s of nanoseconds in a rugged and reliable platform [48]. The CO_2 absorption line used in this mission was chosen near 1572nm for several reasons. As compared to other absorption lines of CO_2, this absorption line has a relatively less insensitivity to temperature changes, no absorption interference by other atmospheric species, and an appropriate absorption line strength allowing space-borne column measurements without a concern about absorption saturation. Robust and alignment-free single frequency all-fiber master oscillator power amplifier (MOPA) system has been identified as a promising transmitter option due to many potential advantages for the space application in the NASA mission.

Several research groups have made efforts to generate high energy pulses at 1572 nm from an Er-doped fiber amplifier (EDF) [49] or an Er-Yb co-doped fiber amplifier (EYDF) MOPA system [50]. When it comes to an all-fiber MOPA system at 1572 nm, however, most demonstrations have been limited to a maximum achievable pulse energy only at several hundreds of micro-joule level to date. The fundamental challenge is that the available gain at 1572 nm is typically much poorer than the gain at short wavelengths in standard Er-doped fiber amplifiers, therefore, requiring a relatively long length of gain fibers for pulse energy amplification against predominant amplified spontaneous emission (ASE) background at short wavelengths. A long length of gain fiber in the 1572 nm fiber amplifier system could also easily result in pulse energy clamping either induced by stimulated Brillouin scattering (SBS) in the case of a strong seed input or by parasitic lasing of out-of-band ASE in the case of a weak seed input, thereby limiting the maximum extractable pulse energy level to ~ 100-200 μJ at 1572 nm.

A good strategy to attain mJ-level pulse energy at 1572 nm from the all-fiber MOPA system is to use multi-stage fiber amplifier for efficient pulse energy amplification in company with a rigorous out-of-band ASE filtering. The prerequisite is to use gain fibers with a sufficient gain per unit length at 1572 nm, which can be benefitted from using heavily Er-doped gain fibers. Nonetheless, it still needs to do a trade-off between several amplifier parameters, including gain fiber length, optical gain, and ASE noise, for the optimization of laser system. Each stage fiber amplifier in the system can employ a short and a minimum required length of the gain fiber for pulse energy amplification, thereby reducing the detrimental out-of-band ASE noise. The operation level of each fiber amplifier needs to be optimized for best signal-to-noise ratio (SNR) rather than maximum pulse energy until the last stage power amplifier, since improperly high-gain operation of a typical fiber pre-amplifier can cause fast ASE growth which results in gain reduction at 1572 nm or even parasitic lasing at short wavelengths in its subsequent fiber amplifiers if without any spectral filtering. In order to avoid a SBS-induced pulse energy limit, core size of the gain fibers at each stage of the amplifiers needs to be increased in accordance with the pulse energy growth.

Multi-component silicate glasses not only offer the capability of high doping concentrations for rare-earth ions (e.g., Er-Yb co-doping) – which in turn enables high pump absorption and high optical gain within a short piece of gain fiber, but also modify their spectral behaviors as well, when compared with silica-glass-based gain fiber, due to both its unique glass compositions and high doping concentrations (thereby modifying the interaction between ions and host, and ions and ions) [51]. With the spectral modification in heavily Er-Yb co-doped silicate glasses, it turns out that their gain spectrum profile is shifted toward long wavelength side – which is beneficial to pulse amplification at long wavelengths.

Fig. 34 shows the cross-section view of two different kinds of Er-Yb co-doped fibers. The fiber in the left is a single-mode double-cladding heavily Er-Yb (1%wt Er and 5%wt Yb) co-doped silicate glass fiber with 8μm core diameter, 145 μm inner glass cladding, and 167 μm outer glass cladding, which has 0.148 NA for the fiber core and 0.5 NA for the inner cladding. The fiber in the right is a typical commercial Er-Yb-doped silica glass double-cladding fiber with 10 μm core diameter, 125 μm inner glass cladding, and 250 μm outer polymer cladding (not shown in the picture).

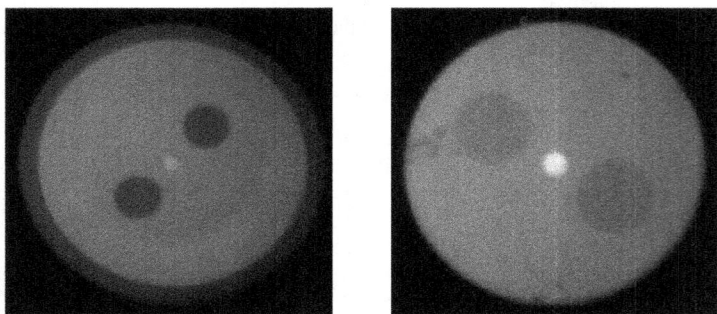

Figure 34. Cross-section view of Er/Yb co-doped silicate glass fiber laser (Left) and commercial Er/Yb co-doped silica glass fiber laser (Right).

Rare Earth - A tribute to the late Mr. Rare Earth, Professor Karl Gschneidner   Materials Research Forum LLC
Materials Research Foundations 164 (2024) 1-66   https://doi.org/10.21741/9781644903056-1

Optical performance of the Er/Yb co-doped silicate glass fiber and commercial Er/Yb co-doped silica glass fiber has been characterized and compared, when used for small-signal amplification in a clad-pumped fiber amplifier. The clad-pumped fiber amplifier was built by either a 1 m long of heavily-doped silicate glass gain fiber or a 2 m long of commercial silica glass gain fiber, seeded with a tunable laser source at an input power of – 6 dBm.

Fig. 35 shows the measured small-signal gain and the signal-to-noise ratio (SNR), as a function of wavelength. At shorter wavelength region below 1545 nm, the commercial silica glass gain fiber exhibits excellent performance in both net gain and SNR. However, at the longer wavelength region from 1565 to 1575 nm, particularly at 1572 nm, the heavily doped silicate glass gain fiber performs significantly better than the commercial silica glass gain fiber in both net gain and SNR. Specifically, when seeded at 1572 nm, the heavily doped silicate glass gain fiber exhibits about 6 dB higher gain as well as 10 dB higher SNR at 1572 nm than the commercial silica glass gain fiber.

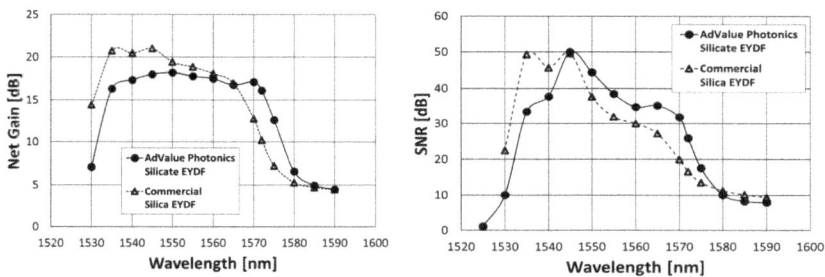

Figure 35. Performance comparison of Er/Yb co-doped silicate glass fiber and commercial Er/Yb co-doped silica glass fiber: wavelength-dependent net gain (a) and signal-to-noise ratio (SNR) (b). Copied from [14].

Fig. 36 shows laser output spectra of the clad-pumped fiber amplifier with the two different gain fiber. Obviously, silicate glass gain fiber performs significantly better than the commercial silica glass gain fiber at 1572 nm.

To produce high energy pulses at 1572 nm, a pulsed seed laser source was built, which consisted of a continuous-wave (CW) DFB laser diode with a MHz-level spectral linewidth centered at 1572 nm, fiber preamplifiers, a fiber-pigtailed acousto-optic modulators (AOM) for pulse generation as well as pulse shaping, and a spectral filter with a 3 dB bandwidth of 0.2 nm to remove ASE noise in the pulsed seed laser source. Ultimately, the pulsed seed laser source at 1572 nm generated ~ 500 ns pulses at the repetition rate of 2.5 kHz with a pulse energy > 6 µJ and a SNR > 50 dB.

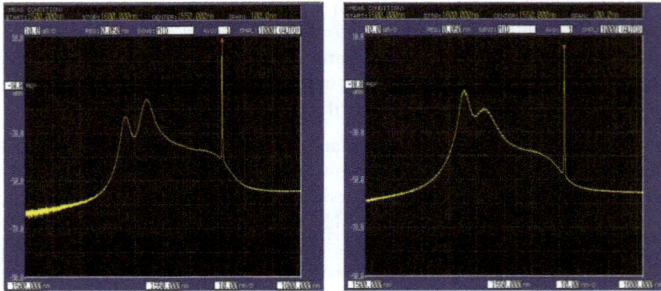

Figure 36. Laser output spectrum from a clad-pumped EDFA with 1 m silicate glass gain fiber (Left) 2 m commercial silica glass gain fiber (Right) seeded with 6 dBm input power at 1572 nm.

By using an AdValue proprietary heavily Er-Yb co-doped silicate glass fiber amplifiers (EYDFAs), we have demonstrated 1.8 mJ, peak power of 3.5kW at the repetition rate of 2.5 kHz single frequency optical pulses at 1572 nm from an all-fiber MOPA system. To our best knowledge, this is the highest pulse energy of single frequency at 1572 nm from all-fiber amplifier system. We also have achieved 1.3 mJ, peak power of 2.5 kW at 7.5 kHz.

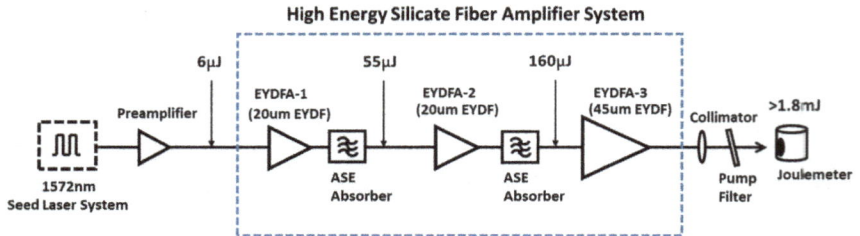

Figure 37. Schematic of the high energy all-fiber MOPA system at 1572nm based on single-mode large-core polarization-maintaining double-cladding silicate Er-Yb co-doped fiber.

To boost the pulse energy at 1572 nm up to mJ-level, we have developed a high energy multi-stage all-fiber amplifier system, with a schematic diagram shown in Fig. 37, by using our proprietary single-mode large core polarization maintaining double-cladding silicate glass Er-Yb co-doped fiber. The entire amplifier system was constructed with 3-stage cascaded Er-Yb doped fiber amplifiers (EYDFA-1 to 3). First two amplifiers (EYDFA-1 and EYDFA-2) were configured with a 28 cm and a 33 cm long single-mode large core silicate Er-Yb co-doped fiber with a core diameter of 20 μm and the last stage amplifier (EYDFA-3) was configured with a 55 cm long single-mode large core silicate Er-Yb co-doped fiber with a core diameter of 45 μm. All of those Er-Yb co-doped fibers were made of silicate glass with co-doping of 1wt% of Er and 5wt% of Yb.

From the first two EYDFAs (EYDFA-1 and -2), we could be able to obtain the output pulse energy of > 70 μJ and > 200 μJ at the pump power of ~10W at 976 nm through a PM (2 + 1) x1 pump combiner. In order to achieve highest pulse energy from the last stage amplifier, obtaining high quality input pulses with a high SNR is very crucial. For this reason, the operation level of the EYDFAs were optimized by slightly lowering pumping power for a better SNR. At the same time, a short piece of un-pumped gain fiber with the same core diameter was inserted after each of the first two EYDFAs (EYDFA-1 and -2) output as an ASE absorber, based on the intrinsic three-level characteristics of rare earth element doped fiber amplifier, to absorb undesired ASE background from the silicate glass EYDFs. As it can be seen from the optical performance of the Er-Yb co-doped silicate glass fiber, as shown in Fig. 35 (a), the signal gain at 1572 nm is > 20 dB smaller than that at 1545 nm. Therefore, absorption of ASE around 1545 nm through the un-pumped gain fiber is substantially stronger than the absorption of the signal at 1572 nm. Owing to the un-pumped gain fiber-based ASE absorber, a SNR of > 50 dB at the output pulse energy of 55 μJ from the first stage high energy amplifier (EYDFA-1) and a SNR of > 40 dB at the output pulse energy of 160 μJ from the second stage (EYDFA-2) can be obtained.

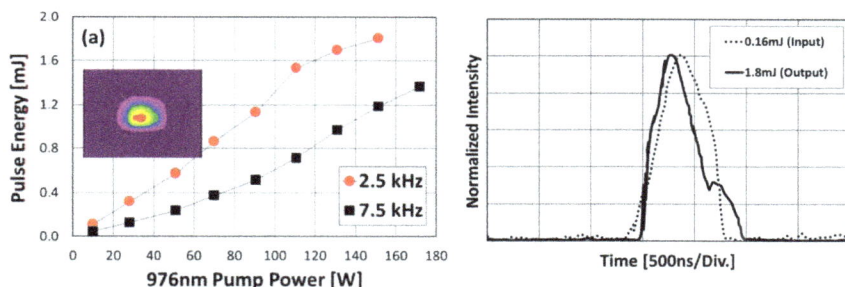

Figure 38. The measured output pulse energy at 2.5 kHz and 7.5 kHz versus pump power (a) and the pulse traces at 2.5 kHz at the pulse energy of 1.8 mJ (b). The inset is a typical beam profile of the last stage power amplifier. Copied from [14].

Continuously, the 160 μJ output pulses at 2.5 kHz were fed into the last stage amplifier (EYDFA-3) based on the 45 μm core diameter fiber. The fiber ends of the last stage amplifier (EYDFA-3) were angle-cleaved with an angle of ~10 degree to avoid the unnecessary signal feedback. The output pulse energy at 1572 nm was recorded using a commercial joulemeter (Gentec, Mach5) after a pump filter in free space and the pulse trace was recorded using an InGaAs photodetector (ThorLabs, PDA10D) and an oscilloscope (Tektronix, TDS 754A).

The output pulse energy from the last stage amplifier (EYDFA-3) pumped at 976 nm is shown in Fig. 38 (a). At the pulse repetition rate of 2.5 kHz, we have obtained a highest pulse energy of 1.8 mJ at the pump power of 151 W delivered through a PM (2 + 1) x1 pump combiner by two of 976 nm wavelength-locked 85 W pump diodes. The output pulse trace from the last stage EYDFA at the pulse energy level of 1.8 mJ at 2.5 kHz is shown in Fig. 38 (b). The output pulse width was measured to be ~ 510 ns FWHM. The corresponding peak power of the pulse energy of 1.8mJ was estimated to be >3.5 kW from the measured pulse width. When the pump power increased to the output pulse energy of >1.7 mJ, an energy saturation behavior was observed. Furthermore, the output pulses become unstable at the pulse energy of 1.8 mJ. It should be noted that the system output is not limited by SBS at this energy level since the theoretical SBS threshold of the last stage amplifier (EYDFA-3) with a core diameter of 45 μm is estimated to be > 5 mJ [46].

Evidently, we have measured the onset of parasitic lasing at 1535 nm at the pulse energy of around 1.8 mJ.

The spectral evolution of the output pulses from the last stage amplifier (EYDFA-3) at the pulse energy of > 1 mJ was measured with an optical spectrum analyzer (Yakogawa, AQ6370) and the corresponding SNR measurement versus output pulse energy are shown in Fig. 39 (a) and (b), respectively. At 1535 nm, ASE growth rate became faster above pulse energy level of 1.5 mJ as it can be seen from the SNR measurement. From the output energy of 1.7 mJ to 1.8 mJ, the SNR suddenly dropped by 6 dB due to the parasitic lasing at 1535 nm.

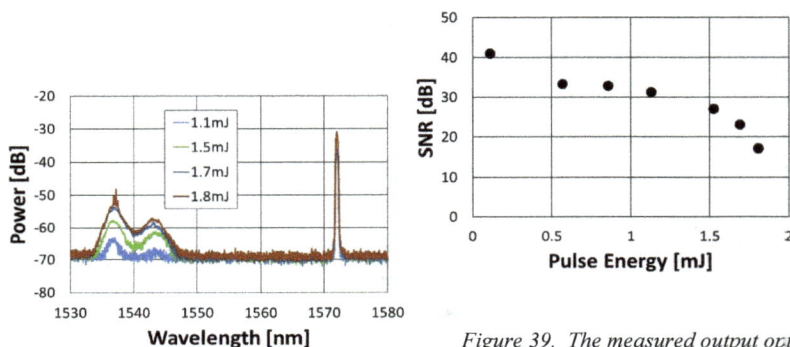

Figure 39. The measured output optical spectra from the last stage amplifier at 2.5 kHz (a) the SNR versus the output pulse energy (b). Copied from [14].

Spectral purity of the output signal at 1572 nm was investigated at the finest resolution bandwidth of 0.05 nm. The zoom-in view of the output signal spectrum at 1572 nm from the last stage amplifier is shown in Fig. 40 (a). There was no apparent sign of spectral distortion from the measured output signal at 1572 nm. We have verified the single frequency operation of our all-fiber MOPA system using a scanning Fabry-Perot interferometer (Micron Optics, FFP-SI) with a free spectral range (FSR) of 1 GHz and a resolution of <10 MHz, which is shown in Fig. 40 (a).

Based on the measured spectra in Fig. 39 (a), we believe that the last stage amplifier (EYDFA-3) in our current system is quite yet to be optimized so that the system output energy could be further improved through another stage of an optimized large signal amplification process from ~ 500 µJ to > 3 mJ utilizing a short length of AdValue Photonics proprietary super large core single-mode polarization maintaining silicate glass EYDF with a core diameter of ~60 µm. In this case, we'll need to lower the pump power for EYDFA-3 to obtain the pulse energy of ~500 µJ at the gain level of ~ 6 dB free from parasitic lasing. Consequently, some of the under-pumped section of the gain fiber in the EYDFA-3 could contribute to obtaining a better SNR from the amplifier acting as an ASE absorber since the length of the gain fiber in the EYDFA-3 is long enough to produce the pulse energy >1.8 mJ.

Rare Earth - A tribute to the late Mr. Rare Earth, Professor Karl Gschneidner   Materials Research Forum LLC
Materials Research Foundations 164 (2024) 1-66   https://doi.org/10.21741/9781644903056-1

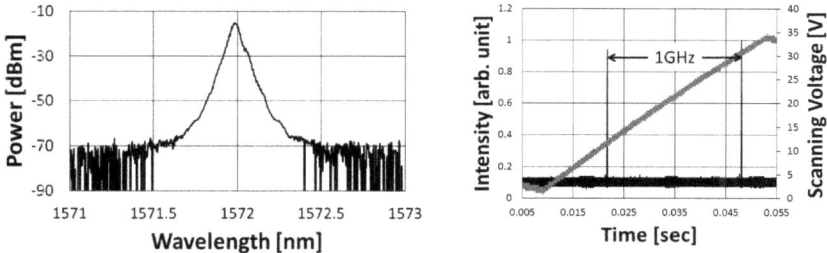

Figure 40. Zoom-in view of the optical output spectrum at 1572 nm from the last stage amplifier (resolution bandwidth: 0.05 nm) (a) the Fabry-Perot interferometer scan of the output optical spectrum at 1572 nm from the last stage amplifier for confirming single frequency operation (b). Copied from [14].

Similarly, we have also operated the entire high energy MOPA system at the pulse repetition rate of 7.5 kHz. It should be noted that the AOM gating pulse window in the seed laser system was re-optimized for the system operation at 7.5 kHz to achieve sufficient pulse energy to saturate the first stage high energy EYDFA. The measured output pulse energy at the pulse repetition rate of 7.5 kHz is shown in Fig. 40 (a). We have obtained the highest pulse energy of 1.36 mJ, a peak power of ~ 2.5 kW with a pulse width of 505 ns at the pulse repetition rate of 7.5 kHz. At 7.5 kHz operation, the measured pulse energy curve didn't show clear indication of energy saturation behavior. No parasitic lasing issue was found from the system at the 7.5 kHz operation and the maximum output pulse energy was only limited by available pump power from the two 80 W diodes at 976 nm.

Demonstration of single frequency optical pulses at 1572 nm is confirmed with a Fabry-Perot scan shown in Fig. 40 (b). Maximum pulse energy of 1.8 mJ (peak power of 3.5 kW) at 2.5 kHz, as well as 1.3 mJ (peak power of 2.5 kW) at 7.5 kHz is the highest pulse energy of single frequency at 1572 nm from all-fiber MOPA system. This illustrates a very promising technology for the NASA Active Sensing of CO_2 Emissions over Nights, Days, and Seasons (ASCENDS) mission. By taking the advantage of the Er/Yb co-doped silicate glass fibers, the system output pulse energy can further be improved up to 3 mJ utilizing a short length of silicate glass gain fiber with a core diameter of ~ 60 μm.

7. Applications of fiber lasers

The development of fiber lasers will continue to have a significant impact on the applications in the future. This section will consider some of the applications of fiber lasers and the advantages they have over discrete component laser systems. We will look at three broad application areas: materials processing, medical applications and atmospheric remote sensing in relation to how fiber lasers described in this chapter could substantially impact these application fields. The applications will consider both continuous and pulsed fiber lasers.

Most glass and crystal laser systems use discrete laser components to modify the laser radiation to tailor it toward a specific application. For instance, an acousto-optic Q-switch can change continuous to pulsed operation, gratings can narrow the laser line emission or cavity mirrors can determine the laser brightness. Fiber lasers now allow the integration of all these and other laser components, which are usually physically discrete devices, to be incorporated into the fiber itself, resulting in a laser system made of a single optical structure. This has significant advantages. The design, construction and operation of laser systems with discrete components usually required the expertise associated with an advanced degree and significant cost to maintain successful laser system operation. This is usually due to individual laser system components having different alignment sensitivity to room temperature and stress, which results in constant system realignment. The mass and size of typical bulk crystal or glass laser systems makes stable long-term operation problematic in environments of changing vibration, temperature and stress.

Fiber lasers have reduced these impediments by allowing all laser components to be placed within the fiber, resulting in a lower mass and volume system that can be more easily controlled environmentally for long term stable operation. In addition, such lasers can now be more easily mass- produced which allows reduced cost and thus broader applications. Some of these applications will now be discussed in detail.

7.1   Materials processing

Lasers have been used for many years for materials processing. Cutting, drilling, welding, and ultrafast modification of a variety of materials with lasers is well known and widely used. The primary issue in materials processing is to absorb the laser energy on the surface, creating heat before conduction, emission and ablation can remove the heat. The major concerns for such laser applications are average power, operational cost, complexity and use by non-specialists. The laser wavelength is not a major driver for this application. "The highest single-mode powers yet achieved in the laboratory, about 2-kW continuous wave (cw), have come from ytterbium-doped fiber lasers emitting near 1.1 µm. But powers have also been increasing from erbium-doped fiber lasers emitting at 1.5 to 1.6 µm and from thulium-doped systems emitting at 1.8 to 2.2 µm. Commercial fiber lasers emitting a couple of hundred watts already are used for materials working, and military developers are taking a hard look at the prospect of laser weapons based on high-power fiber lasers." [52]

Fiber lasers can have high brightness resulting from diffraction limited focusing of single transverse mode lasers. Using computer-controlled laser positioning, materials are processed rapidly, lowering production cost. If fiber lasers are mass-produced, then the working area can be increased by combining many single fiber lasers into a unit capable of processing large material areas. The operational costs can be low due to the high wall plug efficiency of these lasers. Complexity is reduced by having all optical components imbedded into the fiber laser, reducing the need for critical optical alignment in a rugged package that is compact and mobile. Fiber lasers have continued to grow as a technology for many industrial laser solutions.

7.2   Medical applications

Medical applications have surged in recent years allowing patients to elect for outpatient procedures as opposed to complicated surgery. Laser emission can pinpoint intense energy on small areas of tissue, causing tissue cutting and ablation with little bleeding. The laser is used to precisely ablate bone and cartilage, with many applications in orthopedics for arthroscopy, urology for lithotripsy (removal of kidney stones), ENT (ear, nose and throat) for endoscopic sinus surgery, and spine surgery for endoscopic disc removal. As Kincaide wrote in Laser Focus World: "In fiber lasers, the optical fiber itself acts as the resonator cavity. These devices typically comprise a single-mode fiber core doped with erbium, ytterbium (or a combination thereof), or thulium. Energy from a solid-state source is coupled into the fiber's cladding, then moves into the core and pumps the dopant. In the medical field, the most desirable wavelengths are in the 1.3-μm range for imaging and the 1.5-μm to 4-μm range for surgical applications, at powers ranging from milliwatts to more than 100 W." [53]

If a particular wavelength is absorbed by the target tissue type, and transmitted, reflected or scattered by other surrounding tissues, the therapeutic effect may be achieved. The absorption characteristics of different tissue types are determined by tissue chromophores, the main absorbing components, which are hemoglobin in the blood, melanin in the skin and, in general, the water that is present in all tissue types. Hemoglobin and melanin absorb visible light, and water absorbs infrared light. Water absorption depth versus wavelength is shown in Fig. 24. Generally, the laser energy may cause thermal, mechanical and chemical effects.

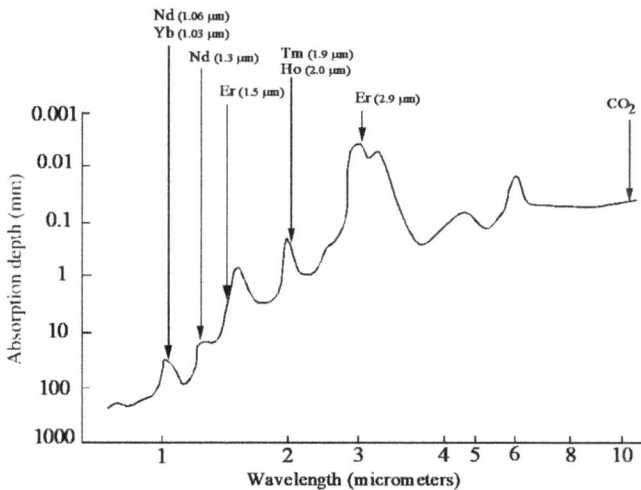

Figure 24. Logarithmic plot of absorption depth vs. wavelength with various laser system operating wavelengths shown.

All tissue contains water, and this can be used to fine-tune the amount and depth of absorption of laser light in the tissue. If the fiber laser is line narrowed and slightly tunable at 2-μm, then tissue light absorption can be controlled precisely by wavelength tuning the laser emission on and off a particular water absorption line near 2-μm, while leaving the laser emission intensity constant. When tissue is heated, water vapor is given off and if the laser is tuned to the water vapor absorption line, then significant absorption will occur in the vapor region. Conversely, if the fiber laser is tuned off the water vapor absorption line, the dominant absorption will occur at the tissue surface. Thus, water amount in the tissue and wavelength of the laser determines tissue absorption. This may open a new field of specialized laser surgery. While operational cost of such lasers is not a major driver, reliability and focus ability are. Fiber lasers that are integrated and modular, thus having greater reliability, will gain greater use in the medical field.

7.3   Atmospheric remote sensing

There are a variety of methods for remote sensing of the atmosphere. Passive remote sensing relies on spectrophotometric techniques, which can provide wide area coverage, but with limited resolution. Active remote sensing relies on lasers utilizing laser radar, or *lidar* (light detection and ranging), which is useful on smaller spatial scales, but with greater resolution. Ouellette wrote in The Industrial Physicist: "There are three basic types of lidar, each operating at different wavelengths. Rangefinder lidar measures the distance from the instrument to a solid target. Differential absorption lidar determines the concentrations of chemicals in the atmosphere—such as ozone, water vapor, and pollutants—and enables underwater mapping in shallow areas. Doppler lidar measures the velocity of a target by using the Doppler shift effect." [54] Some molecular constituents in the atmosphere that are of interest due to their impact on the environment include greenhouse gasses such as carbon dioxide (CO_2), methane (CH_4), carbon monoxide (CO), and water vapor (H_2O). In addition, ozone (O_3) is of importance in solar ultraviolet shielding in the stratosphere. Sulfur dioxide (SO_2) is also of interest because of tropospheric smog pollution, a precursor to acid rain. Tunable, narrowband lasers operating in pulsed mode are ideal for lidar systems in remote sensing on Earth and other worlds.

The very demanding application for fiber lasers will be atmospheric remote sensing on Earth. Here the goal will be to have many laser systems gathering atmospheric data on molecular constituents at diverse locations around the world without human presence. These lasers would be pulsed, emitting at least 1 mJ per pulse at 1000 Hz (1 Watt), resulting in atmospheric profiles of a wide variety of atmospheric constituents. For instance, water vapor is the major atmospheric constituent contributing to weather patterns. A weather forecasting goal of predicting the weather at every square kilometer of the Earth's surface for as far in advance as 30 days would require thousands of lidar stations around the world continuously measuring water vapor profiles and inputting them into large global atmospheric models. These stations would have to operate with high reliability and have long life.

On Mars there is a need to measure atmospheric water vapor profiles at various locations to understand the transport of water vapor from the poles and to determine the locations of sinks and sources of surface water vapor. This is important for missions that will determine if life is present on the Martian surface, since water is a precursor to life as we know it. [55]

The DIAL technique is an active remote sensing technique that takes advantage of the absorption and scattering of pulsed narrow-band laser light along the beam path to obtain the concentration of the molecular species that causes the selective absorption. In practice, two laser pulses are transmitted near simultaneously with one at the peak of the water vapor absorption line called the "*on-line*" and another away from the absorption line called the "*off-line*". As the pulse propagates, some of its power is absorbed and some is scattered back to the telescope receiver. Using the DIAL method, the average water vapor number density n between atmospheric ranges R_1 and R_2 is calculated using the relation

$$n = \frac{1}{2\Delta s(R_2 - R_1)} \ln \left[\frac{P_{on}(R_1)P_{off}(R_2)}{P_{on}(R_2)P_{off}(R_1)} \right] \tag{40}$$

where P is the returned power from the on-line and off-line laser wavelengths and Δs is the differential water vapor absorption cross section between the on-line and off-line wavelengths. The advantage of the DIAL method is that it can be used to obtain range-resolved profiles of atmospheric water vapor with high vertical resolution. In addition to measuring gas concentration profiles, high spatial resolution aerosol backscattering distributions are simultaneously obtained as part of the DIAL measurement using the off-line lidar signal return. DIAL offers the advantage of adjusting vertical resolution by averaging the lidar data that are collected at a very high resolution and lidar measurements can be made during day or night background conditions. The DIAL method has been most successfully demonstrated for ground-based atmospheric ozone [56], [57] and water vapor [58], [59], [60] measurements.

7.4   NASA laser/lidar technology applications

The laser/lidar technology is considered for a wide range of applications that can enable NASA's achievement of its scientific and space exploration goals. These applications fall into five general areas: (1) Earth Science: global monitoring of the atmosphere and surface; (2) Planetary Science: orbiting or land-based instruments providing geological and atmospheric data; (3) Landing Aid: sensors providing hazard avoidance and navigation data; (4) Rendezvous and Docking Aid: sensors providing spacecraft bearing, distance, and approach velocity; (5) Data Transmission: high speed free-space optical communication.

References

[1] W.M. Yen, Optical Spectroscopy of Ions in Inorganic Solids, I. Zschokke, ed. (Kluwer, Netherlands, 1986), pp. 23. https://doi.org/10.1007/978-94-009-4650-7_2

[2] K. Patek, Glass Lasers, (Iliffe, London, 1970).

[3] M. Mortier and D. Vivien, Ann. Chim. - Sci. Mat., Vol. 28 (2003), pp. 21.

[4] C.A. Morrison, D.E. Wortman, and N. Karayanis., J. Phys. C: Solid State Phys., Vol. 9 (1976), pp. L191. https://doi.org/10.1088/0022-3719/9/8/001

[5] J. H. Van Vleck, J. Phys. Chem., Vol. 41 (1937), pp. 67. https://doi.org/10.1021/j150379a006

[6] B.R. Judd, J. Phys. Rev B, Vol. 127 (1962), pp. 750. https://doi.org/10.1103/PhysRev.127.750

[7] G.S. Ofelt, J. Chem Phys., Vol. 37 (1962), pp. 511. https://doi.org/10.1063/1.1701366

[8] B.M. Walsh, Advances in Spectroscopy for Lasers and Sensing, B. Di Bartolo and O. Forte, eds. (Springer, Netherlands, 2006), pp. 403.

[9] H. Warren Moos, J. Lumin, Vol. 1,2 (1970), pp.106-121. https://doi.org/10.1016/0022-2313(70)90027-X

[10] M. J. Weber, Phys. Rev. B, Vol 8 (1973), pp. 54. https://doi.org/10.1103/PhysRevB.8.54

[11] W. H. Dumbaugh, Phys. Chem. Glasses, Vol 27 (1986), pp. 119.

[12] A. Polman, J. Appl. Phys, Vol 82 (1997), pp. 1-39. https://doi.org/10.1063/1.366265

[13] E. Snitzer and R. Woodcock, Appl. Phys. Lett. 6, (1965), pp. 45-46 https://doi.org/10.1063/1.1754157

[14] H.W. Gandy, R.J. Ginther, and J.F. Weller, Phys. Lett. 16 (1965), pp. 266-267. https://doi.org/10.1016/0031-9163(65)90842-5

[15] I. Pocsik, Physica A, Vol. 201 (1993), pp. 34. https://doi.org/10.1016/0378-4371(93)90397-M

[16] R. Reisfeld, and C.K. Jorgensen, Handbook on the Physics and Chemistry of Rare Earths, K.A. Gshneidner, Jr. and L. Eyring, eds., Vol, IX, (Elsevier, Amsterdam, 1987), Chapter 58.

[17] K. Ouannes, K. Lebbou, Brian M. Walsh, M. Poulain, G. Alombert-Goget, Y. Guyot, J. Alloys Compd,, Vol 649 (2015), pp. 564-572. https://doi.org/10.1016/j.jallcom.2015.07.113

[18] B.M. Walsh, N.P. Barnes, and B. Di Bartolo, J. Appl. Phys., Vol. 83 (1998), pp. 2772. https://doi.org/10.1063/1.367037

[19] Animesh Jha, Billy Richards, Gin Jose, Toney Teddy-Fernandez, Purushottam Joshi, Xin Jiang, Joris Lousteau, Prog. Mater. Sci., Vol 57 (2012), pp.1426. https://doi.org/10.1016/j.pmatsci.2012.04.003

[20] Dhiraj K. Sardar, John B. Gruber, Bahram Zandi, J. Andrew Hutchinson and C. Ward Trussell, Appl. Phys, Vol 93 (2003), pp. 2041. https://doi.org/10.1063/1.1536738

[21] Norman P. Barnes, Brian M. Walsh, Farzin Amzajerdian, Donald J. Reichle, George E. Busch, and William A. Carrion, , IEEE J Quant. Elec., Vol. 49 (2013), pp. 238. https://doi.org/10.1109/JQE.2012.2226146

[22] E. Snitzer, Appl. Opt., Vol. 5 (1966), pp. 1487. https://doi.org/10.1364/AO.5.001487

[23] H. Zellmer, U. Willamowski, A. Timmermann, H. Welling, S. Unger, V. Reichel, H.R. Muller, J. Kirchhof, and P. Albers, Opt. Lett., Vol. 20 (1995), pp. 578. https://doi.org/10.1364/OL.20.000578

[24] S.D. Jackson and Y. Li, IEEE J. Quant. Elec., Vol. 39 (2003), pp. 1118. https://doi.org/10.1109/JQE.2003.816094

[25] D.Y. Shen, J.K. Sahu, and W.A. Clarkson, Opt. Express, Vol. 13 (2005), pp. 4916. https://doi.org/10.1364/OPEX.13.004916

[26] J. Jeong, J.K. Sahu, D.N. Payne, and H. Nilsson, Opt. Express, Vol. 12 (2004), pp. 6088. https://doi.org/10.1364/OPEX.12.006088

[27] D.C. Hanna, R.M. Percival, R.G. Smart, and A.C. Tropper, Opt. Comm., Vol. 75 (1990), pp. 283. https://doi.org/10.1016/0030-4018(90)90533-Y

[28] W.L. Barnes and J.E. Townsend, Electron. Lett., Vol. 26 (1990), pp. 746. https://doi.org/10.1049/el:19900487

[29] S.D. Jackson and T.A. King, Opt. Lett., Vol. 23 (1998), pp. 1462. https://doi.org/10.1364/OL.23.001462

[30] R.A. Howard, W.A. Clarkson, P.W. Turner, J. Nilsson, A.B. Grudinin, and D.C. Hanna, Electron. Lett., Vol. 36 (2000), pp. 711. https://doi.org/10.1049/el:20000577

[31] W.A. Clarkson, N.P. Barnes, P.W. Turner, J. Nilsson, and D.C. Hanna, Opt. Lett., Vol. 27 (2002), pp. 1989. https://doi.org/10.1364/OL.27.001989

[32] P. Myslinski, X. Pan, C. Barnard, B.T. Sullivan, J.F. Bayon, Opt. Eng., Vol. 32 (1993), pp. 2025. https://doi.org/10.1117/12.143911

[33] N.P. Barnes, W.A. Clarkson, D.C. Hanna, P.W. Turner, J. Nilsson, B.M. Walsh, in Advanced Solid State Lasers, vol. 50 of OSA Trends In Optics and Photonics Series (Optical Society of America, Washington, D.C., 2001), pp. 88.

[34] J.N. Carter, R.G. Smart, D.C. Hanna, and A.C. Tropper, Elec. Lett., Vol. 26 (1990), pp. 599. https://doi.org/10.1049/el:19900394

[35] J.Y. Allain, M. Monerie, and H. Poignant, Elec. Lett, Vol. 25 (1989), pp. 1660. https://doi.org/10.1049/el:19890221

[36] B.M, Walsh and N.P. Barnes, Appl. Phys. B., Vol. 78 (2004), pp. 325. https://doi.org/10.1007/s00340-003-1393-2

[37] B.M. Walsh, N.P. Barnes, D.J. Reichle, and S. Jiang, J. Non-Cryst. Solids, Vol. 352 (2006), pp. 5344. https://doi.org/10.1016/j.jnoncrysol.2006.08.029

[38] D. J. Richardson, J. Nilsson, and W. A. Clarkson, J. Opt. Soc. Am. B, Vol. 27 (2010), pp. 63-92. https://doi.org/10.1364/JOSAB.27.000B63

[39] Y. Jeong, S. Yoo, C.A. Codemard, J. Nilsson, J. K. Sahu, D. N. Payne, R. Horley, P. W. Turner, L. Hickey, A. Harker, M. Lovelady, and A. Piper, IEEE J. Sel. Top. Quantum Electron., Vol. 13 (2007), pp. 573-579. https://doi.org/10.1109/JSTQE.2007.897178

[40] Y. Jeong, J.K. Sahu, D. N. Payne, and J. Nilsson, Opt. Express, Vol. 12 (2004), pp. 6088-6092. https://doi.org/10.1364/OPEX.12.006088

[41] S.D. Setzler, M. P. Francis, Y. E. Young, J. R. Konves, and E. P. Chicklis, IEEE J. Sel. Top. Quantum Electron., Vol. 11 (2005), pp. 645-657. https://doi.org/10.1109/JSTQE.2005.850249

[42] M. Dubinskii, J. Zhang, and V. Ter-Mikirtychev, Electron. Lett., Vol. 45 (2009), pp. 400-401. https://doi.org/10.1049/el.2009.0505

[43] J. Zhang, V. Fromzel, and M. Dubinskii, Opt. Express, Vol. 19 (2011), pp. 5574-5578. https://doi.org/10.1364/OE.19.005574

[44] L. Sun, S. Jiang, and J. R. Marciante, Opt. Express, Vol. 18 (2010), pp. 12191-12196. https://doi.org/10.1364/OE.18.012191

[45] L. Sun, S. Jiang, J. D. Zuegel, J. R. Marciante, Opt. Lett., Vol. 35, (2010), pp. 706-708. https://doi.org/10.1364/OL.35.000706

[46] Z. Qiang, J. Geng, T. Luo, J. Zhang, and S. Jiang, Appl Opt., Vol. 53 (2014), 643-647. https://doi.org/10.1364/AO.53.000643

Materials Research Foundations 164 (2024) 1-66   https://doi.org/10.21741/9781644903056-1

[47] NASA Science Definition and Planning Workshop Report, "Active Sensing of CO2 Emissions Over Nights, Days, and Seasons (ASCENDS) Mission", (2008), https://cce.nasa.gov/ascends/

[48] A.W. Yu, J.B. Abshire, M. Storm, and A. Betin, Proc. SPIE, Vol. 9342 (2015), pp. 93420M-5.

[49] J. W. Nicholson, A. DeSantolo, M. F. Yan, P. Wisk, B. Mangan, G. Puc, A. W. Yu, and M. A. Stephen, Opt. Exp., Vol. 24 (2016), 19961-19968. https://doi.org/10.1364/OE.24.019961

[50] D. Engin, B. Mathason, M. Stephen, A.Yu, H. Cao, J. Fouron, and M. Storm, Proc. SPIE, Vol. 9728 (2016), pp. 97282S.

[51] Wangkuen Lee, Jihong Geng, Shibin Jiang, and Anthony Yu, Opt. Lett., Vol. 43 (2018), pp. 2264-2267. https://doi.org/10.1364/OL.43.002264

[52] J. Hecht, Laser Focus World, Vol. 41, No. 8 (2005), pp. 66.

[53] K. Kincade, Laser Focus World, Vol. 41, No. 9 (2005), pp. 76.

[54] J. Ouellette, The Industrial Physicist, Vol. 8, No. 2 (2002), pp. 16.

[55] M.D. Smith, J. Geophys. Res., Vol. 107, No. E11 (2002), pp. 25-1. https://doi.org/10.1029/2001JE001522

[56] G. Megie, and R.T. Menzies, Appl. Opt., Vol. 19 (1980), pp. 1173. https://doi.org/10.1364/AO.19.001173

[57] O. Uchino, M. Tokunago, and Y. Miyazoe, Opt. Lett., Vol. 8 (1983), pp. 347. https://doi.org/10.1364/OL.8.000347

[58] E.V. Browell, T.D. Wilkerson, and T.J. McIlrath, Appl. Opt., Vol. 18 (1979), pp. 3474. https://doi.org/10.1364/AO.18.003474

[59] C. Cahen, G. Megie, and P. Flamant, J. Appl. Meteorol., Vol. 21 (1982), pp. 1506. https://doi.org/10.1175/1520-0450(1982)021<1506:LMOTWV>2.0.CO;2

[60] J. Bosenberg, OSA Tech. Digest Series, Vol. 18, No. 2 (1987), pp. 22.

Rare Earth - A tribute to the late Mr. Rare Earth, Professor Karl Gschneidner
Materials Research Foundations 164 (2024) 67-142

Materials Research Forum LLC
https://doi.org/10.21741/9781644903056-2

Chapter 2

A Luminescent Pathway for Anti-Counterfeiting of Currency and Forensic Applications

Payal P. Pradhan, D. Haranath*

Luminescent Materials and Devices Group, Department of Physics, National Institute of Technology, Warangal 506004, Telangana, INDIA

haranath@nitw.ac.in

Abstract

Nowadays, issues regarding counterfeit and forensic applications are gradually increasing. For instance, banking, insurance sectors, the drug industry, and university degrees, including different expensive commercial products, are facing counterfeit hitches with fake products. Duplication of the products could be done by simple methods and the replicated product looks genuine that nobody could doubt. So, to prevent these kinds of offensive activities, latent security ink has a great advantage to write secret codes or symbols. Similarly, the effective identification of latent fingerprints in forensic science is crucial for investigations at the crime scene. Latent fingerprints are generally accomplished on many surfaces. However, extraction of these from porous, non-porous, and colored surfaces is not so easy. To resolve the issue, organic/inorganic dyes, pigments, and nano phosphor materials have been used conventionally, due to the characteristics of nano-size, variable colour emission, afterglow, and high chemical stability, which consent the fingerprint recognition on any type of surface for forensic applications. This comprehensive book chapter includes a brief introduction to luminescence, properties of lanthanides, and lanthanide complexes. Extensive literature survey including synthesis and application of lanthanide-based nano phosphors in the areas of anticounterfeit and latent fingerprint extractions. Amongst the numerous techniques, which have been developed to combat counterfeits, visualization of latent fingerprints, security ink, and printing, innovative luminescent materials have been broadly utilized for various applications not only due to high-throughput production, facile design and simple operation but also due to exceptional security properties. Moreover, this chapter provides a systematic comprehensive overview of the latest developments in Ln-doped luminescent materials, plasmonic nanomaterials, quantum dots (QDs) and metal-organic frameworks for their possible use in the above high in demand applications.

Keywords

Photoluminescence, Latent Fingerprint, Anti-Counterfeiting, Security Ink

Contents

1.   Introduction

1.1   Introduction to Luminescence

1.1.1   History

Luminescence is a technique which is First observed by Nicolas Monardes in an extract of a wood 'Lignum nephriticum' in 1565 and in 1819 Edward D. Clarke discussed a unique property of highly transparent fluorite crystal that had dichroic behavior, reflected and transmitted light showed sapphire blue and bright green respectively. But he was not able to explain that mechanism. Likewise, In Rene Hauy reported the dichroic behavior of some fluorspar crystals in 1822. [1][2]–[5]

Later, Sir George Gabriel Stokes described the phenomena of emission and absorption in 1852. In 1888, the term 'luminescence' for both phosphorescence and fluorescence phenomena were introduced by physicist Eilhard Wiedemann. The name "luminescence" is derived from the Latin word "lumen," which meaning light, and describes the phenomena of light spontaneously emitted from an electronically/vibrationally stimulated state that is not in thermal equilibrium with its environment. [2][6] A wide range of applications could be found in areas as diverse as physics, chemistry, biology, biochemistry, medicine, materials science, environmental science, microelectronics, toxicology, and pharmaceuticals. [7]–[10] The phenomenon where light is emitted by a material that has not arisen from an increase in its temperature is called Luminescence, or "cold light". When materials increase their emission of radiation as their temperature increases, this phenomenon is called incandescence or "hot light". [9]

Luminescence involves at least two phases, the excitation of the electronic system of the material and the subsequent emission of photons. Luminescence differs from various types of scattering in luminescence transitional processes, whose time period is more than the period of a light wave that arises between absorption and emission. As a result, the connection between the oscillation phases of absorbed and radiated light is lost. [10] If the absorbed energy comes from rubbing or crushing crystals, the phenomenon is known as Triboluminescence. Chemiluminescence involves the emission of light on an exo-energetic chemical reaction. [7], [10] In luminescence, an electron of an atom absorbs some energy from the source then it jumps from a higher energy level of the valence band to the conduction band, then the electron comes down to the ground state by emitting energy in the form of radiation. [9]

1.1.2 Mechanism of Luminescence

The luminescence mechanism of a solid substance was explained by the band theory for solids. Each single atom carries a variety of electrons in orbitals near the nucleus. Additionally, some electrons should occupy orbitals farther from the nucleus as the lower energy orbitals closest to it are already filled, [7], [8] for this reason, negatrons will only ever have one alternate electron of the opposite spin in their orbitals. A pulse of radiation, such as that produced by a gauge boson, a fast negatron, or another means, is frequently used to displace the occupied negatron in occupied orbitals thus creating vacancies. As soon as this occurs, associate negatron from the associated outer level might descend and retake the inner, lesser energy, level. A gauge boson is a form in which the extra energy is radiated [3]. Since there can only be two electrons in a level at a time, the energy levels begin to break down into smaller levels. Instead of relating to a single atom, these new orbitals do so in relation to the entire lump. The energy levels of the various sub-levels that form when the outer orbitals overlap and split are marginally different. The characteristics of the crystal structure play a role in how light the solids are [11]. The difference of energy between the valence band's highest energy and the conductivity band's lowest value is known as the band gap energy. When the conductivity band is completely empty and the valence band is completely filled with electrons, the fabric is associated material because in order for an object to conduct electricity, its electrons need to gain energy and migrate to a higher level. The distance between the conductivity band's empty space and the fully occupied valence band can be very small in some materials. [12][11]

1.1.3 Characteristics of Luminescence

Fig. 1 provides a visual representation of the luminescence process. Two different methods of return to the ground state one radioactive and the other non-radioactive are shown. The luminescence process takes place through the former. The other is unrelated to luminescence but is associated with radioactive emission because phonons are transformed into lattice vibrations that carry heat energy.

Solid-state lighting, advanced optical displays, scintillation, X-ray intensification, security applications, and many more fields have discovered numerous uses for luminescent materials

Luminescence is further divided into sub-categories based on the length of the emission.:

• Fluorescence: Upon excitation removal, an exponential afterglow with a lifespan of less than 10^{-8} seconds is seen, which is independent of excitation intensity and temperature [12].

• Phosphorescence: After excitation is removed, there is an additional phenomenon known as an afterglow that has a life of more than 10^{-8} seconds and is frequently depending on the intensity of the excitation. Since heat activation of the metastable activator or trap is a need for emission, the metastable states produced by the impurities, activators, defect centers, and electron or hole traps present in the structure may prolong the luminous emission creating this effect [9].

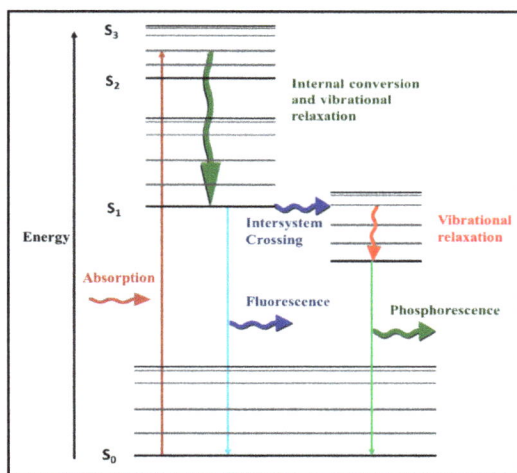

Figure 1. Schematic representation of the difference between fluorescence and phosphorescence.

1.2   Introduction of Lanthanides

Exciting developments in material science and technology over the past ten years have raised the demand for dependable and strong systems for observing, measuring, and regulating a wide range of technical issues [13]. Adopting special qualities of some metal ions, such as those lanthanide family members which can display rather effective long-lived luminescence when raised to their excited states, is a favorable strategy for moving forward with technology in material science. Essential conditions are normally recognized to notice these phenomena. The benefit of having such persistent emissions is the ability to apply time-resolved techniques, which measure the emission after a lengthy pause to allow for the disappearance of all background fluorescence and light scattering.

Other benefits of using lanthanides include their highly distinctive emissions, which have significant Stokes shifts, and the fact that because they involve the excitation of a shielded f-level electron, they are less susceptible to oxygen and related substances quenching than conventional fluorescent dyes [14][15]. The lanthanides, commonly referred to as "rare-earth metals," are made up of elements from the 4fblock (La to Lu). They are comparable in many ways, including ionic radius and oxidation state (III). They have extremely special luminous qualities that can be used in a wide range of devices, including lasers, plasma displays, and liquid crystals [16]. Nanoparticles with lanthanide bases have demonstrated excellent luminous material promise. The effective application of highly luminous rare-earth based nanoparticles, which combines the diverse characteristics of doping ions and the nanoparticle matrix, has assured the widespread use of nanoparticles. Multifunctional nanoparticles can be designed to perform both diagnostic and therapy [17], facilitate the separation and detection of bio-molecules or detect or track tagged biomolecules using a variety of complementary approaches [18][19]. An intriguing method for creating effective luminous multifunctional Nanoprobes is to substitute organic dyes with inorganic fluorophores, such as lanthanide compounds because these substances have beneficial optical properties. Lanthanide ions are good choices for the development of highly luminous and photostable Nano probes without emission intermittency since they have great photostability, narrow emission bands, a long luminescence lifespan (1 ms), and a large absorption band [18]. In comparison to other luminous materials, lanthanide-doped nanomaterials have various benefits including improved chemical stability, a longer lifetime, strong emission, strong thermal stability, broad excitation spectra, low toxicity, and great photochemical stability. Depending on the stimulus, light can be emitted from the ultraviolet to the near-infrared (NIR) region by Ln-doped luminescent materials [20]. With the exception of actinides, the lanthanides are remarkable among the elements which resemble their chemical characteristics, specifically in their oxidation states. This is easily demonstrated by the electronic configuration of the atoms and their generated ions, which effectively exist in their trivalent state Ln (III) in aqueous solutions [21][22].

1.2.1 Why lanthanides?

The majority of transition metal ions are UV/visible light absorbers, but very few of them re-emit even a tiny portion of the energy they have already absorbed as UV or visible emissions. This is a result of their d-electron excited states' high connection to the surrounding environment via the ligand field, which offers an effective de-excitation mechanism [13]. However, all of the trivalent lanthanide ions above lanthanum, denoted by the formula Ln(III), are well known for their luminescence, especially in the solid state under anhydrous circumstances. Because these excited states entail the promotion of one of the 4f electrons, their complexes likewise exhibit this feature. Many monographs have addressed the fundamental excited state characteristics of lanthanide ions [23]. The existence of electrons in the outer 5th shell and for some the 6th shell also shields the 4f electrons, which is the main distinction between this transition metal and other transition metals. The limited bonding activity of these outer electrons protects the inner 4f electrons from the usual chaos of chemical reactions. Electronic transitions between these f orbitals' energy levels are only slightly perturbed as a result. Because of this, radiation-free deactivation mechanisms in Ln (III) are often ineffective and can often be completed by energy emission as light. In a coordinating environment, the ligand field, spin-orbit coupling (SO), and interelectronic repulsion all work together to define the energy of the 4f electronic configuration of a lanthanide ion. The Judd-Ofelt theory can be used to explain the intensity of the f + f absorption spectra [24]. The Laporte rule forbids electric dipole transitions between 4f levels that do not involve a change in parity, and they only take place as a result of relatively strong spin-orbit interactions and the interaction of the ligand field, which mixes the electronic states and is known as the "J-mixing effect." Forced electric dipole transitions are the transitions between these changed states. Molar extinction coefficients are typically weak because it makes theoretically forbidden transitions possible and keeps their probability low. The associated emissions are distinguished by typically lengthy radioactive lifetimes, radioactive lifetimes in the millisecond timescale, and line-like emission bands. The behavior of organic fluorescent compounds, which frequently have radioactive lifetimes in the millisecond range and molar extinction values of the tens of thousands, contrasts significantly with this [25]. Europium (Eu), gadolinium (Gd), samarium (Sm), dysprosium (Dy) and terbium (Tb) are the five central lanthanides, and their energy levels are shown in Figure 6. In chelated forms, all of these can glow while in solution. The main cause of the other lanthanides' infrequent brightness in the solid state is that their excited states and acceptor levels of the ground state manifold do not have very wide gaps, and their minor energy level differences can be effectively bridged by non-radioactive processes [26][27]. (Figure 7). Some of the transitions, which are highly sensitive to the environment's fine details, are reflected in the pattern of emission as probabilities for various transitions. For Eu^{3+} ions, the strongest transitions from the $^5D_0$ to the ^7F manifold are the $^5D_0$ to $^7F_2$, and $^7F_1$ transitions. The ligand environment has a significant impact on the relative intensities of these two emissions. Some other transitions are $^5D_0$ to $^7F_{0,3,5}$. which are either very faint or unobservable [28] and which are expressly prohibited. Since the decay of the excited states is mostly controlled by dark, non-radioactive transitions, these ions often display low emission quantum yields in water [29]–[32].

1.2.2 Applications of lanthanides

One of the most exciting and difficult fields in business is rare-earth technology. The public and the scientific community have both become more aware of the uniqueness of this industry as a result of the current geopolitical instability. The sector is frequently upended by significant changes in supply and demand, the emergence of new uses, the entry of new competitors, and even the introduction of new raw materials. While the rare-earth industry has traits of a commodity business in some parts, it might be termed a premium business in others. Rare-earths are employed as additives in numerous current industrial applications, usually at low mass ratios. Rare-earth recycling is still not done extensively in practice since it is difficult to recycle them profitably. However, things are changing because of new rules governing the handling and disposal of home and industrial waste as well as the recent sharp rise in the price of rare earth mined materials. Since industrial items don't include thorium, recycling has an advantage over mine-produced goods, and this may be a key element in recycling's growth. Rare-earths are recycled by businesses either from scrap created on-site during manufacturing or from commercial items that have reached the end of their useful lives [33]. Over the past century, rare earths have been used in a wide variety of fairly minor technical applications. Huge amounts of the lighter lanthanides, from lanthanum to samarium, were left over after thorium was removed from monazite sand. Cerium (IV), which was frequently employed in chemical analysis, and CeO_2 as an optical polishing powder, is simple to extract from such mixtures. The remaining components were greatly reduced to an alloy called "Mischmetall" that was used for flints in cigarette lighters and is still used extensively in steel production today. Alloys like $SmCo_5$ are used to make permanent magnets [30]. In addition to immunoassays and nucleic acid hybridizations, there is a strong interest in using lanthanides as non-isotopic labels in other areas. However, lanthanide labeling invariably modifies the labeled compounds' molecular structure; as a result, it is reasonable to predict that these compounds' biological functions will change in some way [34]. Recently, there has been a lot of interest in

Rare Earth - A tribute to the late Mr. Rare Earth, Professor Karl Gschneidner   Materials Research Forum LLC
Materials Research Foundations 164 (2024) 67-142   https://doi.org/10.21741/9781644903056-2

using lanthanide nanoparticles as luminous probes. Lanthanide Nanoparticles offer superior sensitivity, biocompatibility, photo-stability, and thermal stability when compared to lanthanide complex-based luminous probes. According to their structural similarities, three general categories of lanthanide nanoparticles that have been used as luminous probes for biological molecules are Nanoparticles made of lanthanide coordination polymers [34]. The ongoing need for more advanced luminescent materials has sparked technological and scientific attempts to enhance the qualities of currently available luminescent materials and to foster the creation of brand-new, efficient luminous nanomaterials with the desired form, size, morphology, optical properties with other characteristics. [35]

2.   Introduction to luminescent material

2.1   Inorganic lanthanide (Ln) doped luminescent nanomaterials

Recent advances in biomedicine, display technologies, photovoltaics, forensic science, solid-state lighting, and other fields have made lanthanide-based luminescent nanostructures well-known on a global scale for their excellent optical and chemical properties resulting from their distinctive electronic structures. [35]–[43] Nanophosphors are another name for luminous nanomaterials based on lanthanides. Trivalent Ln-ions are typically doped in the host lattice of nano phosphors in one or even more quantities. Scandium and yttrium are two of the 15 rare-earth elements that can be employed in lanthanide-based luminous nanomaterials as trivalent ions. Because of their 4f-5d electronic transitions, lanthanide-based luminous materials exhibit acute emission. In comparison to other luminescent substances (quantum dots, plasmonic materials, organic dyes, etc.), Ln-doped luminescent nanomaterials have many benefits, including broad excitation, sharp emission, longer lifetime, high photochemical stability, well chemical stability, good thermal stability, low toxicity, etc [43].

Depending on the excitation, emission from UV to NIR region is displayed by the Ln-based luminescent materials. The Ln-doped luminescent nanomaterials based on their light-emitting properties are divided into two categories as follows:

2.1.1   Down-conversion luminescence

A photon with high energy is changed into a photon with low energy during this luminescence process. The Stokes shift refers to the energy difference between the photons that are emitted and those that are absorbed. Fig. 2 displays a schematic representation of the down-conversion luminescence procedure.

The host lattice is often doped with trivalent lanthanide ions. Lanthanide-based host lattices come in a variety of forms, including phosphates, oxides, vanadates, fluorides, oxy-sulfides, and others. A rare-earth host lattice doped with trivalent Ln-ions serves as an activator. Then the job of the host lattice is to take up the photons' energy and send them on to the activator. The host lattice also serves as a sensitizer, primarily in down-conversion luminescent materials. For instance, O^{2-} is the sensitizer in the case of the Y_2O_3 host lattice, but in the case of $GdVO_4$ VO_4^{3-} is the sensitizer. Due to the effective transfer of energy from the host to the activator, the emission efficiency is high in both situations. The host lattice's crystal structure is preserved with or without doping by optimizing the lanthanide ion doping concentration. In a report, we found that 5 mol% of Eu^{3+} is the ideal amount to dope the Y_2O_3 host lattice with. [44] In addition to the dopant concentration, a number of other variables, including the host lattice, synthesis circumstances, particle size, temperature, calcination, and environment, can affect the PL intensity.[43] [44] [45]–[48] Another lanthanide ion that serves as the sensitizer can be doped to increase the down-conversion's efficiency. The luminescence efficiency is increased due to the sensitizer's adequate energy transfer to the activator from the host lattice. There are, however, few reports of down-conversion-related co-doping in the literature. The most popular co-doping ion for Ln^{3+} (Ln = Dy, Sm, Eu, Tb) to boost luminescence efficiency is Ce^{3+}.

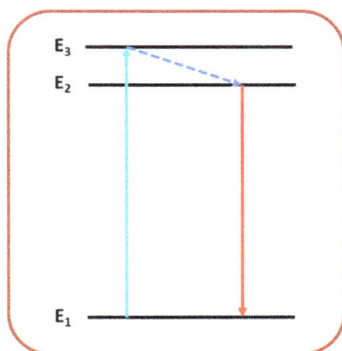

Figure 2. Diagrammatic representation of the downconversion luminescence procedure.

Numerous studies on down-conversion luminescent nanomaterials have been published in the literature; these include studies on $Sr_3Y_2(BO_3)_4$:Ce,Dy, GdF_3:Ce,Ln, $BaYF_5$:Ce,Tb, $NaGdF_4$:Ce, Ln, and YPO_4:Ce, Tb. [48]–[54] It is important to note that in all of these luminous materials, the 4f–5d absorption transition of Ce^{3+} ions dominates the excitation spectrum, whilst the emission spectrum is typical of trivalent Ln-ions, suggesting the transfer of energy from Ce^{3+} to Ln^{3+} ions. [43]

2.1.2   Up-conversion luminescence.

A non-linear process transforms a photon with low energy into a photon with high energy in up-conversion (UC) luminescence. The up-conversion process theoretically involves the absorption of two or more photons and their conversion into higher energy photons (e.g., UV, visible, and NIR). In order to produce higher energy photons, up-conversion typically employs the sequential absorption of numerous photons using trivalent Ln-ions that have a long lifetime and genuine ladder-like energy levels implanted in the suitable inorganic host lattice. The anti-Stokes shift refers to the energy difference between the photons that are emitted and those that are absorbed. [49] The trivalent lanthanide ions Ho^{3+}, Tm^{3+}, Yb^{3+}, and Er^{3+} (emitters) are doped into the inorganic host lattice of upconverting phosphors to generate emission when they are in excitation modes. Trivalent Ln-ions should be carefully chosen in order to adjust the emission wavelength. The up-conversion process has three primary mechanisms: up-conversion energy transfer, photon avalanche, and excited state absorption. [55] In Fig. 3a–g, the schematic representation for the up-conversion procedure is displayed.

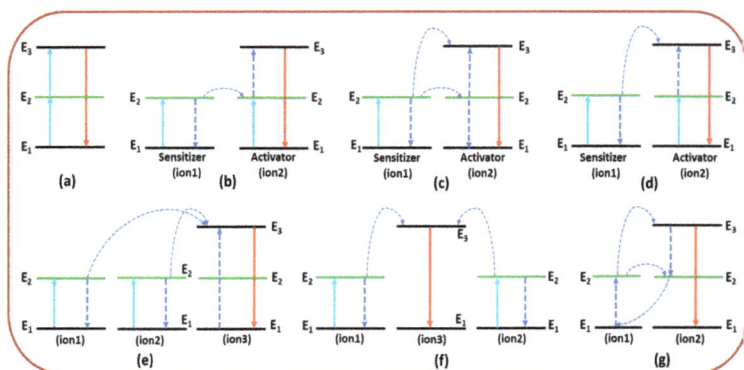

Figure 3. Schematic diagram for the up-conversion process.

Fig. 3a shows a schematic representation of the up-conversion mechanism for materials doped with a single lanthanide ion. As seen in Fig. 3a, two photons are sequentially absorbed as part of the excited state absorption mechanism. Another photon raises the excited ion to the higher energy level (E3) when the Ln ion is sufficiently stimulated by energy photons from the lower energy level (E1) to that level (E3). When photons transition occurs from the state (E3) to the state (E1), up-conversion luminescence takes place. The most effective process is energy transfer up-conversion. Figure 3b–f illustrates the five distinct kinds of energy-transfer up-conversion processes. Sensitizer ions go from the state (E1) to the (E2) by absorbing photons (Fig. 3b). Sensitizer ions transmit energy to activator ions, which are stimulated, throughout this process. From the level (E2) to (E3), the activator ions are promoted by the energy transfer. Only the sensitizer ion absorbs photons, therefore the activator ion is propelled to energy levels (E2) and (E3) by two more energy transfers in sequential energy transfer (Fig. 3c). The activator and sensitizer ions are both of a similar kind When it comes to cross-relaxation up-conversion (Fig. 3d). The photons are absorbed while they are stimulated in an excited state (E2) by the activator and sensitizer ions. Then, as a result of the energy transfer mechanism, the sensitizer returns to the ground state and the activator ion is promoted to the excited state (E3). [55] The cooperative effect plays a role in the energy transfer up-conversion as seen in Fig. 3e and f. In these situations, the luminescence or sensitization process involves luminescent nanomaterials with multiple ions. Cooperative sensitization occurs when ion 1 and ion 2 both sensitizers absorb photons and then transmit energy to the activator (ion 3), which elevates ion 3 to an excited state (E3). Cooperative luminescence, on the other hand, involves two excited ions (ion 1 and ion 2) that work together to absorb photons and produce emission (Fig. 3f). In Fig. 4g, the photon avalanche is seen. The threshold value is the minimal excitation power that is necessary for this operation. The intensity of the luminescence increases by an order of magnitude above the threshold value. The photon avalanche, which generates feedback as seen in Fig. 3g, is fundamentally a looping process coupled to efficient cross-relaxation (CR) and the processes of excited state absorption (ESA) for excitation light. When the process of looping starts, elevated ions 2 at level E1 are raised to level E2, where they emit. Ion 2 is first stocked at level E1 by non-resonant weak ground state

Rare Earth - A tribute to the late Mr. Rare Earth, Professor Karl Gschneidner   Materials Research Forum LLC
Materials Research Foundations 164 (2024) 67-142   https://doi.org/10.21741/9781644903056-2

absorption. This is followed by a successful CR process between ions 1 and 2, consisting of E2 (ion 2) + G (ion 1) → E1 (ion 2) + E1 (ion 1). In order to populate level E1 and complete the loop, ion 1 finally transmits its energy to ion 2. The looping process results in the production of two ions 2 in the metastable E1 state from a single ion 2. The looping process produces an avalanche effect for populating ion 2 in its E1 state, which is followed by two ion 2s creating four ions in their E1 state, and a further four producing eight ions. This results in the photo avalanche (PA) UC, which is produced by the avalanche effect, at the emission level of E2. Since PA typically requires a pump threshold and takes a while (in seconds) to build up, it is simple to identify. Additionally, near the threshold pumping power, the UC photoluminescence's dependence on the pump power intensifies significantly. [55]

2.2 Quantum dots

Quantum dots are often referred to as spherical nanoparticles which are smaller than the exciton in terms of physical dimensions. Exciton Bohr radius confines themselves in all three spatial directions. [56]–[58] The distinguished properties of Quantum dots helped them to play a pivotal role in the field of research. These properties are (i) long wavelength emission lifetime and high fluorescent quantum yield (QY), [59] (ii) use of the single light source for synchronized excitation of numerous QDs, [60] (iii) symmetrical, tapered [61] and (iv) wide spectral windows ranging from the UV to IR region. [62]. QDs materials possess high emission properties in UV regions considering the size of QDs nanoparticles [63]. When QDs are photoexcited in a specific wavelength and they generate an electron-hole pair whose recombination does light emission. In the semiconductor, the excitation of an electron from the valence band to the conduction band by absorbing a photon of equal or greater energy generates an electrostatically neutral electron-hole pair studied as an exciton. The exciton Bohr radius is the mean distance between photogenerated electrons and holes. When the dimensions of the particles approaching the exciton (E.g., about 6 nm is the exciton Bohr radius of CdSe), then due to the quantum confinement effect, the material's optical and electrical properties rely on its physical dimensions. [64] Semiconductor's band structure changes into discrete energy levels under these above mentioned conditions as shown in Fig. 4. and with decreasing particle size the difference of energy between the lowest unoccupied level and the highest occupied level gets widened. In an elementary model, particles in the box approximation restricting the motion of the electron in all 3 dimensions by impenetrable walls can be used to describe quantum confinement. The contribution to the energy gap based on size is inversely proportional to the square of the radius, as stated by Klimov, in a semiconductor-based spherical QD model with radius R, indicating that the gap widens as the QD's size decreases. Due to quantum confinement, the uninterrupted energy levels of electrons discriminate from the bulk materials compared to the nanocrystals. Moreover, the experimental result reveals the confined wave functions of electrons, they become a discrete set of energy levels resulting in the discrete absorption spectrum of QDs. As shown in Figure 5. [65], with the reduction in particle size, there is an increase in the blue shift in their luminescence.

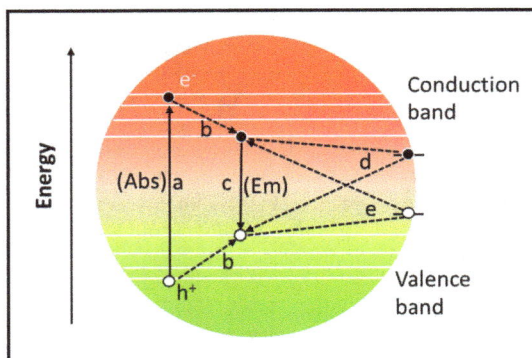

Figure 4. An electrical and optical process in a semiconductor nanocrystal is shown in a schematic diagram. Radiative and non-radiative processes are represented by dashed and solid lines, respectively. Here a- optical excitation, b-thermal relaxation of the excited electrons, c- radiative recombination, d- non-radiative recombination by electron trap, e- non-radiative recombination by hole trap. (Reproduced with permission from Ref [64])

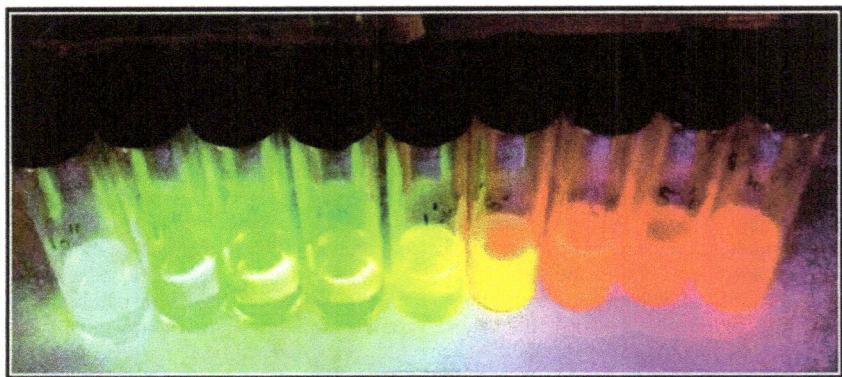

Figure 5. Photograph of CdSe-ZnS QDs in chloroform solutions of varying sizes (smaller to larger from left to right). (Reproduced with permission from Ref [64])

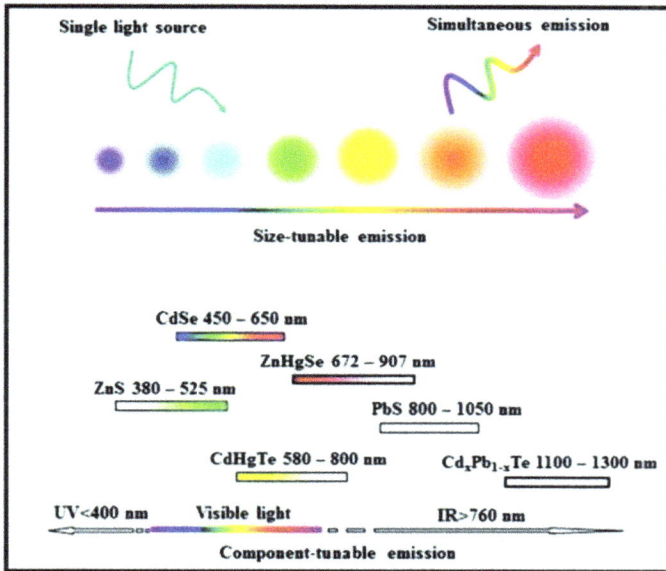

Figure 6. Size adjustable Broad-spectrum windows from the UV to IR region are displayed by the simultaneous fluorescence emission and excitation of several QDs by a single light source. (Reproduced with permission from Ref. [62])

The size of QDs due to the quantum effect affects their optical and electronic properties. Therefore, QDs of various size show variety fluorescence colors. The change in semiconductor QDs' radiation, as well as changes to their size and the range of their emissions' wavelengths, are shown in Fig. 6. [62] There is considerable research on various groups like II-VI (e.g., ZnSe, CdSe, CdS, CdTe), III-V (e.g., InAs, InP), and IV-VI (e.g., PbSe) and their alloys with a restricted nanoscale size for several implementations circulated in the publication. With reference to its dimension, every QD has its limitations for regulating the color emission as in Fig. 6. Several reports are there that describes the correlation between the dimension, configuration, and electronic properties of quantum dots. [66] Lately, Carbon dots (CDs) or Carbon quantum dots (CQDs), which also include graphene quantum dots (GQDs), are a type of carbon-based quantum dot that has undergone extensive scrutiny. [58][67]–[69] CQDs possess wide optical absorption because of π- plasmon covering a wide UV/VIS spectral range up to NIR. The errors of the photoluminescent emissions of these QDs are linked to the configuration and exterior sites. [58][70] In 2004, Xu et al first discovered the carbon quantum dots. [71] But, Sun et al invented the term for fluorescent carbon nanoparticles in 2006. [72] The mechanism of fluorescence emission in the case of CQDs or GQDs could be described in the following two ways: (i) fluorescence emission linked with errors in GQDs surface (ii) fluorescence emission caused by the transition of band gap due to combined π-domains. In the above mechanism, the transitions of bandgap derive from the π-domains. These π-domains are cloistered with the generation of sp2 hybridized islands which is abundant in π-electrons by reducing graphene oxides. [73] The synthesis process of GQDs shows no π-connections between the sp2 islands in the graphitic network as there will be interisland quenching of the intended fluorescence emission if there are any π-connections between the sp2 island. [74][75] To avoid interlayer quenching between dual layers, single-layer graphene sheets could be used for these kinds of bandgap transitions. [76] For optical absorption and fluorescence emissions in GQDs, single-layer graphene sheets are employed as a predecessor for splitting them electronically into isolated π-conjugated domains, which is analogous to large aromatic molecules with elongated π-conjugated of the specified electronic energy bandgap.11 Such electronic transitions reveal weak or no fluorescence emissions but show strong absorption in the ultraviolet (UV) region as depicted in Fig. 7a. Due to the absorption of light by a huge amount of highly dense π-electrons in the sp^2 hybridized islands leads to strong absorption, while due to quenching during migration of exciton through radiation less relaxations into the ground state is the probable cause of the weak emissions. [77] Surface-related defect sites generate the fluorescence mechanism of the second class. In general, this type of surface site consists of incomplete sp2 domains serving as surface energy traps. Both hybridized carbon atoms sp2 and sp3 domains along with other operationalized defects of surface such as carbonyl-linked bounded electronic states [70][78] located in GQDs/CQDs provide their many-color emissions that are acute in the visible light's green and blue regions. These surface defects act like arenes which are discretely integrated into solid moderators that display many-color emissions due to various surface defects with divergent emission and excitation properties. [70][76], [79] According to Robertson and O'Reilly, the sp2 site's π-states regulate the optical characteristics of carbon nanomaterial that contain sp2 and sp3 links. [80] Thus, the Recombination of electron-hole pairs creates the vibrant surface defect-derived fluorescence of GQDs/CQDs in the firmly confined π and π^* electronic sp2 domains. The energy gap of the σ and σ^* states of the sp3 domains is the location of these sites, [81] which emits strong visible emissions. Strong emissions in the visible region and weak absorption in the near UV region are exhibited by such electronic transition as depicted in Fig. 7b. The fluorescent emission of CDs may show the effects from fluctuating sizes of a particle in the sample along with the distribution of different points of emission of each CD. [82]Chua et al. reported the two types of functionalized GQDs: hydrazide-reduced GQDs and NH_2OH functionalized, which revealed red-shift and blue-shift phenomena, respectively [83]. The fascinating tunable PL feature of CDs arises from the effects of quantum confinement [84]. Fluorescence QY (ϕ) possesses a high significant value; if the value increases, the intensity of the fluorescence signal gets stronger and the fluorophore turns brighter [85]. Due to the emissive surface traps, the photoluminescence quantum yield (PLQY) of bare CNDs is frequently low and their brightness could be increased by passivating the surface. The CQDs and GQDs normally show higher quantum yield than that of CNDs due to their functional groups, layered structure, better crystallinity, impurities, and defects of surface [84][86]. The QY, chemical reactivity, electron density, and bandgap of CDs are modified by doping heteroatoms. Good photostability, PL emissions shifted to red and 24% quantum yield in GQDs because of the narrow energy gap assigned to the synergetic result of π–$\pi*$ transitions of the large π-electron system and high nitrogen content of graphite was reported by Lyu et al. 85% QY was displayed by Lin et al. passivating the surface with nitrogen doping and amino groups [87]. Sun et al. reported similar red-shifted PL for fluorine-doped GQDs, caused by increased fluorination resulting in high defects in the surface, as a result more excitons get trapped [88]. The intensity of stable fluorescence emission after continued excitation is referred to as Photostability. Fluorescent CDs generally shows magnificent photostability because of their large

Rare Earth - A tribute to the late Mr. Rare Earth, Professor Karl Gschneidner   Materials Research Forum LLC
Materials Research Foundations 164 (2024) 67-142   https://doi.org/10.21741/9781644903056-2

π-conjugated structure and they display nonblinking PL after surface passivation. Das et al. prepared fluorescent CDs which were nonblinking at the single-particle level and had good photostability [89]. Sun et al. synthesized fluorinated GQDs that showed good photostability, as the electron density of the aromatic framework decreased attached with fluorine having high electron-withdrawing feature [90] In addition, by passivating the surface defects are made more steady to make smooth and productive radiative recombination of electrons and holes in a confined surface, resulting brighter fluorescence emissions. [80] Without any surface modification the emission spectra of CQDs/GQDs show tunability as in semiconductor quantum dots. The quantum yield of GQDs/CQDs is very low due to the defects of low stability. Therefore, modification of surfaces of these QDs doped with polymers gives rise to strong emission and stable defects. Surface modification (passivation) helps in tuning the emission color of CQDs as shown in Fig. 8.

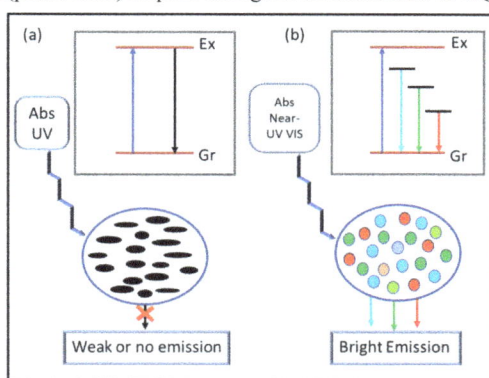

Figure 7. Diagrammatic representation of (a) CQDs with high absorption in the UV range but low emissions and (b) CQDs with low absorption in the near UV-VIS range but high multicolor emission in the visible range.

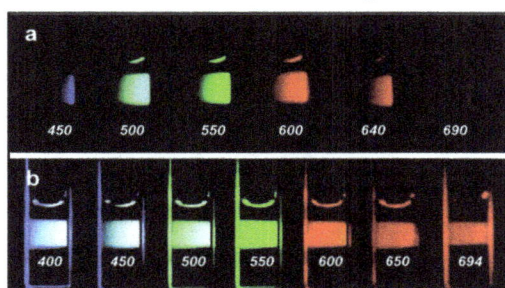

Figure 8. Aqueous solutions of the PEG1500N-attached CDs (a) were excited at 400 nm and were then photographed using the appropriate band-pass filters, and in (b), they were excited at

the stated wavelengths and were then directly captured. (Reproduced with permission from Ref. [91])

2.3   Luminescent nanoscale metal–organic framework

An extensive type of organic-inorganic hybrid crystalline porous materials manufactured from metal ions and organic linkers is known as metal-organic framework (MOFs). [92][93]   MOFs belong to a sub-class of coordination polymers (CPs). As MOFs have both inorganic and organic constituents, these materials of MOF are favorable as multipurpose materials. Their composition can be designed according to the required properties considering the coordination modes of the inorganic metal ions and the configuration of the organic ligands. [93][94] As a result, impressive numerous applications in many fields including optical luminescence, barcode security, storage of gas, delivery of drugs, sensing, diagnostics, exchange of ions, catalysis, and magnetism are being developed. [95][92] However, metal–organic standard bulk crystalline materials frequently do not satisfy the requirements for these mentioned activities. Considering the preferred application, these metal-organic materials need fabrication as bulk crystalline solids and it is essential to miniaturize them on the nanometre size scale and paralyze them at precise sites on surfaces. [93], [96] For instance, nanoscale MOFs are more suitable to fabricate luminescent security links because they can be easily dispersed uniformly into the ink medium with high stability as compared to bulk MOFs. In the above applications, Nanoscale MOFs' high surface area is a more beneficial aspect than their macroscopic counterparts as the former reduces the number of luminescent MOFs necessary for the fabrication of ink. Assessing their physical properties, usually when relatively one dimension of material becomes the size of a nanometre scale then there is an emergence of special physical properties. Thus, Size-dependent electrical, magnetic, and chemical properties are also beneficial aspects of the nanoscale metal-organic materials. [96] The tunable luminescence features of the MOFs cover a broad spectral range from ultraviolet to NIR. In MOFs, luminescence is produced by both inorganic metal ions and organic ligands.

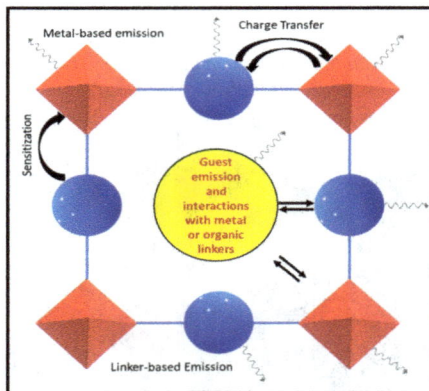

Figure 9. Graphic illustration of the emission potential in a porous MOF, with metal clusters (red octahedral) connected by organic linkers (blue spherical), each with a guest molecule (yellow circle). (Reproduced with permission from Ref. [97]

Rare Earth - A tribute to the late Mr. Rare Earth, Professor Karl Gschneidner   Materials Research Forum LLC
Materials Research Foundations 164 (2024) 67-142   https://doi.org/10.21741/9781644903056-2

Furthermore, the transfer of charge between the ligand and the metal ion is also responsible for luminescence in metal-organic frameworks. In addition, introducing guest molecules or ions will also cause luminescence. The majority of recent research on luminescent MOFs has focused on examining the luminescence mechanism, the fundamental synthesis, emerging properties of luminescent MOFs, and upconverting MOFs with greater credibility for luminescent security applications as well as in display and light-emitting devices. Additionally, the reversible storage of some luminous MOFs has been made possible by the stable porosity of these materials, which releases guest substrates and serves as the host for their differential identification with sensing species. [97], [98] Fig. 9 demonstrates the various methods of luminescence production as detailed below:

2.3.1 Ligand-based luminescence in metal-organic frameworks

In this method, commonly combined organic compounds that absorb incident radiation are known as linkers. Henceforth, the charge gets transferred with coordinated metal ions or clusters by the linkers. Occasionally, there is a possibility of direct emission from the organic linkers. The photophysical process is described in a diagram as shown in Fig. 10. When an organic molecule radiatively changes from its first singlet state, S1, to its ground singlet state, S0, this is referred to as molecular fluorescence. The phosphorescence of organic molecules is the radiative transition from the triplet state T1 to the singlet state S0, which has a lifespan of several microseconds to seconds. The MOFs along π-conjugated organic molecules as linkers exhibit ligand-based luminescence. For the production of MOFs, a wide variety of organic ligands are used, all of which have rigid backbones functionalized with several carboxylate groups for the coordination of the metal and the ligand. [98] The transition from the lowest excited singlet state to the singlet ground state corresponds, in general, to the fluorescence emission of organic ligands, and transitions are either $\pi \rightarrow \pi^*$ or $n \rightarrow \pi^*$. In addition, stabilization of the organic linkers within MOFs results in a reduction in the rate of nonradioactive decay causing growth in quantum efficiencies, fluorescent intensity, and lifetimes. In the solid state, the molecules get adjoining by molecular interactions, enabling the transfer of charge among the organic ligands/linkers, broadening the emission, resulting in a shift of spectra and loss of fine structure. It is possible to change the fluorescence properties of the organic linkers by causing intramolecular or intermolecular interactions between the organic linkers depending on the size and type of the metal ions, the arrangement and orientation of the linkers, and the coordination environment within the MOF. [97], [98].

2.3.2 Lanthanide-doped luminescence in metal-organic frameworks

The ligand-to-metal charge transfer (LMCT) mechanism results in lanthanide complexes and rare earth metals, which are linked to ln-doped luminescence. Narrow and featured 4f-4f electronic transitions are shown in each lanthanide ion. Excluding La^{3+} and Lu^{3+} all Ln^{3+} ions may produce luminescent f–f emissions from a wide range covering ultraviolet (UV) to visible and near-infrared (NIR). Eu^{3+} emits red colour, Tb^{3+} emits green colour, Sm^{3+} emits orange colour and Tm^{3+} emits blue colour while Er^{3+}, Yb^{3+} and Nd^{3+} exhibit near-infrared (NIR) emissions. The sharp and intense emission band of the transition metal originates from f-f transitions. But, due to the confined f–f transitions lanthanide ions suffer from weak absorption. This issue may be solved by the luminescence sensitization process or by creating the antenna effect. [100] The antenna effect mechanism includes three steps. In the first step, the photon energy is absorbed by organic ligands, which exist around Ln^{3+} ions, in the second step the absorbed energy is transferred to Ln^{3+} ions from the organic ligand, and in the third step there is emission from Ln^{3+} ions as shown Fig. 10. The actinide metal ions seen in MOFs containing rare earth metals exhibit the antenna effect as well. [101]

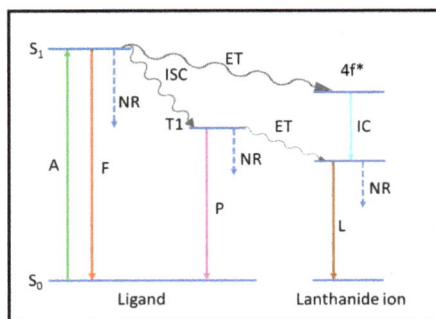

Figure 10. Representation of energy absorption, migration, emission, and process in MOFs. Plain arrows indicate radiative transitions; dotted arrows indicate nonradiative transitions. Abbreviations: A= absorption; L= Lanthanide-centered luminescence; P= Phosphorescence; F=fluorescence; S=Singlet; T=Triplet; ET=Energy transfer; IC= internal conversion; ISC= intersystem crossing. (Reproduced with permission from Ref. [99])

2.3.3   Charge transfer luminescence in MOFs

When a transition is allowed from the charge-transfer excited state to the ground state then the generation of luminescence takes place. MOFs consist of two types of charge transfer which are metal-to-ligand charge transfer (MLCT) and Ligand-to-metal charge transfer (LMCT). Metal-to-ligand charge transfer is the term for an electrical transition that takes place from a metal-clustered orbital to an organic ligand-cantered orbital. A ligand-to-metal transfer is a name for an electronic transition that takes place from an orbital with an organic ligand at its center to one with a metal cluster. MLCT for MOFs occurs less than LMCT, and it is observed that $d10$ Ag^1 and Cu^1- based MOFs if there is a chance of transfer of d-electron. For example, $MOFCu_3(C_7H_2NO_5)_2.3H_2O$, $(C_7H_2NO_5=2,6\text{-dicarboxylate-4-hydroxypyridine4-hydroxy}$ pyridine that is chelidamic acid) which displays blue fluorescence at 398 and 478 nm upon excitation at 333 nm, but $CuAg_2(C_7H_2NO_5)_2$ with free linker 2,6-dicarboxylate-4-hydroxy pyridine both shows green emission when excited at 515 and 526 nm and at 358 and 365 nm, respectively. [102]

2.3.4   Guest-induced luminescence in MOFs

MOFs have controllable pore sizes with regular channel structure that works as flexible/rigid hosts for the encapsulation of lanthanide ions. MOFs feature large porosity that ranges from certain angstroms to certain nanometers which perform as the host lattice for lanthanides ions, organic molecules, and fluorescent dyes. The featured emission of lanthanide ions could surely be detected in liquid surroundings. The quantum yields concerning lanthanide fluorescence are extreme in a liquified environment. MOFs doped with Eu^{3+} and Tb^{3+} prepared by Luo et al. show tunable emission features. As seen in Fig. 11, the Ln-doped MOFs exhibit various colours based upon the doped ion when undergoes excitation at 365 nm. This guest-induced luminescence produces MOFs that help in environmental probing and molecular detection.[103-104].

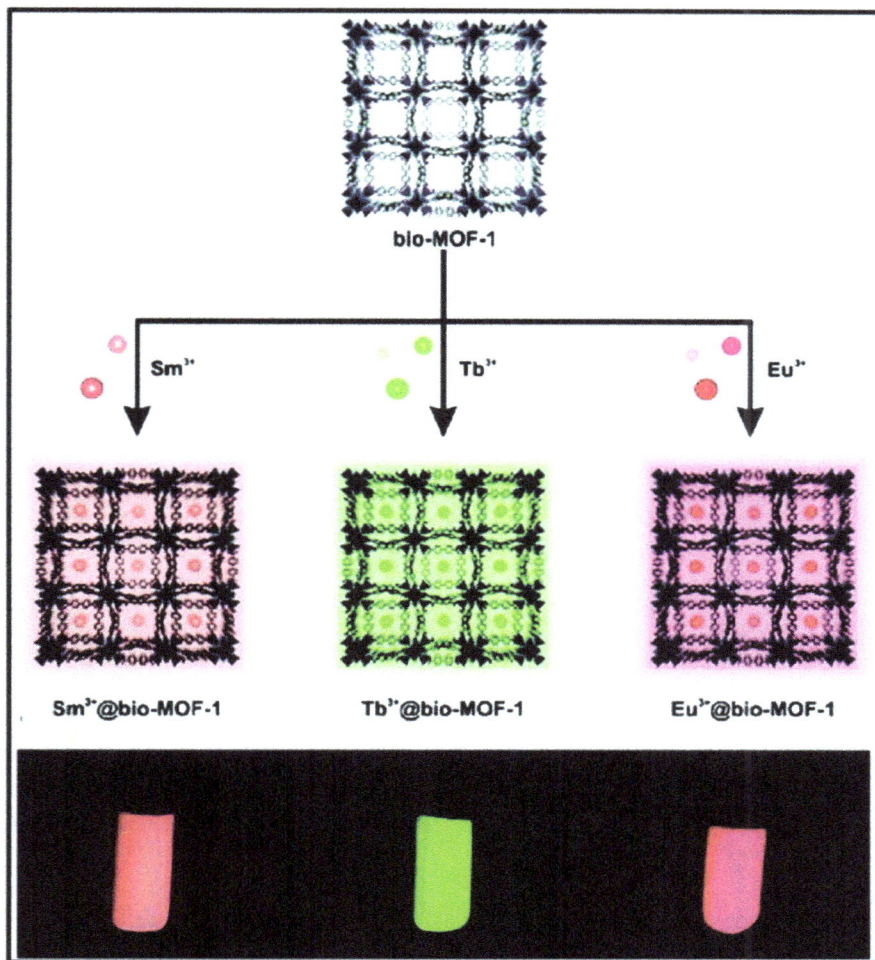

Figure 11. Ln-ions doped in bio-MOF-1. (Reproduced with permission from Ref. [91])

2.4   Plasmonic materials

Collective oscillations of the conduction electrons in metals due to the incident of light form a quasiparticle known as plasmon. The oscillations of conduction electrons can be strongly confined amid a dielectric interface of metal. Localized surface plasmon resonance (LSPR) is the coherent oscillation of electrons of a conduction band in small spherical metallic nanoparticles. Conceptually, the localized surface plasmon is detected in conductive nanoparticles which have a dimension smaller than the incident wavelength. [105][106] Interaction of metal nanoparticles with incident light collectively excites the electron of a conduction band which then goes through a displacement from its initial position with respect to nuclei. Therefore, the Coulombic interaction between the nuclei and electrons leads to rising in restoring forces resulting coherent oscillations of the electron cloud corresponding to nuclei. Oscillations with a frequency depends on four factors: the density of electron, the dimension of the charge distribution and the effective mass of the electron. These collective plasmon oscillations for a spherical nanoparticle are explained in Fig. 12. [107]

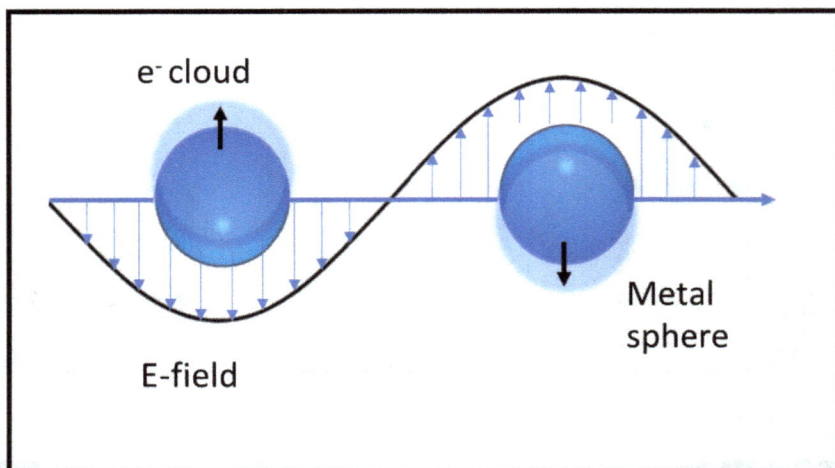

Figure 12. The displacement of the conduction electron charge cloud from the nuclei is depicted schematically in the plasmon oscillation for a sphere. (Reproduced with permission from Ref. [107])

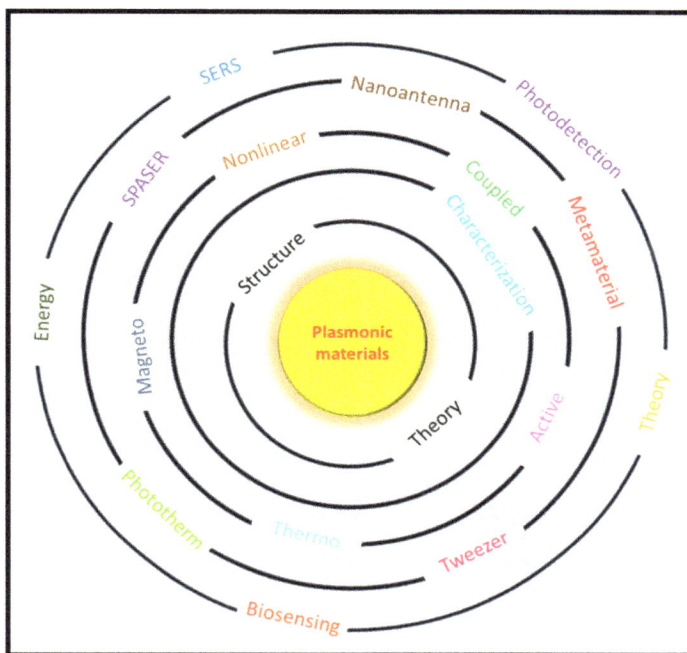

Figure 13. Several subfields of nanoplasmonics that are currently being researched are represented schematically. From the outer to inner sides are (1) the applications closely related to our daily life, (2) the applications in optics and optoelectronics, (3) the properties of hybrid plasmonic nanostructures, and (4) the basics, which cover the materials, theoretical basis, characterisation, and fabrication. (Reproduced with permission from Ref. [105])

Plasmonic nanomaterials, in particular, are those particles whose density of electron can be coupled by using electromagnetic waves with wavelengths significantly greater than the particle, which is possible due to the metal-dielectric contact between the particles and the medium. [108] According to the geometries and relative positions the plasmonic nanoparticles show properties like coupling, absorbance, and scattering. Furthermore, plasmonic nanoparticles have phenomenal properties like high photothermal conversion efficiencies, high electromagnetic field enhancements, and high spectral responses. So, plasmonic nanostructures are a multifaceted group of nanomaterials that have numerous types of functions. [109]–[111] Current field of research studied as 'nanoplasmonics' has got attention globally from various scientific groups including engineering, chemistry, biomedicine and optics. The schematic indication of the numerous domains in nanoplasmonics is shown in Fig. 13. These days, aluminium, copper, graphene, and highly doped oxides are being used for the study of nanoplasmonics. The tunability range of the plasmon for aluminium is greater than silver and gold as shown in Fig. 14a. However, it has been reported by research about the plasmonic nature of noble metals based on their high chemical stability, visible–NIR tunable plasmon and a huge number of chemical processes for their synthesis with regulated shape and size. Metallic nanoparticles have optical properties that defer with the slight increase in dimension for gold nanosphere samples as shown in Fig. 14b. Due to the anisotropic nature, the optical properties of the nanoparticles change remarkably along the growth of nanorods. [112]

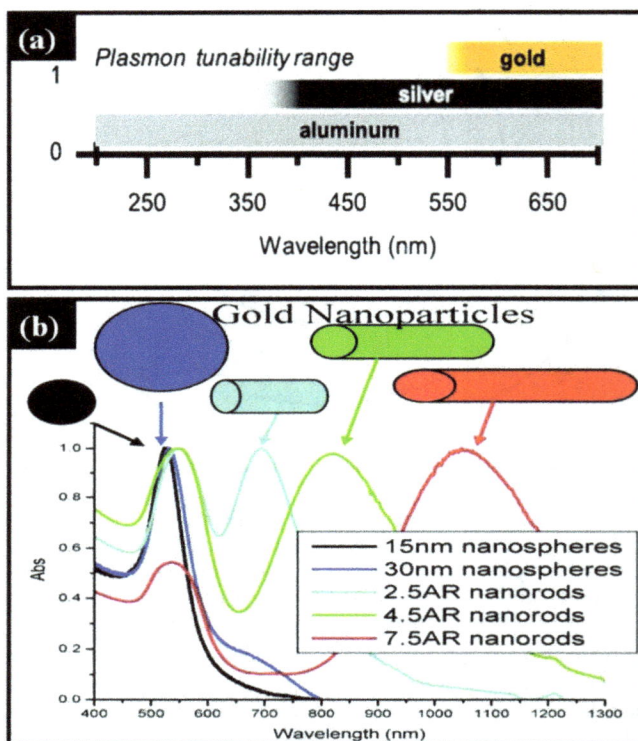

Figure 14. (a) Plasmon tuning ranges of gold and silver in comparison to Al [113], as well as (b) the absorption of different-sized and-shaped gold nanoparticles. (Reproduced with permission from Ref. [108])

3. Luminescent materials for forensic application

3.1 Introduction

The study, evaluation, and detection of evidence at crime scenes that can be used to quickly and easily apprehend suspects is known as forensic technology. Prior to making a decision in court, forensic evidence is an essential tool for identifying culprits through inspections at crime scenes. Additionally, it is a helpful tool for compiling reliable evidence against the accused [114][115]. The majority of forensic technology focuses on collecting biological evidence for criminal identification. toxicology, entomology, Psychiatry, pathology, anthropology, and odontology are among the many scientific disciplines that have been used in criminal investigations. [116][117]. The study of fingerprints, sound profiles, earprints, and handwriting also uses forensic tools. Forensic technology is a different field of science that has been extensively used in the method of law enforcement and to locate criminals along with solving crimes. As a result, it is one of the fundamental sciences utilised to lock up offenders and ensure that justice is served promptly. Bloodstains, textiles, fibre, glass, hair, gunshot residues, paint, and other common evidence substrates are regularly seen in forensic examination [118]. The most common types of forensic evidence used to identify criminals are bloodstains and fingerprints taken using spectroscopic techniques. The elevated areas of the epidermis of a person's fingertips, palms, and soles create the patterns of friction ridge skin. They can be used for person identification in a variety of real-world settings, including law enforcement, authority control, and information encryption, because they are genotypically defined, distinct, and durable [117]. In general, when a finger contacts something, the skin's epidermal secretions are transferred to the surface, leaving an imprint of the friction ridge pattern [114][118]. There are three different forms of deposited fingerprints (FPs): visible, indented (plastic/molded), and latent/invisible prints. Among these, Extraction of information from latent prints is difficult in these cases because of the surface's contrast issues or because pollutants from the environment have deposited on the ridges of the finger or palm of natural secretions. LFPs contain diverse chemical components in addition to reproducing the physical properties of a person's ridge patterns [119][120]. LFPs typically have a combination of intrinsic and extrinsic components. Natural secretion glands, like the exocrine, eccrine, sebaceous, and apocrine secrete the majority of the intrinsic components of a fingerprint. Extrinsic components primarily come from the different pollutants that are present, including oil, blood, make-up, and food-borne pathogens. The sweat produced by the eccrine glands, which are dispersed throughout the body, has more than 98% water. Every person's FPs produce eccrine and sebaceous gland secretions during the deposition process. When minerals and organic compounds, perspiration has the largest residue on finger pores of all of them. Eccrine sweat is linked to lactic acids, urea creatine, amino acids, uric acids, choline and carbohydrates. Squalene, sterol ester, fatty acids, glycerides and wax esters are all mixed into sebaceous sweat. The changes in fingerprint composition can be caused by donor variables such as gender, medication use, metabolism, diet, age, and environment [120]. The ability to discover latent fingerprints is also impacted by time, which in highly dependent on the hazards given by the compounds used, moisture, light, warmth and air. These fingerprint compositions can reveal a great deal about the donor when analysed. LFPs can therefore be used by users as a framework for information storage and be crucial in forensic analysis, medical diagnosis, and health evaluation.

The fundamental idea behind LFP imaging is to create contrast between the release residues and the substrates that contain them [118]. Detailed features for individual identification and direct visualisation of an LFP pattern are made possible by LFP imaging, which is crucial in forensic and crime scene applications. Most prominently, it is possible to get useful data for applications like metabolism analysis and safety inspection by imaging substances of interest in LFPs, such as metabolites or pollutants. For LFP imaging, many of contrast materials and imaging methods have been created over the last few decades [117][121].

How a fingerprint is revealed using a luminescent powder?

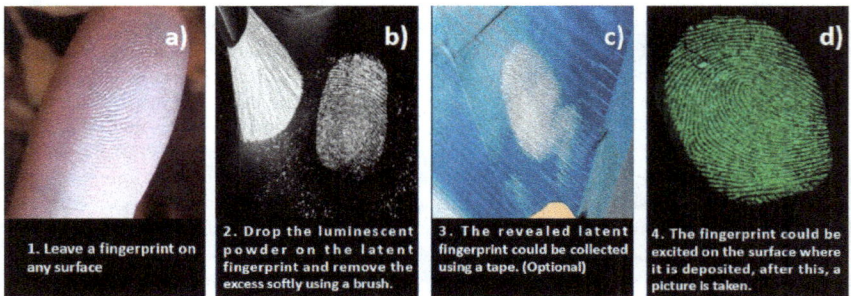

Figure 15. General procedure to lifting fingerprints from a substrate. (Reproduced with permission from Ref. [122])

3.2   Principle and determination of latent fingerprint detection

The primary approach for identifying a person in forensic science is latent fingerprint identification. In order to identify a fingerprint, the intricacies of the ridge pattern must match, and the control fingerprint must be compared to the fingerprint from the crime scene. Numerous level-1 (arch, whorl, and loop), level-2 (minutiae), and level-3 (sweat pores) categories could be seen in the ridge patterns [123].

Figure 16 [124] depicts the specifics of fingerprint traits. Since each person's fingerprint is distinct, it is impossible for two persons to have the same fingerprints. However, Galton concluded that most human fingerprints share the same fingerprint pattern [125]. Human fingerprints are detected by a pattern of ridges and furrows on the hands, feet, and toes that perseveres until death. It is possible for two persons to have the same fingerprint ridge pattern. Ink is typically applied to the left-hand finger and then placed on paper, plastic, or aluminium plates in order to capture a fingerprint. The fingerprint ridge, which includes loops, whorls, and arches, is formed when friction is produced between the finger and the surface which it has been applied to. The form and kind of coating have a significant impact on the finger's ridges. There are a few whorls and arches among the loops that make up the majority of human fingerprint ridges. Typically, the finger's fingerprint ridges are packed with arches. However, there are four categories of fingerprint whorls: major, twin, lateral and accidental. The ridge patterns of fingerprint explain why fingerprint detection is necessary.

The entire finger does not have ridges. Quit ridges are the ridges that abruptly appear at the end. As dividing ridges, certain other ridges are referred to. Additionally, there are three other types of ridges: remote, diversified, and discrete. Large dots are used to represent ridges [126]. Its characteristics are based on the type of ridge, which also reveals important details about the person's identity.

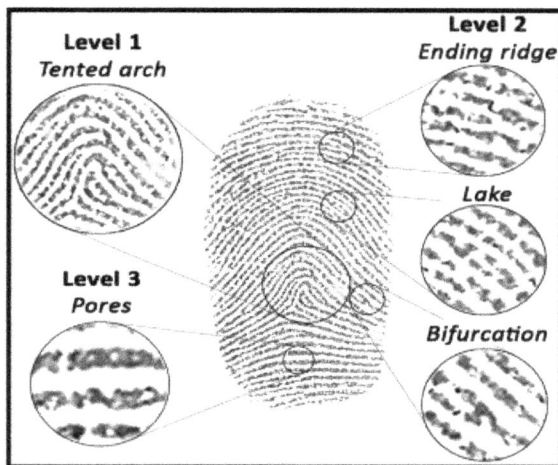

Figure 16. Level 1 (class of mark e tented arch), Level 2 (minutiae e terminating ridge, lake, and bifurcation), and Level 3 (minutiae e lake and bifurcation) are a few examples of fingermark ridge characteristics that have been mentioned (pores). (Reproduced with permission from Ref.[124]).

The detection of latent fingerprints has been improved by the application of several analytical techniques. For the purpose of detecting fingerprints, physical techniques such as vacuum metal deposition, powdering and small particle reagents have been used [126][127]. Cyanoacrylate, iodine, and multi-metallic deposition are examples of physical/chemical methods that have been employed to enhance fingerprint recognition [128][129]. The utilisation of ninhydrin and its derivatives, genipin, deferoxamine, steel complexation with ninhydrin and 1, 2- indanedione has also been studied for the purpose of detecting fingerprints [130].

For enhancing picture quality and easing latent fingerprint identification, all methods necessitate improved chemical reagents. Additionally, equipment has been used to improve fingerprint recognition. These include optical techniques including fluorescence spectroscopy, mirrored images, diffusion-mirrored pictures and absorption, UV-visible absorption [131]. While detecting latent fingerprint, the identification of both new and old fingerprints is done utilising an electrochromic polymer that electrochemically depositing material on stainless steel by employing electrochemical techniques [132]. Thin layer chromatography has been used to identify amino acid and lipid components for detection of latent fingerprint [133]. Due to its solubility in water, 1,8-diazafluoren-9-one and ninhydrin are frequently utilised for detection of latent fingerprint. There have been reports of latent fingerprints developing on porous and non-porous substrate surfaces employing solid ninhydrin [134]. Compared to other approaches, it displayed clear fingerprint patterns and images. Almong et al. examined fingerprints on paper surfaces using ninhydrin and the metal salts of zinc and cadmium chloride [135]. Shorter wavelength reagents with higher resolution than genipin were also found. Genipin was employed to detect latent fingerprints on the paper substrates, and better fingerprint image quality than ninhydrin consisting a high contrast backdrop were noted [136]. For the creation of latent purple-brown fingerprint images, a surface of red luminescence paper was coated with naphthoquinones. In order to reduce the masking effect of fingerprint residues on paper, ninhydrin and ink were utilised [137]. On white bond paper, the donor left their fingerprints, which were then damaged at level 3 by the addition of ninhydrin. For the purpose of finding latent fingerprints, ninhydrin and 1,2-indanedione were concentrated on the fingerprint deposit on thermal paper [138]. In contrast, the majority of materials that are insoluble in water are used to identify latent fingerprints using physical means. Gas chromatography-mass spectrometry (GCMS) has been created to identify fingerprints on porous and non-porous surfaces, and it has also been used to analyse the time of application and ageing of fingerprints. Liquid chromatography-mass spectrometry has been used to separate and quantify amino acids from the detection of latent fingerprints [139]. Papers containing pores that contain ninhydrin and 1,2-indanedione-zinc chloride have also been used. These aid in separating amino acids from the detection of latent fingerprints [140]. An important method for identifying visual fingerprints is Surface-Enhanced Raman Spectroscopy (SERS), which is reliant on the amount of lipids and amino acids present on the residues of the fingerprint detection. SERS uses metal nanoparticles to interact with sweat residues in order to clearly distinguish fingerprint images. SERS is used in conjunction with 4-mercaptobenzoic acids which is a functional complement on nanoparticles, to produce fingerprint pictures, functionalized nanoparticles assist in obtaining crisper fingerprint images. SERS spectroscopy has also been used to identify illicit substances including drugs and explosives in secretions; however, it has no effect on the cyanoacrylate fume technique of latent fingerprint identification.[141]

3.2.1 Identification of fingerprints

The initial stage involves pressing a finger on any surface, as shown in Fig. 15a. To simulate an accidentally left fingerprint, very little pressure should be applied [142]. Second, the latent fingerprint is manually covered in luminous powder (see Fig. 15b), and any extra powder is removed using a brush. As a result, the latent fingerprint's finer details may now be seen with the unaided eye. The final stage could involve collecting the fingerprint using transparent tape (see Fig. 15c). More fingerprint features may be retained on the tape if the tape is removed properly. If the fingerprint is left on the surface, UV light can be used to stimulate it, and the powder used to cover the fingerprint will emit a strong amount of light which will make visible the fingerprint (see Fig. 15d). In this point, a high-resolution shot need to be taken in order to capture the fingerprints' more intricate patterns.

The Automated Fingerprint Identification System (AFIS), as a result was developed to make fingerprint searches and comparisons easier. By digitising the fingerprint cards from advanced databases, the AFIS computer system assists in the method of matching, conserving, and exploring fingerprints [143], [144]. Before the fingerprints are registered in the system of AFIS, the following features must be there:

• Size. Arches, loops, and whorls important fingerprint traits should be discernible in the images.

• Contrast. To effectively discern between the shapes of fingerprints and the surface they were taken from, there should be a high of contrast.

• Sharpness. The impression must have numerous identifying characteristics, such as whorls, loops and arcs, and must be simple to evaluate with the naked eye.

The execution of latent fingerprint identification onto porous and non-porous substrates can be estimated using a number of variables, including background interference, resolution, stability, and brightness.

Fig. 16 depicts the basic ridges that give a fingerprint its uniqueness. Each person has a distinctive fingerprint pattern that distinguishes them from other people because even identical twins have different ridge patterns [143-144].

3.3   Inorganic nanomaterials for fingerprint application

3.3.1   Literature survey

Since the effectiveness and dependability of current procedures are deteriorating, new innovations in forensic analysis are required. This is true even if study on forensic technology skills has stabilised its forensic analytical method. Numerous evaluations that place a focus on LFP imaging have so far been published. For instance, Russell et al. have emphasised how fingerprinting technology has advanced to the point where it can now provide both chemical information and personal identity [145]. The chemical techniques for obtaining chemical data on LFPs have been compiled by Zhang et al. in their study [146]. Using forensic analysis and biological/chemical considerations, Bécue et al. [147] have introduced the techniques for LFP detection. These evaluations largely concentrate on the information extraction and LFP analysis. There isn't much discussion about LFP imaging techniques that can remove background interference. Conducting polymer electrodes, gold(Au) or silver(Ag) metal depositions, electrochemiluminescence, multi-metal depositions [MMD-I and II/SMD], gold-aspartic acid deposition [148], metal oxide depositions such as ZnO, SiO2, and semiconductor quantum dots [149] have all been tested as nanomaterials. The forensic examination of numerous samples and crime scenes has involved the use of numerous analytical tools. The most popular techniques among them are TEM, DLS, SERs, SEM, AFM, UV-Vis, MALDI-MSI, and FTIR.

The microstructure and density of the covering of the NPs are assessed by SEM. Similarly, methods like FT-IR for determining the functional group of both exogenous materials and ligands, UV-Vis for improving LFPs with high performance, DLS for determining particle size, SERs for FP analysis with locating chemical and biological components in the ridge, and AFM for three-dimensional (3D) image detection of FPs ridges [150].

Utilising the hydrothermal co-precipitation technique, solid state synthesis, and combustion synthesis, a phosphorescent material strontium aluminate is simple to synthesise at a reasonable cost. [151][152] This has proven the effectiveness to detect fingerprints. In order to detect fingerprints, Strontium aluminate powders (SAP) doped with Eu^{2+} ions were made by Liu et al. [149]. Under the stimulation of ultraviolet (UV) radiation (526 nm), this substance produced green light.

No extra equipment was needed to visualise the fingerprint utilising the SAP, and the outcomes showed that the fingerprints could be identified for recent or old latent finger prints placed on porous or non-porous surfaces. Fig. 17 displays four fingerprints obtained under identical circumstances from non-porous surfaces like porcelain, glass, foil and plastic bags (the fingerprints were stimulated for two minutes with UV light at 365 nm in a dark atmosphere). The prints of paper, cloth, leather and wood respectively, are represented by the fingerprints in Figs. 17A, 17B, 17C, and 17D. When comparing the images in Figs. 17 and 18, it can be seen that the fingerprints obtained from non-porous surfaces exhibit greater lucidity and resolution than the fingerprints obtained from porous surfaces. The fingerprints left on paper and fabric surfaces (see Fig. 18A and D) have a high definition, but those found on wood and leather surfaces (see Fig. 18C and D) have the poorest quality. In fact, the fingerprints appear to have hundreds of brilliant spots in those last images, making it impossible to visualize them. These spots are not visible in the fingerprints shown in Figure 17. Additionally, fresh photos of finger mark were taken after 7 days. According to the findings, SAP phosphor powders may successfully identify fingerprints on non-porous surfaces with the same precision and description as those seen in prior to one week. Even though the light the fingerprints released after being excited by UV light was less intense, they still exhibited sufficient resolution for collection. After a week, the plastic bag yielded the best-preserved fingerprint.

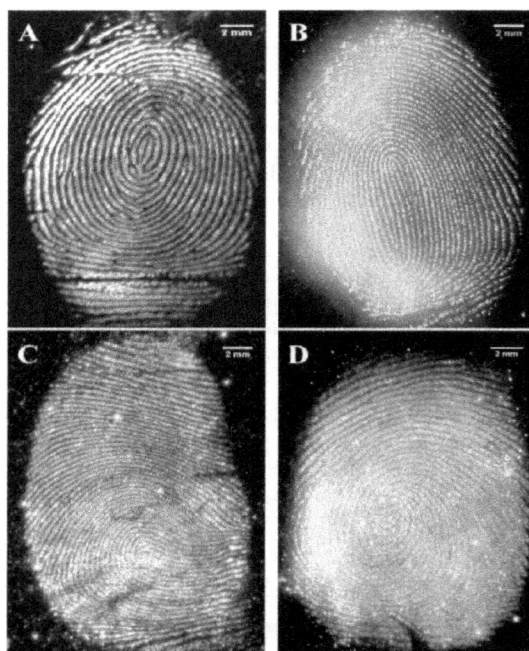

Figure 17. Pictures of fingerprints found with the SAP powder on non-porous materials, including (A) foil, (B) glass material, (C) porcelain, and (D) plastic bag. (Reproduced with permission from Ref. [149])

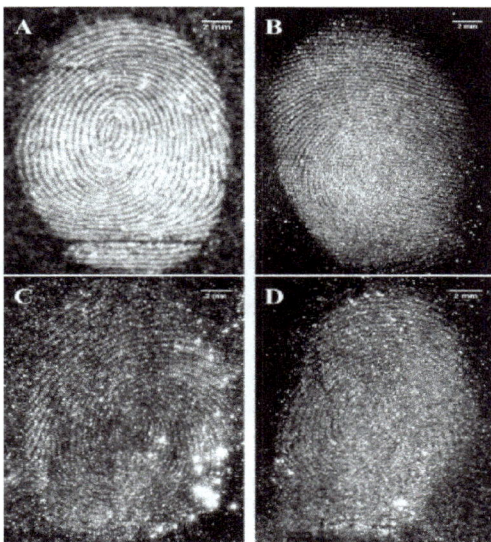

Figure 18. Pictures of fingerprints made on paper were captured when ESA powder detected them on a variety of semi-porous and porous materials. (A) Paper; (B) fabric material; (C) wood material; and (D) leather material. These tests were all conducted using new fingerprints. Following a 2-minute UV light excitation of the labelled prints, all of the photos were captured in complete darkness. (Reproduced with permission from Ref. [149])

To identify latent fingerprints, Strontium aluminate powders including rare earths ($Sr_4Al_{14}O_{25}$ Eu^{2+}, Dy^{3+}) were used by Sharma et al. According to the JCPDS card 74-1810, the SAP that was created by the group is orthorhombic in shape. Figure 19 displays the SEM image of these powders obtained by scanning electron microscopy.

Figure 19. TEM and SEM photographs of the $Sr_4Al_{14}O_{25}$: Eu^{2+}, Dy^{3+} nanophosphor. (Reproduced with permission from Ref. [153])

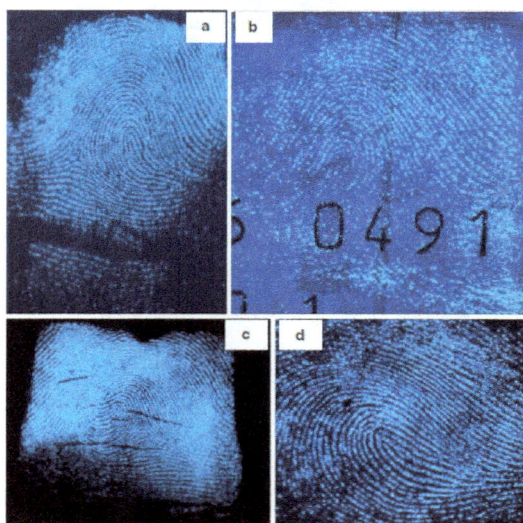

Figure 20. The advanced fingerprints with arranged $Sr_4Al_{14}O_{25}$: Eu^{2+}, Dy^{3+} nanophosphor on (a)compact disk, when UV light is switched off (excited in UV for 5sec only), (b)credit card, under UV radiation, and (c)aluminium foil without UV radiation(d) and with UV radiation. (Reproduced with permission from Ref. [153])

96

Rare Earth - A tribute to the late Mr. Rare Earth, Professor Karl Gschneidner   Materials Research Forum LLC
Materials Research Foundations 164 (2024) 67-142   https://doi.org/10.21741/9781644903056-2

As can be seen, the synthesized powders exhibit an uneven morphology and a porous structure as a result of the gases produced during their combustion during the process of their synthesis. Porous microparticles range in size from 50 to 150 μm. According to the TEM picture, micrometric particles are composed of smaller particles as shown in the inset of Fig. 5 that are between 50 and 100 nm in size. These materials produced greenish-blue emission at CIE coordinates (0.138, 0.359) following 365 nm excitation. At 495 nm the emission peaked. Fingerprint images obtained from the surfaces of aluminium foil, credit cards, and compact discs (CDs) are shown in Fig. 20 both with UV stimulation and without UV stimulation. It is interesting that following UV excitation, the fingerprint on the CD is clear and that the rainbow colour created by light reflection on the Carbon Dots does not interfere with the fingerprint picture (see Fig. 20a). Compared to the fingerprint in Fig. 20a, the credit card fingerprint in Fig. 20b has few features regarding cyan colour. This has most likely happened for SAP's less stringent credit card adherence. Since the picture mentioned in Fig. 20b was captured using UV light, the card's purple reflection can also be seen. In contrast, when the fingerprint is found on the aluminium foil, a clear image is seen (see Fig. 20c). Surprisingly, when fingerprints are activated with UV light, poor contrast and resolution is seen in the fingerprint ridge, as seen in Fig. 20d; continuous arcs and loops are therefore difficult to differentiate. Other groups have reported using $SrAl_2O_4$: Eu^{2+}, Dy^{3+} with monoclinic phase for detection of fingerprint due to the noteworthy advantages of SAP for this purpose [154]. This substance displayed a 365 nm excitation peak and an emission wavelength of 518 nm. This substance emits light with the CIE coordinates (0.238, 0.502), which are associated with the colour green. As seen in Figs. 21A–21B, particles with erratic forms are create the $SrAl_2O_4$: Eu^{2+}, Dy^{3+}. In Fig. 21C, a less magnified picture reveals that the SAP is made up of erratic surfaces with embedded particles [155]. According to Fig. 21D, the implanted nanoparticles had diameters ranging from 5 to 40 nm, with a 10.97 nm average. Due to their ease of dispersion on rough surfaces for fingerprint detection, particles with nanometric diameters are appropriate. On many surfaces, including: glass, aluminium foil, colourful paper, a compact disc, chocolate wrappers, and an optical mouse, numerous fingerprints were also discovered. Due to its lack of contrast in the barcode area, the wrapper presented the most challenging surface. Due to its more intense persistent emission that works well for detecting fingerprints at criminal scenes, In general, SAP nanophosphor with an orthorhombic phase outperforms SAP phosphor with a monoclinic phase for latent finger prints. The ability of $SrAl_2O_4$: Eu^{2+}, Dy^{3+} (SAO: Eu, Dy) with monoclinic phase for the recognition of fingerprints was also examined by D. Chavez et al. [142]. This substance emits green light at a wavelength of 522 nm, and its CIE coordinates were (0.228, 0.646). With the help of a 390 nm UV laser, this substance was stimulated to disclose fingerprints.

$SrTiO_3$ (STO): Pr^{3+}, which is based on strontium and is doped with additional ions such A^+ (A = Na, Li and K), is another substrate that is utilized to identify fingerprints [156]. Figure 22 depicts the single cubic phase and flower-like hierarchical structures that were present in this substance. The average size of these structures was 27 nm. After being excited for 30 seconds by 442 nm, $SrTiO_3$ (STO): Pr^{3+} displayed a sustained emission band at 607 nm and at 500 nm. According to the STO: Pr^{3+} co-doping ion (Na, Li, or K), the STO: Pr^{3+} phosphors accumulated on the fingerprints displayed colour changing from red to pink, as shown in Figure 23. Without a co-doping ion, the STO: Pr^{3+} phosphor gave off a faint red glow (see Fig. 23a). If the STO is co-doped with Na or Li ions, as shown in Figures 23b and c, the pink colour is visible; but, when co-doped with K, the pink colour is less apparent as seen in Fig. 23d. On the surfaces made of ceramic tiles, aluminium foil, metal scales, staplers, stamp pads, marble, granite and tablespoons, fingerprints

Rare Earth - A tribute to the late Mr. Rare Earth, Professor Karl Gschneidner   Materials Research Forum LLC
Materials Research Foundations 164 (2024) 67-142   https://doi.org/10.21741/9781644903056-2

were identified using STO: Pr3+. Because of their carved patterns that prevented the consistent accumulation of the revealing substance (STO: Pr^{3+}) and cause the deposition of powder on their surface, the stapler and stamp pad presented the greatest challenges for the identification of fingerprints. Compared to those found by SAP, the fingerprints recognized by STO: Pr^{3+} showed less contrast, less resolution, and less detail.

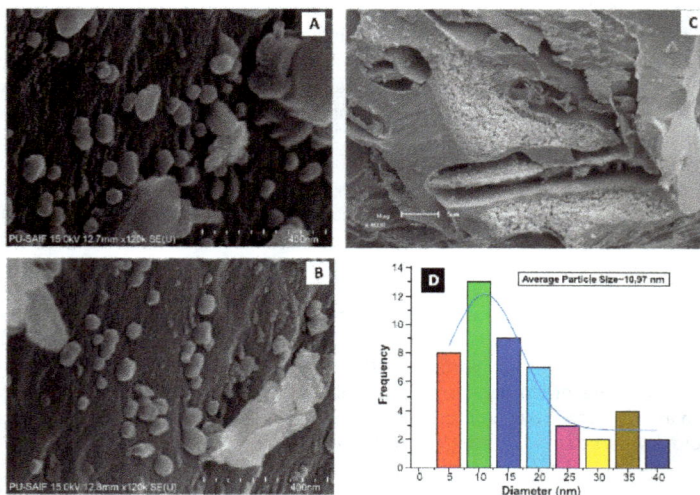

Figure 21. (A) and (B) FE-SEM micrograph of the $SrAl_2O_4$:Eu^{2+}, Dy^{3+} nanophosphor depicting the nanoparticles, (C) the surface morphology of the powder particles is depicted in a SEM micrograph (D) and their average size of particle (Reproduced with permission from Ref. [154])

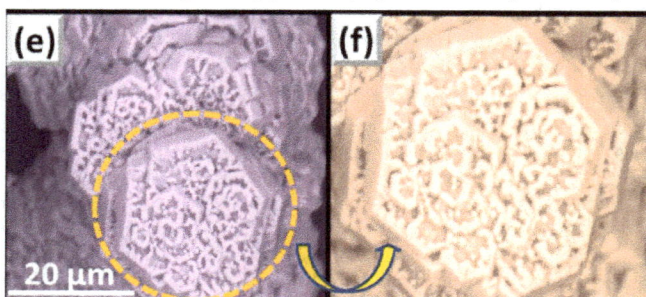

Figure 22. Pr^{3+} (5 mol%) doped SrTiO3 NPs generated by modifying the ultrasonic irradiation period are shown in SEM pictures (e) 5 h and (f) Zoomed part of fig. (e). (Reproduced with permission from Ref.[156])

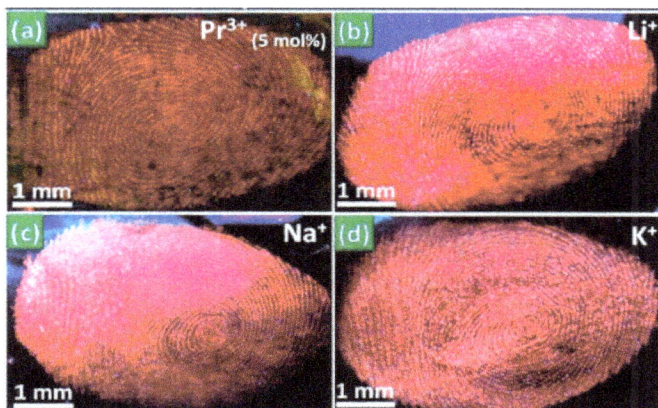

Figure 23. On the surface of aluminium foil, LFPs (Latent Fingerprints) were visible using different monovalent metal ions (Na+, K+, and Li+). (Reproduced with permission from Ref. [156])

Blue phosphor $CaAl_2O_4$ (CALO): Eu^{2+}, Dy^{3+} with monoclinic phase and space group P 21/n was described by Sharma et al. in 2018 [157]. As seen in Fig. 24a, the TEM photographs of this nano phosphor depicted spherical particles with a mean size of 33 nm. Under the stimulation of 245 nm, the nano phosphor produced blue coloured light at 447 nm. When an exposed fingerprint is found on a mobile screen surface, it emits blue light as in Fig. 24b. Because the ridges, cores, enclosures, and bifurcations of a fingerprint can be seen clearly, the authors concluded that the fingerprint has a strong definition, enabling accurate acquisition and recognition. On surfaces including mobile screens, black-coloured papers, steel, and aluminium foil, the gathering of latent fingerprints was accomplished successfully utilizing the CALO nano phosphor. As a result of its nano size, this demonstrated remarkable adhesion characteristics on surfaces including steel and paper. As a result, the nano phosphor was distributed uniformly over the latent fingerprints, preventing the concentration of these substances on fingerprints. It is inappropriate to utilize excessive nano phosphor because it obstructs the ability to see fingerprint details like ridges and grooves.

For the purpose of detecting fingerprints, however, the $CaGdAlO_4$ (CGA): Eu^{3+} nano phosphor was created by Park et al. [158]. According to the JCPDS 24-0192 card, this red substance had a tetragonal crystal structure with space group I4/mmm. The CGA particles have an average size of 48 nm, and its agglomerated dendrite-like particles can be seen in Fig. 25. This substance was labeled as phosphorescent since its emission had a decay time of 1.326–1.646 ms, but the material CGA needs to be activated continually in order to see the fingerprint as it is being collected from any surface. It is observed for a large number of fingerprint characteristics to be recognised, there must be more contrast between the exposed fingerprint and the surface, preventing CGA from releasing a bright red light at 620 nm after being excited at 283 nm. As a result, it is unable to register the fingerprint in the AFIS system. Since the passport surface is rough and has imprinted types of patterns on that surface (see Fig. 26f), it is challenging to disclose a distinct fingerprint picture. This happened as a result of the CGA particles' uneven distribution and accumulation in the latent fingerprint's deep zones, which led to oversaturated bright zones in the image of the fingerprint with the emission of red coloured light. In the contour of the fingerprint the papillary ridges of the fingerprints can be seen even when the interior area of the fingerprint is not clearly defined. Stainless steel, glass, a conical tube, and an aluminium foil all produced fingerprints that were not oversaturated and had enough information to be recorded for the AFIS system (Fig. 26a'–26e'). According to the results reported here, CaGdAlO4: Eu3+ nano phosphors can be used to detect latent fingerprints [158], even though, the contrast and degree of detail weren't as good as they were for the earlier materials (SrTiO$_3$:Pr^{3+} or SrAl$_2$O$_4$: Eu^{2+}, Dy^{3+} nano phosphors; CaGdAlO$_4$: Eu^{3+}).

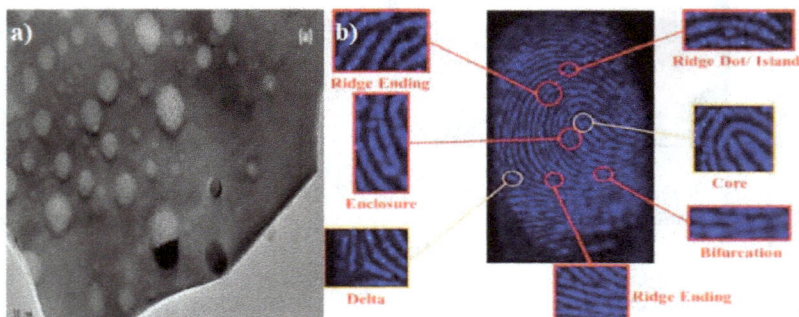

Figure 24. a) TEM micrograph. b) The development of a fingerprint using tagged nanoparticles of manufactured nanophosphor activated by UV light. (Reproduced with permission from Ref. [157])

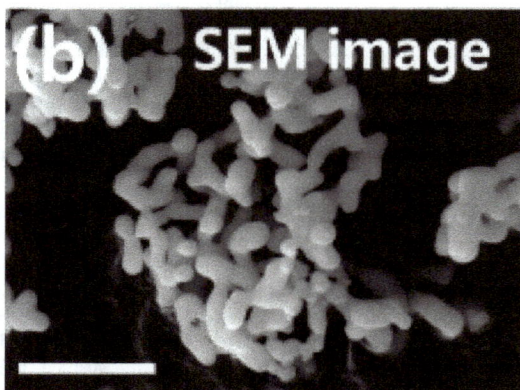

Figure 25. SEM picture of CGA:7Eu^{3+} (doped with 7 mol% of Eu^{3+}) nanophosphors. (Reproduced with permission from Ref. [158])

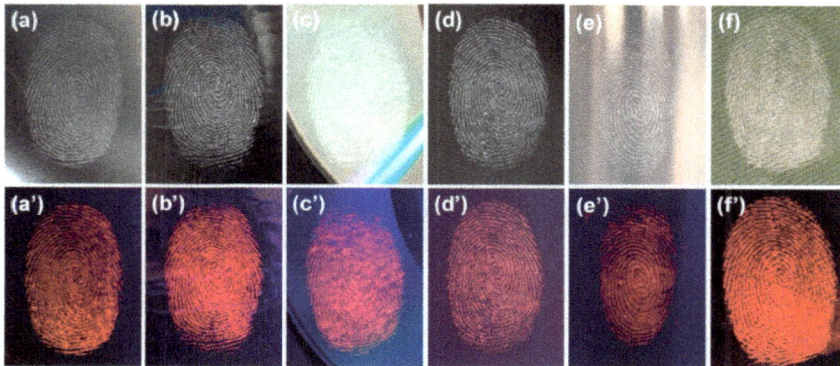

Figure 26. Fluorescence pictures of latent fingerprints created with CGA:7Eu³⁺ nanophosphors on various substrates are captured digitally; (a–f) in bright field, (a'- f') under irradiation (254 nm), (a,a') stainless steel, (b,b') aluminum foil, (c,c') compact disc, (d,d') glass, (e,e') conical tube, (f,f') passport. (Reproduced with permission from Ref. [158])

$Zn_2TiO_4:Sm^{3+}$ is another luminous substance that has been utilised to detect latent fingerprints [159]. According to JCPDS card No. 77-0014, this substance had a cubic phase and space group Fd-3m. Spherical nanoparticles with diameters between 27 and 40 nm form this substance, $Zn_2TiO_4:Sm^{3+}$, when excited by UV radiation, emit red light at 611 nm. Visible light reveals that the powder is white. Fig. 27b shows the red colour that results from UV light excitation of the powder using a lamp (254 nm, 4W). Fingerprints on a glass plate shown in Fig. 27c are without UV irradiation and in Fig. 27d are with radiation respectively. Fig 27a shows the TEM image which have 27-40 nm of spherical nano particles. $Zn_2TiO_4:Sm^{3+}$ nano phosphors revealed great adherence to the fingermark ridges due to their nano size. This was linked to the prepared materials' porousness because higher porosities are known to have stronger adhesion forces. Additionally, high porosity substrates are better for the diffusion of chemicals into the fingerprint for its identification. The findings of this study imply that $Zn_2TiO_4:Sm^{3+}$ functions as a sensor for the detection of latent fingerprint on various surfaces.

Despite having low contrast, the fingerprint displayed red emission when illuminated by UV light. In contrast to $SrAl_2O_4$ strontium aluminates, which had phosphorescence of the order of hours, this material's decay period was shorter at 6.2–6.48 ms [160]. As a result, this research revealed that $Zn_2TiO_4:Sm^{3+}$ nano phosphors might be used in forensic science for person recognition, however, their luminescence needs to be enhanced. A $Zn_3Ga_2Ge_2O_{10}:0.5\%Cr^{3+}$ phosphor with persistent emission in the NIR was created by King et al. [161]. Although Pan et al. [162] already performed the optical characterization of this material, Kim et al. investigated the possibility of using it to identify fingerprints. This team showed that the powder's NIR phosphorescence makes it possible to capture latent fingerprints quickly and easily without interfering with the substrate's UV-induced fluorescence (254, 312, 365 nm). The utilization of NIR persistent phosphors for latent fingerprint detection is described for the first time in this report.

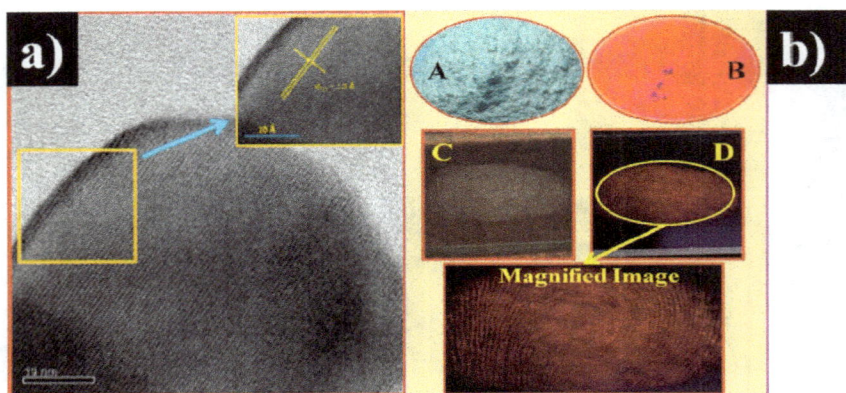

Figure 27. a) TEM image of Sm^{3+} (3 mol%) doped Zn_2TiO_4 nanophosphor b) Latent fingerprint development of Zn_2TiO_4:Sm^{3+} (3 mol %) nanophosphor. (A) Powder under natural light (B) powder under UV light (C) Fingerprint under visible light on glass plate (D)Under UV light. (Reproduced with permission from Ref. [159])

Numerous types of nanoparticles with varied morphologies have been created in the last 20 years and used in forensic investigations. Silica metal oxide, europium oxide, zinc oxide, titanium dioxide, iron oxide nanoparticles, as well as gold and silver nanoparticles, cadmium selenide quantum dots (CdSeQDs), cadmium telluride quantum dots (CdTeQDs), cadmium sulfide quantum dots (CdSQDs), polymeric dots, conjugated-polyelectrolyte dots (CPEDs), carbon dots (CDs), fluorescent silica nanoparticles, luminous mesoporous silica nanoparticles, aggregation-induced emission luminous molecules integrated nanomaterials, and rare earth metals in the forensic application are the method to detect fingerprints has been developed using fluorescent nanoparticles [163][164].

The following is a description of the reason behind using these nanoparticles to identify fingerprints: Target molecules of metabolites and explosives are primarily addressed by the silver and gold nanoparticles, which are corresponding to antibodies and aptamers target molecules of metabolites and explosives, respectively. They have additionally demonstrated a variety of characteristics, including a large surface area, a compact size, electrical conductivity, and thermal conductivity [165]. These compounds have been employed in the detection of fingerprints and are used at crime scenes to positively identify culprits. Latent fingerprint detection has been applied to metal oxide nanoparticle powders including europium oxide nanoparticles, titanium dioxide, iron oxide, silica, and zinc oxide nanoparticles. Due to their strong contrast, superior coating improved fingerprint image resolution, and ridge properties in daylight, these various metal oxide powders are employed on surfaces where fingerprint residue has been left behind. These characteristics support the use of latent fingerprints to identify criminal activity. Quantum dots (QDs) are generally employed in liquid crystal displays (LCDs), optical devices, bioimaging, and as biomarkers due to their strong fluorescence properties. Better quantum yields, larger surface areas, smaller sizes, and improved photo-stability have all been induced [166] Due to their small size, in forensic applications, QDs were superior nanoparticles for creating fingerprints because they had a powerful interaction with the fingerprint residue, they have shown improved fluorescence properties to research the detection of fingerprints on diverse porous and non-porous materials. Because of the high-quality pictures, greater background contrast, improved visibility, good ridge features, higher selectivity, and increased sensitivity of these types of nanoparticles for fingerprint identification. Due to their excellent optical and chemical properties, rare earth fluorescent upconverters nanomaterials (UCNMs) and fluorescent QD powders have become new tools for generating latent fingerprints. Good surface modification, good chemical characteristics, high fluorescence intensity, photostability, and minimal toxicity are only a few of their advantages [167].

In general, QDs have superior water solubility, easier functionalization, and reduced toxicity because they don't include heavy metals like cadmium. QDs have so far been used in a variety of medication delivery, bioimaging, and sensing applications. Additionally, compared to conventional organic dyes, these materials have demonstrated good emission activity, wide absorption, and less photobleaching [168]. These fluorescent powders have demonstrated excellent uses in thermometry, biomedicine, and lighting. On a variety of substrates with multicolored surfaces, latent fingerprints can be developed using fluorescent powders. Due to the emission characteristics of the powdered materials, fingerprints may be easily seen under UV irradiation to quickly identify the offender. To improve the detection of latent fingerprints and aid in the identification of criminals, materials like aluminium foil and carbon-based compounds are frequently employed.

Two different kinds of hydrothermally generated rare-earth fluorescent nanomaterials, YVO_4: Eu (purple emission) and $LaPO_4$: Ce, Tb (green emission), have been used successfully in LFP detection, according to Wang et al. [169]. On specific surfaces like marbles, paper, glass, tiles, aluminium sheets, timber materials, and plastic materials, with regard to LFP detection, these fluorescent materials have shown excessive sensitivity and better bright marks. $LaPO_4$: Ce and Tb luminescent materials have subsequently been used in forensic research for LFP detection. Wang et al. [170] created novel fluorescent compounds such as $NaYF_4$: Yb and Er to address the issues of poor assessment like background noise, low assessment, autofluorescence interference, and low sensitivity in fingerprint recognition at real crime scenes on various materials. The fluorescence intensity of $NaYF_4$: Yb, Er compounds, which varies on temperature, oleic acid volume, phase composition, and time of response has been improved. The effectiveness of these fluorescent materials in detecting LFPs on unusual substrates such as plastic plates, black and multicolored marbles, ceramics, Chinese paper money, notice sheets, and glass was demonstrated. When $NaYF_4$: Yb and Er powders are added, this luminescence powder is also capable of generating high-quality fingerprint pictures on unique materials because it has better sensitivity, better assessment, less background interruption, and less autofluorescence interference than commonly used powders like magnetic powders, bronze powder, and low-cost emission powders. When silica nanoparticles are added to radical purple luminescence powder to create hybrid materials, as demonstrated by Saif et al. [171], high-quality LFP images are produced. One of the rarest rare earth elements, europium (Eu^{3+}) ion exhibits crimson emission and also has non-toxic impacts on users in LFP detection. Sol-gel techniques have been used to create hybrid materials composed of Eu^{3+}: $Y_2Ti_2O_7$ embedded onto silica. An in vivo method was used to investigate the non-toxicity of this nano matrix Eu^{3+}: $Y_2Ti_2O_7$ embedded into silica hybrid nanomaterials. There is no hazardous potential for this matrix component. It has also been found

Materials Research Foundations 164 (2024) 67-142   https://doi.org/10.21741/9781644903056-2

that the nano phosphor in Eu^{3+}: $Y_2Ti_2O_7/SiO_2$ hybrid compounds is essentially harmless. As can be shown in Fig. 28, this matrix cloth has thus been successfully employed in LFP on various non-porous and porous surfaces.

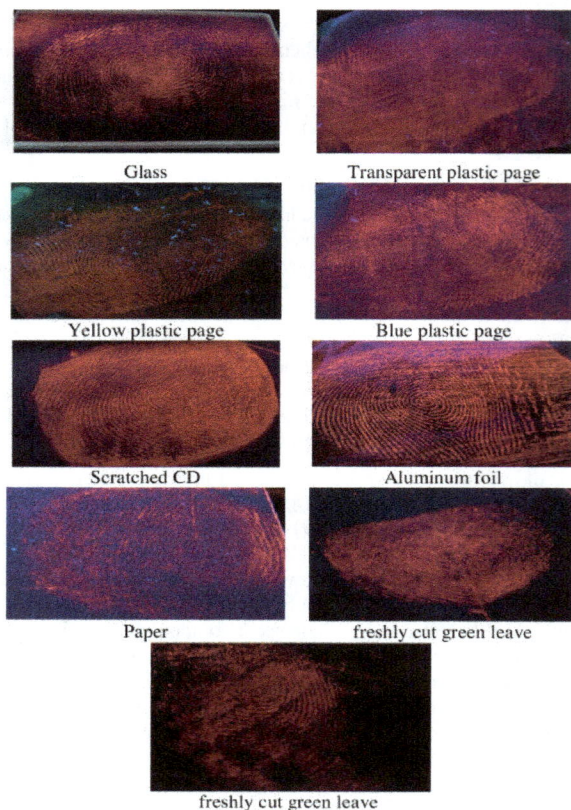

Glass

Transparent plastic page

Yellow plastic page

Blue plastic page

Scratched CD

Aluminum foil

Paper

freshly cut green leave

freshly cut green leave

Figure 28. Photographs of Eu^{3+}: $Y_2Ti_2O_7$ embedded in silica matrix powders showed the development of recent fingerprints from various surfaces triggered by a mercury lamp. (Reproduced with permission from Ref. [171])

The photochemical characteristics and quantum efficiency were enhanced by the luminous materials of Eu^{3+} ion-based metal oxide nanoparticles. When alumina (Al_2O_3) nanoparticles are doped with the Eu^{3+} ion, they become Eu^{3+}: Al_2O_3 nanoparticle powder, which has unique qualities such as chemical inertness, high melting point, and photochemical stability [172]. Under UV light irradiation, the rare earth metal-doped aluminium oxide nanoparticles were utilized as labeling agents at various nonporous and porous surfaces for LFP detection. With glass slides, highlighter pens, aluminium foil, stainless steel plates, and stainless bowls, these produced images with higher contrast. The enhanced fluorescence activity of our Eu^{3+} doped Al_2O_3 nanopowder made it more appropriate for LFP detection when exposed to UV light than commercial powders. Under UV light irradiation, this substance improved the ability to identify fingerprints on the surface of porous and non-porous materials.

3.3.2   Effect of surface

The kind of surface on which the fingerprints are "printed" affects how easy or difficult it is to discover them. The smoothness, softness, and porosity of the surface where fingerprints are deposited, as well as whether they are visible or not, have all been taken into consideration by analysts when dividing fingerprints into two categories. [144]. The surface's porosity, structure, and temperature can all change when a fingerprint is applied to it, which affects how the fingerprints are deposited. The surface porosity is critical to how fingerprint residues behave and greatly affects the development technique that can be used. Based on this characteristic, the surfaces are divided into three main groups: porous, semi-porous, and nonporous, according to how fingerprints develop on them. These porosity groups provide direction for selecting the right kind of fingerprint development sequence to be implemented since the contact between the fingerprint and the surface can directly affect the interaction with chemical, optical, and physical properties. For instance, a chemical treatment should be used to increase contrast if the surface is porous rather than using a powder. Additionally, a fluorescent chemical treatment might make the surface more noticeable against the background. Understanding the surface is essential for the proper generation of latent fingerprints; if an odd surface is found, research on a sample surface should be done to identify the substrate.

Non-porous surfaces include polished metal, glass, aluminium foil, and polyethylene plastic bags. These surfaces do not absorb any of the fingerprint residues and their exposure to corrosive environments and chemicals will retain FPs residue on top of it. Ink stains left by fingerprints on non-porous surfaces are delicate and readily cleaned using a cloth or organic solvents. In contrast, porous materials like cardboard or paper can absorb the FP material because the water-soluble elements are absorbed into the surface shortly after deposition. Aluminum foil, glass, polished metal, and plastic bags are examples of non-porous surfaces. No fingerprint residue is absorbed by these surfaces. FPs residue will remain on a non-porous surface that has been exposed to corrosive conditions and chemicals. Fingerprint ink stains are fragile and easily removed with a cloth or organic solvent from non-porous surfaces. Any insoluble substance will be left behind after exposure to water, which will wash away the water-soluble components. Porous materials, however, such as paper or cardboard, can take up the fingerprint substance because it is swiftly incorporated into the surface after the deposition of the water-soluble components.[173]

The leftover eccrine material, such as amino acids and salts, retains the mark and is absorbed into the paper, depending on the relative humidity in the area. Lipophilic material can persist on porous surfaces for a long time and can be targeted by more delicate approaches, although typically being inadequate to be handled by powdering procedures. These surfaces that don't fit into the non-porous or porous categories include polymer banknotes, varnished woods, and painted surfaces. These surfaces usually contain more layers and are more sophisticated. Designing a fingerprint method requires testing it on many surfaces to see which ones perform best. Studies of existing detection systems have looked into the best detection sequence or the order in which fingerprints develop using a range of methodologies. ninhydrin, iodine spray, Iodine fuming, and 1,8-

diazafluoren-9-one (DFO) are a few agents used to get the LFPs of porous surfaces [174]. Additionally, AgNO₃ is frequently used to uncover LFPs because it combines with the elements of sebaceous sweat to create a silver-grey deposit that is simple to see with the unaided eye. Latent fingerprints could be found on flat, smooth surfaces using silver nitrate, cyanoacrylate fuming, and powders. Metal powders including fluorescent organic powders, brass, carbon, and aluminium are frequently used to identify fingerprints. Rhodamine 6G, rhodamine B, and fluorescein are the conventional organic dyes used to detect LFPs. Paper, polished metal, and glass surfaces are frequently used in crime scenes to detect fingerprints [175]. Additionally, there are luminous inorganic powders for fingerprint identification that provide greater contrast than regular organic powders.

4.1 Luminescent materials for anti-counterfeiting application

4.1 Introduction

Counterfeiting is the illicit duplication of something precious specially to deceive. It is a growing issue that creates a risk to both an individual and a whole society globally. It has the potential to have major negative implications for the public and society can even harm economic growth through the spread of unlawful or counterfeit documents, goods, and cash [176][177]. Counterfeiting of vital papers, cash, and items is a severe issue with negative security, health, and economic consequences for governments, corporations, and people worldwide. It is evaluated that the process of counterfeiting represents a multibillion-dollar underground business, with millions of counterfeit goods made each year. Counterfeiting is a growing advanced crime and requires advanced solutions in order to counter and prevent counterfeiting acts. An additional severe threat to public health and safety is the counterfeiting of pharmaceutical products. Altogether the acts of counterfeiting cause threats in various fields including health sectors, trading sectors, industrial sectors, and service sectors. Researchers from all over the world are being pushed by these alarming increases in fabrication to come up with secure and cutting-edge anti-counterfeiting solutions to the problem [178]. Universally, security is the largest menace in the case of identification of cash, verification of valuable documents such as passports, certificates, identity cards, and bank accounts, and recognition of branded products. Instead of the top security set for the safeguarding of valuable documents, the availability of advanced gadgets and modern technologies have smoothened counterfeiting acts [179][180]. Growing counterfeiting activities are responsible for breaching the rights of copyright owners with damages to society. Therefore, anti-counterfeiting measures are crucial for the protection of brands and priceless documents because they make genuine goods more difficult to replicate and simpler to validate. Remarkable steps to develop anti-counterfeiting technologies have been taken to address this universal issue [181]. Numerous anti-counterfeiting plans have been developed that include holograms, barcodes, and watermarks. [182] Recently, various anti-counterfeiting and verification techniques have been improved to protect valuable products and documents such as RFID (Radio-Frequency Identification), machine-readable codes, security holograms, thermochromic inks, DNA coding, ultra-resistant labels and luminescence printing. But these old-fashioned anti-counterfeiting plans have a high chance of duplication. Luminescence printing is a broadly used technology to combat counterfeiting due to the simplistic composition of luminescent materials and high output [183]. Secured security printing needs a suitable selection of materials followed by basic features like eco-friendly high-yielding synthesis along with durabilit. Currently, luminescence printing techniques based on fluorescent-resistant labels are used in currencies. Most optical data storage and security applications use luminescent materials as anti-counterfeiting materials. It is extremely difficult to duplicate the optical characteristics of luminous materials. As a result, many kinds of luminescent materials, such as Ln-doped nanomaterials, CDs, QDs, MOFs, etc., are frequently found in security printing [184]. Escalating the need for advanced luminous materials has stimulated scientific research and study to upgrade the features of these materials and to prepare the latest efficient luminescent nanomaterials with the desired dimensions and optical properties etc. Various applications in the fields of display technologies, bio-medicine, forensics, anti-counterfeiting, solid-state lighting, etc are shown by luminescent nanoparticles. To secure high-value products, important documents, banknotes, and currency nanomaterials are broadly used as security inks. For the protection of genuine merchandise luminescent tags are used through the incorporation of luminescent nanomaterials, where these materials are used as security inks to print them. and these printed elements commonly contain single-ink colour in halftone patterns for example, banknotes showing luminescent parts under UV radiation. Lanthanide compounds are used for making luminescent security inks due to their distinctive fingerprint spectrum. Lanthanide-doped luminescent inks are used as security labels and tags. Anti-counterfeiting images or patterns characterized with multicolor and multimode photoluminescence provide resistance to counterfeiting activities. The term 'multimode' refers to multiple activities that occur in response to diverse modes of stimulation, essentially affecting a number of factors, including lifetime and colour, leading to changes in its fluorescence characteristics that are dependent on the stimulation. Ordinary fluorescent materials essentially exhibit static fluorescence, which implies that patterns don't change while a fixed stimulus is present. This means that the static luminescent materials used in the data encryption security pattern are susceptible to duplication by substitute materials with similar luminescent properties, which is quite challenging considering their reasonable authentication in the anti-counterfeiting fields for secure and sensitive encryption purposes. Most of these static luminous photoluminescent patterns are invisible in normal light and are only seen under UV/NIR radiation, which is regarded as one of the most crucial encrypting features. However, various static luminescent materials combined under modulation lead to the development of multicolor complex patterns resulting in high-end sensitive data encryption including procurement of data for anti-counterfeiting activities. It is obvious that a fixed set of colored ink is necessary for a multi-hued pattern, which consists of red-green-blue (RGB) emitting static luminous materials. Andres et al. reported lanthanide ions with tris-dipicolinate lanthanide complexes for green- and red-emitting luminous security inks to create a multi-colored image that is not visible under normal light in this area. [185] When exposed to UV light, the ink releases RGB colors. In this study, a commercially available blue ink was combined with processed red and green emitting luminous inks to produce a full-color image with a multicolored pattern using inkjet printing. Recent developments of different materials like labels, security inks, and tags including new classes of materials such as security resins have been made to prevent counterfeiting. Whatever the use, each material will be endowed with a difficult-to-replicate feature, thus increasing the efforts required to precisely reproduce the protected objects. Anti-counterfeiting elements have broad applications across different industries and provide a crucial layer of protection for culturally significant objects. Food, pharmaceutical, apparel, banking, and art industries all have a vested interest in developing innovative anti-counterfeiting.

4.2   Literature Survey

Following a brief conversation, we delved into the specifics of various luminous material-based anti-counterfeiting solutions.

4.2.1   Inorganic material-based security ink for anti-counterfeiting application

Here we have discussed the various kinds of lanthanide-doped luminescent security inks based on nanomaterials and the dual mode of emission-based luminescent nanomaterials, up-conversion based luminescent nanomaterials, and down-conversion based luminescent nanomaterials that are utilized to make security inks are also explained in this part.

Down-conversion luminescent security inks emitting the red and green were described by Andres et al.. [186] The full-colour images created by an inkjet printer using the mentioned green and red emitting luminescent inks emit green, red, and blue colours when exposed to an excited source of UV light and obscures their visibility when lit by white light. To print full-color graphics in the visible spectrum, commercial blue ink was blended with green and red emitting luminous ink.

To safeguard trademarks and essential documents like legal tender, currencies, and certificates against counterfeiting, security ink manufacturers utilize luminous nanorods. Additionally, in another research, Chen et al. reported that the phosphor of $(Y, Gd) VO_4:Bi^{3+}, Eu^{3+}$ can be used for anti-counterfeiting applications. [187] Temperature and wavelength affect the colour emitted by $(Y, Gd) VO_4:Bi^{3+}, Eu^{3+}$ phosphors. In addition, by varying the concentrations of Bi^{3+} and Eu^{3+}, the emitted colour can be changed from green to orange.

Blumenthal et al. described upconverting nanophosphors in the creation of security inks for counterfeit prevention. [188] Upconverting toluene-doped nano phosphors, oleic acid as a capping agent, Perspex as a binding agent, and methyl benzoate constitute the upconverting ink solution. White light does not make these printed features visible, but light with a wavelength of 980 nm can excite them and make them visible. The printed feature patterns are depicted in Figures 29a and b under white light and when stimulated at 980 nm, respectively.

In additional research, Meruga et al. reported a method to formulate security inks for anti-counterfeiting activities using up-conversion nano phosphor. [189] They have printed QR codes using two distinct colour upconverting inks i.e., blue and green by aerosol jet printing. Fig. 29c and d illustrate the use of green ink for printing ink that has been upconverted. When stimulated at 980 nm, the QR is not visible in white light but displays a bright green colour in NIR. Fig. 29e shows a printed QR code that is stimulated at a 980 nm wavelength. The image shows a printed upconverting QR code excited at wavelength of 980 nm. In a different study, Meruga et al. revealed that QR codes are made of three colours of upconverting inks (red, green, and blue). 300 When stimulated at 980 nm, the tricolor QR code generated by an aerosol jet printer displays a distinct multicolor image. Red, green, and blue inks were also printed tightly and overlapped on one another to generate blue, green, red, cyan, yellow, white, and magenta up-conversion luminescence.

In Fig. 29f, a printed QR code is displayed both in natural light and after being excited at 980 nm. Wang et al. also successfully detected hidden fingerprints using lanthanide-doped up-conversion luminescent nanomaterials, which are not visible in white light but emit green when activated by light with a 980 nm NIR wavelength. 308 They have found latent fingerprints on a variety of items, including marble, floor leather, glass and ceramic tiles with varying polymeric materials, wood

Rare Earth - A tribute to the late Mr. Rare Earth, Professor Karl Gschneidner   Materials Research Forum LLC
Materials Research Foundations 164 (2024) 67-142   https://doi.org/10.21741/9781644903056-2

materials, metallic materials, surface textures, and various papers. Hidden fingerprints on the surface of numerous paper materials, which were blotted with NaYF$_4$: Yb, Er UCNPs, are shown in Fig. 29g–J. These fingerprints were later recognized by NIR–induced fluorescent imaging after being excited at a wavelength of 980 nm. As a result, this study offers fresh ideas for using lanthanide-doped nanomaterials in forensic science to find hidden fingerprints. Zhang et al. described how multicolor barcoding was used in single up-conversion crystals. [190] In their study, they used a simple hydrothermal procedure to create multicolor emitting up-conversion micro rods and determined their anti-counterfeiting uses.

The application of multicolor barcoding in single up-conversion crystals was described by Zhang et al. [190] They created multicolor emitting up-conversion micro rods using a straightforward hydrothermal procedure and identified uses for them in anti-counterfeiting research. The end-on growth of β-NaYF$_4$ micro rods employing β-NaYF$_4$ nanoparticles as the precursor material was also addressed. The multicolor emitting microrods' fluorescent picture is shown in Fig. 29k when they are excited by light at a wavelength of 980 nm. In dimethyl sulphoxide solvent the multicolor emitting up-conversion micro rods were well distributed, which is transparent to natural light and exhibits vivid colour following stimulation at 980 nm wavelength under NIR. The enlarged luminous image of a stamped letter S that is printed with microrod-based inks may be seen in Fig. 29l. In Fig. 29l, the emission spectra of micro rods are also depicted. The use of fluorescent barcodes for anti-counterfeiting activities is thus made possible by these multicolor-emitting micro rods.

Additionally, You et al. pioneered a technique for flexible or digital printing that makes use of up-conversion nanoparticle-based ink to produce high-resolution security patterns for anti-counterfeiting operations. [195] They created hydrophilic and hydrophobic luminous security inks employing up-conversion nano phosphors for their investigation, and they used inkjet printing to produce anti-counterfeiting graphics. Additionally, the exact domain in which the up-conversion and down-conversion nano phosphors were mixed produced a variety of patterns that could be seen upon stimulation. In Fig. 29m and n, multicolor fluorescent picture patterns of the blue down-conversion nano phosphor (NaYF$_4$: Yb, Tm) and the green up-conversion nano phosphor (NaYF$_4$: Yb, Er) are shown by directly writing on a black surface with a pen. As a result, the technique indicated is extremely advantageous to combat counterfeiting due to its simple production and low likelihood of imitation. Sangeetha et al. used atomic force microscopy (AFM) nano xerography using up-conversion nano phosphors formulated three-dimensional (3D) QR codes. [194] In this work, it was found that the polarizability of the nanocrystals, concentration of NC, charge patterns of the surface potential, and polarity of the dispersion solvent all have a role in the formation of 3D patterns utilising an accumulation of nanocrystals of NaYF. Both a QR code having a changeable width and a QR code with two colours that produce up-conversion nano phosphors have been developed. Using a photoluminescence mapping technique, the produced QR code may be quickly captured. Figures 29o and p display the atomic force microscopy (AFM) topographical pictures of the QR code that were produced utilising the two separate upconverting nanocrystals. After sending the light through a 500 nm short pass filter, Fig. 29q shows blue emission with 485 nm up-conversion PL mapping. In Fig. 29r, the QR code hides the number "31," but when light is passed via a long pass filter (530 nm) and a short pass filter (650 nm), it emits green at a wavelength of 545 nm, revealing the number. As a result, the AFM nano xerography result indicates that it may be the best option for premium counterfeit protection. For the detection and recognition of orthodontic adhesives, which display apparent luminescence upon IR radiation as illustrated in

Fig. 29s and t. Ramos et al. proposed 3D security-ink stamps or luminous tags. [194] When Baride et al. created NIR to NIR upconversion security inks using Yb^{3+}/Tm^{3+} based β-$NaYF_4$ nano phosphors, they reported another intriguing finding. The printed QR codes in Fig. 30 are not visible in natural light, but their NIR luminous pictures (800 nm) can be easily photographed with a charge-coupled device (CCD) camera. [197]

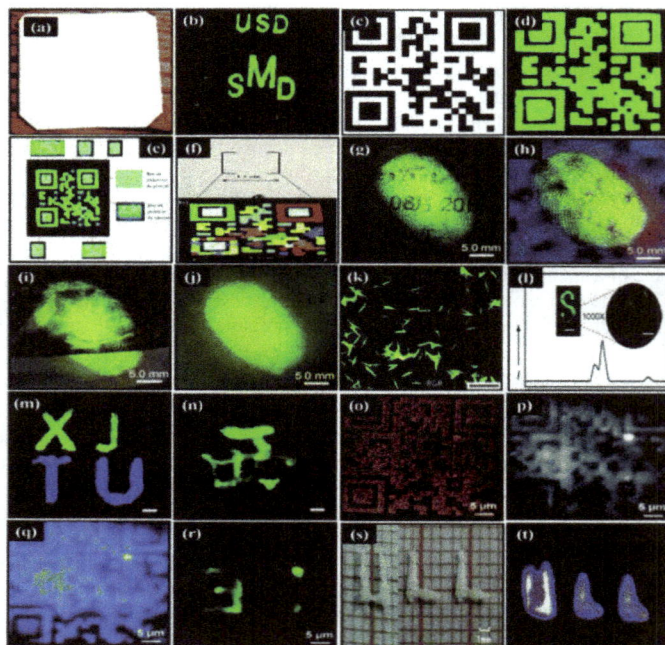

Figure 29. (a) & (b) Images in natural light and 980 nm UV light, [188] (c) and (d) the QR code in natural light and (980 nm) UV light, [191] (e) a photograph of the QR code and the literal letters has been written in blue upconverting ink. [191] (f) An RGB upconverting ink was used to print a QR code (excitation at 980 nm) [192] (g–j) fluorescence pictures of fingerprints on several materials, [193] (k) and (l) fluorescence pictures of dual-mode microrods and letter S written with the same microrods, [194] (m) and (n) multicolor patterning of upconverting inks, [195] (o) An AFM topography image of a binary NC assembly formed of upconverting NCs, (p) mapping of this QR code's entire up-conversion photoluminescence, (q) image of blue emitting QR code by up-conversion PL mapping, and (r) the QR code's buried number "31" is only visible in green (545 nm) after passing the PL light through 530 nm long pass and 650 nm short pass filters [194]. (s & t) image of the 3D-printed letters ULL in both natural and infrared lighting conditions. (Reproduced with permission from Ref. [196])

The ability of up-conversion based security inks or down-conversion based security inks to use one-stage coding means that the created luminous ink can be activated by a single wavelength and then exhibits only that particular wavelength. This strategy of coding seems to be significantly less efficient for document security. Heavy-performance different stage coding with multiple excitations and at least double colour emission is therefore predicted. A big challenge in this field that has recently been addressed in a few publications is creating phosphors with all of the aforementioned desirable qualities at an affordable price. The creation of multistage excitable ink enables multistage coding in a single host lattice, which is advantageous for high-end document protection to prevent document forgery in addition to being cost-effective. A method to create a dual-mode security ink based on Ln-based NCs was described by Lu et al.. The fake ink was applied on printing stamps that were designed to prevent counterfeiting. [198] As illustrated in Fig. 31a, the printed stamps (in Chinese characters) produce brilliant green and blue colors under excitation at 980 and 365 nanometers, respectively.

The screen-printing method can be used to print ink on paper. Fig. 31b displays the printed design in natural light at various excited wavelengths (379 nm and 980 nm). Fig. 31c illustrates the recommended application of this dual-mode luminous security ink for the purpose of preventing monetary fraud. Therefore, new prospects for dual-mode protection against counterfeiting are presented by this multistage excitable invisible ink.

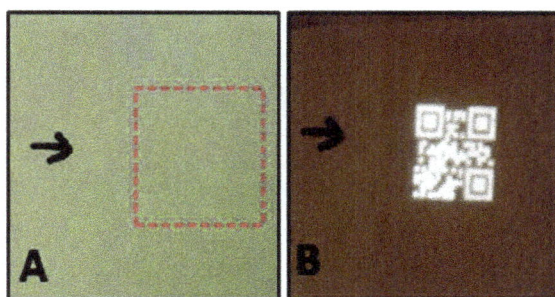

Figure 30. (a) and (b) codes written using NIR-to-NIR up-conversion ink in natural lighting using a 980 nm source. (Reproduced with permission from Ref. [197])

A hybrid inorganic-organic nanostructure that generated visible emissions through both Up-conversion and Down-conversion processes was used by Singh et al. to demonstrate how to make a dual-mode security ink. [194] While the Gd_2O_3:Er^{3+}/Yb^{3+} nano phosphor (an inorganic material) aids in the up-conversion process in the hybrid structure, the Eu (DBM)3Phen complex (an organic material) aids in the down-conversion process. In order to create a stable security ink that may be utilized to print security codes, this hybrid substance was dispersed in ethanol. Fig. 31d–g displays the printed pattern and ink solutions at observable UV and NIR excitation wavelengths.

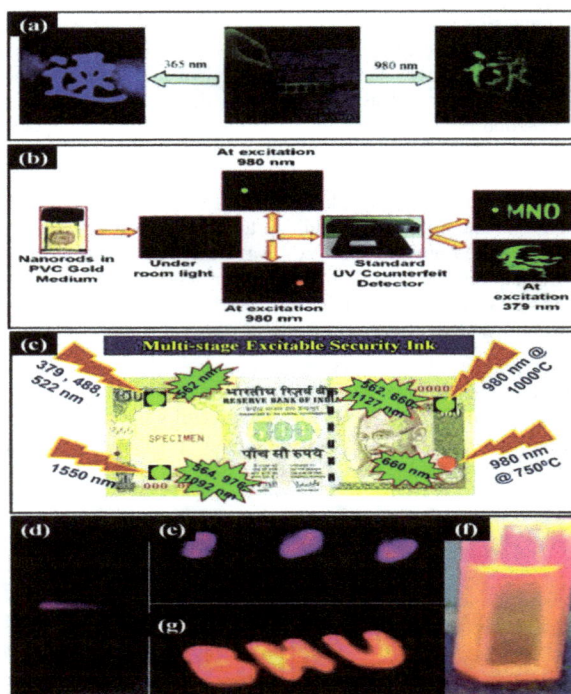

Figure 31. (a) Fluorescence pictures of the characters imprinted on the transparent material in bright field (centre), UC (right), and DC (left) orientations, [198] (b) printed patterns were examined in both natural light and at two different wavelengths for excitation (379 nm and 980 nm), [199] (c) an application of upconverting ink (d-g) UV and NIR excitation wavelength up-conversion of ink solutions and printed patterns. (Reproduced with permission from Ref. [44])

4.2.2 Quantum dot-based security ink for anti-counterfeiting applications

Quantum dots (QDs) exhibit variable size-dependent light emission. QDs can be used to create security inks because of their exceptional photoluminescence activity. Rewritable multicolor fluorescent patterns based on QDs were revealed by Lu et al. [200]. In this rewritable luminous system, the Branched polyethyleneimine (BPEI) serves as both a remover and a writer. Fig. 32a depicts the procedure for deleting or writing the fluorescent patterns. As a result, this method can create multistate memory chips with extremely high levels of data security and storage for a variety of anti-counterfeiting applications.

Another study by Bao et al. used inkjet printing to create luminous CdS QD-polymer nanocomposite patterns that were then exposed to hydrogen sulfide gas. The patterns were created using an ink that was loaded with cadmium as the source. [201] When patterning nanocomposites for optoelectronic devices, the fabrication method is simple, adaptable, and has a wide range of uses. Photographs and fluorescence pictures of CdS QD-polymer composite patterns made from dot-matrix technology are displayed in Fig. 43b. Potential applications for this CdS QD-polymer composite include anti-counterfeiting measures. Gao et al. developed the multifunctional m-SiO2/CdTe/Ag fluorescent/antibacterial nanomaterials for usage in anti-counterfeiting techniques. [202] In their investigation, the mesoporous-SiO2 nanospheres were filled with CdTe QDs, which inhibit QD agglomeration. A composite ink with antimicrobial properties was also created using Ag nanoparticles. In Fig. 32c–f, the painted objects with the composite ink are displayed in both natural and ultraviolet light. Producing fluorescent/antibacterial inks for anti-counterfeiting applications may be made possible by the m-SiO2/CdTe/Ag composite inks.

By using an ionic liquid (IL)-assisted solution approach, the ability of water-soluble silica QDs to be used in anti-counterfeiting applications was demonstrated by Zhou et al. [203] The synthetic silica QDs have temperature-sensitive characteristics and display excitation-dependent photoluminescence. Fluorescence microscope images of Fig. 32g–j show the patterns created under UV light from silicon QD solutions (triangle) and easily available green fluorescent inks (circle).

Using the excellent photothermal conversion efficiency of Cu_7S_4 nanocomposites, Cui et al. proposed a strategy for the photothermal latent fingerprint (LFP) imaging technique in another report. [204] On several substrates with various backgrounds, they used photothermal LFP imaging in their investigation. The proposed photothermal imaging method allowed for the recognition of intricate details on a fingerprint, such as termination, bifurcation, crossover, and whorl, in high-magnification pictures, as illustrated in Fig. 32k. In addition, effectively created LFP ridge patterns with good separation between furrows and ridges and distinct ridge patterns were another accomplishment. [204]

Additionally, Wang et al. developed an ink using the double mode (up-conversion and down-conversion) luminescent carbon dot fluid (CDF) direct thermal breakdown technique. [205] The produced ink showed stability at room temperature for up to six months and is suitable for inkjet printing. Pictures of the inkjet-printed symbol on weighing paper were captured with carbon dot fluid and shown in Fig. 33a under normal, NIR, and UV radiation.

Through pyrolysis and microwave treatment, Wen et al. created brilliant inks based on green fluorescent CDs using cotton as the precursor. [206] Figures 33b and c illustrate the colourful designs produced by CD-based ink and its advantageous use in counterfeit detection.

The one-pot MW synthesis of luminous CNDs from citric acid and urea in distilled water is described by Qu et al. to produce biocompatible fluorescent ink. The presence of functionalities of O-H and N-H surface resulted in the spherical dimension (size = 1–5 nm and zeta potential = +88.1 mV) and increased water stability and hydrophilicity of the carbon-rich nanodots.

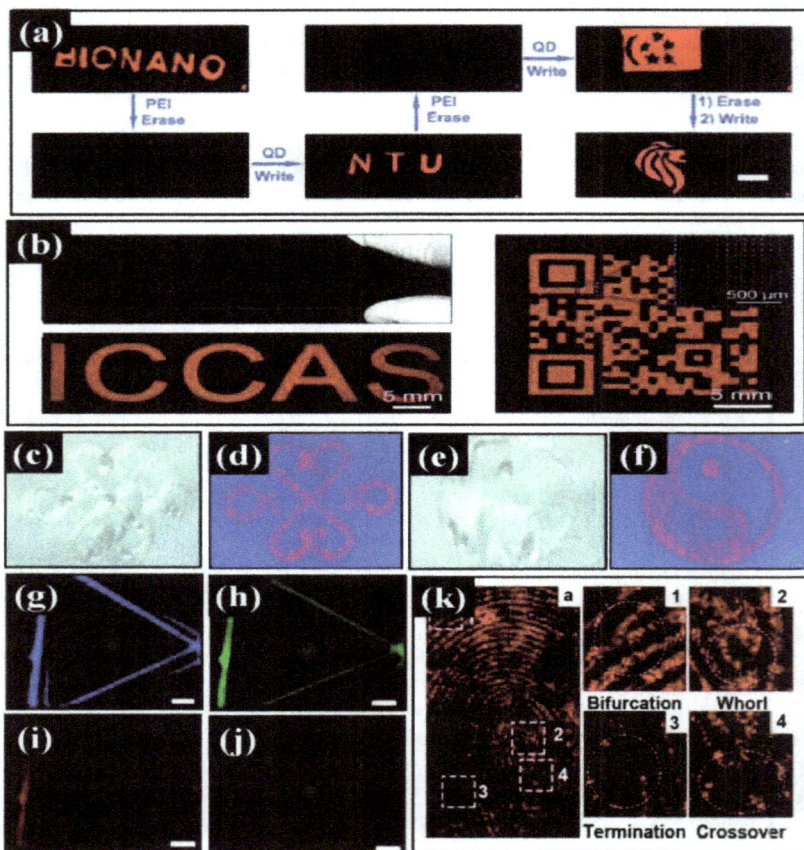

Figure 32. (a) QD-based fluorescent patterns with reversible writing and erasing, [200] (b) a fluorescent image 2D code pattern along with photograph of a text pattern [201] (c-f, anti-counterfeit labels coated with security ink exposed to visible and UV light, [202] (g-j) Fluorescent pictures of a pattern made of either commercially available green ink (circle) or silica nanodot solution (triangle) can be seen when exposed to UV light. [203] (k) photothermal images of LFPs. (Reproduced with permission from Ref. [204])

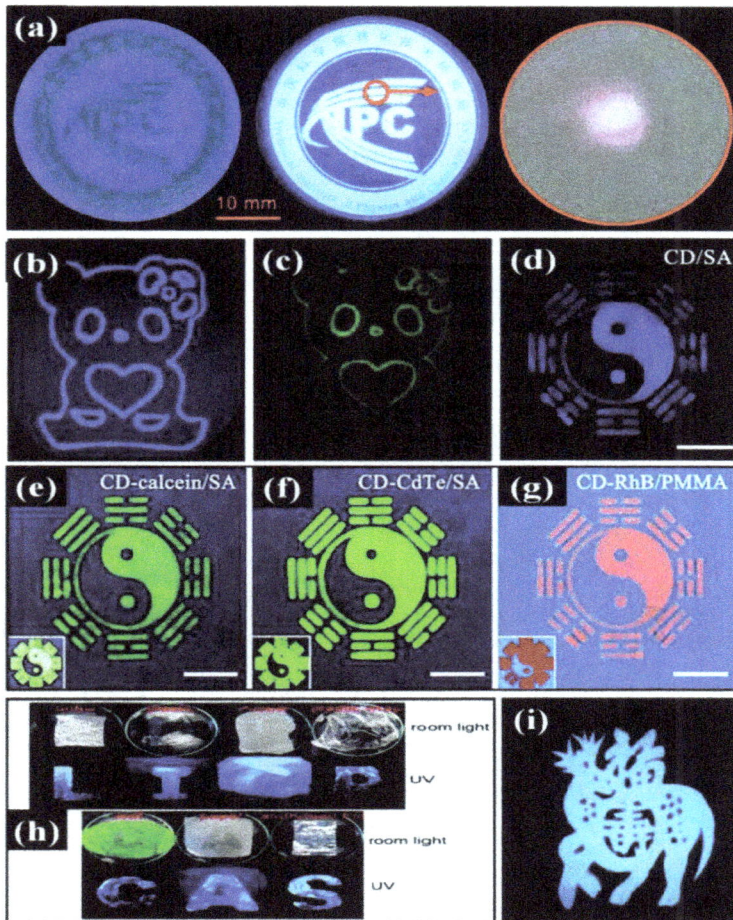

Figure 33. (a) Images of the logo printed using inkjet on weighing paper taken with CDF in natural, UV, and NIR light, [207] (b, c) Images of the patterns taken with CDF excited in UV light (470 nm), [206](d–g) fluorescence pattern images under UV light utilising CD-RhB/PMMA, CD/SA, CDCdTe/SA, and CD-calcein/SA, [208] (h) photos of various letters printed on various surfaces applying CQD fluorescent, [209] (i) Under UV light, the supernatant is used as the ink to create a fluorescent image of a printed object. (Reproduced with permission from Ref. [210])

Similar to this, urea and diammonium hydrogen citrate and green-emitting CNDs with a quantum yield of 46.4% detailed by Khan et al. were used as biocompatible fluorescent ink. [211].

An amphiphilic fluorescent carbon dot (CD)-based luminous security ink was created utilising natural eggs of chicken as precursor materials, according to Wang et al., and the quick plasma-induced method has been disclosed. [212] The synthesized carbon dots (CDs) possessed the amphiphilic character and are simply dissolved in water and most organic solvents. Both silk-screen printing and inkjet can be done with fluorescent ink derived from CDs. The images of fluorescent patterns created with fluorescent ink on CDs are shown in Fig. 33d-g.

In a different study, Wang et al. described the pyrolysis-based synthesis of a luminous security ink based on carbon quantum dots utilizing citric acid as a precursor and a capping agent based on imidazolium (1-aminopropyl-3-methylimidazolium bromide [APMIm][Br]). [213] Anion exchange and phase transfer were used to adjust the amphiphilicity of the synthesized CQDs. In addition, the author showed how these hydrophobic CQDs may be used to create luminous inks for use as anti-counterfeiting writing materials on a variety of surfaces. Fig. 33h displays images of numerous letters that were written using CQD fluorescence on various surfaces.

Ran et al. [214] reported on the preparation of nitrogen-based carbon nanodots (NCNDs) utilizing the hydrothermal technique through self-polymerization, and carbonization with allylamine as the carbon and nitrogen source. The optimal reaction temperature for the synthesis of 10.7% N-containing NCNDs (dimension = 2.88 0.4 nm) was 493 K sustained for 1.5 hours. These NCNDs had a QY of 15%, and their peak blue emission occurred at 350 nm for excitation.

Gao et al. unveiled a different technique for producing nitrogen-based carbon dots (N-CDs) through hydrothermal processing. [210] When the supernatant is subjected to UV light, it emits a hyperfine blue emission and can also be utilized to create fluorescent ink. Under UV light, the printed item photo employing the supernatant as ink is depicted in Fig. 33i. These NCNDs could also be used as fluorescent pens. Zhou et al. performed a similar study to create NCNDs without the use of organic solvents by employing allylamine and citric acid as N and carbon precursor molecules. [211] Graphitic Carbon Nitride (g-C3N4) was chemically oxidized by HNO3 to produce graphitic CNQDs by Song et al. [215] before being heated hydrothermally. Aqueous ink writing that is based on these CNQDs provides three levels of security for preserving data secrecy, allowing the coding information to be kept completely private. The carbon framework's heteroatom content impacts the PL characteristics, while the surface functional groups, precursors or specific elements that are doped in it significantly influence the structural properties of CNDs.

Utilizing these qualities, Zhang et al. described the S and N-based CNDs utilizing citric acid and cysteine as precursors during hydrothermal processing [216]. The average lifetime of the CNDs' bright blue fluorescence was 11.61 ns. On filter paper, a printed pattern made from an aqueous solution of CNDs was utilized as fluorescent ink; the pattern was invisible in natural light but fluoresced blue under 365 nm UV radiation, indicating its potential in anti-counterfeiting activities. In a unique technique described by Yang et al. [217], intramolecular rotation is prohibited and excitation energy is transferred to fluorescence in order to maintain the solid-state fluorescence in CNDs. Melamine and a di-thiosalicylic acid (DTSA)/acetic acid solution were used as precursors in a solvothermal procedure to create S and N-based hydrophobic CNDs (SNCNDs). When the SNCNDs were dissolved in an aqueous solution, they displayed blue FL. As more water (more than 50%) is added, the solution becomes turbid and has a suspension. Because they are hydrophobic, the CNDs continuously assemble, turning on the red solid-state fluorescence while turning off the blue FL. These S and N doped CNDs were used in two-switch mode fluorescent ink that was printed on filter paper and that appeared clear in natural light but exhibited blue fluorescent when exposed to 365 nm UV light. [218]

4.2.3 Luminescent MOFs for anti-counterfeiting applications

The family of luminous materials includes the amazing class of functional materials known as luminescent metal-organic frameworks (LMOFs). The nanoscale processability, specified porosity, simple synthesis, predicted orderly arrangements, and good photophysical features are only a few of their distinctive qualities. [219] By carefully choosing the constituent units, specifically the metal ions and struts (organic linker), a wide range of functional and structural tunability in LMOFs for various applications that includes sensing, gas storage & separation, drug administration, and catalysis can be accomplished. The intriguing characteristics of metal-organic frameworks (MOFs) and their potential for usage in a variety of applications make them a subject of significant interest. Recent years have seen the use of metal-organic frameworks (MOFs) in anti-counterfeiting activities. [220]

The creation of luminous inks based on lanthanide-MOFs was described by Luz et al. [221] Using an inkjet printer, fluorescent MOFs were produced on flexible substrates (plastic foil and paper). According to the author, MOF-doped security inks may be utilized to print secret security codes or labels for anti-counterfeiting activities. Under UV light, Fig. 34a and b display the vivid red and green coloured emission patterns that were coated on the polyethylene terephthalate (PET) material by utilising lanthanide MOF inks (R and G-MOFs) (254 nm).

Li et al. introduced a unique method for producing patterned luminous metal-organic-framework (LMOF) films utilising microwave deposition with electrochemical assistance in another paper. [222] A patterned lanthanide hydroxide was created on conducting glass using the electrochemical deposition process. In addition, the patterned lanthanide hydroxide was transformed into MOFs using a microwave method. This approach might be the best option for spatially locating MOFs on the surface of the substrate. Thus, these LMOFs could be utilised in applications for counterfeit detection. The images of the LMOF-based barcodes (Eu-MOFs and Tb-MOFs) are displayed in Fig. 34c and d.

A metal-organic-framework (MOF) film that has been photo functionalized with lanthanides and has various emission bands that correspond to lanthanide ions was described by Zhou et al. as a simple method for creating a barcode system. [224] Forecasting and adjusting the intensity of various emission bands is made possible by managing the filtered dye loading, that generates distinct radiometric optical codes. This technique can therefore be utilised to create barcodes with lanthanide doped MOFs and may have uses in anti-counterfeiting technology. The two-dimensional radiometric code matrix and Figure 34e displays the Eu^{3+}/Tb^{3+}@MIL-100 (ln) film emission spectra with different combinations of dye loading.

A novel technique for luminous barcodes based on nanoscale MOFs was reported by Lu et al (MOF-253). [226] By introducing lanthanide ions (Eu^{3+}, Tb^{3+}, and Sm^{3+}) via the post-synthetic method (PSM), the barcoding was achieved. Each lanthanide ion emits light at a proportionate rate to its concentration in the MOFs, creating distinct luminous barcodes that based on the ratios and compositions of the lanthanide ions. As a result, the ratio of lanthanide ions utilized and their structure affect luminous barcodes. The typical transition between the Eu^{3+} and Tb^{3+} ions following excitation (330 nm) is visible in the emission spectra of MOF-253-Tb_xEu_{1-x}, as illustrated in Fig. 34f. Fig. 34f also depicts the color-coding of the barcode readout in the CIE chromaticity diagram and a photo of the barcoded material dispersed in ethanol during stimulation at 330 nm.

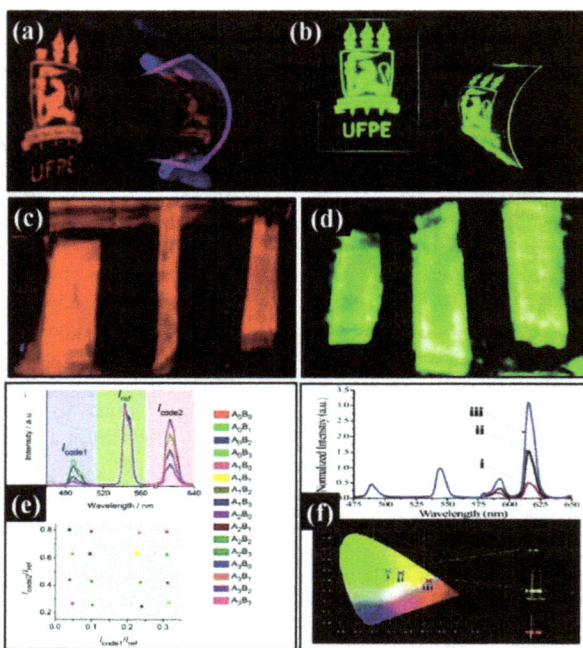

Figure 34. (a) and (b) Fluorescence photographs of R and G-MOFs under UV light, [223] (c) and (d) Tb-MOF and Eu-MOF patterns can be seen in fluorescence pictures, [222] (e) Eu3+/Tb3+@MIL-100 (In) films' emission spectra and a two-dimensional ratio metric coding matrix, both with different dye (FL and MB) loading combinations, [224] (f) Picture of the barcoded material based on MOF-253 (distributed in ethanol) excited at 330nm, normalised to the Tb^{3+} signal, emission spectra of Eu^{3+} and Tb^{3+} excited at 330nm, and color-code of the barcode readings in the CIE chromaticity chart, (i), (ii), and (iii) is MOF-253-Tb0.999Eu0.001, MOF-253-$Tb_{0.995}Eu_{0.005}$ and MOF-253-$Tb_{0.99}Eu_{0.01}$. (Reproduced with permission from Ref. [225])

4.2.4 Plasmonic nanomaterials for anti-counterfeiting applications

For the new-generation security tags in anti-counterfeiting applications, plasmonic nanoparticles hold a lot of potential. However, there have only been a few reports based on the use of plasmonic nanoparticles for anti-counterfeiting purposes published.

Via glutathione-coated gold nanoclusters (GS-AuNCs), which were created using a solution-based microwave technique, Zhang et al. generated luminous inks. [227] By adjusting the time of reaction, the emission colour of glutathione-coated gold nanoclusters (GS-AuNCs) was modified. When the response time was prolonged, the GS-AuNCs' photoluminescence emission became blue-shifted. Fig. 35a-c displays images of text printed on filter paper utilising luminescent inks doped with GS-AuNC in both natural and ultraviolet radiation (365 nm). In Fig. 35d the phrase "NANOTECHNOLOGY," where the letters "NA" have been screen printed with silver nanoparticles, "CH" with gold nanorods, and "GY" with a commercial green ink; the first letters "NO" have two overlapping layers, the first printed with magnetic nanoparticles and the second with the same printing with silver nanoparticles; Moreover, the second letter "NO" has two overlapping layers: one above printed with gold nanorods and one bottom printed with magnetic nanoparticles.

Another approach for integrating optical and magnetic substances for anti-counterfeiting purposes was revealed by Campos-Cuerva et al. [228]. They select single-phase spherical nanoparticles (magnetite, Au and Ag) for their study and screen-printed them on a variety of substrates. A single spot can produce optical and magnetic signals that can be identified separately when optical and magnetic materials are combined. The printed signals of these codes remain steady under accelerated conditions offering a high multi-stage security inks production method for anti-counterfeiting techniques.

For anti-counterfeiting purposes, there are additional security inks based on electroluminescence and chemiluminescence. A method for visualizing and identifying latent fingerprints on metal surfaces using ECL was described by Xu et al. They described an additional method of using electrochemiluminescence to detect protein/polypeptide residues in latent fingerprints (ECL). [229]

4.3   Various types of printing techniques

As an example of an anti-counterfeit application, while installing a security feature by hand is relatively effective in a simple fashion, it is not especially appealing to execute in large-scale automated manufacturing processes. However, it should be noted that research describing these procedures may be in an early stage of method development, with more complicated ways of implementing these techniques (such as ink-jet printing) becoming the subject of future reports once optimization has been achieved. Next-generation ink developments have the potential to open up new doors for academics working on anti-counterfeiting applications. As a result, using luminous material-based security ink in the creation of classified papers will unquestionably be a step toward safeguarding a country's integrity and economic progress. In this sense, the selection of luminous materials for anti-counterfeiting and encryption applications is essential, since it is largely determined by the security pattern's output signal, design, and complexity.

The following sections provide a quick overview of the three primary ways for printing anti-counterfeiting objects, patterns, and security codes:

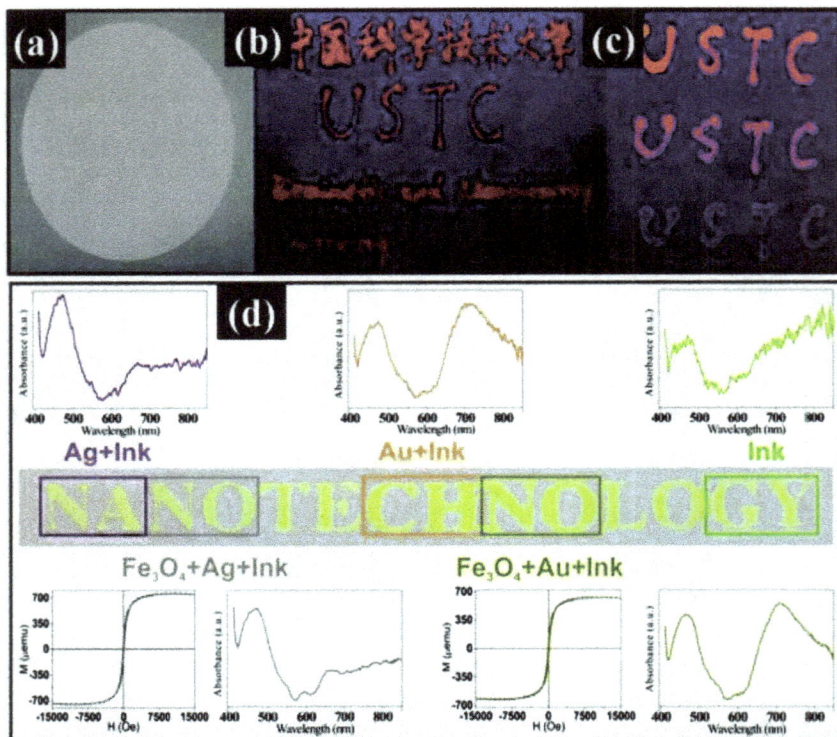

Figure 35. Photos of ordinary filter paper with invisible words printed on it in AuNC-3 h ink (a) in natural light and (b, c) in UV light (365 nm), [230] (d) and a photo of the printed word "Nanotechnology" utilising ink, (Ag + ink), and (Au + ink), (Reproduced with permission from Ref. [228])

4.3.1 Screen printing techniques

In the method of screen printing, ink is applied to a surface through a mesh, with the exception of areas where a blocking stencil has created an ink resistance. For the purpose of placing ink into the open mesh holes, a squeegee is dragged across the screen. After that, the screen can quickly touch the substrate with a reverse stroke along the line of contact. Due to the screen being able to spring back after the blade has passed, the ink can moisten the substrate and be drawn out through the mesh holes. In previous articles, we described how to use manufactured ink for printing graphics and security codes utilising the method of screen-printing. The method for printing of screen is shown step-by-step in Fig. 36.

As was already mentioned in the previous section, Campos-Cuerva et al. presented a method for screen printing plasmonic nanoparticle-doped inks on various types of paper. [228]

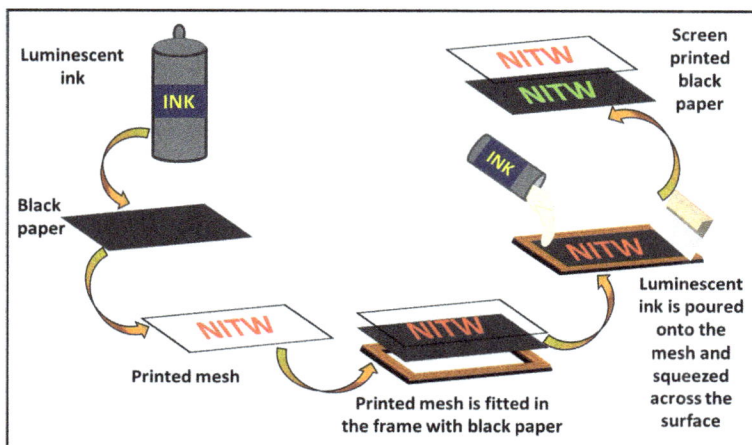

Figure 36. Process flow diagram for screen printing. (Reproduced with permission from Ref. [199])

4.3.2 Aerosol jet printing techniques

The aerosol jet printing technique utilise aerodynamic focusing to precisely distribute colloidal suspensions and/or chemical precursor solutions. A planar or three-dimensional substrate is exposed to a focussed, deposited, and patterned aerosol stream of the deposition material. The fundamental system consists of two essential parts: (i) liquid raw materials atomization module and (ii) The additional aerosol-focusing and droplet-depositing module. Meruga et al. showed how to create inks based on green and blue up-conversion nanoparticles and use them to print QR codes on paper utilizing aerosol jet printing process. [231] Fig. 37 depicts the procedure to print QR codes utilizing an aerosol jet printer. To transfer the QR code to the aerosol jet printer, it was initially created using a QR code creator and saved in an Auto-CAD file. After that, an aerosol jet printer was used to print the QR code on paper. The printing parameters were a 35 V atomizer power, a 125 cm sheath, a 30 cm atomizer, a 250 mm nozzle size, with a 5 mm/s deposition speed. Similar procedures were used by Meruga et al. in a different study to create and print luminous inks based on red-green-blue up-conversion nanoparticles. [231]

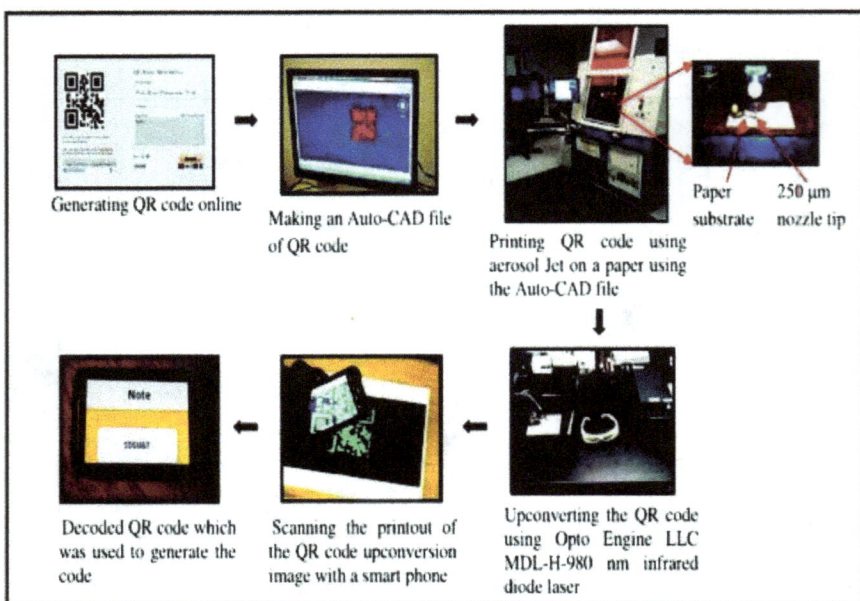

Figure 37. Guidelines for producing a QR code with the help of an aerosol jet printing machine. (Reproduced with permission from Ref. [231])

4.3.3 Inkjet printing techniques

Employing a quick, quasi-adiabatic contraction of the chamber volume brought on by piezoelectric activity, a fixed amount of ink is ejected from a nozzle into a chamber in the inkjet printing process. [232] A liquid-filled chamber constricts in consequence of an applied external voltage. A liquid drop discharges from the nozzle as a result of the sudden decrease creating a shockwave in the liquid. This approach has been examined and expanded upon in recent review studies. [233] The discharged drop is propelled downward until it hits the substrate by gravity and air resistance, at which point it spreads as a result of motion momentum and surface tension-enabled flow along the surface. As shown in Fig. 34, the solvent then evaporates, causing the drop to dry.

Recent studies have shown that the viscosity, which is reliant on the molar mass of the polymer, has a significant impact on how drops spread and the final shape of the printed object. [234]

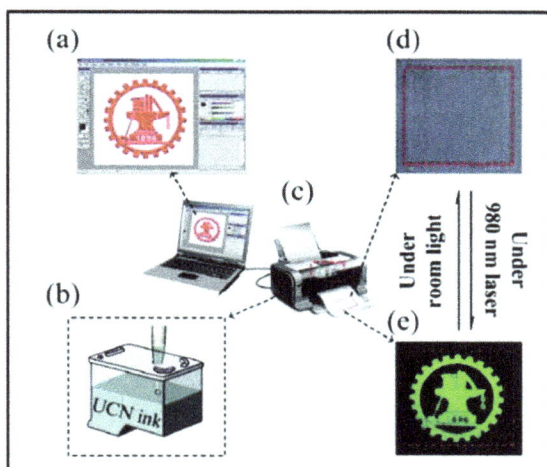

Figure 38. *schematic Flowchart for inkjet printing. (Reproduced with permission from Ref. [235])*

Inkjet printers' usefulness for producing security codes can be attributed to their appealing properties such as (i) The potential for entirely extra operation, wherein depending on the requirement the corresponding inks are only applied. (ii) Fig. 38 illustrates the adaptability in the preference of structural patterns for complex security designs, where alterations may be done immediately utilizing printer-control systems based on software. (iii) Its suitability for usage with materials covering a vast area. (iv) Possibility of mass production and high spatial resolution (based on DPI). [235] For the fabrication of high resolution and high brightness security patterns, You et al. developed an adaptable and digital inkjet printing process. [235] The inkjet printing concept for up-conversion inks are shown in Fig. 38.

Conclusions

This comprehensive chapter has been allocated an indication of two significant applications of Ln-doped nanomaterials, those being emissive in the luminescent constituents of LFPs developers (solutions, vapors, powders, and suspensions) and anti-counterfeiting measures (dopants, inks, resins, and labels). This discussion generally focused on the imperative nanomaterials that have been used for the recognition and development of LFPs and anti-counterfeiting. Presently, most of the forensic issues are carried out with the help of well-known RE-doped phosphors such as $NaYF_4$, LaF_3, and Y_2O_3 doped with Ln-elements, that are likely to be used as security inks, dyes, and labels. In addition to that, these phosphors are also used as LFPs developers, powders, colloids, and solutions. This chapter starts with the history of luminescent nanomaterials, elementary perceptions of mechanism, and their classification. Followed by these things, it discusses the development of required Ln-doped inorganic luminescent nanomaterials, MOFs, quantum dots, and plasmonic nanomaterials in detail. Additionally, we discussed their utilization in LFPs and anti-counterfeiting activities by recognizing fingerprints on different surfaces and printing the security codes/images. After that, we explained some techniques of printing like inkjet printing, screen printing and aerosol jet printing for security purposes.

The most important consequence of the chapter provides a way to resolve the addressed issues which consist of a solution series associated with different types of luminescent nanomaterials, and their utilization in ink preparation for exclusive security printing to protect banknotes, medicine, documents, branded items, etc. from counterfeiting. Despite the enormous progress that has been done in the LFPs labels and security inks depending on Ln-doped nanomaterials to date, several issues and challenges still remain to be determined such as the synthesis of Luminescent nanomaterials, the design of ink formulations and different types of labels to develop LFPs. Full exploitation of luminescent nanomaterials is the foremost challenge in the present scenario, which are free of these limitations. Another major task in this area is to eliminate background reflection from different types of non-porous/porous materials, the desired size of the particles, develop a medium for fabricating the luminescent ink in which luminescent materials should be dispersed with good colloidal stability, a long lifetime and cost-effective. This discussion will reignite the interest of researchers to carry out further research to improve the limitations of the current methods.

Acknowledgments

The authors PPP and DH thank the Director, NIT Warangal and the Department of Science and Technology (DST) of the Government of India for the financial support under the project #CRG/2021/007142.

References

[1] J. R. Partington, "Lignum nephriticum," *Ann. Sci.*, vol. 11, no. 1, pp. 1–26, Mar. 1955, https://doi.org/10.1080/00033795500200015

[2] B. Valeur and M. N. Berberan-Santos, "A Brief History of Fluorescence and Phosphorescence before the Emergence of Quantum Theory," *J. Chem. Educ.*, vol. 88, no. 6, pp. 731–738, Jun. 2011, https://doi.org/10.1021/ed100182h

[3] A. U. Acuña and F. Amat-Guerri, "Early History of Solution Fluorescence: The Lignum nephriticum of Nicolás Monardes BT - Fluorescence of Supermolecules, Polymers, and Nanosystems," M. N. Berberan-Santos, Ed. Berlin, Heidelberg: Springer Berlin Heidelberg, 2008, pp. 3–20. https://doi.org/10.1007/4243_2007_006

[4] A. U. Acuña, "More thoughts on the narra tree fluorescence [1]," *J. Chem. Educ.*, vol. 84, no. 2, p. 231, 2007, https://doi.org/10.1021/ed084p231

[5] "J. Chem. Educ. 2006, 83, 655–661," *J. Chem. Educ.*, vol. 83, no. 8, p. 1138, Aug. 2006, https://doi.org/10.1021/ed083p1138.2

[6] S. E. Braslavsky, "Glossary of terms used in photochemistry, 3rd edition (IUPAC Recommendations 2006)," vol. 79, no. 3, pp. 293–465, 2007, https://doi.org/doi:10.1351/pac200779030293

[7] P. R. Selvin, T. M. Rana, and J. E. Hearst, "Luminescence Resonance Energy Transfer," *J. Am. Chem. Soc.*, vol. 116, no. 13, pp. 6029–6030, Jun. 1994, https://doi.org/10.1021/ja00092a088

[8] S. Schietinger, L. Menezes, B. Lauritzen, and O. Benson, "No Title," *Nano Lett.*, vol. 9, p. 2477, 2009

[9] A. Cha, G. E. Snyder, P. R. Selvin, and F. Bezanilla, "Atomic scale movement of the voltage-sensing region in a potassium channel measured via spectroscopy," *Nature*, vol. 402, no. 6763, pp. 809–813, 1999, https://doi.org/10.1038/45552

[10]   A. Beeby *et al.*, "Luminescence imaging microscopy and lifetime mapping using kinetically stable lanthanide(III) complexes.," *J. Photochem. Photobiol. B.*, vol. 57 2–3, pp. 83–89, 2000

[11]   K. V. R. Murthy and H. Virk, "Luminescence Phenomena: An Introduction," *Defect Diffus. Forum*, vol. 347, pp. 1–34, Dec. 2013, https://doi.org/10.4028/www.scientific.net/DDF.347.1

[12]   H.-S. Qian and Y. Zhang, "Synthesis of Hexagonal-Phase Core–Shell NaYF4 Nanocrystals with Tunable Upconversion Fluorescence," *Langmuir*, vol. 24, no. 21, pp. 12123–12125, Nov. 2008, https://doi.org/10.1021/la802343f

[13]   A. G. Macedo *et al.*, "Effects of phonon confinement on anomalous thermalization, energy transfer, and upconversion in Ln3+-doped Gd2O3 nanotubes," *Adv. Funct. Mater.*, vol. 20, no. 4, pp. 624–634, 2010, https://doi.org/10.1002/adfm.200901772

[14]   "Advanced Materials - 2010 - Liu - A Strategy to Achieve Efficient Dual-Mode Luminescence of Eu3 in Lanthanides Doped.pdf."

[15]   Y. Liu, D. Tu, H. Zhu, and X. Chen, "Lanthanide-doped luminescent nanoprobes: Controlled synthesis, optical spectroscopy, and bioapplications," *Chem. Soc. Rev.*, vol. 42, no. 16, pp. 6924–6958, 2013, https://doi.org/10.1039/c3cs60060b

[16]   W. J. Kim, M. Nyk, and P. N. Prasad, "Color-coded multilayer photopatterned microstructures using lanthanide (III) ion co-doped NaYF4 nanoparticles with upconversion luminescence for possible applications in security," *Nanotechnology*, vol. 20, no. 18, p. 185301, 2009, https://doi.org/10.1088/0957-4484/20/18/185301

[17]   D. E. Wortman and C. A. Morrison, "Energy levels and predicted absorption spectra of rare-earth ions in rare-earth arsenides. Interim report, 1 July-30 September 1992," United States, 1992. [Online]. Available: https://www.osti.gov/biblio/6941498

[18]   C. Jiang, F. Wang, N. Wu, and X. Liu, "Up- and down-conversion cubic zirconia and hafnia nanobelts," *Adv. Mater.*, vol. 20, no. 24, pp. 4826–4829, 2008, https://doi.org/10.1002/adma.200801459

[19]   S. Heer, K. Kömpe, H. U. Güdel, and M. Haase, "Highly efficient multicolour upconversion emission in transparent colloids of lanthanide-doped NaYF4 nanocrystals," *Adv. Mater.*, vol. 16, no. 23–24, pp. 2102–2105, 2004, https://doi.org/10.1002/adma.200400772

[20] P. Ghosh and A. Patra, "Tuning of Crystal Phase and Luminescence Properties of Eu3+ Doped Sodium Yttrium Fluoride Nanocrystals," *J. Phys. Chem. C*, vol. 112, no. 9, pp. 3223–3231, Mar. 2008, https://doi.org/10.1021/jp7099114

[21] O. Ehlert, R. Thomann, M. Darbandi, and T. Nann, "A four-color colloidal multiplexing nanoparticle system.," *ACS Nano*, vol. 2 1, pp. 120–124, 2008

[22] I. M. Clarkson *et al.*, "Experimental assessment of the efficacy of sensitised emission in water from a europium ion, following intramolecular excitation by a phenanthridinyl group," *New J. Chem.*, vol. 24, no. 6, pp. 377–386, 2000, https://doi.org/10.1039/B001319F

[23] D. Hreniak *et al.*, "Enhancement of luminescence properties of Eu3+:YVO4 in polymeric nanocomposites upon UV excitation," *J. Lumin.*, vol. 131, no. 3, pp. 473–476, 2011, https://doi.org/https://doi.org/10.1016/j.jlumin.2010.10.028

[24] T. V. U. Gangan, S. Sreenadh, and M. L. P. Reddy, "Visible-light excitable highly luminescent molecular plastic materials derived from Eu3+-biphenyl based β-diketonate ternary complex and poly(methylmethacrylate)," *J. Photochem. Photobiol. A Chem.*, vol. 328, pp. 171–181, 2016, https://doi.org/https://doi.org/10.1016/j.jphotochem.2016.06.005

[25] X. Chen, H. Z. Zhuang, G. K. Liu, S. T. Li, and R. S. Niedbala, "Confinement on energy transfer between luminescent centers in nanocrystals," *J. Appl. Phys.*, vol. 94, pp. 5559–5565, 2003

[26] F. J. Steemers, W. Verboom, D. N. Reinhoudt, E. B. van der Tol, and J. W. Verhoeven, "Diazatriphenylene complexes of Eu3+ and Tb3+; promising light-converting systems with high luminescence quantum yields," *J. Photochem. Photobiol. A Chem.*, vol. 113, no. 2, pp. 141–144, 1998, https://doi.org/https://doi.org/10.1016/S1010-6030(97)00324-9

27] S. I. Klink *et al.*, "A Systematic Study of the Photophysical Processes in Polydentate Triphenylene-Functionalized Eu3+, Tb3+, Nd3+, Yb3+, and Er3+ Complexes," *J. Phys. Chem. A*, vol. 104, no. 23, pp. 5457–5468, Jun. 2000, https://doi.org/10.1021/jp994286+

[28] E. Brunet, O. Juanes, R. Sedano, and J.-C. Rodríguez-Ubis, "Synthesis of Novel Macrocyclic Lanthanide Chelates Derived from Bis-pyrazolylpyridine," *Org. Lett.*, vol. 4, no. 2, pp. 213–216, Jan. 2002, https://doi.org/10.1021/ol0169527

[29] J.-F. Mei *et al.*, "A novel photo-responsive europium(iii) complex for advanced anti-counterfeiting and encryption," *Dalt. Trans.*, vol. 45, no. 13, pp. 5451–5454, 2016, https://doi.org/10.1039/C6DT00346J

[30] M. R. Ganjali, P. Norouzi, T. Alizadeh, and M. Adib, "Application of 8-amino-N-(2-hydroxybenzylidene)naphthyl amine as a neutral ionophore in the construction of a lanthanum ion-selective sensor," *Anal. Chim. Acta*, vol. 576, no. 2, pp. 275–282, 2006, https://doi.org/https://doi.org/10.1016/j.aca.2006.06.037

[31] N. Sabbatini, M. Guardigli, and J.-M. Lehn, "No Title," *Coord. Chem. Rev.*, vol. 123, p. 201, 1993

[32] A. P. S. Samuel, J. Xu, and K. N. Raymond, "Predicting Efficient Antenna Ligands for Tb(III) Emission," *Inorg. Chem.*, vol. 48, no. 2, pp. 687–698, Jan. 2009, https://doi.org/10.1021/ic801904s

[33]   M. Kawa and J. M. J. Fréchet, "Self-Assembled Lanthanide-Cored Dendrimer Complexes: Enhancement of the Luminescence Properties of Lanthanide Ions through Site-Isolation and Antenna Effects," *Chem. Mater.*, vol. 10, pp. 286–296, 1998

[34]   X. Bai *et al.*, "Size-Dependent Upconversion Luminescence in Er3+/Yb3+-Codoped Nanocrystalline Yttria: Saturation and Thermal Effects," *J. Phys. Chem. C*, vol. 111, no. 36, pp. 13611–13617, Sep. 2007, https://doi.org/10.1021/jp070122e

[35]   L. Xu, X. Jiang, K. Liang, M. Gao, and B. Kong, "Frontier luminous strategy of functional silica nanohybrids in sensing and bioimaging: From ACQ to AIE," *Aggregate*, vol. 3, no. 1, p. e121, Feb. 2022, https://doi.org/https://doi.org/10.1002/agt2.121

[36]   X. Li, F. Zhang, and D. Zhao, "Lab on upconversion nanoparticles: optical properties and applications engineering via designed nanostructure," *Chem. Soc. Rev.*, vol. 44, no. 6, pp. 1346–1378, 2015, https://doi.org/10.1039/C4CS00163J

[37]   H. Dong, L.-D. Sun, and C.-H. Yan, "Energy transfer in lanthanide upconversion studies for extended optical applications," *Chem. Soc. Rev.*, vol. 44, no. 6, pp. 1608–1634, 2015, https://doi.org/10.1039/C4CS00188E

[38]   L. Prodi, E. Rampazzo, F. Rastrelli, A. Speghini, and N. Zaccheroni, "Imaging agents based on lanthanide doped nanoparticles.," *Chem. Soc. Rev.*, vol. 44, no. 14, pp. 4922–4952, Jul. 2015, https://doi.org/10.1039/c4cs00394b

[39]   M.-K. Tsang, G. Bai, and J. Hao, "Stimuli responsive upconversion luminescence nanomaterials and films for various applications," *Chem. Soc. Rev.*, vol. 44, no. 6, pp. 1585–1607, 2015, https://doi.org/10.1039/C4CS00171K

[40]   S. V Eliseeva and J.-C. G. Bünzli, "Lanthanide luminescence for functional materials and bio-sciences.," *Chem. Soc. Rev.*, vol. 39 1, pp. 189–227, 2010

[41]   J.-C. G. Bünzli and S. V Eliseeva, "Lanthanide NIR luminescence for telecommunications, bioanalyses and solar energy conversion," *J. Rare Earths*, vol. 28, no. 6, pp. 824–842, 2010, https://doi.org/https://doi.org/10.1016/S1002-0721(09)60208-8

[42]   B. M. Van Der Ende, L. Aarts, and A. Meijerink, "Lanthanide ions as spectral converters for solar cells," *Phys. Chem. Chem. Phys.*, vol. 11, no. 47, pp. 11081–11095, 2009, https://doi.org/10.1039/b913877c

[43]   S. Gai, C. Li, P. Yang, and J. Lin, "Recent progress in rare earth micro/nanocrystals: Soft chemical synthesis, luminescent properties, and biomedical applications," *Chem. Rev.*, vol. 114, no. 4, pp. 2343–2389, Feb. 2014, https://doi.org/10.1021/CR4001594/ASSET/IMAGES/LARGE/CR-2013-001594_0026.JPEG

[44]   S. K. Singh, A. K. Singh, and S. B. Rai, "Efficient dual mode multicolor luminescence in a lanthanide doped hybrid nanostructure: A multifunctional material," *Nanotechnology*, vol. 22, no. 27, 2011, https://doi.org/10.1088/0957-4484/22/27/275703

[45]   M. N. Luwang, R. S. Ningthoujam, S. K. Srivastava, and R. K. Vatsa, "Disappearance and recovery of luminescence in Bi3+, Eu 3+ codoped YPO4 nanoparticles due to the presence of water molecules Up to 800 °c," *J. Am. Chem. Soc.*, vol. 133, no. 9, pp. 2998–3004, Mar. 2011, https://doi.org/10.1021/JA1092437

[46]   J.-C. Boyer and F. C. J. M. van Veggel, "Absolute quantum yield measurements of colloidal NaYF4: Er3+, Yb3+ upconverting nanoparticles," *Nanoscale*, vol. 2, no. 8, pp. 1417–1419, 2010, https://doi.org/10.1039/C0NR00253D

[47]   N. Bogdan, F. Vetrone, G. A. Ozin, and J. A. Capobianco, "Synthesis of ligand-free colloidally stable water dispersible brightly luminescent lanthanide-doped upconverting nanoparticles.," *Nano Lett.*, vol. 11, no. 2, pp. 835–840, Feb. 2011, https://doi.org/10.1021/nl104192

[48]   H.-X. Mai, Y.-W. Zhang, L.-D. Sun, and C.-H. Yan, "Highly Efficient Multicolor Up-Conversion Emissions and Their Mechanisms of Monodisperse NaYF4:Yb,Er Core and Core/Shell-Structured Nanocrystals," *J. Phys. Chem. C*, vol. 111, no. 37, pp. 13721–13729, Sep. 2007, https://doi.org/10.1021/jp073920d

[49]   J. L. Sommerdijk, A. Bril, and A. W. de Jager, "Two photon luminescence with ultraviolet excitation of trivalent praseodymium," *J. Lumin.*, vol. 8, no. 4, pp. 341–343, 1974, https://doi.org/https://doi.org/10.1016/0022-2313(74)90006-4

[50]   M. N. Luwang, R. S. Ningthoujam, S. K. Srivastava, and R. K. Vatsa, "Preparation of white light emitting YVO4: Ln3+ and silica-coated YVO4:Ln3+ (Ln3+ = Eu3+, Dy3+, Tm3+) nanoparticles by CTAB/n-butanol/hexane/water microemulsion route: Energy transfer and site symmetry studies," *J. Mater. Chem.*, vol. 21, no. 14, pp. 5326–5337, 2011, https://doi.org/10.1039/C0JM03470C

[51]   T.-W. Kim, P.-W. Chung, I. I. Slowing, M. Tsunoda, E. S. Yeung, and V. S.-Y. Lin, "Structurally Ordered Mesoporous Carbon Nanoparticles as Transmembrane Delivery Vehicle in Human Cancer Cells," *Nano Lett.*, vol. 8, no. 11, pp. 3724–3727, Nov. 2008, https://doi.org/10.1021/nl801976m

[52]   C. Liu, H. Wang, X. Zhang, and D. Chen, "Morphology- and phase-controlled synthesis of monodisperse lanthanide-doped NaGdF4nanocrystals with multicolor photoluminescence," *J. Mater. Chem.*, vol. 19, no. 4, pp. 489–496, 2009, https://doi.org/10.1039/B815682D

[53]   P. Hartnagel, S. Ravishankar, B. Klingebiel, O. Thimm, and T. Kirchartz, "Comparing Methods of Characterizing Energetic Disorder in Organic Solar Cells," *Adv. Energy Mater.*, vol. n/a, no. n/a, p. 2300329, Mar. 2023, https://doi.org/https://doi.org/10.1002/aenm.202300329

[54]   B. Fan, C. Chlique, O. Merdrignac-Conanec, X. Zhang, and X. Fan, "Near-Infrared Quantum Cutting Material Er3+/Yb3+ Doped La2O2S with an External Quantum Yield Higher than 100\%," *J. Phys. Chem. C*, vol. 116, pp. 11652–11657, 2012

[55]   J. Zhou, Q. Liu, W. Feng, Y. Sun, and F. Li, "Upconversion luminescent materials: Advances and applications," *Chem. Rev.*, vol. 115, no. 1, pp. 395–465, Jan. 2015, https://doi.org/10.1021/CR400478F

[56]   L. E. Brus, "Electron-electron and electron-hole interactions in small semiconductor crystallites: The size dependence of the lowest excited electronic state," *J. Chem. Phys.*, vol. 80, pp. 4403–4409, May 1984, https://doi.org/10.1063/1.447218

[57]   M. G. Bawendi, M. L. Steigerwald, and L. E. Brus, "The Quantum Mechanics of Larger Semiconductor Clusters ('Quantum Dots')," *Annu. Rev. Phys. Chem.*, vol. 41, no. 1, pp. 477–496, Oct. 1990, https://doi.org/10.1146/annurev.pc.41.100190.002401

[58]   K. A. S. Fernando *et al.*, "Carbon Quantum Dots and Applications in Photocatalytic Energy Conversion," *ACS Appl. Mater. Interfaces*, vol. 7, no. 16, pp. 8363–8376, Apr. 2015, https://doi.org/10.1021/acsami.5b00448

[59]   I. L. Medintz, H. T. Uyeda, E. R. Goldman, and H. Mattoussi, "Quantum dot bioconjugates for imaging, labelling and sensing," *Nat. Mater.*, vol. 4, no. 6, pp. 435–446, 2005, https://doi.org/10.1038/nmat1390

[60]   M. Han, X. Gao, J. Z. Su, and S. Nie, "Quantum-dot-tagged microbeads for multiplexed optical coding of biomolecules," *Nat. Biotechnol.*, vol. 19, no. 7, pp. 631–635, 2001, https://doi.org/10.1038/90228

[61]   W. C. W. Chan, D. J. Maxwell, X. Gao, R. E. Bailey, M. Han, and S. Nie, "Luminescent quantum dots for multiplexed biological detection and imaging," *Curr. Opin. Biotechnol.*, vol. 13, no. 1, pp. 40–46, 2002, https://doi.org/https://doi.org/10.1016/S0958-1669(02)00282-3

[62]   J. Zhou, Y. Yang, and C. Zhang, "Toward Biocompatible Semiconductor Quantum Dots: From Biosynthesis and Bioconjugation to Biomedical Application," *Chem. Rev.*, vol. 115, no. 21, pp. 11669–11717, Nov. 2015, https://doi.org/10.1021/acs.chemrev.5b00049

[63]   J. Zhou, Z. Sheng, H. Han, M. Zou, and C.-X. Li, "Facile synthesis of fluorescent carbon dots using watermelon peel as a carbon source," *Mater. Lett.*, vol. 66, pp. 222–224, 2012

[64]   S. Silvi and A. Credi, "Luminescent sensors based on quantum dot-molecule conjugates," *Chem. Soc. Rev.*, vol. 44, no. 13, pp. 4275–4289, 2015, https://doi.org/10.1039/c4cs00400k

[65]   V. I. Klimov, "Spectral and dynamical properties of multiexcitons in semiconductor nanocrystals.," *Annu. Rev. Phys. Chem.*, vol. 58, pp. 635–673, 2007, https://doi.org/10.1146/annurev.physchem.58.032806.104537

[66]   C. Carrillo-Carrión, S. Cárdenas, B. M. Simonet, and M. Valcárcel, "Quantum dots luminescence enhancement due to illumination with UV/Vis light," *Chem. Commun.*, no. 35, pp. 5214–5226, 2009, https://doi.org/10.1039/b904381k

[67]   S. Kang *et al.*, "Simple preparation of graphene quantum dots with controllable surface states from graphite," *RSC Adv.*, vol. 9, no. 66, pp. 38447–38453, 2019, https://doi.org/10.1039/C9RA07555K

[68]   Y. Wang and A. Hu, "Carbon quantum dots: synthesis, properties and applications," *J. Mater. Chem. C*, vol. 2, no. 34, pp. 6921–6939, 2014, https://doi.org/10.1039/C4TC00988F

[69]   P. G. Luo *et al.*, "Carbon-based quantum dots for fluorescence imaging of cells and tissues," *RSC Adv.*, vol. 4, no. 21, pp. 10791–10807, 2014, https://doi.org/10.1039/C3RA47683A

[70]   L. Cao, M. J. Meziani, S. Sahu, and Y.-P. Sun, "Photoluminescence Properties of Graphene versus Other Carbon Nanomaterials," *Acc. Chem. Res.*, vol. 46, no. 1, pp. 171–180, Jan. 2013, https://doi.org/10.1021/ar300128j

[71]   X. Xu *et al.*, "Electrophoretic Analysis and Purification of Fluorescent Single-Walled Carbon Nanotube Fragments," *J. Am. Chem. Soc.*, vol. 126, no. 40, pp. 12736–12737, Oct. 2004, https://doi.org/10.1021/ja040082h

[72]   Y.-P. Sun *et al.*, "Quantum-Sized Carbon Dots for Bright and Colorful Photoluminescence," *J. Am. Chem. Soc.*, vol. 128, no. 24, pp. 7756–7757, Jun. 2006, https://doi.org/10.1021/ja062677d

[73]   S. Y. Lim, W. Shen, and Z. Gao, "Carbon quantum dots and their applications.," *Chem. Soc. Rev.*, vol. 44, no. 1, pp. 362–381, Jan. 2015, https://doi.org/10.1039/c4cs00269e

[74]   K. Krishnamoorthy, M. Veerapandian, K. Yun, and S. Kim, "The chemical and structural analysis of graphene oxide with different degrees of oxidation," *Carbon N. Y.*, vol. 53, pp. 38–49, 2013

[75]   G. Eda *et al.*, "Blue Photoluminescence from Chemically Derived Graphene Oxide," *Adv. Mater.*, vol. 22, no. 4, pp. 505–509, Jan. 2010, https://doi.org/https://doi.org/10.1002/adma.200901996

[76]   T. Gokus *et al.*, "Making Graphene Luminescent by Oxygen Plasma Treatment," *ACS Nano*, vol. 3, no. 12, pp. 3963–3968, Dec. 2009, https://doi.org/10.1021/nn9012753

[77]   A. P. Demchenko and M. O. Dekaliuk, "Novel fluorescent carbonic nanomaterials for sensing and imaging.," *Methods Appl. Fluoresc.*, vol. 1, no. 4, p. 42001, Aug. 2013, https://doi.org/10.1088/2050-6120/1/4/042001

[78]   J. Shen, Y. Zhu, X. Yang, J. Zong, J. Zhang, and C. Li, "One-pot hydrothermal synthesis of graphene quantum dots surface-passivated by polyethylene glycol and their photoelectric conversion under near-infrared light," *New J. Chem.*, vol. 36, no. 1, pp. 97–101, 2012, https://doi.org/10.1039/c1nj20658c

[79]   K. Krishnamoorthy, M. Veerapandian, R. Mohan, and S.-J. Kim, "Investigation of Raman and photoluminescence studies of reduced graphene oxide sheets," *Appl. Phys. A*, vol. 106, no. 3, pp. 501–506, 2012, https://doi.org/10.1007/s00339-011-6720-6

[80]   S. Y. Lim, W. Shen, and Z. Gao, "Carbon quantum dots and their applications," *Chem. Soc. Rev.*, vol. 44, no. 1, pp. 362–381, 2015, https://doi.org/10.1039/c4cs00269e

[81]   C. Mathioudakis, G. Kopidakis, P. C. Kelires, P. Patsalas, M. Gioti, and S. Logothetidis, "Electronic and optical properties of a-C from tight-binding molecular dynamics simulations," *Thin Solid Films*, vol. 482, no. 1, pp. 151–155, 2005, https://doi.org/https://doi.org/10.1016/j.tsf.2004.11.133

[82]   A. L. Himaja, P. S. Karthik, and S. P. Singh, "Carbon Dots: The Newest Member of the Carbon Nanomaterials Family," *Chem. Rec.*, vol. 15, no. 3, pp. 595–615, Jun. 2015, https://doi.org/https://doi.org/10.1002/tcr.201402090

[83]   T. A. Tabish and S. Zhang, "Graphene quantum dots: Syntheses, properties, and biological applications," *Compr. Nanosci. Nanotechnol.*, vol. 1–5, pp. 171–192, 2019, https://doi.org/10.1016/B978-0-12-803581-8.04133-3

[84]   X. T. Zheng, A. Ananthanarayanan, K. Q. Luo, and P. Chen, "Glowing Graphene Quantum Dots and Carbon Dots: Properties, Syntheses, and Biological Applications," *Small*,

vol. 11, no. 14, pp. 1620–1636, Apr. 2015,
https://doi.org/https://doi.org/10.1002/smll.201402648

[85]   Z. G. Khan and P. O. Patil, "A comprehensive review on carbon dots and graphene quantum dots based fluorescent sensor for biothiols," *Microchem. J.*, vol. 157, p. 105011, 2020, https://doi.org/https://doi.org/10.1016/j.microc.2020.105011

[86]   A. Cayuela, M. L. Soriano, C. Carrillo-Carrión, and M. Valcárcel, "Semiconductor and carbon-based fluorescent nanodots: the need for consistency," *Chem. Commun.*, vol. 52, no. 7, pp. 1311–1326, 2016, https://doi.org/10.1039/C5CC07754K

[87]   H. Lin, J. Huang, and L. Ding, "Preparation of Carbon Dots with High-Fluorescence Quantum Yield and Their Application in Dopamine Fluorescence Probe and Cellular Imaging," *J. Nanomater.*, vol. 2019, p. 5037243, 2019, https://doi.org/10.1155/2019/5037243

[88]   H. Sun, H. Ji, E. Ju, Y. Guan, J. Ren, and X. Qu, "Synthesis of Fluorinated and Nonfluorinated Graphene Quantum Dots through a New Top-Down Strategy for Long-Time Cellular Imaging," *Chem. – A Eur. J.*, vol. 21, no. 9, pp. 3791–3797, Feb. 2015, https://doi.org/https://doi.org/10.1002/chem.201406345

[89]   A. Das, V. Gude, D. Roy, T. Chatterjee, C. K. De, and P. K. Mandal, "On the Molecular Origin of Photoluminescence of Nonblinking Carbon Dot," *J. Phys. Chem. C*, vol. 121, no. 17, pp. 9634–9641, May 2017, https://doi.org/10.1021/acs.jpcc.7b02433

[90]   H. Sun, H. Ji, E. Ju, Y. Guan, J. Ren, and X. Qu, "Synthesis of fluorinated and nonfluorinated graphene quantum dots through a new top-down strategy for long-time cellular imaging," *Chem. - A Eur. J.*, vol. 21, no. 9, pp. 3791–3797, 2015, https://doi.org/10.1002/chem.201406345

[91]   H. X. Mai *et al.*, "No Title," *J. Am. Chem. Soc.*, vol. 128, p. 6426, 2006

[92]   S. L. Li and Q. Xu, "Metal-organic frameworks as platforms for clean energy," *Energy Environ. Sci.*, vol. 6, no. 6, pp. 1656–1683, 2013, https://doi.org/10.1039/c3ee40507a

[93]   Q. L. Zhu and Q. Xu, "Metal-organic framework composites," *Chem. Soc. Rev.*, vol. 43, no. 16, pp. 5468–5512, 2014, https://doi.org/10.1039/c3cs60472a

[94]   T.-H. Chen *et al.*, "Thermally robust and porous noncovalent organic framework with high affinity for fluorocarbons and CFCs.," *Nat. Commun.*, vol. 5, p. 5131, Oct. 2014, https://doi.org/10.1038/ncomms6131

[95]   Y. Ikezoe, G. Washino, T. Uemura, S. Kitagawa, and H. Matsui, "Autonomous motors of a metal–organic framework powered by reorganization of self-assembled peptides at interfaces," *Nat. Mater.*, vol. 11, no. 12, pp. 1081–1085, 2012, https://doi.org/10.1038/nmat3461

[96]   A. Carné, C. Carbonell, I. Imaz, and D. Maspoch, "Nanoscale metal–organic materials," *Chem. Soc. Rev.*, vol. 40, no. 1, pp. 291–305, 2011, https://doi.org/10.1039/c0cs00042f

[97]   M. D. Allendorf, C. A. Bauer, R. K. Bhakta, and R. J. T. Houk, "Luminescent metal-organic frameworks.," *Chem. Soc. Rev.*, vol. 38, no. 5, pp. 1330–1352, May 2009, https://doi.org/10.1039/b802352m

[98]   Y. Cui, Y. Yue, G. Qian, and B. Chen, "Luminescent Functional Metal–Organic Frameworks," *Chem. Rev.*, vol. 112, no. 2, pp. 1126–1162, Feb. 2012, https://doi.org/10.1021/cr200101d

[99]   W. Bi, M. Zhou, Z. Ma, H. Zhang, J. Yu, and Y. Xie, "CuInSe2ultrathin nanoplatelets: Novel self-sacrificial template-directed synthesis and application for flexible photodetectors," *Chem. Commun.*, vol. 48, no. 73, pp. 9162–9164, 2012, https://doi.org/10.1039/c2cc34727j

[100] E. G. Moore, A. P. S. Samuel, and K. N. Raymond, "From Antenna to Assay: Lessons Learned in Lanthanide Luminescence," *Acc. Chem. Res.*, vol. 42, no. 4, pp. 542–552, Apr. 2009, https://doi.org/10.1021/ar800211j

[101] C. L. Cahill, D. T. de Lill, and M. Frisch, "Homo- and heterometallic coordination polymers from the f elements," *CrystEngComm*, vol. 9, no. 1, pp. 15–26, 2007, https://doi.org/10.1039/B615696G

[102] Y. Qin *et al.*, "Efficient ambipolar transport properties in alternate stacking donor–acceptor complexes: from experiment to theory," *Phys. Chem. Chem. Phys.*, vol. 18, no. 20, pp. 14094–14103, 2016, https://doi.org/10.1039/C6CP01509C

[103] Y. Xiao, P. Weidler, S. Lin, C. Wöll, Z.-G. Gu, and J. Zhang, "Chiral Metal–Organic Cluster Induced High Circularly Polarized Luminescence of Metal–Organic Framework Thin Film," *Adv. Funct. Mater.*, vol. 32, Aug. 2022, https://doi.org/10.1002/adfm.202204289

[104] J. An, C. M. Shade, D. A. Chengelis-Czegan, S. Petoud, and N. L. Rosi, "Zinc-Adeninate Metal−Organic Framework for Aqueous Encapsulation and Sensitization of Near-infrared and Visible Emitting Lanthanide Cations," *J. Am. Chem. Soc.*, vol. 133, no. 5, pp. 1220–1223, Feb. 2011, https://doi.org/10.1021/ja109103t

[105] L. M. Liz-Marzán, C. J. Murphy, and J. Wang, "Nanoplasmonics," *Chem. Soc. Rev.*, vol. 43, no. 11, pp. 3820–3822, 2014, https://doi.org/10.1039/C4CS90026J

[106] M. L. Brongersma, "Introductory lecture: nanoplasmonics," *Faraday Discuss.*, vol. 178, no. 0, pp. 9–36, 2015, https://doi.org/10.1039/C5FD90020D

[107] K. L. Kelly, E. Coronado, L. L. Zhao, and G. C. Schatz, "The Optical Properties of Metal Nanoparticles: The Influence of Size, Shape, and Dielectric Environment," *J. Phys. Chem. B*, vol. 107, no. 3, pp. 668–677, Jan. 2003, https://doi.org/10.1021/jp026731y

[108] S. Eustis and M. A. El-Sayed, "Why gold nanoparticles are more precious than pretty gold: Noble metal surface plasmon resonance and its enhancement of the radiative and nonradiative properties of nanocrystals of different shapes," *Chem. Soc. Rev.*, vol. 35, no. 3, pp. 209–217, 2006, https://doi.org/10.1039/B514191E

[109] K. M. Mayer and J. H. Hafner, "Localized Surface Plasmon Resonance Sensors," *Chem. Rev.*, vol. 111, no. 6, pp. 3828–3857, Jun. 2011, https://doi.org/10.1021/cr100313v

[110] P. Nagpal, N. C. Lindquist, S.-H. Oh, and D. J. Norris, "Ultrasmooth patterned metals for plasmonics and metamaterials.," *Science*, vol. 325, no. 5940, pp. 594–597, Jul. 2009, https://doi.org/10.1126/science.1174655

Materials Research Foundations 164 (2024) 67-142   https://doi.org/10.21741/9781644903056-2

[111]   S. K. Ghosh and T. Pal, "Interparticle Coupling Effect on the Surface Plasmon Resonance of Gold Nanoparticles: From Theory to Applications," *Chem. Rev.*, vol. 107, no. 11, pp. 4797–4862, Nov. 2007, https://doi.org/10.1021/cr0680282

[112]   P. R. Sajanlal, T. S. Sreeprasad, A. K. Samal, and T. Pradeep, "Anisotropic nanomaterials: structure, growth, assembly, and functions.," *Nano Rev.*, vol. 2, 2011, https://doi.org/10.3402/nano.v2i0.5883

[113]   M. W. Knight, N. S. King, L. Liu, H. O. Everitt, P. Nordlander, and N. J. Halas, "Aluminum for plasmonics.," *ACS Nano*, vol. 8, no. 1, pp. 834–840, Jan. 2014, https://doi.org/10.1021/nn405495q

[114]   P. Hazarika and D. A. Russell, "Advances in fingerprint analysis.," *Angew. Chem. Int. Ed. Engl.*, vol. 51, no. 15, pp. 3524–3531, Apr. 2012, https://doi.org/10.1002/anie.201104313

[115]   P. Rastogi and K. R. Pillai, "A study of fingerprints in relation to gender and blood group," *J. Indian Acad. Forensic Med.*, vol. 32, pp. 11–14, 2010

[116]   Q. Wei, M. Zhang, B. Ogorevc, and X. Zhang, "Recent advances in the chemical imaging of human fingermarks (a review)," *Analyst*, vol. 141, no. 22, pp. 6172–6189, 2016, https://doi.org/10.1039/C6AN01121G

[117]   A. V Ewing and S. G. Kazarian, "Infrared spectroscopy and spectroscopic imaging in forensic science," *Analyst*, vol. 142, no. 2, pp. 257–272, 2017, https://doi.org/10.1039/C6AN02244H

[118]   A. Bécue, "Emerging fields in fingermark (meta)detection – a critical review," *Anal. Methods*, vol. 8, no. 45, pp. 7983–8003, 2016, https://doi.org/10.1039/C6AY02496C

[119]   T. J. Comi, S. W. Ryu, and R. H. Perry, "Synchronized Desorption Electrospray Ionization Mass Spectrometry Imaging.," *Anal. Chem.*, vol. 88, no. 2, pp. 1169–1175, Jan. 2016, https://doi.org/10.1021/acs.analchem.5b03010

[120]   S. Cadd, M. Islam, P. Manson, and S. Bleay, "Fingerprint composition and aging: A literature review.," *Sci. Justice*, vol. 55, no. 4, pp. 219–238, Jul. 2015, https://doi.org/10.1016/j.scijus.2015.02.004

[121]   F. Cortés-Salazar, D. Momotenko, H. H. Girault, A. Lesch, and G. Wittstock, "Seeing big with scanning electrochemical microscopy.," *Anal. Chem.*, vol. 83, no. 5, pp. 1493–1499, Mar. 2011, https://doi.org/10.1021/ac101931d

[122]   D. Chávez, C. R. Garcia, J. Oliva, and L. A. Diaz-Torres, "A review of phosphorescent and fluorescent phosphors for fingerprint detection," *Ceram. Int.*, vol. 47, no. 1, pp. 10–41, 2021, https://doi.org/https://doi.org/10.1016/j.ceramint.2020.08.259

[123]   E. Prabakaran and K. Pillay, "Nanomaterials for latent fingerprint detection: a review," *J. Mater. Res. Technol.*, vol. 12, pp. 1856–1885, 2021, https://doi.org/https://doi.org/10.1016/j.jmrt.2021.03.110

[124]   K. Scotcher and R. Bradshaw, "The analysis of latent fingermarks on polymer banknotes using MALDI-MS," *Sci. Rep.*, vol. 8, no. 1, p. 8765, 2018, https://doi.org/10.1038/s41598-018-27004-0

[125] S. Of and C. Vision, "Stereomicroscopic Gender Determination From Fingerprint Ridge Stereomicroscopic Gender Determination From Fingerprint Ridge Density and Fingerprint," *Proc. 24th Myanmar Mil. Med. Conf.*, no. February, 2017, https://doi.org/10.13140/RG.2.2.13274.39368

[126] S. B. Nikam and S. Agarwal, "Ridgelet-based fake fingerprint detection," *Neurocomputing*, vol. 72, no. 10, pp. 2491–2506, 2009, https://doi.org/https://doi.org/10.1016/j.neucom.2008.11.003

[127] B. J. Jones, A. J. Reynolds, M. Richardson, and V. G. Sears, "Nano-scale composition of commercial white powders for development of latent fingerprints on adhesives," *Sci. Justice*, vol. 50, no. 3, pp. 150–155, 2010, https://doi.org/https://doi.org/10.1016/j.scijus.2009.08.001

[128] A. A. Cantu, "Silver Physical Developers for the Visualization of Latent Prints on Paper.," *Forensic Sci. Rev.*, vol. 13, no. 1, pp. 29–64, Jan. 2001

[129] B. Schnetz and P. Margot, "Technical note: latent fingermarks, colloidal gold and multimetal deposition (MMD): Optimisation of the method," *Forensic Sci. Int.*, vol. 118, no. 1, pp. 21–28, 2001, https://doi.org/https://doi.org/10.1016/S0379-0738(00)00361-3

[130] C. Lennard, P. A. Margot, M. Sterns, and R. N. Warrener, "Photoluminescent Enhancement of Ninhydrin Developed Fingerprints by Metal Complexation: Structural Studies of Complexes Formed Between Ruhemann's Purple and Group IIb Metal Salts," *J. Forensic Sci.*, vol. 32, pp. 597–605, 1987

[131] J. Friesen, "Forensic Chemistry: The Revelation of Latent Fingerprints," *J. Chem. Educ.*, vol. 92, pp. 497–504, Mar. 2015, https://doi.org/10.1021/ed400597u

[132] A. L. Beresford and A. R. Hillman, "Electrochromic Enhancement of Latent Fingerprints on Stainless Steel Surfaces," *Anal. Chem.*, vol. 82, no. 2, pp. 483–486, Jan. 2010, https://doi.org/10.1021/ac9025434

[133] S. K. Bramble, "SEPARATION OF LATENT FINGERMARK RESIDUE BY THIN-LAYER CHROMATOGRAPHY," *J. Forensic Sci.*, vol. 40, pp. 969–975, 1995

[134] R. Yang and J. Lian, "Studies on the development of latent fingerprints by the method of solid–medium ninhydrin," *Forensic Sci. Int.*, vol. 242, pp. 123–126, 2014, https://doi.org/https://doi.org/10.1016/j.forsciint.2014.06.036

[135] J. Almog, G. Levinton-Shamuilov, Y. Cohen, and M. Azoury, "Fingerprint Reagents with Dual Action: Color and Fluorescence," *J. Forensic Sci.*, vol. 52, no. 2, pp. 330–334, Mar. 2007, https://doi.org/https://doi.org/10.1111/j.1556-4029.2007.00383.x

[136] G. Levinton-Shamuilov, Y. Cohen, M. Azoury, A. Chaikovsky, and J. Almog, "Genipin, a novel fingerprint reagent with colorimetric and fluorogenic activity, part II: optimization, scope and limitations.," *J. Forensic Sci.*, vol. 50, no. 6, pp. 1367–1371, Nov. 2005

[137] S. Coughlan, "Using acetone to increase visualization of ninhydrin-developed fingerprints obscured by common pen ink," vol. 62, pp. 330–333, Jul. 2012

[138] O. P. Jasuja, M. A. Toofany, G. Singh, and G. S. Sodhi, "Dynamics of latent fingerprints: The effect of physical factors on quality of ninhydrin developed prints — A preliminary

study," *Sci. Justice*, vol. 49, no. 1, pp. 8–11, 2009,
https://doi.org/https://doi.org/10.1016/j.scijus.2008.08.001

[139] M. F. Mangle, X. Xu, and M. de Puit, "Performance of 1,2-indanedione and the need for sequential treatment of fingerprints," *Sci. Justice*, vol. 55, no. 5, pp. 343–346, 2015, https://doi.org/https://doi.org/10.1016/j.scijus.2015.04.002

[140] M. de Puit, M. Ismail, and X. Xu, "LCMS Analysis of Fingerprints, the Amino Acid Profile of 20 Donors," *J. Forensic Sci.*, vol. 59, no. 2, pp. 364–370, Mar. 2014, https://doi.org/https://doi.org/10.1111/1556-4029.12327

[141] W. Song *et al.*, "Detection of protein deposition within latent fingerprints by surface-enhanced Raman spectroscopy imaging," *Nanoscale*, vol. 4, no. 7, pp. 2333–2338, 2012, https://doi.org/10.1039/C2NR12030E

[142] D. Chavez, C. R. Garcia, I. Ruiz-Martinez, J. Oliva, E. Rivera-Rosales, and L. A. Diaz-Torres, "Fingerprint detection on low contrast surfaces using phosphorescent nanomaterials," *AIP Conf. Proc.*, vol. 2083, no. March 2019, 2019, https://doi.org/10.1063/1.5094304

[143] C. Huynh and J. Halámek, "Trends in fingerprint analysis," *TrAC Trends Anal. Chem.*, vol. 82, pp. 328–336, 2016, https://doi.org/https://doi.org/10.1016/j.trac.2016.06.003

[144] N. Singla, M. Kaur, and S. Sofat, "Automated latent fingerprint identification system: A review," *Forensic Sci. Int.*, vol. 309, p. 110187, 2020, https://doi.org/https://doi.org/10.1016/j.forsciint.2020.110187

[145] A. H. Malik, N. Zehra, M. Ahmad, R. Parui, and P. K. Iyer, "Advances in conjugated polymers for visualization of latent fingerprints: A critical perspective," *New J. Chem.*, vol. 44, no. 45, pp. 19423–19439, 2020, https://doi.org/10.1039/d0nj04131a

[146] G. Ren *et al.*, "Nitrogen-doped carbon dots for the detection of mercury ions in living cells and visualization of latent fingerprints," *New J. Chem.*, vol. 42, no. 9, pp. 6824–6830, 2018, https://doi.org/10.1039/c7nj05170k

[147] A. Bécue, "Emerging fields in fingermark (meta)detection-a critical review," *Anal. Methods*, vol. 8, no. 45, pp. 7983–8003, 2016, https://doi.org/10.1039/c6ay02496c

[148] Y.-P. Luo, Y.-B. Zhao, and S. Liu, "Evaluation of DFO/PVP and its application to latent fingermarks development on thermal paper," *Forensic Sci. Int.*, vol. 229, no. 1, pp. 75–79, 2013, https://doi.org/https://doi.org/10.1016/j.forsciint.2013.03.045

[149] L. Liu, Z. Zhang, L. Zhang, and Y. Zhai, "The effectiveness of strong afterglow phosphor powder in the detection of fingermarks," *Forensic Sci. Int.*, vol. 183, no. 1, pp. 45–49, 2009, https://doi.org/https://doi.org/10.1016/j.forsciint.2008.10.008

[150] S. Das *et al.*, "Molecular Fluorescence, Phosphorescence, and Chemiluminescence Spectrometry," *Anal. Chem.*, vol. 84, no. 2, pp. 597–625, Jan. 2012, https://doi.org/10.1021/ac202904n

[151] D. Chavez, C. R. Garcia, I. Ruiz-Martinez, J. Oliva, E. Rivera-Rosales, and L. A. Diaz-Torres, "Fingerprint detection on low contrast surfaces using phosphorescent nanomaterials," *AIP Conf. Proc.*, vol. 2083, no. 1, p. 20001, Mar. 2019, https://doi.org/10.1063/1.5094304

[152] Q. Xiao, L. Xiao, Y. Liu, X. Chen, and Y. Li, "Synthesis and luminescence properties of needle-like SrAl2O4:Eu, Dy phosphor via a hydrothermal co-precipitation method," *J. Phys. Chem. Solids*, vol. 71, no. 7, pp. 1026–1030, 2010, https://doi.org/https://doi.org/10.1016/j.jpcs.2010.04.017

[153] V. Sharma, A. Das, V. Kumar, O. M. Ntwaeaborwa, and H. C. Swart, "Potential of Sr4Al14O25: Eu2+,Dy3+ inorganic oxide-based nanophosphor in Latent fingermark detection," *J. Mater. Sci.*, vol. 49, no. 5, pp. 2225–2234, 2014, https://doi.org/10.1007/s10853-013-7916-2

[154] V. Sharma, A. Das, and V. Kumar, "Eu2+,Dy3+ codoped SrAl2O4 nanocrystalline phosphor for latent fingerprint detection in forensic applications," *Mater. Res. Express*, vol. 3, no. 1, p. 15004, 2016, https://doi.org/10.1088/2053-1591/3/1/015004

[155] W. Shan, L. Wu, N. Tao, Y. Chen, and D. Guo, "Optimization method for green SrAl2O4:Eu2+,Dy3+ phosphors synthesized via co-precipitation route assisted by microwave irradiation using orthogonal experimental design," *Ceram. Int.*, vol. 41, no. 10, Part B, pp. 15034–15040, 2015, https://doi.org/https://doi.org/10.1016/j.ceramint.2015.08.050

[156] S. Yeshodamma, D. V Sunitha, R. B. Basavaraj, G. P. Darshan, B. D. Prasad, and H. Nagabhushana, "Monovalent ions co-doped SrTiO3:Pr3+ nanostructures for the visualization of latent fingerprints and can be red component for solid state devices," *J. Lumin.*, vol. 208, pp. 371–387, 2019, https://doi.org/https://doi.org/10.1016/j.jlumin.2018.12.044

[157] V. Sharma, A. Das, V. Kumar, V. Kumar, K. Verma, and H. C. Swart, "Combustion synthesis and characterization of blue long lasting phosphor CaAl2O4: Eu2+, Dy3+ and its novel application in latent fingerprint and lip mark detection," *Phys. B Condens. Matter*, vol. 535, pp. 149–156, 2018, https://doi.org/https://doi.org/10.1016/j.physb.2017.07.019

[158] J. Y. Park, J. W. Chung, S. J. Park, and H. K. Yang, "Versatile fluorescent CaGdAlO4:Eu3+ red phosphor for latent fingerprints detection," *J. Alloys Compd.*, vol. 824, p. 153994, 2020, https://doi.org/https://doi.org/10.1016/j.jallcom.2020.153994

[159] K. M. Girish, S. C. Prashantha, R. Naik, and H. Nagabhushana, "Zn2TiO4: A novel host lattice for Sm3+ doped reddish orange light emitting photoluminescent material for thermal and fingerprint sensor," *Opt. Mater. (Amst).*, vol. 73, pp. 197–205, 2017, https://doi.org/https://doi.org/10.1016/j.optmat.2017.08.009

[160] R. E. Rojas-Hernandez, F. Rubio-Marcos, M. Á. Rodriguez, and J. F. Fernandez, "Long lasting phosphors: SrAl2O4:Eu, Dy as the most studied material," *Renew. Sustain. Energy Rev.*, vol. 81, pp. 2759–2770, 2018, https://doi.org/https://doi.org/10.1016/j.rser.2017.06.081

[161] R. S. P. King and D. A. Skros, "Sunlight-activated near-infrared phosphorescence as a viable means of latent fingermark visualisation," *Forensic Sci. Int.*, vol. 276, pp. e35–e39, 2017, https://doi.org/https://doi.org/10.1016/j.forsciint.2017.04.012

[162] Z. Pan, Y.-Y. Lu, and F. Liu, "Sunlight-activated long-persistent luminescence in the near-infrared from Cr3+-doped zinc gallogermanates," *Nat. Mater.*, vol. 11, no. 1, pp. 58–63, 2012, https://doi.org/10.1038/nmat3173

[163] X. Ran, Z. Wang, Z. Zhang, F. Pu, J. Ren, and X. Qu, "Nucleic-acid-programmed Ag-nanoclusters as a generic platform for visualization of latent fingerprints and exogenous

substances," *Chem. Commun.*, vol. 52, no. 3, pp. 557–560, 2016, https://doi.org/10.1039/C5CC08534A

[164]   L. K. Bharat, G. S. R. Raju, and J. S. Yu, "Red and green colors emitting spherical-shaped calcium molybdate nanophosphors for enhanced latent fingerprint detection.," *Sci. Rep.*, vol. 7, no. 1, p. 11571, Sep. 2017, https://doi.org/10.1038/s41598-017-11692-1

[165]   V. Prasad, S. Lukose, P. Agarwal, and L. Prasad, "Role of Nanomaterials for Forensic Investigation and Latent Fingerprinting—A Review," *J. Forensic Sci.*, vol. 65, no. 1, pp. 26–36, Jan. 2020, https://doi.org/https://doi.org/10.1111/1556-4029.14172

[166]   T. J. Bukowski and J. H. Simmons, "Quantum Dot Research: Current State and Future Prospects," *Crit. Rev. Solid State Mater. Sci.*, vol. 27, no. 3–4, pp. 119–142, Jul. 2002, https://doi.org/10.1080/10408430208500496

[167]   J. Zhu *et al.*, "Emitting color tunable carbon dots by adjusting solvent towards light-emitting devices," *Nanotechnology*, vol. 29, no. 8, p. 85705, 2018, https://doi.org/10.1088/1361-6528/aaa321

[168]   R. B. Basavaraj, G. P. Darshan, B. Daruka Prasad, S. C. Sharma, and H. Nagabhushana, "Rapid visualization of latent fingerprints using novel $CaSiO_3$:Sm^{3+} nanophosphors fabricated via ultrasound route," *J. Rare Earths*, vol. 37, no. 1, pp. 32–44, 2019, https://doi.org/https://doi.org/10.1016/j.jre.2018.04.019

[169]   M. Wang, M. Li, A. Yu, J. Wu, and C. Mao, "Rare Earth Fluorescent Nanomaterials for Enhanced Development of Latent Fingerprints," *ACS Appl. Mater. Interfaces*, vol. 7, no. 51, pp. 28110–28115, Dec. 2015, https://doi.org/10.1021/acsami.5b09320

[170]   M. Wang, Y. Zhu, and C. Mao, "Synthesis of NIR-Responsive $NaYF_4$:Yb,Er Upconversion Fluorescent Nanoparticles Using an Optimized Solvothermal Method and Their Applications in Enhanced Development of Latent Fingerprints on Various Smooth Substrates," *Langmuir*, vol. 31, no. 25, pp. 7084–7090, Jun. 2015, https://doi.org/10.1021/acs.langmuir.5b01151

[171]   M. Saif *et al.*, "Novel non-toxic and red luminescent sensor based on Eu^{3+}:$Y_2Ti_2O_7$/SiO_2 nano-powder for latent fingerprint detection," *Sensors Actuators B Chem.*, vol. 220, pp. 162–170, 2015, https://doi.org/https://doi.org/10.1016/j.snb.2015.05.040

[172]   A. Das and V. Shama, "Synthesis and characterization of Eu^{3+} doped α-Al_2O_3 nanocrystalline powder for novel application in latent fingerprint development," *Adv. Mater. Lett.*, vol. 7, no. 4, pp. 302–306, 2016, https://doi.org/10.5185/amlett.2016.6310

[173]   G. S. Sodhi and J. Kaur, "Physical developer method for detection of latent fingerprints: A review," *Egypt. J. Forensic Sci.*, vol. 6, no. 2, pp. 44–47, 2016, https://doi.org/https://doi.org/10.1016/j.ejfs.2015.05.001

[174]   B. Stojanović, O. Marques, and A. Nešković, "Latent overlapped fingerprint separation: a review," *Multimed. Tools Appl.*, vol. 76, no. 15, pp. 16263–16290, 2017, https://doi.org/10.1007/s11042-016-3908-y

[175]   V. Sharma, A. Das, and V. Kumar, "Eu^{2+}, Dy^{3+} codoped $SrAl_2O_4$ nanocrystalline phosphor for latent fi ngerprint detection in forensic applications," *Mater. Res. Express*, vol. 3, no. 1, p. 15004

[176] Y. Liu *et al.*, "Inkjet-printed unclonable quantum dot fluorescent anti-counterfeiting labels with artificial intelligence authentication," *Nat. Commun.*, vol. 10, no. 1, p. 2409, 2019, https://doi.org/10.1038/s41467-019-10406-7

[177] H. Zhang *et al.*, "Materials and Technologies to Combat Counterfeiting of Pharmaceuticals: Current and Future Problem Tackling," *Adv. Mater.*, vol. 32, no. 11, p. 1905486, Mar. 2020, https://doi.org/https://doi.org/10.1002/adma.201905486

[178] G. Schirripa Spagnolo, L. Cozzella, and C. Simonetti, "Banknote security using a biometric-like technique: a hylemetric approach," *Meas. Sci. Technol.*, vol. 21, no. 5, p. 55501, 2010, https://doi.org/10.1088/0957-0233/21/5/055501

[179] X. Li and Y. Hu, "Luminescent films functionalized with cellulose nanofibrils/CdTe quantum dots for anti-counterfeiting applications," *Carbohydr. Polym.*, vol. 203, pp. 167–175, 2019, https://doi.org/https://doi.org/10.1016/j.carbpol.2018.09.028

[180] Z. Zhang *et al.*, "Switchable up and down-conversion luminescent properties of Nd(III)-nanopaper for visible and near-infrared anti-counterfeiting," *Carbohydr. Polym.*, vol. 252, p. 117134, 2021, https://doi.org/https://doi.org/10.1016/j.carbpol.2020.117134

[181] R. Huang *et al.*, "Tunable upconversion of holmium sublattice through interfacial energy transfer for anti-counterfeiting," *Nanoscale*, vol. 13, no. 9, pp. 4812–4820, 2021, https://doi.org/10.1039/D0NR09068A

[182] D. Wang *et al.*, "Achieving Color-Tunable and Time-Dependent Organic Long Persistent Luminescence via Phosphorescence Energy Transfer for Advanced Anti-Counterfeiting," *Adv. Funct. Mater.*, vol. 33, no. 1, p. 2208895, Jan. 2023, https://doi.org/https://doi.org/10.1002/adfm.202208895

[183] K. Jiang, L. Zhang, J. Lu, C. Xu, C. Cai, and H. Lin, "Triple-Mode Emission of Carbon Dots: Applications for Advanced Anti-Counterfeiting," *Angew. Chemie Int. Ed.*, vol. 55, no. 25, pp. 7231–7235, Jun. 2016, https://doi.org/https://doi.org/10.1002/anie.201602445

[184] Kanika, P. Kumar, S. Singh, and B. K. Gupta, "A Novel Approach to Synthesise a Dual-Mode Luminescent Composite Pigment for Uncloneable High-Security Codes to Combat Counterfeiting," *Chem. – A Eur. J.*, vol. 23, no. 67, pp. 17144–17151, Dec. 2017, https://doi.org/https://doi.org/10.1002/chem.201704076

[185] "Eur J Inorg Chem - 2022 - Salaam - Tris-dipicolinate Lanthanide Complexes Influence of the Second Hydration Sphere on the.pdf."

[186] J. Andres, R. D. Hersch, J. E. Moser, and A. S. Chauvin, "A new anti-counterfeiting feature relying on invisible luminescent full color images printed with lanthanide-based inks," *Adv. Funct. Mater.*, vol. 24, no. 32, pp. 5029–5036, 2014, https://doi.org/10.1002/adfm.201400298

[187] L. Chen *et al.*, "The temperature-sensitive luminescence of (Y,Gd)VO4:Bi3+,Eu3+ and its application for stealth anti-counterfeiting," *Phys. status solidi – Rapid Res. Lett.*, vol. 6, no. 7, pp. 321–323, Jul. 2012, https://doi.org/https://doi.org/10.1002/pssr.201206234

[188] T. Blumenthal *et al.*, "Patterned direct-write and screen-printing of NIR-to-visible upconverting inks for security applications," *Nanotechnology*, vol. 23, no. 18, p. 185305, 2012, https://doi.org/10.1088/0957-4484/23/18/185305

[189] W. Gao, W. Ge, J. Shi, Y. Tian, J. Zhu, and Y. Li, "Stretchable, flexible, and transparent SrAl2O4:Eu2+@TPU ultraviolet stimulated anti-counterfeiting film," *Chem. Eng. J.*, vol. 405, p. 126949, 2021, https://doi.org/https://doi.org/10.1016/j.cej.2020.126949

[190] Y. Zhang *et al.*, "Multicolor Barcoding in a Single Upconversion Crystal," *J. Am. Chem. Soc.*, vol. 136, no. 13, pp. 4893–4896, Apr. 2014, https://doi.org/10.1021/ja5013646

[191] J. M. Meruga, W. M. Cross, P. Stanley May, Q. Luu, G. A. Crawford, and J. J. Kellar, "Security printing of covert quick response codes using upconverting nanoparticle inks," *Nanotechnology*, vol. 23, no. 39, 2012, https://doi.org/10.1088/0957-4484/23/39/395201

[192] J. M. Meruga, A. Baride, W. Cross, J. J. Kellar, and P. S. May, "Red-green-blue printing using luminescence-upconversion inks," *J. Mater. Chem. C*, vol. 2, no. 12, pp. 2221–2227, 2014, https://doi.org/10.1039/c3tc32233e

[193] M. Wang *et al.*, "NIR-induced highly sensitive detection of latent fingermarks by NaYF4:Yb,Er upconversion nanoparticles in a dry powder state," *Nano Res.*, vol. 8, no. 6, pp. 1800–1810, 2015, https://doi.org/10.1007/s12274-014-0686-6

[194] N. M. Sangeetha *et al.*, "3D assembly of upconverting NaYF4 nanocrystals by AFM nanoxerography: Creation of anti-counterfeiting microtags," *Nanoscale*, vol. 5, no. 20, pp. 9587–9592, 2013, https://doi.org/10.1039/c3nr02734a

[195] M. You, J. Zhong, Y. Hong, Z. Duan, M. Lin, and F. Xu, "Inkjet printing of upconversion nanoparticles for anti-counterfeit applications," *Nanoscale*, vol. 7, no. 10, pp. 4423–4431, 2015, https://doi.org/10.1039/c4nr06944g

[196] J. Méndez-Ramos, J. C. Ruiz-Morales, P. Acosta-Mora, and N. M. Khaidukov, "Infrared-light induced curing of photosensitive resins through photon up-conversion for novel cost-effective luminescent 3D-printing technology," *J. Mater. Chem. C*, vol. 4, no. 4, pp. 801–806, 2015, https://doi.org/10.1039/c5tc03315b

[197] A. Baride *et al.*, "A NIR-to-NIR upconversion luminescence system for security printing applications," *RSC Adv.*, vol. 5, no. 123, pp. 101338–101346, 2015, https://doi.org/10.1039/c5ra20785a

[198] Y. Liu, K. Ai, and L. Lu, "Designing lanthanide-doped nanocrystals with both up- and down-conversion luminescence for anti-counterfeiting," *Nanoscale*, vol. 3, no. 11, pp. 4804–4810, 2011, https://doi.org/10.1039/c1nr10752f

[199] P. Kumar, J. Dwivedi, and B. K. Gupta, "Highly luminescent dual mode rare-earth nanorod assisted multi-stage excitable security ink for anti-counterfeiting applications," *J. Mater. Chem. C*, vol. 2, no. 48, pp. 10468–10475, 2014, https://doi.org/10.1039/C4TC02065K

[200] Z. Lu, Y. Liu, H. Wu, X. W. Lou, and C. M. Li, "Rewritable multicolor fluorescent patterns for multistate memory devices with high data storage capacity," *Chem. Commun.*, vol. 47, no. 34, pp. 9609–9611, 2011, https://doi.org/10.1039/c1cc13448e

[201] B. Bao *et al.*, "Patterning Fluorescent Quantum Dot Nanocomposites by Reactive Inkjet Printing," *Small*, vol. 11, no. 14, pp. 1649–1654, Apr. 2015, https://doi.org/https://doi.org/10.1002/smll.201403005

[202]   Y. Gao *et al.*, "Decorating CdTe QD-Embedded Mesoporous Silica Nanospheres with Ag NPs to Prevent Bacteria Invasion for Enhanced Anticounterfeit Applications," *ACS Appl. Mater. Interfaces*, vol. 7, no. 18, pp. 10022–10033, May 2015, https://doi.org/10.1021/acsami.5b02472

[203]   L. Zhou, A. Zhao, Z. Wang, Z. Chen, J. Ren, and X. Qu, "Ionic Liquid-Assisted Synthesis of Multicolor Luminescent Silica Nanodots and Their Use as Anti-counterfeiting Ink," *ACS Appl. Mater. Interfaces*, vol. 7, no. 4, pp. 2905–2911, Feb. 2015, https://doi.org/10.1021/am5083304

[204]   J. Cui, S. Xu, C. Guo, R. Jiang, T. D. James, and L. Wang, "Highly Efficient Photothermal Semiconductor Nanocomposites for Photothermal Imaging of Latent Fingerprints," *Anal. Chem.*, vol. 87, no. 22, pp. 11592–11598, Nov. 2015, https://doi.org/10.1021/acs.analchem.5b03652

[205]   B. Wang, W. Liang, Z. Guo, and W. Liu, "Biomimetic super-lyophobic and super-lyophilic materials applied for oil/water separation: A new strategy beyond nature," *Chem. Soc. Rev.*, vol. 44, no. 1, pp. 336–361, 2015, https://doi.org/10.1039/c4cs00220b

[206]   X. Wen *et al.*, "Green synthesis of carbon nanodots from cotton for multicolor imaging, patterning, and sensing," *Sensors Actuators B Chem.*, vol. 221, pp. 769–776, 2015, https://doi.org/https://doi.org/10.1016/j.snb.2015.07.019

[207]   F. Wang, Z. Xie, B. Zhang, Y. Liu, W. Yang, and C. Y. Liu, "Down- and up-conversion luminescent carbon dot fluid: Inkjet printing and gel glass fabrication," *Nanoscale*, vol. 6, no. 7, pp. 3818–3823, 2014, https://doi.org/10.1039/c3nr05869g

[208]   J. Wang, C. F. Wang, and S. Chen, "Amphiphilic egg-derived carbon dots: Rapid plasma fabrication, pyrolysis process, and multicolor printing patterns," *Angew. Chemie - Int. Ed.*, vol. 51, no. 37, pp. 9297–9301, 2012, https://doi.org/10.1002/anie.201204381

[209]   B. Wang *et al.*, "Tunable amphiphilicity and multifunctional applications of ionic-liquid-modified carbon quantum dots," *ACS Appl. Mater. Interfaces*, vol. 7, no. 12, pp. 6919–6925, 2015, https://doi.org/10.1021/acsami.5b00758

[210]   S. Gao *et al.*, "A green one-arrow-two-hawks strategy for nitrogen-doped carbon dots as fluorescent ink and oxygen reduction electrocatalysts," *J. Mater. Chem. A*, vol. 2, no. 18, pp. 6320–6325, 2014, https://doi.org/10.1039/c3ta15443b

[211]   K. Muthamma, D. Sunil, and P. Shetty, "Carbon dots as emerging luminophores in security inks for anti-counterfeit applications - An up-to-date review," *Appl. Mater. Today*, vol. 23, p. 101050, 2021, https://doi.org/https://doi.org/10.1016/j.apmt.2021.101050

[212]   J. Wang, C.-F. Wang, and S. Chen, "Amphiphilic Egg-Derived Carbon Dots: Rapid Plasma Fabrication, Pyrolysis Process, and Multicolor Printing Patterns," *Angew. Chemie Int. Ed.*, vol. 51, no. 37, pp. 9297–9301, Sep. 2012, https://doi.org/https://doi.org/10.1002/anie.201204381

[213]   B. Wang *et al.*, "Tunable Amphiphilicity and Multifunctional Applications of Ionic-Liquid-Modified Carbon Quantum Dots," *ACS Appl. Mater. Interfaces*, vol. 7, no. 12, pp. 6919–6925, Apr. 2015, https://doi.org/10.1021/acsami.5b00758

[214] L. Li and T. Dong, "Photoluminescence tuning in carbon dots: Surface passivation or/and functionalization, heteroatom doping," *J. Mater. Chem. C*, vol. 6, no. 30, pp. 7944–7970, 2018, https://doi.org/10.1039/c7tc05878k

[215] M. L. Brongersma, "Introductory lecture: Nanoplasmonics," *Faraday Discuss.*, vol. 178, pp. 9–36, 2015, https://doi.org/10.1039/c5fd90020d

[216] R. Santonocito, M. Intravaia, I. M. Caruso, A. Pappalardo, G. Trusso Sfrazzetto, and N. Tuccitto, "Fluorescence sensing by carbon nanoparticles," *Nanoscale Adv.*, vol. 4, no. 8, pp. 1926–1948, 2022, https://doi.org/10.1039/d2na00080f

[217] A. Tan, G. Yang, and X. Wan, "Ultra-high quantum yield nitrogen-doped carbon quantum dots and their versatile application in fluorescence sensing, bioimaging and anti-counterfeiting," *Spectrochim. Acta Part A Mol. Biomol. Spectrosc.*, vol. 253, p. 119583, 2021, https://doi.org/https://doi.org/10.1016/j.saa.2021.119583

[218] C. Ma, Z. Wang, Z. Hu, Y. Wang, Y. Zhao, and J. Shi, "Preparation of submicron monodisperse melamine resin microspheres and nitrogen-doped carbon microspheres derived from them," *New Carbon Mater.*, vol. 35, no. 3, pp. 269–285, 2020, https://doi.org/https://doi.org/10.1016/S1872-5805(20)60489-9

[219] A. Hazra, U. Mondal, S. Mandal, and P. Banerjee, "Advancement in functionalized luminescent frameworks and their prospective applications as inkjet-printed sensors and anti-counterfeit materials," *Dalt. Trans.*, vol. 50, no. 25, pp. 8657–8670, 2021, https://doi.org/10.1039/d1dt00705j

[220] K. A. White, D. A. Chengelis, K. A. Gogick, J. Stehman, N. L. Rosi, and S. Petoud, "Near-infrared luminescent lanthanide MOF barcodes.," *J. Am. Chem. Soc.*, vol. 131, no. 50, pp. 18069–18071, Dec. 2009, https://doi.org/10.1021/ja907885m

[221] L. L. da Luz *et al.*, "Inkjet Printing of Lanthanide-Organic Frameworks for Anti-Counterfeiting Applications.," *ACS Appl. Mater. Interfaces*, vol. 7, no. 49, pp. 27115–27123, Dec. 2015, https://doi.org/10.1021/acsami.5b06301

[222] W. J. Li *et al.*, "Patterned growth of luminescent metal-organic framework films: A versatile electrochemically-assisted microwave deposition method," *Chem. Commun.*, vol. 52, no. 20, pp. 3951–3954, 2016, https://doi.org/10.1039/c6cc00519e

[223] L. L. Da Luz *et al.*, "Inkjet Printing of Lanthanide-Organic Frameworks for Anti-Counterfeiting Applications," *ACS Appl. Mater. Interfaces*, vol. 7, no. 49, pp. 27115–27123, 2015, https://doi.org/10.1021/acsami.5b06301

[224] Y. Zhou and B. Yan, "Ratiometric multiplexed barcodes based on luminescent metal-organic framework films," *J. Mater. Chem. C*, vol. 3, no. 32, pp. 8413–8418, 2015, https://doi.org/10.1039/c5tc01311a

[225] Y. Lu and B. Yan, "Luminescent lanthanide barcodes based on postsynthetic modified nanoscale metal–organic frameworks," *J. Mater. Chem. C*, vol. 2, no. 35, pp. 7411–7416, 2014, https://doi.org/10.1039/C4TC01077A

[226] M. You, J. Zhong, Y. Hong, Z. Duan, M. Lin, and F. Xu, "Inkjet printing of upconversion nanoparticles for anti-counterfeit applications.," *Nanoscale*, vol. 7, no. 10, pp. 4423–4431, Mar. 2015, https://doi.org/10.1039/c4nr06944g

[227] J. Zhang *et al.*, "A microwave-facilitated rapid synthesis of gold nanoclusters with tunable optical properties for sensing ions and fluorescent ink," *Chem. Commun.*, vol. 51, no. 52, pp. 10539–10542, 2015, https://doi.org/10.1039/C5CC03086B

[228] C. Campos-Cuerva *et al.*, "Screen-printed nanoparticles as anti-counterfeiting tags," *Nanotechnology*, vol. 27, no. 9, p. 95702, 2016, https://doi.org/10.1088/0957-4484/27/9/095702

[229] Y. Chen *et al.*, "Chirality-activated mechanoluminescence from aggregation-induced emission enantiomers with high contrast mechanochromism and force-induced delayed fluorescence," *Mater. Chem. Front.*, vol. 3, no. 9, pp. 1800–1806, 2019, https://doi.org/10.1039/C9QM00312F

[230] J. Zhang *et al.*, "A microwave-facilitated rapid synthesis of gold nanoclusters with tunable optical properties for sensing ions and fluorescent ink," *Chem. Commun.*, vol. 51, no. 52, pp. 10539–10542, 2015, https://doi.org/10.1039/c5cc03086b

[231] J. M. Meruga *et al.*, "Stable Inks Containing Upconverting Nanoparticles Based on an Oil-in-Water Nanoemulsion.," *Langmuir*, vol. 34, no. 4, pp. 1535–1541, Jan. 2018, https://doi.org/10.1021/acs.langmuir.7b03415

[232] M. Singh, H. M. Haverinen, P. Dhagat, and G. E. Jabbour, "Inkjet printing-process and its applications.," *Adv. Mater.*, vol. 22, no. 6, pp. 673–685, Feb. 2010, https://doi.org/10.1002/adma.200901141

[233] E. Tekin, P. J. Smith, and U. S. Schubert, "Inkjet printing as a deposition and patterning tool for polymers and inorganic particles," *Soft Matter*, vol. 4, no. 4, pp. 703–713, 2008, https://doi.org/10.1039/B711984D

[234] J. Perelaer, P. J. Smith, C. E. Hendriks, A. M. J. van den Berg, and U. S. Schubert, "The preferential deposition of micro-particles at the boundary of inkjet printed droplets.," *Soft Matter*, vol. 4, no. 5, pp. 1072–1078, Apr. 2008, https://doi.org/10.1039/b715076h

[235] M. You, J. Zhong, Y. Hong, Z. Duan, M. Lin, and F. Xu, "Inkjet printing of upconversion nanoparticles for anti-counterfeit applications," *Nanoscale*, vol. 7, no. 10, pp. 4423–4431, 2015, https://doi.org/10.1039/c4nr06944g

Rare Earth - A tribute to the late Mr. Rare Earth, Professor Karl Gschneidner   Materials Research Forum LLC
Materials Research Foundations 164 (2024) 143-176   https://doi.org/10.21741/9781644903056-3

Chapter 3

Overview of Treating Skin Diseases and Rejuvenating Skin using Light Sources

Aachal A. Sharma, D. Haranath*

Luminescence Materials and Devices (LMD) Group, Department of Physics, National Institute of Technology Warangal, 506004, Telangana, India

haranath@nitw.ac.in

Abstract

The electromagnetic spectrum of sunlight, which includes infrared (IR, 52–55%), visible (VIS, 42–43%), and ultraviolet radiation (UV, 3-5%), is a naturally occurring source of light. Out of all these radiations, UV radiations with specific wavelengths have been shown to have clinically important biological effects that are both efficient and safe for human skin. UV radiation exposure primarily aids in the formation of vitamin D synthesis and has the potential to suppress immunological response, making it a more acceptable wavelength range for the diagnosis and treatment of various skin dermatoses. While comparing natural sunlight which contains less percentage of UV radiation with broadband range artificial UV therapy is an efficient, reliable, affordable, and, nonsurgical therapeutic option available for skin dermatoses. Recent investigations and literature reviews demonstrate that various types of phototherapies such as UVB narrowband, PUVA, etc. have incompatible efficacy toward the management of dermatoses that often depend upon the action, for instance, and absorption spectra of the biological molecule (chromophores) contained in the human skin. We attempt to categorize human skin conditions by their symptoms, topological facts, and dosimetry prescribed for the cures by the minimal erythemal dosage (MED) of phototherapy recommended by the dermatologist. Descriptive analysis of types of therapies carried out on narrowband (311 ± 2 nm) and broadband (295-315 nm) ultraviolent radiation B (UVB) therapy, light-sensitive organic drug (Psoralen) in combination with UVA (PUVA) therapy and laser therapy. This review article aims at the development in the field of dermatology by reviewing historical data, therapeutic options available, a detailed understanding of skin dermatoses, and the availability of light sources for the clearance of various skin dermatoses.

Keywords

Ultraviolet Radiation, Skin Diseases, Phototherapy, Human Skin, Psoralene, Action Spectra, Absorption spectra

Contents

Rare Earth - A tribute to the late Mr. Rare Earth, Professor Karl Gschneidner   Materials Research Forum LLC
Materials Research Foundations 164 (2024) 143-176   https://doi.org/10.21741/9781644903056-3

1. Introduction

Disability-adjusted life years (DALYs) for a disease or health condition are the sum of the years of life lost due to premature mortality (YLLs) and the years lived with a disability (YLDs) due to prevalent cases of the disease or health condition in a population. According to DALYs data, skin disorders contributed 1.79% to the worldwide burden of disease[1]. Nearly one-third of the world's population suffers from various skin disorders such as psoriasis, vitiligo, lichen sclerosis, atopic dermatitis, etc., making them the fourth most common cause of other human diseases[2]. Despite their widespread prevalence and visibility, the burden of skin disorders is frequently underestimated. Skin diseases affect individuals of all ages, gender, and, complexation. All skin disorders are climate sensitive and include all conditions that clog, irritate, or inflame the human skin. The common cause of skin disease involves bacteria or fungus or parasite traps, virus infection, immune system imbalance, genetic condition, etc. However, most researchers believe that severe skin diseases (Psoriasis, vitiligo, etc.) are mainly autoimmune in nature. From antiquity, the first light source used for the treatment of such autoimmune diseases was natural sunlight. The use of sunlight for the treatment of various skin diseases is effective due to its Ultraviolet (UV) component. A specific range of the wavelength of the UV band has biological advantages on the human skin. This specific UV radiation induces a variety of immune system changes inside the human skin that lead to skin rejuvenation. The introduction of this chapter includes detailed information about the natural and artificial sources of light, Human skin, the role of UV radiation in the treatment of skin diseases, its historical aspects, the nature and mechanism of UV radiation on human skin, and lastly based on the action and absorption spectra choice of a UV wavelength for specific skin diseases.

1.1 Sunlight: A natural source of light

A natural source of light available all over the world is sunlight. Mostly from the surface of the sun, it takes 8.3 minutes to get to the earth[3]. The sun's unique characteristic is that it generates electromagnetic radiation spanning the majority of the electromagnetic spectrum. In the solar system, the sun emits electromagnetic radiation covering a major portion of the electromagnetic spectrum. It includes gamma rays [10^{-4} nm to 10^{-2} nm], X-rays [10^{-2} nm to 10 nm], ultraviolet (UV) rays [10 to 400 nm], visible rays [400 to 750 nm], infrared(IR) rays [750 to 1 mm], microwaves [1 mm to 1 m], and, radio waves [$\geq$ 1 m][4] as shown in Figure 1.

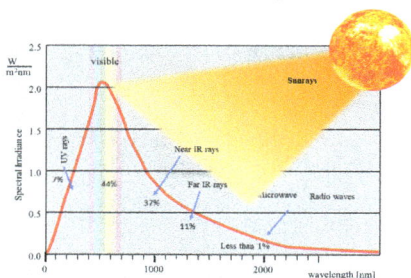

Figure 1: According to the solar radiation spectrum, UV rays make up a smaller percentage of the radiation that reaches the earth's surface than visible and infrared rays.

Since longer as well as shorter wavelength radiation output is less through the sun's surface, sometimes the short wavelength X-rays come from the hottest and active part of the sun and Gamma rays emitted during solar flares. As gamma rays generated during the fusion process in the sun's core are continuously absorbed by the solar plasma, they always reemit at the higher wavelength mostly UV, Visible, and IR through its surface. Hence, when the sun's surface sunlight strikes the earth's atmosphere, some portion of it gets absorbed while the rest proceeds toward the earth's atmosphere. This concludes that most of the sunlight includes infrared, visible, and ultraviolet parts of the electromagnetic spectrum. Despite making up a very small amount of sunlight, UV radiation is a very significant part of solar radiation. It has both positive and negative biological impacts[5]. Nearly half of the total radiation that the Earth's surface is exposed to is visible radiation. The Sun's total irradiance spectrum includes the visible band, which has the strongest output. The primary advantage of infrared radiation is that it generates heat. At the Earth's surface, infrared radiation makes up over half of all solar radiation. Sunlight has several health benefits, for instance, the formation of vitamin D, the ability to reduce high blood pressure, subsidiary bone health, regulate immune system functioning, etc. [6]. The three bands of sunlight namely UV, visible, and IR have various applications depending on the field. In this chapter, we are dealing with the light sources useful for the treatment of skin diseases. Out of three bands of sunlight, UV radiation has more potential and ability to cause changes in the immune system of human skin and helps in skin rejuvenation. The Importance of UV radiation is discussed below.

1.2   Ultraviolet (UV) radiation

Ultraviolet (UV) radiation was first invented by Johann Wilhelm ritter on 22 Feb 1801. The radiation is referred to be UV because its frequency is higher than that of visible light. It is a form of electromagnetic radiation that is non-ionizing in nature. Sunlight is the only naturally occurring source of UV radiation. It contains 5-6% of UV rays and the rest is visible and IR rays. These UV rays are strongly attenuated from the surface of the earth by its protecting shield commonly called the ozone layer. Thus, the amount of UV varies significantly with latitude and has contributed to several biological adaptations, such as variations in human skin tone around the globe[7]. As UV light has binary effects of positive and negative impact on human skin it helps in the making of vitamin D and a mutagen but is also responsible for sunburn, and sun tanning, and mainly causes skin cancer. Also, UV photons are antibacterial, they are utilized to disinfect water and instruments. Some man-made UV sources are tanning beds, mercury vapor lamps, halogen lamps, fluorescent lamps, and incandescent lights, as well as some types of lasers. International Commission on illumination classification subdivides the UV wavelength band (100-400 nm) into three categories namely UV vacuum, UVA, UVB, and, UVC, and their interaction with the earth's atmosphere is shown in Figure 2.

UV vacuum spans a range of 100 to 200 nm and the UV range below 200 is known as a vacuum was made by the German physicist Victor Schumann. The oxygen in the air substantially absorbs this wavelength range.

UVC range covers a UV spectrum from 200 to 280 nm. The most active wavelength value for UVC rays is 265 nm. Due to the small wavelength value the ozone layer absorbs the majority of UVC radiation before it reaches the earth's surface. Hence artificial sources are the only way to use UVC rays. As these rays possess more energy and hence penetrate less into human skin. Due to its characteristic germicidal effect, it is utilized in germicidal lamps to purify water, air from the industries where harmful and toxic chemicals are produced.

Rare Earth - A tribute to the late Mr. Rare Earth, Professor Karl Gschneidner Materials Research Forum LLC
Materials Research Foundations 164 (2024) 143-176 https://doi.org/10.21741/9781644903056-3

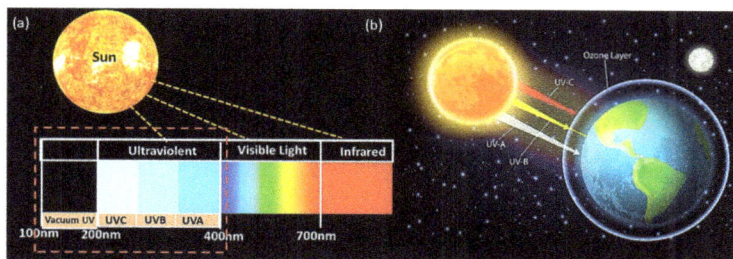

Figure 2: a) Classification of UV spectrum b) Behaviour of UV rays when passing through the Ozone layer.

UVB wavelength lies between 280 to 315 nm most of the UVB rays get absorbed by the atmosphere and contribute to the photochemical process that creates the ozone layer. Since it has higher energy than UVA rays, it can penetrate human skin to the epidermis. It directly damages DNA and causes sunburn, and suntan [5]. Moreover, it also promotes skin aging and considerably develops the risk of skin carcinoma. However, it is also necessary for the production of vitamin D through the activation of ergosterol in the epidermal layer of the skin [5].

UVA radiations cover the wavelength range of 315 to 400 nm. UVA radiation has lower energy in the spectrum hence it penetrates deeply into the human skin to the dermis. These radiations cause indirect damage to the DNA by forming free radicals and reactive oxygen species. The main effects of UVA radiation on the skin include wrinkles and skin cancer.

DNA integrity is crucial for the efficient operation of all cellular activities. UV light exposure changes DNA structure, which has an impact on all biological systems, including humans and microbes[8].

1.3 Human Skin

Human skin is the most sensitive organ of the body. The UV, visible, and infrared radiations that come from sunlight are mostly exposed to human skin. IR rays penetrate deeper into the skin, its large percentage is transmitted into the subcutaneous tissues and most of the part reaches till dermis. The IR rays are mostly absorbed by the water molecules present inside the skin and those molecules convert rays into kinetic energy which comes up on the skin as heat. Visible radiation helps to increase the amount of proteins, oxygen-active form, and DNA mutation in human skin. The UV rays are not visible to our eyes but human skin is most sensitive towards these radiation. The skin is made up of water, proteins, fats, and, minerals. The primary function of the skin is to shield the body from temperature, chemicals, bacteria, and pathogens[9]. Three layers of tissues make up the skin, they are the epidermal, the dermal, and, the hypodermal layer as shown in Figure 3. The epidermal layer, the uppermost portion of the skin is enriched completely with cells and thin in nature. It is composed of 5 layers of its own namely stratum basale, stratum spinosum, stratum granulosum, stratum lucidum, and stratum corneum.

Rare Earth - A tribute to the late Mr. Rare Earth, Professor Karl Gschneidner   Materials Research Forum LLC

Materials Research Foundations 164 (2024) 143-176   https://doi.org/10.21741/9781644903056-3

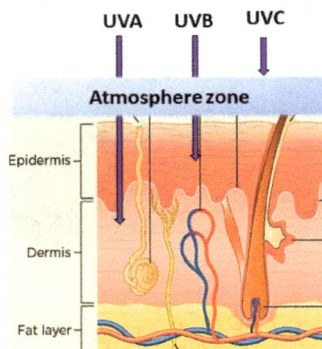

Figure 3: Cross-section view of human skin with UV effects.

The stratum basal cell layer is the deepest and usually lies above the dermal layer. T In this basal cell layer new cells namely keratinocytes, and melanocytes formed. These cells produce the protein keratin and melanin so the change in their metabolism is responsible for biological effects on human skin[10]. The stratum spinosum, stratum granulosum, and stratum lucidum make up the middle section and contribute to the flexibility and strength of the skin as well as contain keratinocyte granules[11]. The top layer stratum corneum is visible to the naked eye and comprises corneocytes, or dead keratinocytes, which serve as a shed for new keratinocyte cells. Collagen, elastic tissues, and other extracellular elements like veins, nerve endings, hair follicles, and glands make up the fibrous structure of the dermis layer. The dermis' primary functions are to maintain and shield the skin's deeper layers, aid in thermoregulation, and enhance sensation. The bottom layer of the skin is the hypodermis, which has a yellowish texture due to the presence of carotene pigment. Its thickness varies between individuals based on the amount of fatty tissue that accumulates in various body regions as a result of a person's hormones and heredity. The hypodermis layer connects the dermis layer of skin to the body's muscles and bones, insulating and safeguarding the body from any harm[12]. It also aids in the storage of energy.

Different wavelengths of the electromagnetic spectrum have different penetration inside the human skin as shown in Figure 4. As depicted in fig. lower and higher wavelengths have less penetration inside the skin over other wavelengths[13]. This concludes that the depth of the penetration increases with an increase in the wavelength. As IR rays have high wavelengths compared to visible and UV rays hence successfully penetrate the deep layers of the skin. In the UV regions, UVA and UVB also penetrate differently depending on their wavelength value[7]. UVB penetrates the epidermis but due to higher wavelength value, UVA penetrates deep into the dermis layer.

Rare Earth - A tribute to the late Mr. Rare Earth, Professor Karl Gschneidner

Materials Research Forum LLC

Materials Research Foundations 164 (2024) 143-176

https://doi.org/10.21741/9781644903056-3

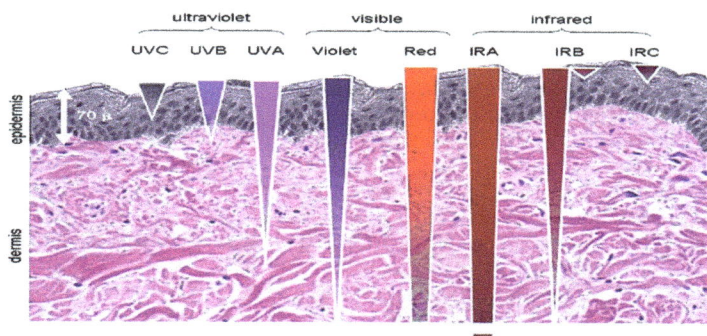

Figure 4: Variation in the penetration depth of UV, Visible, and IR rays into human skin. (Reproduced from the source: SCENIHR, Health effects of artificial light, 19 March 2012)

The distribution and activity of a few important immune system genomic components are altered by UVA radiation. The body's capacity to fight certain diseases is decreased as a result of changes in immunological metabolism. Also, UVA rays activate melanin pigment that exists in the upper layer of the skin that generates long-lasting tan also its deep penetration leads to losing the elasticity of the skin and causes wrinkles[14]. Even though UVB rays are universally accepted as a proven and efficient treatment option for an abundance of skin diseases. The excess amount of UVB exposure on human skin turns out into several hazardous symptoms also such as sunburn, suntan, the cloudy lens in the eye, wrinkle formation, immune suppression, direct DNA damage, and in the worst condition skin cancer[15]. Skin carcinomas are the most frequent disease seen in many countries with fair-skinned people. A fair-skinned individual has less amount of protective pigment called melanin in their skin. Compared to darker skin tones individuals have less chance of getting skin cancer due to the excess presence of melanin in their skin. Caucasian people have pale, poorly browned skin that burns frequently, hence the consequences of UVR on human skin are more obvious in them[16].

Vitamin D3 production is one of several positive health benefits of UVR exposure in humans. Vitamin D synthesis is encouraged by UVB exposure having a peak wavelength at 295 nm and an abrupt cut-off at 315 nm. A wavelength above 315 nm is not useful in the formation of vitamin D. In addition to supporting healthy bones, vitamin D is also thought to control cell proliferation and guard against cancer[17].

1.3.1 Classification of Fitzpatrick skin phototypes

The Classification of Fitzpatrick Skin Phototype (FSP) or Fitzpatrick scale is a numerical classification scheme for the skin tone of individuals. FSP was first established by Harvard University dermatologist Dr. Thomas Fitzpatrick in 1975. This scheme tabulates human Skin type determined by the quantity of melanin pigment present and how responsive the skin is to sun exposure[18]. This scheme helps in predicting an individual's overall disadvantage of skin damage and skin carcinoma. This well-known method's validity and reliability are highly acknowledged, and it has been used in numerous research projects. In the epidermal layer of skin, melanocytes formed the pigment melanin which is responsible for natural skin pigmentation in all human beings. The capacity of human skin to withstand UV radiation is subject to the amount of melanin present in the skin. Although the number of melanocytes in healthy skin is constant, the amount of melanin that is synthesized varies depending on genetics, which causes variable amounts of skin pigmentation. Patients with type I skin scores between 0 and 6 are flammable even in low UV intensity, while those with type VI skin scores between 35 and 36 will often not be injured but will tan rapidly given in Table 1[19]. To determine a patient's susceptibility to UV radiation and to guide specific treatments, such as phototherapy and laser operations, the doctor will need to assess the patient's skin type. During phototherapy, UVA/TL-01 starting dosages and increments can be chosen based on Fitzpatrick skin type[20].

Table 1: Classification of Fitzpatrick skin type

Sr. No.	Types	Skin pigmentation	Numerical score	Effects
1.	Type I	very fair skin/freckled/red hair	0-6	always burns, never tans.
2.	Type II	fair skin	7-13	usually burns, tans eventually
3.	Type III	fair to olive skin	14-20	Occasionally burns, tans easily
4.	Type IV	brown skin	21-27	rarely burns, tans easily
5.	Type V	dark brown skin	28-34	very rarely burns, tans easily
6.	Type VI	black skin	35-36	never burns, tans easily

Rare Earth - A tribute to the late Mr. Rare Earth, Professor Karl Gschneidner   Materials Research Forum LLC
Materials Research Foundations 164 (2024) 143-176   https://doi.org/10.21741/9781644903056-3

1.4   Light therapy or Phototherapy

Light therapy is an effective therapeutic invention for the treatment of plenty of skin diseases. This therapy includes a direct and controlled regime of non-ionizing radiations such as ultraviolet and visible light of different wavelengths on the damaged tissue to cause a change in its immune response[21]. Dates back Greeks people used sunrays for the first time as a treatment source for various therapeutic conditions. At the beginning of the 20th century, the epoch of confined phototherapy began. The father of modern phototherapy Niels Finsen discovered that cutaneous TB can be successfully treated with ultraviolet radiation from a carbon arc and natural sunshine[22]. For his ground-breaking achievement, he was awarded the Nobel Prize in medicine in 1903. Finally, in 1923, the use of artificial light sources in the care of individuals with psoriasis and other skin disorders gained acceptance throughout the world. Artificial light sources include fluorescent lights, halogen lamps, light-emitting diodes (LEDs), etc[23]. The basic needs for phototherapy devices are:

- Handy lamps or LEDs with a focused beam of light at one or a specific area of skin.

- Special units have a small opening for treating the hands and feet of patients respectively

- Booths or large panels that used for exposure to light

- Combs for scalp treatment

- Excimer lasers for highly intense beam treatment

These light devices emit light in the appropriate UV region so that patients get cured of particular diseases with minimal erythemal doses. The immune system cells in the skin under this UV light were inhibited. UV light helps reduce skin conditions caused by an overreaction of the immune system[24]. Regularly exposing the affected area to these lights helps in the reduction of inflammation and also prevents skin cells from growing quickly. To achieve an effective reaction without experiencing any negative side effects, each phototherapy session needs to be customized and adjusted by several factors[25]. Based on the mode of emission range UV therapy is subdivided into types namely UVB broadband and UVB narrowband phototherapy, photochemotherapy or PUVA, photodynamic therapy, and laser therapy. UVB broadband phototherapy exploits a broad range of UV radiation of wavelength 280–320 nm. UVB narrowband phototherapy wavelength ranges from 311±2 nm. UVB exposure can be applied alone or in conjunction with other topical treatments. Photochemotherapy or PUVA phototherapy is the amalgamation of a photosensitized drug and UV radiation of the appropriate wavelength range[26]. In this therapy, Psoralen generally acts as a photosensitizing agent along with UVA radiation. As UVB narrowband phototherapy is less carcinogenic than PUVA, hence effective in the treatment of psoriasis. Also, UVB narrowband phototherapy causes a less erythemal effect on the human skin during treatment than broadband UVB, so it is preferable. Due to all its characteristics and features UVB narrowband phototherapy has gained a lot of popularity in recent studies. Several specialized phototherapies exist, which include lasers, photodynamic therapy (PDT), bath-PUVA, and extracorporeal photochemotherapy.

1.4.1   Historical context

Since antiquity, people are suffering from several skin diseases without knowing liable treatment for them. Then outdoor sunlight therapy or heliotherapy came into the picture as an effective treatment course of action with ample biological benefits for the management of several skin diseases[27]. The ancient Egyptians, Greece, and, Rome has the earliest records of a connection between sun exposure and skin disease going back five millennia[28]. They used direct sun exposure to irradiate the afflicted skin portion of the individuals. The sunshine was also venerated as a god of wellness by the Inca, Assyrian, and early Germanic tribes[29]. Later in 525 BC Herodotus et. al. related the magnitude of sun irradiation to the vigor of the human skull[22]. Indian literature review dating to 1500 BC, for treating patients affected with several skin diseases used natural leafy herbs Bakuchi, more popularly known as Psoralea corylifolia contains black seeds to apply on patchy, non-pigmented areas of skin earlier than being exposed to sunlight. Buddhist writings dating back to 200 AD and Sung dynasty records from the 10th century also contain similar references[30]. Ibn al-Bitar in 1100 AD wrote a book *Mofradat El-Adwiya*, in this book he gave one more herbal panacea namely Ammi majus in combination with sunlight for the treatment of leukoderma effectively. This mentioned fact is the very first in the field of particular skin disease recovery, which is now frequently referred to as PUVA or photochemotherapy. According to a literature review, in 2000 BC people of Egypt generally used boiling extract of leaves, seeds, or roots of umbelliferous plants, such as bullwort (Ammi majus Linnaeus), to treat the affected areas of individuals suffering from vitiligo after letting the patient lie in the sun[31].In the short period, this beneficial panacea has been followed for the management of other diseases too in antiquity as far back as 1200–2000 BC by physicians and herbalists with specially designed UVA units. El Mofty uses ammoidin (8- methoxy psoralen (8-MOP)) and bergapten (5- methoxy psoralen (5-MOP)) extracted from the leaves of Ammi majus Linnaeus to treat vitiligo successfully[32]. A new era of 8-MOP photosensitizers used in conjunction with exposure to UVA radiations for the treatment of psoriasis was reported by Pathak, Fitzpatrick, Tanenbaum, and Parrish and they called it "Photochemotherapy or PUVA." The first use of oral 8-MOP in the treatment of psoriasis was reported in 1967. In the 19th century addition to this, Downes and Blunt demonstrated the antibacterial power of light in 1877. They demonstrated through their experiment that the growth of foreign invaders like bacteria and fungi is hindered when exposed to light[33]. In 1890 Theobald A. Palm reported that the sun could play a therapeutic role in rickets[34]. With successful results, phototherapy replaces heliotherapy at the end of the 19th century. Modern phototherapy techniques with remarkable outcomes are a legacy of the 20th century. In the early 20th century light therapy was promoted by Swiss physician and climatologist Auguste Rollier and American medical doctor John Harvey Kellogg. At the start of the 20th century in 1901 Niels Finsen, a Danish physician did a small experiment in which he demonstrated an amazing fact that refractive rays from sunlight and any electric arc have a restorative effect on human skin tissues[35]. If the irradiation is done in a controlled manner, then damaged tissues can be prevented from inflammation and other side effects. He originally employed the sun as a light source, creating instruments with quartz lenses and filters to block out IR and visible rays while concentrating the antibacterial UV rays on the tissue that needed to be treated. Soon an ineffective light source, for example, a UV radiation source, or a carbon arc lamp took the place of the sunlight[36]. These were soon superseded by sealed quartz mercury vapor lamps that were cooled by air or water, which were much more practical.

In 1903 Finsen received a noble prize for his grand and extraordinary invention of the treatment methodology for the lupus vulgaris. He has cured 80% of patients out of 800 with a well-designed artificial UVR emitting instrument in his lab[22]. Also, artificial UV radiation has been used to treat psoriasis effectively. Rollier was greatly impressed by Finsen's ground-breaking study. In

1903, Rollier established his own Institute of Heliotherapy in Leysin, where he coupled sunbathing with climatic treatments using cold air and high altitude. According to reports after World War I, he helped 1746 out of 2167 patients regain their condition[37]. The next approach made by archaic people was using artificial UV radiation in combination with topical tar and dithranol for various skin disease management. By the end of the first decade of the 20th century, phototherapy was widely used in therapeutic settings in northern European and northern American towns.

John Harvey Kellogg, a physician came up with several types of light-emitting devices made up of tungsten filament mainly for exposing patients locally or throughout their bodies to visible light[38]. Dinshah Ghadiali relates color theory with human psychology and finally, in 1920 he designed a Spectro-chrome instrument to treat various diseases by falling color beams of light on ill patients as shown in Figure 5. The only component of this device was a box with a tungsten light bulb inside and colored filters attached. Color therapy is not effective for treating skin diseases, except neonatal jaundice, because the majority of molecules in the human body are colorless and cannot be absorbed in the visible spectrum[39].

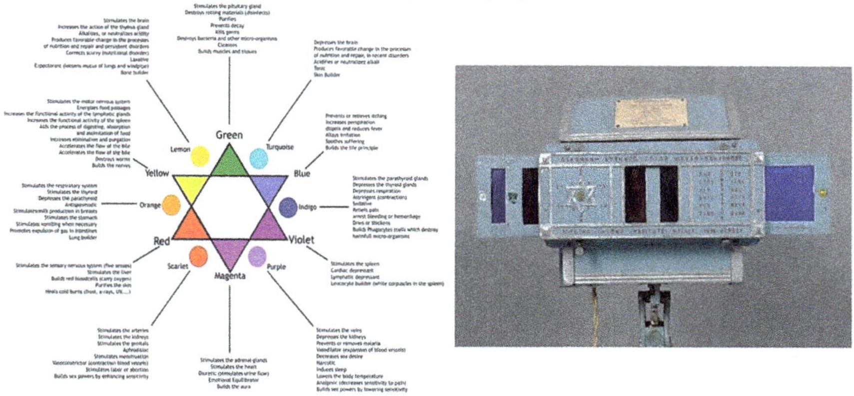

Figure 5: Spectro-chrome instrument designed by Dinshah Ghadiali. (Reproduced with permission from Massart Illustration, Feb 27, 2014)

Also, in the 1960s Wiskemann made Osram ULTRAVITALUX lamps with fluorescent tubes making an important contribution to the field of phototherapy[22]. This lamp specifically designs for the emission of light in the UVA region. The accessibility of fluorescent lamps generating UVA rays at an efficient intensity to enable suitable therapy over realistic irradiation intervals allowed a team from Harvard Medical School to publish a seminal study[40] in 1974, which transformed the practice of dermatology throughout the world. This seminal study immediately expanded to several nations and was utilized as a second-line therapy with positive outcomes. Today, photochemotherapy is a common and well-established kind of treatment, especially for the two most difficult skin conditions, vitiligo, and psoriasis. The majority of modern booths or chambers known as cabinets that emit synthetic UVA and UVB radiation have replaced natural sun therapies[41].

1.4.2   Mechanism of UV therapy

The pathogenesis of UV therapy is multifunctional and complex. It is the direct exposure of radiation of a suitable wavelength UV radiation on an affected area of skin to alter its profile. Depending upon exposure type various disease management has been done by phototherapy. Primarily when UV radiation irradiates human skin it is absorbed by the well-known endogenous chromophore, a UV absorber present inside the skin[42]. Various types of chromophores present inside the skin include deoxyribonucleic acid (DNA), urocanic acid, nucleotides, lipids, amino acids, and melanin causing alterations in their structural and functional characteristics. These changes are regarding the action spectra of incident wavelength and the absorption spectra of the chromophore in the skin[43]. The chromophore shifts into a more energetic electronic state under the right conditions, either the rigid singlet state or the more stable triplet state. The triplet state may last for a few seconds even if the singlet state lasts for an infinitesimally brief period, allowing for photochemical processes to occur. This reaction results in the production of various photo products and mediators that help in the modification of skin pigment[44]. The phototherapy mechanism encompasses four steps (a) Alteration of cytokine profile, certain cells of the immune system secreted several cytokines such as interferon, interleukin, and growth factors that affect other immune cells. UV radiation is responsible for the rise in the release of immunosuppressor cytokines for their profile alteration. (b) Induction of apoptosis planned cell death in response to external factors. (c) Promotion of Immunosuppression, reduction of the activation or efficacy of the immune system. (d) Anti-inflammatory effect[45]. Different wavelength of UV radiation has a different impact on the skin tissues.

The major targeted macromolecules by UVB radiation are nuclear DNA and proteins[45]. UV exposure causes the nucleotides in DNA to become cross-linked with the amino acids in these related proteins. The DNA-protein complex is important in the deactivation of bacteria by UV radiation [46]. DNA absorbing UVB leads to the formation of dimers of nucleic acid such as pyrimidine dimers and includes other photo products. These photoproducts suppress the synthesis of DNA in epidermal cells such as keratinocytes, melanocytes, and, Langerhans cells. This initiated inhibition of DNA transcription in the keratinocytes and T-cells that stops the cell cycle immediately and results in death. It also enhances the conversion of trans-urocanic acid (trans-UCA) into cis-urocanic acid (cis-UCA) and plays an important role in the reduction of the number of natural skin cells in a patient's body. The immune response pattern is altered by UVB irradiation, which increases the formation of reactive oxygen species and increases the synthesis and enzyme activity of the cellular antioxidant system[47]. The tumor suppressor gene p53, which is involved in cell cycle regulation, is upregulated by UVB radiation. Additionally, other macromolecular compounds, such as lipids, and polysaccharides, lack significant UVB absorption bands and are therefore not directly harmed by UVB light. Some micro molecules such as urocanic acid, melanin, etc. can also be a major absorbers of UVB radiation.

Photochemotherapy is the combination of applying the organic compound psoralene and then exploring the skin to the UVA radiation. Psoralen acts as a chromophore for only UVA radiation, as its activity is confined to those layers of skin where UVA radiation penetrates[48]. In the UV region UVA radiation has high energy so it is highly reactive toward human skin. Firstly, psoralene interpolates between the DNA strands. Upon irradiation, psoralene activates and reacts in two ways: (1) an oxygen-independent reaction that shows interaction with the cellular DNA which leads to the formation of mono products with a pyrimidine base. Theremin products are formed-

Rare Earth - A tribute to the late Mr. Rare Earth, Professor Karl Gschneidner   Materials Research Forum LLC
Materials Research Foundations 164 (2024) 143-176   https://doi.org/10.21741/9781644903056-3

cross-linked with DNA that is responsible for the impeding of epidermal cells and T cells causing cell death. (2) an oxygen-dependent reaction shows interaction with molecular oxygen and formed free reactive oxygen species (ROS). These ROS steal electrons from the lipids of the cells accusingly membrane damage. Psoralene interacts much more with proliferating cells than resting cells. The transfer of melanosomes from melanocytes to epidermal cells caused by PUVA's promotion of melanogenesis results in increased pigmentation, although neither the size nor the distribution of melanosomes is altered, the exact mechanism is not known till now[49].

1.4.3 Action spectrum and absorption spectrum

The action spectra refer to the rate of biological responses at each wavelength of light. More precisely it is depending on the absorption spectra of the corresponding substance[50]. It is very difficult to measure the level of biological response, as it is not linear with the incident intensity. It simply represents a specific wavelength at which the response is maximum for the management of the specific disease[51]. The radiation absorption spectrum shows the relation between overall absorption by the substance as a function of the different wavelength ranges. However, a suspected skin chromophore's action and absorption spectra do not usually correspond exactly because scattering and absorption within the skin modify the radiation before it reaches the chromophore. The rate of radiation delivery is the irradiance, expressed as power per unit surface area (W/cm2); whereas the exposure dose, i.e., total radiation energy delivered (radiant exposure), is the product of irradiance and time in seconds (J/cm)[41].

2. Skin diseases

Worldwide 70% of people suffer from skin diseases at some point in their journey of survival. These skin diseases are inflammatory, painful, and sometimes life-threatening. Skin disease classifications are mainly based on the symptoms and severity of the patient's health condition. Generally, all skin conditions are classified based on the duration of the period. According to that some are referred to as short-term or temporary skin diseases, and some as long-term or permanent skin diseases[52]. Some short-term skin diseases include acne, Sunburn, Contact Dermatitis, cold sores, Blister, and hives. Long-term skin diseases include psoriasis, neurodermatitis, vitiligo, palmoplantar pustulosis, mycosis fungoides, etc.

Short-term skin diseases are treated by using antihistamine tablets, steroid injections, ointments, etc. long-term skin diseases such as psoriasis, vitiligo, Ofuji's disease, atopic dermatitis, solar urticaria, atopic dermatitis, etc. are treated by therapies such as UV therapy, photodynamic therapy, PUVA, laser therapy, etc. Remaining skin diseases such as exfoliative dermatitis, seborrheic keratosis, herpes zoster, and skin cancers are treated by other treatment options by dermatologists.

2.1 Various skin diseases and treatments for their management

Here we have discussed several skin diseases that can be explained based on their definition, symptoms, causes, and mechanism along with the available treatment options.

2.1.1 Psoriasis

Psoriasis is now considered to be a genetically determined inflammatory systemic autoimmune disease. The term psoriasis comes from the Greek word 'psora', which means to 'itch'. It is defined as a long-lasting, nontransmissible inflammatory disease that negatively impacts patients' quality of life[53]. It affects any age to any sex but most probably shows the peak at 30-39 and 60-69 years in men and 20-29 and 50-59 years in women[54]. Psoriasis has wide-ranging appearances and it is estimated that 2–4% of the world population is affected, i.e., 125 million people. Each year, more than 2,00,000 additional people get the disease. Psoriasis majorly affects patients' physical, mental, and financial health. This chronic disease is not limited to human skin and has evidence of other severe modalities such as arthritis, type 2 diabetes, cardiovascular disease, depression, and so on. Psoriasis affects various parts of the human body followed by commonly known symptoms outlined in Table 2 along with its cause and trigger factors.

Although the pathophysiology of psoriasis is complicated, it is widely acknowledged that the imbalance of immune cells is the disease's primary mediator. Dysregulation of T-helper cell-1 (Th-1) and T-helper cell-17 (Th-17) plays a key role in the initial stages of psoriasis[55]. As psoriasis cause is multifactorial that also involves multiple potential susceptibility factors that increase the risk for the genetically vulnerable person[56]. The interaction of immune cells and chemical cytokines that affect skin cells and cause disease is the result of susceptibility factors and genetic predisposition. These complex mechanisms initiate due to the invasion of T-helper cells and dendritic cells. This initiation leads to the production of pro-inflammatory cytokines such as tumor necrosis factor-alpha (TNF-α), interferon-gamma (INF-γ), interleukin 17 (IL-17), IL-22/23/12/1β[45]. These inflammatory cytokines activate the most dominant cell of the epidermis i.e., keratinocytes and other fighter skin cells too. Keratinocytes are mainly known to form a protective shield against the internal and external barriers inside the epidermis of human skin. They grow from the innermost layer of the epidermis i.e., stratum Basale and slowly move to the stratum corneum, the upper layer of the epidermis. Keratinocytes become Corneocytes when they reach the surface after losing their nuclei while going upward. While this process takes roughly 23 days in healthy people, it only takes 3-5 days for keratinocytes to fully migrate in the skin of the psoriatic patient. Due to the keratinocytes' accelerated migration, they were unable to shed their nuclei, which prevented them from properly maturing[57]. Due to this they easily scraped off and start forming a large number of dead keratinocytes on the upper surface of the skin leading to edema, redness, and inflammation. One of the causes of psoriasis is the abnormal growth of a large number of sterile keratinocytes at the skin's surface[58]. The immune cells abnormality in the skin occurs in the various parts of the body based on that psoriasis has various classifications as follows:

Plaque psoriasis

Plaque psoriasis corresponds to the attacking of infection-fighting cells misguidedly on healthy skin cells. This causes new skin cells to grow much faster than normal, and they build up in thick patches. These patches are red, inflamed shields with a silvery-white build-up of dead skin cells (scales) which makes plaque psoriasis a common type of disease. Most probably it affects symmetrically on both sides of the body but is majorly seen on the elbows, knees, scalp, etc. The hue of these patches varies depending on how much melanin is present in human skin. Its scientific name is chronic stationary psoriasis or commonly known as psoriasis vulgaris. The cause of plaque psoriasis has not been confirmed yet but it is considered an autoimmune disease or sometimes it seems to be run in the families[59]. The factors that trigger the condition include infection, any skin injury, stress, tobacco, alcohol, etc.

Rare Earth - A tribute to the late Mr. Rare Earth, Professor Karl Gschneidner   Materials Research Forum LLC
Materials Research Foundations 164 (2024) 143-176   https://doi.org/10.21741/9781644903056-3

Nail psoriasis

Psoriasis can affect fingernails and toenails, causing pitting, abnormal nail growth, and discoloration. Psoriatic nails might loosen and separate from the nail bed (onycholysis). Severe disease may cause the nail to crumble.

Guttate psoriasis

It is a rare form of psoriasis identified by the various tiny, scaly, Salmon-colored to pink droplet-like papules on the skin follows by a throat infection. Guttate is a Latin word derived from gutta meaning drop. This form of psoriasis comes suddenly at an early age or in adulthood. Guttate psoriasis accounts for about 2% of all cases of psoriasis. Many papules cover a large portion of the body, starting with the scalp, limbs, and trunk. Scales of dry skin that peel or flake off the red patches also may appear. Genetics and bacterial infection are the two main causes of guttate psoriasis.

Inverse Psoriasis

It is also an immune-mediated disease that appears in the skin folds such as the armpits, belly button, under the breast, and groin. It looks like a shiny, smooth, discolored (brown, red, or purple) rash, and it may feel damp. The build-up of new cells replacing old cells creates a shiny rash. Inverse psoriasis is particularly subject to irritation from rubbing and sweating because of its location in skin folds and tender areas.

Palmo-plantar Pustular psoriasis

Palmo-plantar pustular psoriasis (PPPP) is characterized by multiple sterile blisters on the palms and soles. Blisters first appear as yellowish monomorphic lesions that turn a brown color with chronicity and associated scaling as the pus consists of white blood cells. Most patients with PPPP are smokers. It is also mostly seen in adults.

Erythrodermic psoriasis

This particular type of psoriasis covers almost the entire body with a periodic, prevalent peeling rash that can burn or itch strongly. It can be short-lived (acute) or long-term (chronic). Approximately 1 in 3 people who develop erythrodermic psoriasis already have plaque psoriasis.

Psoriatic arthritis

Patients suffering from psoriasis have a major possibility of getting psoriatic arthritis. Server joint pain, laboriousness, and, inflammation with silvery dead cells are the major symptoms of psoriasis arthritis. Most people experience psoriasis for several years before being notified they have psoriatic arthritis. However, for some people, joint issues start either concurrently with or before skin patches develop. They can range in severity from mild to severe and can affect any area of your body, including your fingertips and backbone. Psoriatic arthritis has the potential to be incapacitating and devastating without treatment. In both psoriasis and psoriatic arthritis, disease flares can alternate with periods of remission[60].

Rare Earth - A tribute to the late Mr. Rare Earth, Professor Karl Gschneidner   Materials Research Forum LLC
Materials Research Foundations 164 (2024) 143-176   https://doi.org/10.21741/9781644903056-3

Table 2: Psoriasis major affected areas, causes, symptoms, and trigger factors.

Affected area	Common symptoms	Cause	Trigger factors
Knees, elbows, trunk, scalp, lower back	Red Rashes, dry, itchy, scaly, small bumps, dry skin, thickened, flakiness, peeling, inflamed tendons, itching, joint stiffness, plaque, or small dents in nails	The pathogenesis of psoriasis is multifactorial and includes Autoimmune-mediates disease. Genetics	Injuries to the skin like cuts, wounds, insect bites, drinking alcohol, smoking, stress, hormonal changes during puberty and menopause, etc.

2.1.2 Vitiligo

A widespread, persistent, noncontagious condition known as vitiligo results in the loss of melanocytes in the skin and mucosa. Differences in skin color among individuals are caused by variations in pigmentation. The uppermost layer of the human skin is the epidermal layer which is enriched with keratinocytes, melanocytes, and Langerhans cells that plays important role in various immunological activities. Out of which melanocytes cells are responsible for the production of pigment commonly known as melanin which is responsible for the human skin color[61]. The formation of melanin by each individual's melanocytes determines the color of the skin, mucosa, and hair on the entire body given in Table 3. Sometimes body treated healthy cells as a foreign bodies like viruses, and bacteria so that it changes their response and starts overreacting towards the production of antibodies. This process mistakenly kills melanocytes that there working stop this leading to cause vitiligo[62]. Genetic issues, stress, and environmental factors can also sometimes cause disease. This disease makes skin color lighten or fully white compared to the natural skin color. If the loss of the skin pigment area is less than 1 centimeter it is termed a macule or if it is more than this it is a patch. It affects individuals of all skin types but is effectively visible in the individual of brown and black color.

Patients with vitiligo have significantly reduced quality of life due to the disease's severe consequences on their personality and social life. Loss of skin pigment can occur in any part of the body specifically hands, face, arm, feet, inner linings of mouth, lips, nose, and opening areas of the body, and these patches grow with time. It also causes early-age whitening or greying of the hair on the scalp, eyelashes, eyebrows, or beard. It affects an individual's mental and social life a lot. In unusual circumstances, vitiligo is irreversible until death, however, it is treatable, and vitiligo treatments can restore color to the damaged skin[63]. The chances of affecting individuals with vitiligo get increase if they are suffering from diseases like Addison's disease, sickle cell anemia, type 1 diabetes, psoriasis, etc.

Although vitiligo can begin at any age, it often manifests before the age of 30. Classification of vitiligo according to the areas of the body that it has affected.

Universal vitiligo:

With this type of vitiligo occurs in nearly all skin surfaces. Skin losses its pigmentation all over the body.

Generalized vitiligo

It is a well-known type and allows symptoms to visible on many parts of the physique. The discolored patches frequently similarly develop on homologous body regions.

Segmental vitiligo

In these cases, symptoms appear only on one side or specific part of the body. This type tends to occur at an adult age, progress for one or two years then stop.

Localized vitiligo

Symptoms of this type can be seen only in a few parts of the body.

Mucosal vitiligo:

It affects the inner lining of the mouth and genitals.

Trichome vitiligo:

This type causes a bullseye with a white or colorless center, then an area of lighter pigmentation, and an area of your natural skin tone.

The face and hands:

This type of vitiligo, also known as acrofacial vitiligo, affects the hands, face, and the skin surrounding body openings like the eyes, nose, and ears. It's challenging to forecast how this illness will develop. Without therapy, the patches may occasionally stop developing. The majority of the time, pigment loss affects the majority of the skin as it progresses[32]. The skin occasionally regains its color.

Table 3: Vitiligo's major affected areas, causes, symptoms, and trigger factors.

Affected areas	Common Symptoms	Cause	Trigger factors
Hands, feet, arms, face, genital areas, mucosa	White or lighten color patches on the various parts of the body	• Autoimmunological change • Stress • Genetical factors • Environmental factors	Injuries to the skin like cuts, wounds, insect bites, drinking alcohol, smoking, stress, hormonal changes during puberty and menopause, etc.

2.1.3 Polymorphic light eruption (PMLE):

A polymorphic light eruption is a typical skin rash brought on by exposure to natural sunlight or artificial UV radiations or genetics. The "Polymorphous" refers to how it appears differently in several individuals. The term "eruption" describes the rash's abrupt onset, which often occurs 30 minutes after UV radiation exposure. UV radiation may alter a compound in your skin and your immune cells combine with the new compound[64]. According to the genome-wide analysis, four genes such as complement 1s subunit (C1s), scavenger receptor B1 (SCARB1) fibronectin (FN1), immunoglobulin superfamily member 3 (IGSF3), caspase-1 (CASP1) and paraoxonase 2 (PON2) are responsible for the formation of the auto-antigen. After being exposed to UV light, an unidentified photo antigen becomes allergenic. It is an autoimmune-mediated photosensitive disease. The T-cell function is not reduced by UV radiation in patients with polymorphic light eruptions until after photo-hardening has taken place. The photo-hardening effect corresponds to where further UV exposure does not cause PMLE. Although there is a chance of involving whipping, hyperkeratosis, and acanthosis nigricans of the epidermis, and/or the development of immunological tolerance[65].

The effect of UV includes the formation of inflammatory cytokine IL 1 and also leads to the preponderance of T-cells. In a genetically predisposed individual, it is believed to be caused by a body-produced type IV delayed-type hypersensitivity to an allergen after exposure to sunlight. Additionally, it is believed that skin microbiome or microbial components may contribute to the pathophysiology of the illness. The rash typically manifests as small, inflammatory pimples or slightly elevated regions of skin that itch and burn. Sometimes tiny bumps may appear bloody and scary. It appears usually on the face neck, forearms, and legs given in Table 4. It is also called sun allergy, prurigo aestivalis, or sun poisoning. The reaction typically occurs when skin is exposed to sun rays often happens in the spring and early summer. Mostly it is visible in lighter skin than brown or black skin. It generally starts in females between the ages of 20 and 40, but it can also occur earlier or later in life.

Table 4: PMLE major affected areas, causes, symptoms, and trigger factors.

Affected areas	Common symptoms	Cause	Trigger factors
face neck, forearms, and legs	Inflamed, raised rough patches, Itching or burning	• Oxidative stress • Photosensitizer • Genetics • Oestrogen effects	insect bites, drinking alcohol, smoking, stress, etc.

2.1.4   Phototherapy in HIV infection

The fact that both phototherapy and photochemotherapy can cause immune system suppression, which might aggravate HIV status, makes this a contentious topic[66]. As the HIV regulator becomes active, virus generation and infectious gene regulation may increase [67]. Additionally, it has been suggested that HIV-induced immune suppression may speed up skin cancer growth as a result of UV radiation[68][69]. On the other side, several studies indicate that UVB and PUVA are useful for treating HIV-related psoriasis and itch. However, to draw the proper conclusions, long-term observations are necessary[70][71]. According to the information at hand, UVB is more likely to be hazardous to HIV-positive individuals than PUVA.

3.   Treatment options for curing skin diseases using Phototherapy

Depending on the severe condition of the individual and the necessity, different types of treatments have been used for the management of skin disease. Primary tests are carried out to determine the types of skin diseases. Based on the intensity and severity of the skin disease secondary treatments are available for the diagnosis.

3.1   Primary tests for the identification of the skin disease type

Visual representation is unable to identify the type of skin disease. Knowing the types of diseases that need to be treated is essential for receiving an accurate diagnosis for management. This can be done by using various primary diagnostic tests for instance:

Rare Earth - A tribute to the late Mr. Rare Earth, Professor Karl Gschneidner   Materials Research Forum LLC
Materials Research Foundations 164 (2024) 143-176   https://doi.org/10.21741/9781644903056-3

Biopsy: a shrill and minor layer of skin is examined underneath the microscope during these tests.

Culture: The skin sample is tested for bacterial, or virus infection.

Skin patch test: Test for any allergic reaction by applying a small amount of any organic compound.

Wood light test: This test includes UV exposure to clearly see the skin's pigmentation.

Diascopy: Press a microscope slide on a patch of skin to observe any color changes.

Dermoscopy: Hand-held device dermatoscope is used to determine skin lesions.

Tzanck test: This test is useful for checking for herpes simplex or herpes zoster in the fluid from a blister.

Several options are available for treating skin diseases after diagnosis they are antibiotics, antihistamines, laser skin resurfacing, medicated creams, ointments or gels, moisturizers, oral medications (taken by mouth), steroid pills, creams or injections, etc, however, some of them are not useful for long term chronic conditions and some of them have adverse side effects hence UV therapy or phototherapy is used for its effective and proven approach towards the management of almost various skin conditions along with some other mortalities. Many systemic and topical therapies are available for the diagnosis of skin diseases as detailed below.

3.2   Ultraviolet (UV) therapy

Ultraviolet (UV) therapy is a therapeutic method for the management of various skin conditions, along with some other conditions such as depression, Jaundice, diabetes, etc. in an individual. It is also called UV phototherapy or UV light therapy. It consists of exposure of the individual to the UV band of various wavelengths depending upon the type of disease to be diagnosed ultraviolet light. It is also said that it works in various ways such as enhancement in the collagenase activity, increase in the formation of INF-γ, and decrease in TGF-β[72]. The UV band of the electromagnetic spectrum has been renowned for its medical application for decades, although the use of artificial light allows the precise dosing of radiation from the past century with UVA in 1974, UVB narrowband therapy in 1984, UVAB in 1985, and UVA$_1$ during 1992. Additionally, this therapeutic modality is well-known for killing bacteria and indorsing wound healing. Hence known as a promising and effective therapeutic option for several modalities. Psoriasis, vitiligo, atopic dermatitis, mood and sleep disorder, diabetes, etc, are normally treated by UV therapy with effective results.

In the electromagnetic spectrum, the UV band lies between x-rays and the visible region that is well known for its non-radiation and high energy (4-12.4 eV) characterization. UV rays have the capability of reflection, scattering, and absorption by endogenous chromophores. Due to its high energy, it has a large degree of absorption compared to the other wavelength of the electromagnetic spectrum[73]. The degree of absorption generally depends upon the width of the epidermal layer and the production of melanin by melanocytes in the human skin. The extent of the reaction or damage depends on the wavelength of ultraviolet and the amount of ultraviolet absorbed. Absorption of UV radiation characterizes by the broadening of the epidermis and the skin color of an individual. Lesions will cause the skin to thicken, which will enhance dispersion and absorption while decreasing penetration.

Each region of the UV band has its biological effect as UV-C is well known for its germicidal effect, UV-B is specifically used for the management of various skin dermatoses with fewer side effects, and UV-A is characteristic of wound healing and infant jaundice. Based on the UV band region UV therapy is broadly classified into the following types such as UVB broadband phototherapy, UVB narrowband phototherapy, UVA phototherapy or PUVA or photochemotherapy, photodynamic therapy, and Laser phototherapy. The dermatologist specifies the type of phototherapy that must be utilized based on each individual's unique skin condition.

3.2.1　Types of phototherapies

The sunlight emits a wide range of UV bands that consist of various ranges such as UVA, UVB, and UVC. Through the ozone layer, the earth is protected from UVA and UVB rays, but UVC is absorbed. The UV spectrum emitted by the sunlight is broad and includes useful as well as hazardous wavelengths to avoid these and to use only specific wavelengths various light devices emitting comparable wavelengths to the sunlight are available in the market for artificial UV therapy.

UV-A phototherapy

UV-A wavelength ranges from 320 to 400 nm, as DNA molecules cannot readily absorb it. It causes indirect damage to the DNA molecule by absorbing photons through non-DNA chromophores. Absorption of UV-A radiations causes the formation of reactive oxygen species (ROS) that leads to the oxidation of DNA and damage to its protein by mutation[73]. The UV-A range is further classified into two types namely UVA_1 ranges from 340 to 400 nm and UVA_2 ranges from 320 to 340 nm.

UVA_1 has the lowest energy within the UV band hence responsible for the caspase-mediated cell death of Langerhans cells, mast cells, etc. Also, prevent the action of various cytokines activities such as IL-5, 13, 31. The epidermis, medium, and deep dermal components, particularly blood vessels, are all reached by UVA_1 (340 to 400 nm). UV therapy with UVA_1 not required any organic compound as a chromophore. UVA_1 is thought to be efficient in the treatment of skin disease by inhibiting various immunological functions. Protocols for applying UVA_1 vary depending on the condition being treated, but treatment is often administered three to five times weekly at dosages beginning at 20 to 30 J/cm^2 and gradually increasing [74]. A useful alternative for the treatment of autoimmune and inflammatory illnesses is UVA_1. This method may be used either alone for treatment or in conjunction with other traditional treatment options. It can be used on youngsters, pregnant women, and those who should not be exposed to any organic compound. As PUVA includes a broad UVA band that includes UVA_1 and UVA_2 both hence UVA_1 therapy has fewer side effects compared to PUVA.

Generally, UVA radiation is useful in the management of Atopic dermatitis, Cutaneous T-cell lymphoma, Urticaria pigmentosa, Lupus erythematosus, etc. Additionally, UVA_1 stimulates fibroblasts to create matrix metalloproteinases, which degrade extracellular matrix excess collagen. This is why it helps treat skin diseases that cause sclerosing (scar-like) issues. This therapy includes metal halide lamps equipped with specially designed optical fiber. To cover the whole-body dosage can be given by lie-down or standing in $UV-A_1$ cabinets.

$UV-A_2$ is the same as UV-B it penetrates superficially till the epidermal layer. In terms of sensitivity, immune regulation, and amounts of radiation, UVA_2 is similar to UVB. $UV-A_2$ has

Rare Earth - A tribute to the late Mr. Rare Earth, Professor Karl Gschneidner   Materials Research Forum LLC
Materials Research Foundations 164 (2024) 143-176   https://doi.org/10.21741/9781644903056-3

more erythemal effects compared to the UV-A_1 wavelength due to its more energy value[75]. Recent research on the meticulous role of the UV wavelength domains in a nickel model of recall contact hypersensitivity has shown that solar-induced immunosuppression peaks at 300 nm for UVB and 370 nm for UVA$_1$ as shown in figure 6. The immunosuppression caused by UVA$_1$ photons, which make up the majority of sunshine and can be absorbed during routine everyday activities, is three times more than that caused by UVB rays at moderate levels[76], although UVA$_2$ has not been demonstrated to inhibit the immune response in humans[77].

UVA wavelength is in combination used with an organic compound psoralene commonly known as PUVA or Photochemotherapy. Psoralene makes skin sensitive to UVA radiations. It interacts with the skin in three steps. In the absence of UV rays, psoralene interpolates with the DNA. Secondly, they take up photons after being exposed to UVA, become triggered, and with mutual sharing attach to the DNA bases. As a result, there is a formation of cross-linked pairs which shows antiproliferative, antiangiogenic, and apoptotic effects. Third, when activates psoralene combines with oxygen results in the formation of ROS that is responsible for damage to the cell membrane[78]. This DNA psoralene cross linkage prevents DNA replication and results in cell apoptosis and alters the cytokine profile. The amount of proliferating epidermal cells can be decreased by PUVA. The unmatured lymphocytes are strongly repressed by photochemotherapy. UVA exposure to the skin also helps in the enhancement of the formation of melanogenesis. This wavelength is well known for Killing T-cells, increasing the formation of collagenase-1 in the dermal layers, and decreasing the production of collagen I and III which is responsible for an anti-fibrotic effect[79]. PUVA involves the use of 8- methoxy psoralen and 5-methoxypsoralene approximately 2 hours before irradiation of UVA wavelength generally two or three times a week. The dosage of radiation increases every week until the visibility of a small erythemal effect on the affected portion [80][74]. After the session, skin and eye photoprotection must be secured for 24 hours.

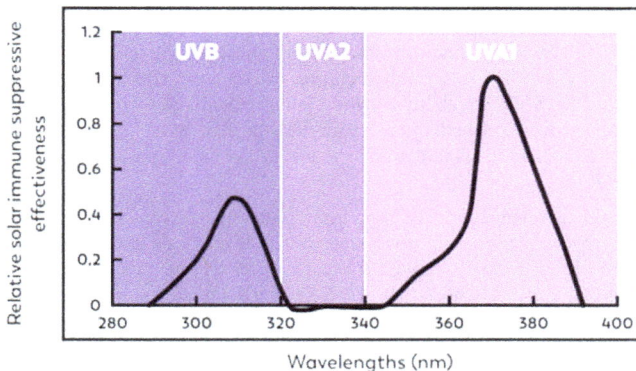

Figure 6: Solar effectiveness for immunosuppression in humans shows that compared to all UVA$_1$ have a predominant immunosuppressive effect. (Reproduced with permission from Ref [75, 97])

PUVA is usually used in the management of diseases like eczema, psoriasis, graft-versus-host disease, vitiligo, mycosis fungoides, large plaque parapsoriasis, and so on.

UVB phototherapy

UVB radiation in the UV band ranges from 285-320 nm, absorbed by the epidermal and upper portion of the dermal layer. Further, it is classified into two types UVB broadband phototherapy, and, UVB narrowband phototherapy. UVB Broadband phototherapy requires more exposure time with a broad range of wavelengths. Initially, broadband UVB (290-320nm) is used for the diagnosis of various skin conditions such as psoriasis, vitiligo, etc., later it is found that wavelength at 311 nm is more effective and superficial in the clearing of psoriasis and vitiligo effectively with fewer side effects[81]. Narrowband refers to the narrow region of UVB radiation ranging from 311 ± 2 nm. Currently, narrowband UVB phototherapy is the most effective component for the management of psoriasis; it is also effective for many other dermatoses, including atopic dermatitis, PMLE, vitiligo, solar urticaria, lichen planus, etc.

The exact mechanism of UVB interaction with skin cells is multifunctional and complex. A lowering in pro-inflammatory cytokines, a reduction in innate immunity inhibiting the activity of Langerhans cells, and a series of biotic events that result in the deactivation of T-cell mediated immune system functioning are all caused by NB-UVB in the treatment of various skin diseases[82]. The possibility of a better psoriasis cure has improved with Philips' creation of UVB narrowband fluorescent lamps (Philips TL-01). The TL-01 produces UV at a wavelength that is therapeutically ideal for psoriasis, allowing for quicker clearance and longer remissions[83]. Due to its shorter treatment time, UVB 311 nm has a much-reduced carcinogenic effect. Being a cost-effective and safe therapeutic option UVB narrowband phototherapy sets a revolutionary controlled delivery route for the diagnosis of several skin alignments[84][85].

Targeted phototherapy

This type of phototherapy also known as Laser therapy includes the use of a 308 nm excimer or lamp. It's been employed in dermatology since nearly its beginning as of its excellent degree of correctness, efficiency, and low frequency of clinical effects. A device that emits electromagnetic radiation is called a LASER, which stands for Light Amplification by Stimulated Emission of Radiation. A laser is distinguished by the emission of light at a specific wavelength that is monochromatic and homogeneous in nature. The use of laser light particularly in the therapeutic field began in 1997.

The monochromatic wavelength of 308 nm emitting excimer laser whose range is close to the narrowband UVB region was used for the management of psoriasis and various skin diseases in various countries. It works well for a variety of other chronic and localized inflammatory dermatoses that affect only 10% of the body's surface. This therapy can specifically be used for locations that are challenging to reach with other therapeutic choices for example the scalp, genital areas, etc [86]. Using a spot-on hint between 14 and 30 mm in diameter, laser phototherapy is applied to the lesion while avoiding healthy skin. Because of this, the treatment only needs a few sessions, which reduces the long-term adverse effects brought on by other types of phototherapies.

Excimer laser emits xenon and chloride gas that in combination produce unstable "excited dimers". The excited dimers break down for the emission of the monochromatic wavelength of value 308 nm, which reaches epidermal cells and fibroblasts[87]. On the other side, the excimer lamp emits light irregularly and takes longer than the laser to emit the same frequency. It has the benefits of being simpler to transport, allowing the treatment of larger regions, and having lower operational expenses[88]. The number of treatments required to completely eradicate psoriasis is drastically reduced by these approaches and systems, from 25 to 5–10. The techniques and technologies applied also minimize the period and storage space for each treatment, which further lowers the price of phototherapy. For instance, since only the associated disorders are treated, the doctor should not worry about sunburn on the skin that is not affected. Hence there is no need to calculate the radiation exposure related to one minimal erythemal dose as in other types of phototherapies.

Photodynamic therapy

Photodynamic therapy (PDT) includes the combination of light and photosensitizer for the production of highly reactive oxygen radicals for the management of skin diseases. UV light causes the PDT drug to be excited into the triplet state, resulting in the formation of free radicals which provide molecular oxygen energy to generate singlet oxygen. PDT produces oxygen intermediates by using longer-wavelength light with greater penetration[89]. Porphyrins, ALA, and benzo porphyrin derivatives are photosensitizing compounds used to treat dermatological diseases. Free radicals' formation causes cell membrane damage that results in the apoptosis of cells. PDT is used for treating Basal cell carcinoma, Squamous cell carcinoma, Cutaneous T-cell lymphoma, and, Kaposi's sarcoma. PDT is also effective for non-malignant illnesses including psoriasis, genital warts, and actinic keratoses. Since tumor cells vary from normal cells in their vasculature, the medication preferentially localizes in them, making it more effective against these disorders[90].

Ultraviolet combs

UV combs are specially designed devices usually use for the diagnosis of a scalp-corresponding disease like scalp psoriasis. Sometimes UV combs also use in the management of seborrheic dermatitis. This therapy corresponds to the direct exposure of UV radiation to the scalp. The device is handy and identical to household combs, at easy to disinfect. Although scientific research on the therapeutic efficacy of the majority of devices is absent, there are no occurrences of acute or long-term negative effects following the appropriate application of the technique[91].

Home phototherapy

In the comfort and privacy of the patient's home, home phototherapy is an alternative non-drug option for the self-treatment of skin conditions. Home phototherapy with hand devices that emits UVB radiation is usually suggested for an exceptional individual, who exhibits good mental capacity and treatment compliance. However, several global elements have a detrimental impact on the recommendation of this therapy, difficulty in handling the device or the amount of dosage, and the specific distance from the phototherapy unit. When traditional phototherapy is not possible for the individual, less typical phototherapy techniques like heliotherapy (exposure to sunshine), whether or not psoralen, have been suggested. Additionally, home LEDs and tanning beds with blue light emission may be effective in curing psoriasis. The use of protective glasses, gloves, and blankets for covering remaining areas and other prevention is strictly necessary during home phototherapy[83].

4.   Light Sources for UV therapy

In the realm of medicine, light-based equipment is frequently used for surgical and diagnostic operations, medical photography, phototherapy, etc. The use of light in dermatology goes back to

Rare Earth - A tribute to the late Mr. Rare Earth, Professor Karl Gschneidner   Materials Research Forum LLC
Materials Research Foundations 164 (2024) 143-176   https://doi.org/10.21741/9781644903056-3

ancient times in Egypt when sunlight was a ground-breaking source that was presumably employed for the first time to treat skin conditions. UV therapy involved the use of light radiation for the diagnosis of the physical and mental health of individuals. Illumination parameters such as wavelength and dose of therapy can be varied to have distinct effects on cells and tissues. Afterward, day-by-day growth in medical science along with the hands of innovative techniques resulted in the development of new and more sophisticated options in the replacement of available light-emitting sources. In the market various light sources are available for the treatment . The detailed information about the light sources is given below:

Arc Lamp

It is commonly known as a gas discharge lamp. It produces light due to an electric arc. It consists of a transparent envelope containing two electrodes with plasma between them. This envelope is filled with gases such as mercury, xenon, etc. When high voltage is applied across the electrode electron started the movement as they reach the excited state. After a known interval of time, they come back to the ground state by the emission of light. The gas used in the lamp decides the spectrum of output[92]. A low-pressure mercury arc lamp is a type of popular fluorescent lamp. Solar simulators employ xenon bulbs that specifically emit white light.

Halogen lamps

It is also called an incandescent lamp, made up of a transparent envelope containing tungsten filament in it. The transparent envelope is occupied with a mixture of inert gas and less amount of halogen including iodine or bromine. The early development of halogen lamps is comparable to that of incandescent light bulbs. In 1882, the method of using chlorine to keep lamps from turning black was patented. Iodine was used as the halogen gas in the halogen lamp developed in 1959 that was commercially successful. The halogen lamp works on the principle of halogen-cycle, hence differentiate from the incandescent bulb. In halogen lamps, halogen gas plays an important role in combining functioning with the filament of the lamp. Halogen gas chemically reacted with the evaporated tungsten filament preventing it from sticking to the glass surface. The evaporated tungsten filament is redeposited back to the filament and increases the life of the halogen lamp[93]. This helps to produce light of efficient luminescence intensity at a higher temperature compared to the normal tungsten lamp. Halogen lamp spectrum ranges from near UV to deep IR radiation. The operating temperature of the halogen lamp is high hence emitted light is of high-intensity value the shifted towards blue. This makes the halogen lamp spectrum equivalent to the spectrum of the sun. Generally, halogen lamps are useful for the treatment of infant jaundice. Due to their high-temperature effect, quartz halogen lights are made to be kept away from young children during the treatment[94].

Mercury vapor lamp

It is a gas-discharge lamp that generates light by passing an electric arc through evaporated mercury. It is invented by peter in 1903. This lamp is specifically designed to emit UV light for medical applications. Mercury vapor lamps are categorized into 2 types:

Alpine sunlamp

It is air cooled mercury vapor pressure lamp. This lamp consists of a U-shaped tube filled with halogen gas like argon at low pressure. U-tube is sealed from both sides and a small amount of mercury is poured into that tube. Also contains a burner made from quartz material connected to the tube. A quartz burner allows UV radiation to flow through it at high temperatures. Electrodes are put in metal caps at the end of glass tubes. A high potential of value 400V is applied across metal caps with the help of a step-up transformer. As voltage supplies from the tube argon gas start ionizing resulting in the electric current flow through a tube. Metal caps consist of positive and negative electrodes hence electron moves towards the negative electrode and vice-versa. When these ions move through the tube, they collide with the argon gas resulting in the ionization of the gas. Ionization of argon gas generated UV radiations. Also, electric current flow through the tube generates heat that makes mercury vaporize. This also causes the ionization of mercury due to the flow of current. In this kind of lamp UV radiation is produced due to argon ionization, mercury ionization, and vaporization. Argon and mercury ions combine again when the lamp is switched off, bringing everything inside the tube back to a neutral state.

Kromayer lamp:

It is a water-cooled mercury vapor lamp. It is UV lamps that reduce the risk of burns and absorb IR rays. Distilled water is flowing all around the lamp to absorb IR. To chill water, a pump and a cooling fan are integrated within the lamp's body. Following use, the burner should be turned off to allow the lamp to cool for five minutes. Water is pumped back and forth between two quartz windows at the lamp's front, allowing UV to shine through.

Fluorescent lamps:

As mercury vapor lamp is known to produce shot-wavelength UV rays that have several erythemal effects to avoid these there is a need for a lamp that produces long UV wavelength. Hence, various types of fluorescent tubes have been designed. Each fluorescent tube is comprised of a long-wave ultra-violet-transmitting glass and is roughly 120 cm long[95]. Each fluorescent tube has a unique phosphor coating on the inside, and the type of coating determines the spectrum of each tube. If any short UV light is generated in this situation, it is absorbed by the phosphor and emits light back at a longer wavelength. The output of the tube may be partially UVB and partially UV-A or UV-A, like in the PUVA system, depending on which specific phosphor is employed, although precise control of the emitted wavelength is feasible.

Theraktin tunnel:

Theraktin tunnel is a semi-cylindrical lamp with four fluorescent bulbs put inside. Each tube is mounted in its reflector, resulting in uniform radiation production that enables treatment of the entire body to be evenly distributed over two halves.

Cold-Light UV-A1

This lamp has a filter inserted to block out light with a wavelength of 530 nm, and it also distributes the high heat load produced by the UVA_1 generator. When it comes to removing lesions and shortening the duration of atopic dermatitis (AD) flare-ups, this form of UV light is more efficient than UV-A alone or standard UVA_1.

PUVA apparatus:

When a lot of UVA is needed to treat psoriasis, it is typically employed for that purpose. In PUVA, the tubes are typically installed in a vertical battery on a wall or on all four sides of a box that encloses the patient. The term PUVA (psoralen ultraviolet A) refers to the type of ultraviolet radiation that is often administered two hours after the patient has taken a photoactive medication like psoralen.

Excimer

The excimer emits UV light due to the spontaneous emission of excimer molecules. Excimers are polyatomic or diatomic molecules that have unbound or weakly bound ground states and stable excited electronic states. It is a source of quasimonochromatic light of wavelength 308 nm. A combination of excited gases called an excimer releases extra energy as UV radiation when it decomposes. It is available in both the form lamp as well as laser. The excimer lamp is incoherent, and monochromatic and emits light in between the range of 306-310 nm. The excimer laser emits light of wavelength 308 nm, which is of a single wavelength, pulsing and delivering as targeted phototherapy. It is commonly used for the curing of psoriasis, atopic dermatitis, and vitiligo[96], and has also been used in the treatment of early-stage mycosis fungoides. In the prurigo variant of atopic dermatitis (AD), excimer laser treatment for 10 weeks has been demonstrated to be more effective than clobetasol propionate. Along with this excimer lamp produces a 207 nm wavelength that is useful for germicidal activity. This excimer lamp/laser has a lot of advantages such as high energy emitted photons, absence of visible and IR radiation, low heating of radiating surface, Eco-friendly, etc.

TL-01 Ultraviolet Lamp

This lamp has been demonstrated to be particularly useful for treating psoriasis since it generates a narrowband UVB spectrum at 311 nm. The narrow-band, TL-01 UVB air-conditioned lamp is a modified version of this lamp that has been designed and tested on several patients suffering from atopic dermatitis. The majority of the patients treated with this UVB narrowband phototherapy experienced long-term benefits, making it a successful and sparing steroid treatment for persistent severe atopic dermatitis.

TL-12 Philips Lamp

This near-UV lamp generates light at a wavelength between 270 and 350 (nm), covering all UVB and a portion of UVA. Use of this lamp is limited.

5. Advantages of UV therapy

Advantages of UV therapy over other treatment options available for skin diseases diagnosis include:

1. With the help of handy UV therapy units and lasers we can expose only the affected area of the skin.

2. Highly intense devices so in less duration and with fewer sessions' treatment can be done.

4. Phototherapy works for whole areas of the body also it permits diagnosis for difficult areas easily such as the scalp, inner lining of the nose, genitals, etc.

5. Simple and effective management for children

6. Needs less space for the Equipment setup

7. Painless and non-surgical treatment

8. Suitable for all skin types and so on.

6.   Side effects of UV therapy

With various advantages, UV therapy has some side effects that we cannot neglect but they are not much dangerous. During diagnosis, there is a chance of getting a fever, burning, itching, dryness of the skin with some erythemal effects that fade within 24 hours, and sometimes nausea. UV therapy results in some physiological effects that are listed below:

1.   Erythema caused by capillary and arteriole enlargement of the capillaries and arterioles.

2.   Tanning or change in the pigment of the skin.

3.   Peeling off and thickening of the skin.

4.   Enhancement in the production of vitamin D due to UV exposure.

5.   As the body's resistance to infection rises as a result of the stimulation of the reticuloendothelial system, the prophylactic effect begins to take effect.

6.   The immunosuppressive effects of ultraviolet light, which kill Langerhans cells and promote the growth of suppressor T cells, may contribute to the development of skin cancer.

7.   Intense amounts of UV-B and UV-C radiation can cause photokeratitis and conjunctivitis, which can irritate the eyes and cause watering, a gritty sensation, and a phobia of light (photophobia). Cataracts may also develop as a result of a high UVA dosage.

8.   Large exposure duration increases the chances of aging, specifically in light-skinned individuals.

Conclusions

Phototherapy uses UV radiation to heal human skin from several skin diseases. UV radiation plays a vital role in the synthesis of vitamin D, which helps in suppressing the activity of skin diseases. Among various categories of UV radiation, UVA and UVB radiation are preferred for phototherapy. The emission of these two types of radiation depends purely on the selection of phosphors used in phototherapeutic devices. For any particular treatment, UVA and UVB light-emitting phosphors within a selected wavelength range are required. It is most significant that the emission spectra of the synthesized phosphor fit well with the action spectra of the disease to be treated. LED-based phototherapy devices will be safer and use modern technology, which will enable them to replace mercury-based phototherapy devices. Future devices could be more effective since the LED-based devices will have additional facilities to control the intensity and area of treatment. Compared to all other treatments, phototherapy shows effective results with fewer side effects and no surgery.

Acknowledgments

The authors AAS and DH thank the Department of Science and Technology (DST) of the Government of India for financial support under projects viz. INSPIRE scheme #IF200233 and #CRG/2021/007142, respectively.

References

[1]   C. Karimkhani *et al.*, "Global skin disease morbidity and mortality an update from the global burden of disease study 2013," *JAMA Dermatology*, vol. 153, no. 5, pp. 406–412, 2017. https://doi.org/10.1001/jamadermatol.2016.5538

[2]   A. M. Dessie, S. F. Feleke, S. G. Workie, T. G. Abebe, Y. M. Chanie, and A. K. Yalew, "Prevalence of Skin Disease and Its Associated Factors Among Primary Schoolchildren: A Cross-Sectional Study from a Northern Ethiopian Town," *Clin. Cosmet. Investig. Dermatol.*, vol. 15, no. April, pp. 791–801, 2022. https://doi.org/10.2147/CCID.S361051

[3]   F. R. S. A. royal Harold spencer jones, "Measuring the distance of the sun from the earth," *Nature*, vol. 153, pp. 181–187, 1944. https://doi.org/https://doi.org/10.1038/153181a0

[4]   J. C. Zwinkels and C. Canada, "Light, Electromagnetic Spectrum," *Encycl. Color Sci. Technol.*, vol. 204, pp. 1–8, 2014. https://doi.org/10.1007/978-3-642-27851-8

[5]   D. I. Pattison and M. J. Davies, "Actions of ultraviolet light on cellular structures.," *EXS*, no. 96, pp. 131–157, 2006. https://doi.org/10.1007/3-7643-7378-4_6

[6]   M. N. Mead, "Benefits of sunlight: a bright spot for human health.," *Environ. Health Perspect.*, vol. 116, no. 4, 2008. https://doi.org/10.1289/ehp.116-a160

[7]   J. D'Orazio, S. Jarrett, A. Amaro-Ortiz, and T. Scott, "UV radiation and the skin," *Int. J. Mol. Sci.*, vol. 14, no. 6, pp. 12222–12248, 2013. https://doi.org/10.3390/ijms140612222

[8]   C. Green, B. L. Diffey, and J. L. M. Hawk, "Ultraviolet radiation in the treatment of skin disease," *Phys. Med. Biol.*, vol. 37, no. 1, pp. 1–20, 1992. https://doi.org/10.1088/0031-9155/37/1/001

[9]   A. L. Byrd, Y. Belkaid, and J. A. Segre, "The human skin microbiome," *Nat. Rev. Microbiol.*, vol. 16, no. 3, pp. 143–155, 2018. https://doi.org/10.1038/nrmicro.2017.157

[10]   D. R. Bergfelt, "Anatomy and Physiology of the Mare," *Equine Breed. Manag. Artif. Insemin.*, pp. 113–131, 2009. https://doi.org/10.1016/B978-1-4160-5234-0.00011-8

[11]   Yagi M and Yonei Y, "Glycative stress and anti-aging: 7.Glycative stress and skin aging," *Glycative Stress Res.*, vol. 5, no. 1, pp. 50–54, 2018

[12]   S. B. Hoath and D. G. Leahy, "The Organization of Human Epidermis: Functional Epidermal Units and Phi Proportionality," *J. Invest. Dermatol.*, vol. 121, no. 6, pp. 1440–1446, 2003. https://doi.org/10.1046/j.1523-1747.2003.12606.x

[13]   Nanda Karmaker *et al.*, "Fundamental characteristics and application of radiation," *GSC Adv. Res. Rev.*, vol. 7, no. 1, pp. 064–072, 2021. https://doi.org/10.30574/gscarr.2021.7.1.0043

[14]   R. P. Gallagher and T. K. Lee, "Adverse effects of ultraviolet radiation: A brief review," *Prog. Biophys. Mol. Biol.*, vol. 92, no. 1, pp. 119–131, 2006. https://doi.org/10.1016/j.pbiomolbio.2006.02.011

[15]   "Exposure to Artificial UV Radiation and Skin Cancer," *Group*, 2006

[16]   K. J. Gromkowska-Kępka, A. Puścion-Jakubik, R. Markiewicz-Żukowska, and K. Socha,

"The impact of ultraviolet radiation on skin photoaging — review of in vitro studies," *J. Cosmet. Dermatol.*, vol. 20, no. 11, pp. 3427–3431, 2021. https://doi.org/10.1111/jocd.14033

[17] R. Nair and A. Maseeh, "Vitamin D: The sunshine vitamin," *J. Pharmacol. Pharmacother.*, vol. 3, no. 2, pp. 118–126, 2012. https://doi.org/10.4103/0976-500X.95506

[18] M. Fors, P. González, C. Viada, K. Falcon, and S. Palacios, "Validity of the Fitzpatrick Skin Phototype Classification in Ecuador," *Adv. Ski. Wound Care*, vol. 33, no. 12, pp. 1–5, 2020. https://doi.org/10.1097/01.ASW.0000721168.40561.a3

[19] V. K. Sharma, V. Gupta, B. L. Jangid, and M. Pathak, "Modification of the Fitzpatrick system of skin phototype classification for the Indian population, and its correlation with narrowband diffuse reflectance spectrophotometry," *Clin. Exp. Dermatol.*, vol. 43, no. 3, pp. 274–280, 2018. https://doi.org/10.1111/ced.13365

[20] V. Gupta and V. K. Sharma, "Skin typing: Fitzpatrick grading and others," *Clin. Dermatol.*, vol. 37, no. 5, pp. 430–436, 2019. https://doi.org/10.1016/j.clindermatol.2019.07.010

[21] A. Malerich, Sarah; Desai, "Phototherapy: A Review of Literature," *Jaocd*, vol. 40, 2018

[22] H. Hönigsmann, "History of phototherapy in dermatology," *Photochem. Photobiol. Sci.*, vol. 12, no. 1, pp. 16–21, 2013. https://doi.org/10.1039/c2pp25120e

[23] Philips, "Effective light therapy"

[24] K. Biniek, K. Levi, and R. H. Dauskardt, "Solar UV radiation reduces the barrier function of human skin," *Proc. Natl. Acad. Sci. U. S. A.*, vol. 109, no. 42, pp. 17111–17116, 2012. https://doi.org/10.1073/pnas.1206851109

[25] F. Html *et al.*, "BioMed Research International Adolescents," vol. 2018, pp. 1–10, 2017

[26] Royal Free Hospital Department of Dermatology, "PUVA Treatment," 2003

[27] C. W. Saleeby, "The advance of heliotherapy," *Nature*, vol. 109, no. 2742, p. 663, 1922. https://doi.org/10.1038/109663a0

[28] I. G. Ferreira, M. B. Weber, and R. R. Bonamigo, "History of dermatology: the study of skin diseases over the centuries," *An. Bras. Dermatol.*, vol. 96, no. 3, pp. 332–345, 2021. https://doi.org/10.1016/j.abd.2020.09.006

[29] K. MCGEE, "a Synopsis of the History of Orthoptics.," *Am. Orthopt. J.*, vol. 14, pp. 95–98, 1964. https://doi.org/10.1080/0065955x.1965.11981449

[30] L. I. Grossweiner, *The Science of Phototherapy*. 2005. [Online]. Available: https://link.springer.com/book/10.1007/1-4020-2885-7#about

[31] M. Bhambri, "International Journal of Botany Studies Ammi majus: A plant with multifunctional medicinal properties," vol. 6, no. 3, pp. 3–4, 2021, [Online]. Available: www.botanyjournals.com

[32] P. With and W. Received, "Vitiligo Photochemotherapy," pp. 3–4, 2015

[33] A. Grzybowski, J. Sak, and J. Pawlikowski, "A brief report on the history of phototherapy," *Clin. Dermatol.*, vol. 34, no. 5, pp. 532–537, 2016. https://doi.org/10.1016/j.clindermatol.2016.05.002

[34] R. W. Chesney, "Theobald palm and his remarkable observation: How the sunshine vitamin came to be recognized," *Nutrients*, vol. 4, no. 1, pp. 42–51, 2012. https://doi.org/10.3390/nu4010042

[35] R. Kropp, "Niels Ryberg Finsen," *Pneumologie*, vol. 70, pp. S180–S183, 2016. https://doi.org/10.1055/s-0042-118376

[36] N. R. Finsen and F. Islands, "Niels Ryberg Finsen," pp. 1–6, 1904

[37] A. hanslmeier M. vazquez, "Ultraviolet radiation in the solar system," p. 380, 2006. https://doi.org/https://doi.org/10.1007/b13626

[38] A. E. Loignon, "Bringing Light to the World: John Harvey Kellogg and Transatlantic Light Therapy," *J. Transatl. Stud.*, vol. 20, no. 1, pp. 103–128, 2022. https://doi.org/10.1057/s42738-022-00092-7

[39] J. Schwarcz and D. Ph, "Colorful Nonsense : Dinshah Ghadiali and His Spectro-Chrome Device," pp. 1–5, 2003

[40] "Parrish1974," 2010

[41] P. S. Hemne, R. G. Kunghatkar, S. J. Dhoble, S. V. Moharil, and V. Singh, "Phosphor for phototherapy: Review on psoriasis," *Luminescence*, vol. 32, no. 3, pp. 260–270, May 2017. https://doi.org/10.1002/bio.3266

[42] F. R. De Gruijl, "Biological action spectra," *Radiat. Prot. Dosimetry*, vol. 91, no. 1–3, pp. 57–63, 2000. https://doi.org/10.1093/oxfordjournals.rpd.a033235

[43] G. Horneck, P. Rettberg, R. Facius, K. Scherer, and A. Medicine, "Quantification of biological effectiveness of uv radiation g. horneck, p. rettberg, r. facius and k. scherer," vol. 9, no. Figure 1, pp. 51–52, 2006

[44] H. Hönigsmann, "Phototherapy for psoriasis," *Clin. Exp. Dermatol.*, vol. 26, no. 4, pp. 343–350, 2001. https://doi.org/10.1046/j.1365-2230.2001.00828.x

[45] J. P. Nater, "Phototherapy in psoriasis," *Geneesmiddelenbulletin*, vol. 12, no. 9, pp. 38–41, 1978. https://doi.org/10.5005/jp/books/12839_38

[46] M. Neves-Petersen, G. Gajula, and S. Petersen, "UV Light Effects on Proteins: From Photochemistry to Nanomedicine," *Mol. Photochem. - Var. Asp.*, pp. 125–158, 2012

[47] N. de M. Barros, L. L. Sbroglio, M. de O. Buffara, J. L. C. e. S. Baka, A. de S. Pessoa, and L. Azulay-Abulafia, "Phototherapy," *An. Bras. Dermatol.*, vol. 96, no. 4, pp. 397–407, Jul. 2021. https://doi.org/10.1016/j.abd.2021.03.001

[48] R. King, K. Paver, and K. Poyzer, "Photochemotherapy in psoriasis.," *Med. J. Aust.*, vol. 1, no. 2, pp. 57–58, 1979. https://doi.org/10.5694/j.1326-5377.1979.tb111984.x

[49] H. Hönigsmann, "Mechanisms of phototherapy and photochemotherapy for photodermatoses," *Dermatol. Ther.*, vol. 16, no. 1, pp. 23–27, 2003.

https://doi.org/10.1046/j.1529-8019.2003.01604.x

[50]   R. B. Setlow, "Action Spectra: Skin," *Sunscreen Photobiol. Mol. Cell. Physiol. Asp.*, pp. 1–10, 1997. https://doi.org/10.1007/978-3-662-10135-3_1

[51]   A. D. Crosswell and K. G. Lockwood, "Best practices for stress measurement: How to measure psychological stress in health research," *Heal. Psychol. Open*, vol. 7, no. 2, 2020. https://doi.org/10.1177/2055102920933072

[52]   B. Zhang *et al.*, "Opportunities and Challenges: Classification of Skin Disease Based on Deep Learning," *Chinese J. Mech. Eng. (English Ed.*, vol. 34, no. 1, 2021. https://doi.org/10.1186/s10033-021-00629-5

[53]   L. Chen *et al.*, "Oncotarget 7204 www.impactjournals.com/oncotarget Inflammatory responses and inflammation-associated diseases in organs," *Oncotarget*, vol. 9, no. 6, pp. 7204–7218, 2018, [Online]. Available: www.impactjournals.com/oncotarget/

[54]   R. Parisi, I. Y. K. Iskandar, E. Kontopantelis, M. Augustin, C. E. M. Griffiths, and D. M. Ashcroft, "National, regional, and worldwide epidemiology of psoriasis: Systematic analysis and modelling study," *BMJ*, vol. 369, 2020. https://doi.org/10.1136/bmj.m1590

[55]   P. Hu *et al.*, "The Role of Helper T Cells in Psoriasis," *Front. Immunol.*, vol. 12, no. December, pp. 1–10, 2021. https://doi.org/10.3389/fimmu.2021.788940

[56]   F. O. Nestle and C. Conrad, "Mechanisms of psoriasis," *Drug Discov. Today Dis. Mech.*, vol. 1, no. 3, pp. 315–319, 2004. https://doi.org/10.1016/j.ddmec.2004.11.005

[57]   F. Benhadou, Di. Mintoff, and V. Del Marmol, "Psoriasis: Keratinocytes or Immune Cells - Which Is the Trigger?," *Dermatology*, vol. 235, no. 2, pp. 91–100, 2019. https://doi.org/10.1159/000495291

[58]   S. C. Weatherhead, P. M. Farr, and N. J. Reynolds, "Spectral effects of UV on psoriasis," *Photochem. Photobiol. Sci.*, vol. 12, no. 1, pp. 47–53, 2013. https://doi.org/10.1039/c2pp25116g

[59]   C. Ni and M. W. Chiu, "Psoriasis and comorbidities: Links and risks," *Clin. Cosmet. Investig. Dermatol.*, vol. 7, pp. 119–132, 2014. https://doi.org/10.2147/CCID.S44843

[60]   C. Balakrishnan and N. Madnani, "Diagnosis and management of psoriatic arthritis," *Indian J. Dermatol. Venereol. Leprol.*, vol. 79, no. SUPPL. 1, pp. 18–24, 2013. https://doi.org/10.4103/0378-6323.115507

[61]   C. Niu, X. Lu, and H. A. Aisa, "Preparation of novel 1,2,3-triazole furocoumarin derivatives via click chemistry and their anti-vitiligo activity," *RSC Adv.*, vol. 9, no. 3, pp. 1671–1678, 2019. https://doi.org/10.1039/C8RA09755K

[62]   A. Arora and M. Kumaran, "Pathogenesis of vitiligo: An update," *Pigment Int.*, vol. 4, no. 2, p. 65, 2017. https://doi.org/10.4103/2349-5847.219673

[63]   M. L. Frisoli, K. Essien, and J. E. Harris, "Vitiligo: Mechanisms of Pathogenesis and Treatment," *Annu. Rev. Immunol.*, vol. 38, pp. 621–648, 2020. https://doi.org/10.1146/annurev-immunol-100919-023531

[64]   C. E. Artz, C. M. Farmer, and H. W. Lim, "Polymorphous Light Eruption: a Review,"

Curr. Dermatol. Rep., vol. 8, no. 3, pp. 110–116, 2019. https://doi.org/10.1007/s13671-019-0264-y

[65] I. An, M. Harman, and I. Ibiloglu, "Polymorphus Light Eruption-An Indian Scenario," *Indian Dermatol. Online J.*, vol. 10, no. 4, pp. 481–485, 2017. https://doi.org/10.4103/idoj.IDOJ

[66] S. E. Ullrich, "Does exposure to UV radiation induce a shift to a Th-2-like immune reaction?," *Photochem. Photobiol.*, vol. 64, no. 2, pp. 254–258, 1996. https://doi.org/10.1111/j.1751-1097.1996.tb02454.x

[67] E. Vicenzi and G. Poli, "Ultraviolet irradiation and cytokines as regulators of HIV latency and expression," *Chem. Biol. Interact.*, vol. 91, no. 2–3, pp. 101–109, 1994. https://doi.org/10.1016/0009-2797(94)90030-2

[68] H. W. Lim, S. Vallurupalli, T. Meola, and N. A. Soter, "UVB phototherapy is an effective treatment for pruritus in patients infected with HIV," *J. Am. Acad. Dermatol.*, vol. 37, no. 3 I, pp. 414–417, 1997. https://doi.org/10.1016/S0190-9622(97)70142-7

[69] A. Alothman and H. S. H. Mohamed, "Recurrent Nasal Sebaceous Carcinoma in Human Immunodefiency Virus / Hepatitis B Virus (HIV / HBV) Coinfection Patient : A Case Report," vol. 3, no. 10, pp. 329–332, 2015. https://doi.org/10.12691/ajmcr-3-10-7

[70] T. Meola, N. A. Soter, R. Ostreicher, M. Sanchez, and J. A. Moy, "The safety of UVB phototherapy in patients with HIV infection," *J. Am. Acad. Dermatol.*, vol. 29, no. 2, pp. 216–220, 1993. https://doi.org/10.1016/0190-9622(93)70171-O

[71] S. Mattinen, A. Lagerstedt, and K. Krohn, "Effect of PUVA on immunologic and virologic findings in HIV-infected patients," *J. Am. Acad. Dermatol.*, vol. 24, no. 3, pp. 404–410, 1991. https://doi.org/10.1016/0190-9622(91)70060-F

[72] J. M. Nuhu, J. Mohammed, and M. Muhammad, "UV therapy: Physiotherapists' perception of therapeutic efficacy and barriers to usage," *Hong Kong Physiother. J.*, vol. 32, no. 1, pp. 44–48, 2014. https://doi.org/10.1016/j.hkpj.2014.02.002

[73] R. P. Rastogi, Richa, A. Kumar, M. B. Tyagi, and R. P. Sinha, "Molecular mechanisms of ultraviolet radiation-induced DNA damage and repair," *J. Nucleic Acids*, vol. 2010, 2010. https://doi.org/10.4061/2010/592980

[74] T. M. Lotti and S. Gianfaldoni, "Ultraviolet A-1 in dermatological diseases," *Adv. Exp. Med. Biol.*, vol. 996, pp. 105–110, 2017. https://doi.org/10.1007/978-3-319-56017-5_9

[75] F. Bernerd, T. Passeron, I. Castiel, and C. Marionnet, "The Damaging Effects of Long UVA (UVA1) Rays: A Major Challenge to Preserve Skin Health and Integrity," *Int. J. Mol. Sci.*, vol. 23, no. 15, pp. 1–33, 2022. https://doi.org/10.3390/ijms23158243

[76] D. L. Damian, Y. J. Matthews, T. A. Phan, and G. M. Halliday, "An action spectrum for ultraviolet radiation-induced immunosuppression in humans," *Br. J. Dermatol.*, vol. 164, no. 3, pp. 657–659, 2011. https://doi.org/10.1111/j.1365-2133.2010.10161.x

[77] D. D. Moyal and A. M. Fourtanier, "Effects of UVA radiation on an established immune response in humans and sunscreen efficacy," *Exp. Dermatology, Suppl.*, vol. 11, no. 1, pp.

28–32, 2002. https://doi.org/10.1034/j.1600-0625.11.s.1.7.x

[78]   I. Furocoumarins, "Yearly Review Recent Advances in Psoralen Phototoxicity," vol. 50, no. 6, pp. 859–882, 1989

[79]   R. Vangipuram and S. R. Feldman, "Ultraviolet phototherapy for cutaneous diseases: A concise review," *Oral Dis.*, vol. 22, no. 4, pp. 253–259, 2016. https://doi.org/10.1111/odi.12366

[80]   J. L. Bolognia, "Rectal suppositories of 8-methoxsalen produce fewer gastrointestinal side effects than the oral formulation," *J. Am. Acad. Dermatol.*, vol. 35, no. 3 PART I, pp. 424–427, 1996. https://doi.org/10.1016/s0190-9622(96)90609-x

[81]   J. Foerster *et al.*, "Narrowband UVB treatment is highly effective and causes a strong reduction in the use of steroid and other creams in psoriasis patients in clinical practice," *PLoS One*, vol. 12, no. 8, pp. 1–14, 2017. https://doi.org/10.1371/journal.pone.0181813

[82]   K. Boswell *et al.*, "Narrowband ultraviolet B treatment for psoriasis is highly economical and causes significant savings in cost for topical treatments," *Br. J. Dermatol.*, vol. 179, no. 5, pp. 1148–1156, 2018. https://doi.org/10.1111/bjd.16716

[83]   P. V Prasad, T. Paari, K. Chokkalingam, and V. Vijaybushanam, "Malignant syphilis (leus maligna) in a HIV infected patient.," *Indian journal of dermatology, venereology and leprology*, vol. 67, no. 4. pp. 192–4, 2001. [Online]. Available: http://www.ncbi.nlm.nih.gov/pubmed/17664738

[84]   D. Bilsland *et al.*, "An appraisal of narrowband (TL-01) UVB phototherapy. British Photodermatology Group Workshop Report (April 1996)," *Br. J. Dermatol.*, vol. 137, no. 3, pp. 327–330, 1997. https://doi.org/10.1111/j.1365-2133.1997.tb03733.x

[85]   S. H. Ibbotson *et al.*, "An update and guidance on narrowband ultraviolet B phototherapy: A British Photodermatology Group Workshop Report," *Br. J. Dermatol.*, vol. 151, no. 2, pp. 283–297, 2004. https://doi.org/10.1111/j.1365-2133.2004.06128.x

[86]   K. Ly, M. P. Smith, Q. G. Thibodeaux, K. M. Beck, W. Liao, and T. Bhutani, "Beyond the Booth: Excimer Laser for Cutaneous Conditions," *Dermatol. Clin.*, vol. 38, no. 1, pp. 157–163, 2020. https://doi.org/10.1016/j.det.2019.08.009

[87]   T. B. M. Abrounk, E. Levin, M. Brodsky, J. R. Gandy, "Excimer laser for the treatment of psoriasis: safety, efficacy, and patient acceptibility," *Psoriasistarget Ther.*, no. 6, pp. 165–173, 2016

[88]   C. Lopes, V. F. M. Trevisani, and T. Melnik, "Efficacy and Safety of 308-nm Monochromatic Excimer Lamp Versus Other Phototherapy Devices for Vitiligo: A Systematic Review with Meta-Analysis," *Am. J. Clin. Dermatol.*, vol. 17, no. 1, pp. 23–32, 2016. https://doi.org/10.1007/s40257-015-0164-2

[89]   C. A. Robertson, D. H. Evans, and H. Abrahamse, "Photodynamic therapy (PDT): A short review on cellular mechanisms and cancer research applications for PDT," *J. Photochem. Photobiol. B Biol.*, vol. 96, no. 1, pp. 1–8, 2009. https://doi.org/10.1016/j.jphotobiol.2009.04.001

[90] S. Kwiatkowski *et al.*, "Photodynamic therapy – mechanisms, photosensitizers and combinations," *Biomed. Pharmacother.*, vol. 106, no. July, pp. 1098–1107, 2018. https://doi.org/10.1016/j.biopha.2018.07.049

[91] M. Nakamura, B. Farahnik, and T. Bhutani, "Recent advances in phototherapy for psoriasis [version 1; referees: 2 approved]," *F1000Research*, vol. 5, pp. 1–8, 2016. https://doi.org/10.12688/F1000RESEARCH.8846.1

[92] A. Carolina Sparavigna, "Carbon-Arc Light as the Electric Light of 1870," *Int. J. Sci.*, vol. 0, no. 10, pp. 1–7, 2014. https://doi.org/10.18483/ijsci.581

[93] R. Jenkins, B. Aldwell, S. Yin, M. Meyer, A. J. Robinson, and R. Lupoi, "Energy efficiency of a quartz tungsten halogen lamp: Experimental and numerical approach," *Therm. Sci. Eng. Prog.*, vol. 13, no. July, p. 100385, 2019. https://doi.org/10.1016/j.tsep.2019.100385

[94] A. Namin, C. Jivacate, D. Chenvidhya, K. Kirtikara, and J. Thongpron, "Construction of tungsten halogen, pulsed LED, and combined tungsten halogen-LED solar simulators for solar cell i - V characterization and electrical parameters determination," *Int. J. Photoenergy*, vol. 2012, 2012. https://doi.org/10.1155/2012/527820

[95] H. Kohli, S. Srivastava, S. K. Sharma, S. Chouhan, and M. Oza, "Design of Programmable LED Based Phototherapy System," *Int. J. Opt.*, vol. 2019, 2019. https://doi.org/10.1155/2019/6023646

[96] H. J. Vreman, R. J. Wong, and D. K. Stevenson, "Phototherapy: Current methods and future directions," *Semin. Perinatol.*, vol. 28, no. 5, pp. 326–333, Oct. 2004. https://doi.org/10.1053/j.semperi.2004.09.003

[97] Beggs, Sarah BS; Short, Jack MD; Rengifo-Pardo, Monica MD; Ehrlich, Alison MD, MHS, "Applications of the Excimer Laser: A Review","American Society for Dermatologic Surgery".,vol. 41, pp. 1201–1211, 2015. https://doi.org/10.1097/DSS.0000000000000485

Rare Earth - A tribute to the late Mr. Rare Earth, Professor Karl Gschneidner   Materials Research Forum LLC
Materials Research Foundations 164 (2024) 177-210   https://doi.org/10.21741/9781644903056-4

Chapter 4

Prospective Potential Applications and Emerging Tendencies in Rare Earth Materials

R. Ramakrishna Reddy

Former UGC Emeritus Professor, Department of Physics, Sri Krishnadevaraya University, Anantapur -515 001, A. P; India

rajururreddy@gmail.com

Abstract

Rare Earth Elements (REE) remain underappreciated despite their numerous applications. High-temperature superconductors, phosphors (for energy-saving lamps, flat-screen monitors, and flat-screen televisions), rechargeable batteries (home and automotive), magnetic resonance image scanning systems, superconductors, laser technology, and very strong permanent magnets all rely on them (used for instance, in wind turbines and hard-disk drives). The article provides a brief overview of the various applications of REES. Medical imaging and therapy using rare earth doped lasers: potential environmental effects. New methods for isolating specific rare earth elements have been highlighted in applications as diverse as metallurgy and recycling.

Keywords

REEs, Potential Applications, Rare Earth Doped Fiber Lasers, Diagnostic Imaging, Environmental Effects

Contents

1.   Indispensable rare earth elements: A brief overview

Awareness and interest in rare earth elements have risen in recent years (REE). Many things we rely on every day contain these various, often foreign parts. Today, Rare Earth Elements (REEs) are employed in nearly every aspect of our technological lives. Since REEs are used in so many cutting-edge technologies, including some that are crucial to national security, there is a constant pressure to increase supply by discovering new sources and developing more efficient methods of extraction.

A miner in Ytterby, Sweden, found a mysterious black rock in 1788; this was the first known instance of rare earth. The ore was given the names "rare" and "earth" because that was the geological word for acid-soluble rocks in the 18th century. The Swedish chemist Johan Gadolin discovered a new "earth" in 1794 and named it yttria after the city it was found in. During the 19th century, naming new elements was a highly regarded yet hotly debated topic among European scientists. In 1803, Jöns Jacob Berzelius isolated cerium, then in 1828, he separated thorium; he called both elements. Swedish chemist Carl Gustaf Mosander began systematic analysis of the mixed rare earths in 1839, and it was during this time that he discovered and named lanthanum, erbium, and terbium. Neodymium and praseodymium were both called after a pupil of Robert Bunsen's, Carl Auer von Welsbach [1-4].

Rare earth elements are studied in geochemistry, marine chemistry, and environmental chemistry (REEs). Due to their unique physical qualities, REEs are indispensable materials in cutting-edge technology, particularly in the semiconductor sector. Many "green" and contemporary conveniences rely on these elements, despite their seemingly out-of-place placement at the bottom of the periodic table. The elements lanthanum through lutetium, scandium through yttrium, and everything in between are known together as rare earth elements. All lanthanides have their valence electrons in 4f-obitals, placing them in the f-block at the very bottom of the periodic table. Lanthanides' unique characteristics, such as light emission and magnetism, come from their non-bonding electrons, which are hidden deep within the atom and shielded by 4d and 5p electrons. When exposed to UV light, the f-electrons of $Eu3+$ are able to momentarily absorb that energy by jumping to a higher energy level, then release that energy as light as they fall back to earth. The capacity of 'excited' f-electrons to emit light is one reason why REEs can be found in lasers, energy-efficient light bulbs, and electronic displays. The electrical arrangements of the elements provide a rationale for rare earths' critical role in so many modern inventions [5, 6]. Rare earth

Rare Earth - A tribute to the late Mr. Rare Earth, Professor Karl Gschneidner   Materials Research Forum LLC
Materials Research Foundations 164 (2024) 177-210   https://doi.org/10.21741/9781644903056-4

elements, or rare earths as they are commonly called, are essential to the operation of many modern technological devices and services essential to our everyday lives in the areas of communication, work, education, safety, energy, and transportation etc., (Figs 1 and 2).

These so-called "rare-earth" elements are actually quite common. They are also plentiful in the Earth's crust, like tin, lead, and zinc. It was in limited mineral deposits that these elements were given the moniker "rare earths" when their discovery in the 19th century prompted the label. Rare earths are never discovered alone or in extremely high quantities; rather, they are always found in a combination with other rare earths or with radioactive elements like uranium and thorium. Rare earth elements are notoriously difficult to isolate from one another and from their natural settings [6-8] due to their unusual chemical characteristics. As a result of these characteristics, they are hard to clean. It takes a lot of ore and creates a lot of toxic waste for today's production methods to extract a relatively modest amount of rare earth metals. The techniques of processing produce waste such as radioactive water, poisonous fluorine, and acids.

Fig.1 Applications of Rare Earths in Several Disciplines: https://www.drishtiias.com/daily-updates/daily-news-analysis/rare-earth-metals-1

Many modern technologies rely heavily on rare earth elements (REEs). The electronics, clean energy, aerospace, automotive, and defense sectors are just some of the many industrial fields that make use of REEs. In 2020, the production of permanent magnets is expected to account for 29 percent of the predicted demand for REEs [9,10]. Permanent magnets are essential to the operation of many electronic equipment, including mobile phones, televisions, computers, vehicles, wind turbines, and aircraft. With their luminous and catalytic capabilities, REEs find widespread application in cutting-edge technology and environmentally friendly products.

Therefore, several scientific disciplines [4,10] can benefit from research into the analytical chemistry of REEs, including geology, oceanography, environmental science, and materials science. Scientists can analyse REEs using inductively coupled plasma mass spectrometry (ICP-MS), X-ray fluorescence (XRF), or laser-induced breakdown spectroscopy (LIBS).

Rare Earth - A tribute to the late Mr. Rare Earth, Professor Karl Gschneidner   Materials Research Forum LLC
Materials Research Foundations 164 (2024) 177-210   https://doi.org/10.21741/9781644903056-4

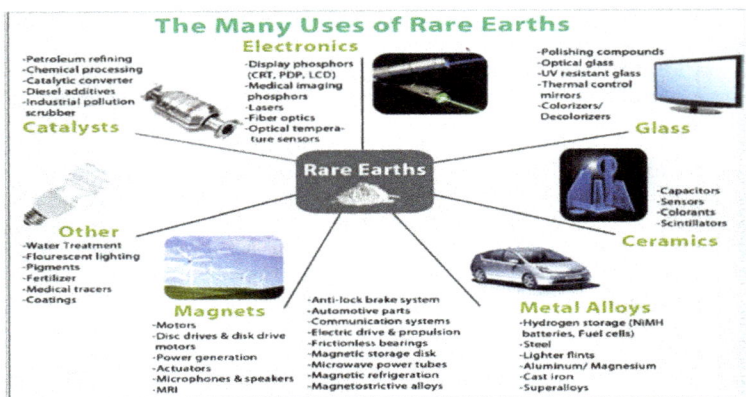

Fig.2. Many uses of Rare earths (The usage of rare earth elements and minerals are detailed in https://www.jxscmachine.com/new/., Ider Kadir.pdf;jsessionid=5D68D19F59E4D7871152EA4B0E2961D3;sequence=1 https://www.theseus.fi/bitstream/handle/10024/105572/Ider Kadir.pdf

2. Exceptional rare earth elements for practical use

Only 17 of the 118 elements in the periodic table are rare earth elements, but these elements play crucial roles in many areas, including defense, renewable energy, environmental protection, and economic development. [11,12].

Elements that are rarely covered in chemistry courses can be found in two rows at the very bottom of the periodic table. Rare earth elements (also known as lanthanides) can be found in the second-to-last row. Perhaps the most common criticism levelled at these so-called "rare earths" is that they are neither earth nor uncommon. "Rare earths" refer to a group of 17 elements that are chemically very similar and have many different technological uses. Despite their frightening names, most of these substances may be found in abundance in nature. Rare earth elements, in general, are useful in more than one setting [13,14]. The chemical characteristics of rare earth elements are shared with 17 other metals on the periodic table (REEs). Rare earths are typically found as homogeneous rocks, making separation into numerous rare earths compounds a technically complex and potentially expensive operation. Although rare earth elements are actually quite common, there are several technological, political, and environmental challenges associated with mining for and refining the metals. The finding of rare earth element sources for mining is becoming increasingly significant and valuable as these elements are essential building blocks for a greener economy.

There is a lot of momentum behind the rare earth metals business right now, and it's only going to pick up steam from here. In 2022, the global market for rare earth elements is expected to be worth \$2821.3 mln, and by 2028, that number is expected to increase to \$3979.4 mln. The industry's explosive growth has created highly fluctuating market conditions. The value of rare earths to the global economy has increased over the past two decades as new technologies have been introduced. Electronics manufacturers rely heavily on rare earths, and most cutting-edge innovations simply couldn't be made without them [14,15].

To fuel our electronics, bring our economy into the digital era, and secure our borders, rare earth elements (REEs) are essential to our daily lives. Nonetheless, a small number of producers once again holds sway: China is the world's largest exporter of rare earths, and Russia is the world's largest exporter of palladium. Existing problems with NATO's military weapon supply lines need to be addressed immediately.

Alternative economic, geopolitical, environmental, and technological policies are addressed in light of projected increases in both supply and demand. products made from rare earth elements, discussing their use in renewable energy and the risks associated with a shortage. Since Praseodymium (Pr) may stand in for Neodymium in high-intensity permanent magnets, it has made the cut as one of the rarest elements. These compounds are "critical" because of their short supply. In light of this, and the fact that we anticipate a rise in demand, we anticipate that the cost of these pricier components will climb over the next decade [11, 12].

There is no other type of these metals that sees as rapid consumption, since they find use in so many different manufacturing and consumer contexts. These chemicals have been called "The Vitamins of Modern Industry" because they are crucial to the functioning of all modern technological devices. Despite the availability of suitable substitutes, rare earth elements continue to find widespread use in numerous technologies. There are many examples of cutting-edge technology in use today, such as lasers, optical glass, fibre optics, masers, radar detecting devices, nuclear fuel rods, mercury-vapor lamps, highly reflective glass, computer memory, nuclear batteries, and high-temperature superconductors. Minerals with high conductivity to electricity often contain rare earth elements. Strongest magnets available today are made from rare earth elements [16].

Rare earth elements are nevertheless increasingly used in numerous technologies, despite the fact that suitable substitutes have been identified for some applications. There are many examples of state-of-the-art technology in use today. Some of them include lasers, optical glass, fibre optics, masers, radar detecting devices, nuclear fuel rods, mercury vapour lamps, highly reflective glass, computer memory, nuclear batteries, and high-temperature superconductors. Elemental rare earths are frequently found in conductive minerals. Magnets produced from rare earth elements are currently the strongest on the market. The manufacture of permanent magnets represents a potentially lucrative market for rare earth elements. Magnets find use in numerous fields, including electronics, transportation, energy, and the medical field [16]. Magnetic resonance imaging (MRI) scanners, computer hard drives, microwave power tubes, anti-lock brakes, and other auto parts are just some of the many places you can find these magnets.

3.   Memory devices that utilize rare metals

Integrated optical memories, and especially on-chip optical memories, are fundamental to the development of scalable and practical applications .The coherent optical memory relies on a type-IV waveguide constructed on the outside of rare-earth ion-doped crystals.Over the past few decades, rising electronic device use and data storage requirements have revealed substantial openings for the development of cutting-edge technologies that may meet the computing and development requirements of a globalised society [17]. After being mined from the ground and processed further, rare earth minerals can be put to a number of distinct uses. Through our daily usage of electronic devices, the vast majority of us are exposed to rare earth materials. Scientists need to look into new and improved methods of energy conversion and storage to keep up with the ever-increasing demands of modern life. Since its oxides are rather stable, perovskites, a well-known energy material with the chemical formula $ABX3$, serve a crucial role as a long-term energy storage material. Using laser-induced breakdown field spectroscopy (LEIBF), we look at the effects of rare-earth ion implantation in $HfO2$ thin films and their use in resistive switching memory systems. Devices for effective quantum conversion between optical photons and microwave photons will be made possible by the development of nano-photonic devices in rare-earth doped crystals. Using atomic frequency comb protocols or controlled reversible inhomogeneous broadening, photons can be stored directly in optical transitions, with storge durations ultimately constrained by the optical coherence time. Nano-photonic resonators, which improve the coupling between photons and a small ensemble of atoms, should be used to embed the rare-earth atoms on the chip in order to obtain high efficiencies on the device. Our products are neodymium and erbium based [18].

Because of the worldwide shortage of microprocessors, or computer chips, the global economy is struggling to recover from the effects of the recent coronavirus outbreak. Manufacturers of integrated circuits are mostly to blame for the current shortfall because they have slashed output. Rare-earth magnets are used in disc drives for reading and writing optical material. For more efficient energy storage systems, rare-earth electrochemistry is essential Energy density, charging speed, and fire safety are all improved by conventional dielectric and electrolytic capacitors compared to standard batteries. Supercapacitors and ultracapacitors are more concentrated, promising energy storage devices [19]. Supercapacitors, ultracapacitors, pseudo capacitors, and electrochemical double-layer capacitors are all names for electrochemical capacitors. The necessity for low power solid-state storage is contributing to the meteoric rise of NAND flash memories, which are used in an ever-expanding number of increasingly resource-intensive applications [18]. A quantum memory having the capability of interacting with the wavelength of light utilised in modern telecommunications systems (1.5 microns). This information is stored in a single erbium atom doped into a crystal.

Rare earth metals and computer chips, which are used in virtually all contemporary technology, are in increasingly short supply. Numerous products, from computer hard drives to digital cameras to optical fibre connections, make use of them. Making our media more portable has been greatly aided by advances in rare earth-based electrode materials for supercapacitors and their composites [17,18]. Memory chips, DVD players, rechargeable batteries, mobile phones, catalytic converters, magnets, compact fluorescent bulbs, and many more fall under this category.

Phase change memory is a potential game-changer for electronic data storage. We were motivated to find the finest dopants and understand how they work because of evidence that doping with rare-earth elements can drastically affect the characteristics of these materials.

4.   Vital function of rare metals in human health

When it comes to metals employed in cancer diagnosis and treatment, rare earth elements are truly special. Multiple studies have looked into the potential of lanthanides for use in cancer treatment and diagnosis. Green energy, information technology, health, the military, and lanthanum carbonate (used to treat hyperphosphatemia) are just some of the many trillion-dollar global businesses that rely on rare earths. The increasing variety of molecular imaging and radiotherapies that rely on rare earth elements (REE) has contributed to the increased demand for these elements [20]. Magnetic resonance imaging is commonly used to make a diagnosis (MRI).

As a result of its unique qualities, REE have been put to use in a wide variety of technological contexts, including manufacturing, medicine, agriculture, and even zootech. For thousands of years, people have relied on metals for their health and wellness. Ayurveda, the ancient Indian medical system. Although these metals are toxic in large doses, they are often necessary in therapeutic and orthopedic devices [21,22]. Anemia can be treated with iron; stomach issues with bismuth, cobalt, and nickel; depression with lithium; leishmaniasis with antimony; cancer with platinum or radioactive metals; and psoriasis with arsenic. Gold is so effective that it is even used to treat rheumatoid arthritis.

The usage of surgical lasers and PET scintillation detectors has increased as a result of recent advancements in medicine. The use of rare earth elements (REE) in medicine appears to be on the rise as scientists learn more about how to take advantage of the special properties of REE alone, in conjunction with other elements, or in a variety of other configurations. Minerals can be found in many forms, and they are essential to the manufacturing of many different types of pharmaceuticals, therapeutics, and medical apparatus. For instance, the extremely porous refractory metal tantalum is great for implants because it encourages bone formation and attachment when used near bone. Specifically, lanthanum has been linked to a variety of beneficial metabolic effects in humans, including decreased cholesterol, blood pressure, appetite, and risk of blood clots. People with renal failure often take lanthanum carbonate to lower their blood phosphate levels. Problems with calcium absorption are a serious health risk when phosphate levels are high [22].

At least two methods exist for employing yttrium in the treatment of cancer, one of which involves directing the element at cancer cells themselves, while the other involves poisoning the blood supply to the cancerous tissue itself. In order to alleviate the suffering caused by osteoblastic metastatic bone lesions, samarium is utilized as a palliative treatment. It can also be administered intravenously (IV). Rare earth elements are highly sought after in the medical device and equipment business. For instance, the element terbium (atomic number 65) is used mostly in electrical equipment alloys and as a green phosphor in electronic displays. Erbium, gadolinium, thulium, ytterbium, and yttrium are essential in medical applications such as x-ray and MRI machines, laser surgery, and cancer therapy. Crucial alloys often include the elements cerium, lanthanum, and neodymium. In the medical manufacturing sector, ceramic technology is growing at an astounding rate, as evidenced by the explosion in the number of electronic implants [23].

Lasers, infrared light filters, and optical fibres are some of the most common applications for erbium (atomic number 68), another rare earth element (REE). It is an essential component in the creation of lasers for use in elective surgeries like dental implant surgery to improve one's appearance. The common medium for solid-state lasers is yttrium-aluminum-garnet (YAG). Focused wavelengths of light produced by erbium-doped YAG lasers are useful for oral surgery and dentistry. Range finders and target designators for guided missiles are only two examples of the various industrial, medical, and military applications of YAG lasers that have benefited from neodymium doping (https://www.sciencehistory.org/learn/science-matters/case-of-rare-earth-elements-science). The intravenous administration of a solution of an organic gadolinium complex or gadolinium compound is used in medical MRI and MR angiography to enhance image quality.

Samarium, often known as Sm (62) or silver, is an extremely valuable metal with many applications in fields as diverse as transportation, defense, and commercial technology. In today's world, REEs are used to power anything from medical equipment and defense systems to alternative energy sources. It can also be used in intravenous radiation treatments for killing cancer cells, in combination with other chemicals. It also aids in the treatment of malignancies of the lung, prostate, breast, and bones [22,23]. The more common name for these magnets is "permanent magnets," yet their uses can be found in everything from stereos and electric vehicle motors to medical tools and weapons.

All actions in mining processes including extraction, separation and after recycling of REEs could produce danger to human health and environment. Rare earth elements are metals that are used for a variety of purposes, some of which provide clear benefits to humanity and the environment, However, there are potential public health risks associated with rare earth element extraction and processing, particularly in relation to radioactivity.

Scintillators made from lanthanides (and REEs more generally) play a crucial role in medical imaging and diagnostics techniques like CT and PET. Since lanthanide-activated phosphors are so crucial to a wide range of physical and biomedical research areas, there has been a flurry of activity in the development of novel synthetic techniques and related instruments for investigating the kinetics of energy transfer processes [24-26]. Magnetic properties of paramagnetic Ln(III) ions have been extensively employed to get structural information using nuclear magnetic resonance (NMR). MRI contrast agents, photodynamic therapy, anticancer drug delivery, enhancing the efficacy and safety of yttrium-90 radioimmunotherapy, and chemoembolization for the treatment of hepatocellular carcinoma are all examples of the applications of nanomaterials doped with lanthanide complexes in cancer diagnosis and treatment.[24].

Nanoparticles that have been doped with rare earth ions have found several applications in biomedical imaging, medication delivery, tumour therapy, and disease diagnosis [24]. Water solubility, biocompatibility, drug-loading capacity, and tumor-specific targeting are just few of the benefits that can be conferred upon rare earth-doped nanoparticles through surface functionalization. The advantages of rare earth doped up conversion nanoparticles are reduced toxicity, increased penetration depth, enhanced imaging sensitivity, and elimination of photobleaching (UCNPs). In biomedical research, namely cancer detection and treatment. Medical applications for metal nanoparticles derived from rare earths and lanthanides have sparked a surge in study.

As a difficult and multidisciplinary process, Yttrium microsphere therapy calls for the close cooperation of the Nuclear Medicine, Conventional and Interventional Radiology, Hepatology, Oncology, and its Hepato Bilio Pancreatic Area. One such rare earth metal with promising cancer-fighting properties is neodymium. When looking for brain tumours, gadolinium is an essential rare earth element [27]. Gadolinium, in particular, favors clinicians who battle against brain tumours, who employ it in MRI as a contrast substance. The use of gadolinium aids in the detection of microscopic cancers. Samarium-153 lexidronam (EDTMP) is used for the treatment of bone metastases. Bone metastases are a major complication of cancer treatment (https://nanografi.com/blog/rare-earth-elements-in-cancer-diagnosis-treatment/9).

Rare Earth - A tribute to the late Mr. Rare Earth, Professor Karl Gschneidner   Materials Research Forum LLC
Materials Research Foundations 164 (2024) 177-210   https://doi.org/10.21741/9781644903056-4

5.   The electronics and semiconductor industries rely critically on rare earth elements

Rare earth elements (REEs) are a family of metals that may be modest in size yet have a significant impact on the modern electrical industry (28). Due to their remarkable magnetic, electrochemical, and luminous properties, rare earth elements (REEs) are being used in a broad range of electronic applications. Smartphones, hard drives, electric vehicles, military defence systems, and even clean energy and medical technologies are just some of the many possible uses. Rare earths are metals with unique properties like great resistance to heat and corrosion, as well as high magnetic and electrical strengths and reflecting features (29-32).

Many of today's most useful conveniences simply wouldn't work without rare earth elements (REE). Metals used in high-tech equipment are often thought of as being uncommon but ubiquitous. The modern world is teeming with high-tech gadgets, from commonplace electronic gadgets to sophisticated medical tools and security systems.

Rare earth elements and tech metals are listed below (https://www.techmetalsresearch.net/what-are-technology-metals/):

Alloys containing lanthanum are utilized in hydrogen automobiles and batteries.

- Magnets and lasers both benefit from the use of neodymium.
- Many different types of aviation engines, fibre optic cables, and magnets all make use of neodymium.
- Pacemakers and precision-guided missiles both rely on promethium.
- Terbium finds its way into electronics, including light bulbs, memory devices, and x-ray machines. A common application for lithium is in battery technology.
- Tantalum has numerous electronic applications, including capacitors, resistors, and more.
- It's common for LCD screens to use indium.
- The semiconductor, LED, and integrated circuit industries all rely on gallium.
- Photovoltaic cells and voltage regulators all require selenium.
- Many common household appliances, such as light bulbs, televisions, and stoves, contain cerium.
- Alloys containing dysprosium are utilised in wind generators, electric vehicles, and nuclear power plants.
- Lasers and optical fibers would be useless without erbium.
- The element europium has a variety of applications, including in LEDs, nuclear power plants, and lasers.

Think about neodymium, one of the most popular rare earths. When combined with iron and boron, neodymium produces magnets that are 12 times stronger than traditional iron magnets, a property that led to its initial application in green laser pointers. For both civilian and military use, there is a significant need for rare earth elements. When combined with tin, indium creates indium tin oxide, an alloy that is widely used in consumer electronics for its high electrical conductivity and optical transparency. This alloy is a key component in the production of flat-screen displays. Most of the rare earths that are essential to electronics are not widely known; erbium, for instance, is an essential component of optical fibres. Erbium is used for amplification, which is necessary for long-distance optical fibre transmission. Excitable erbium ions are driven into a high energy state by irradiation with a laser while they are embedded within brief segments of optical fiber.

Sales of smartphones, laptops, TVs, PCs, etc., are driving the worldwide rare earth elements market as a result of rising consumer expenditure and rising internet penetration. The increased demand for rare earth elements may be traced back to rising sales of electric and hybrid vehicles around the world, which in turn can be attributed to rising consumer awareness and rising government programmers and plans. In 2019, neodymium has the largest market share due to its widespread use in the production of various consumer electronics. These include everything from portable speakers and computer hard drives to hybrid vehicles and automobile batteries. As most of these essential minerals come from China, the supply situation for rare earth elements is complicated by global political conflicts [32,33]. It is projected that the global rare earth metals market would grow from its 2021 revenue of $7,063.8 mln to $15,473.0 mln in 2030, a CAGR of 9.1 %. The rising demand for these materials from the electronics, aerospace, and automotive sectors is the main reason for this prediction.

Due to their widespread application in modern manufacturing and the scarcity of their main production to a handful of locations, rare earths have recently attracted the attention of businesses, governments, and researchers. Sourcing alternatives to primary materials has been recommended as a way to lessen the severity of this problem with 31 essential materials [32-35].. With rare earths being present in many different types of 32 waste (such as electronic and lighting) and industrial byproducts (such as slags and coal combustion 33 products), scientists are investigating extraction at the laboratory scale.

REEs are vital components in a wide range of modern sectors, including electronics, renewable energy, and vehicle production. The extraction of REE from industrial wastes has received a lot of attention as the readily available REE materials dwindle. Scientific research is focusing on methods to recycle key materials, which are indispensable to industry [36,37]. However, there is currently no commercially viable method for recycling pure rare earth elements from magnets found in e-waste [36-39]. With 2.2 billion PCs, tablets, and smartphones scheduled to ship worldwide this year, some surveys have called this a massive squandered opportunity. Rare earth elements that are suitable for recycling are notoriously elusive. Time-consuming purifications, poor extractability, and excessive effluent streams are all issues with the current REE recovery technologies. The extraction rate, as noted by Vishwanath et al. [38], is highly sensitive to the initial REE content in the feed solution.

6.   Rare earths in clean technologies

The need to incorporate renewable production into the energy mix is becoming more pressing as the worldwide demand for petrochemical goods rises and competes for the limited oil resources currently mined as an energy source. The sun, the wind, geothermal heat, and the tidal currents are all renewable energy sources that can be harnessed to produce electricity, but their widespread adoption is limited by a dearth of readily available raw materials. Therefore, a reliable and continuous supply of REE is crucial for the development of renewable energy sources. For every percentage point that renewable energy production rises, REE reserves fall by 0.18 percentage points and greenhouse gas emissions increase by 0.90 percentage points throughout the extraction process. Renewable sources of energy as wind, geothermal, solar, tidal, and electric power are being prioritised [39] because to their widespread availability in nature.

To provide this greener, renewable power, numerous technologies utilise rare earth elements. Earth Elements,' which states that increased availability of rare earth minerals or crucial minerals will be necessary to meet the growing demand in the renewable energy sector, particularly the wind energy sector, to address climate change. Rare earth element and permanent magnet value chains are seen as essential for European industrial development because of their significance in the green transition and the digital transformation. Rare earth is most crucial in the renewable energy sector, and the sector's growth prospects are a deciding factor in the rare earth market [40,41].

94% of the transportation sector's use of renewable energy by 2040 is expected to come from electric vehicles (EVs) fueled by renewable energy, with biofuels accounting for the remaining 6%. By 2040, 1100 Mtoe of SDS's heat will come from renewable sources, up from the current level of 0 Mtoe (https://thinkrcg.com/rare-earth-metals-and-their-role-in-renewable-energy-benefits-and-challenges/). Geological exploration has uncovered 32 million tonnes of neodymium, praseodymium, dysprosium, and terbium; of this total, 1.6 million tonnes are in legally compatible reserves outside of China. Technologies that don't need fossil fuels are the Renewable energy technologies offer a promising new market for rare earths. Mineral resources used are different for each technological application. The density, melting point, conductivity, and thermal conductivity of REEs are all exceptionally high.

Classic mining metals like copper, nickel, manganese, graphite, and zinc all have an important function and will experience a growth in demand, despite the fact that the revolutionary potential of lithium and rare earth elements for energy storage has received new attention. These elements are crucial for the functioning of emission-free vehicles and power plants, such as wind turbines and electric cars, which in turn help to reduce the rate at which global temperatures are rising. The efficiency, life, and power of batteries depend on the presence of elements including lithium, nickel, cobalt, manganese, and graphite. Since the development of NdFeB permanent magnets, rare earths have risen to prominence in the technological world [42,43]. Their initial implementation in magnets prompted the downsizing of communication equipment like computers and mobile phones, and they are now being studied for a wide variety of other uses. In 2018, generators with permanent magnets were employed by nearly all European offshore wind turbines and by over 76% of offshore wind turbines worldwide (6). In onshore applications, however, where the demand for high-output generators in a compact package is not as stringent, permanent magnet generators may be supplanted. Rare earths can be utilised to increase the efficiency of silicon, which could find value in applications such as energy conversion devices [40] that rely on powerful motors.

Copper and aluminium are in great demand because they are used in electrical networks, and copper is the cornerstone of all energy-related technology. Increased demand for rare earths in the energy sector is caused by the sector's transition to new, low-emission technologies, much as the increase in demand seen in the car industry. Although REEs play a significant role in the production and distribution of conventional energy, they are used more in renewable energy technologies (especially wind turbines) [43]. Demand for REE production is expected to rise as more renewable energy sources are put into use. Installed wind power-generating capacity has increased dramatically over the past decade due to government renewable energy legislation and incentives as well as expanding commercial and business interest. Wind's proportion of total U.S. power output went from 2.3 percent to 7.3 percent between 2010 and 2019, and global installed capacity of wind power increased by more than 230 percent during the same time period [44].

As wind power generation has increased, so has the need for rare earths in the industry. Wind turbines are not alone in using rare earth elements (REEs) in magnets; many manufacturers employ permanent magnets with REEs since it is more cost-effective, requires less maintenance, and keeps working even when the wind speed is low. 159 Some wind turbines include as much as a metric tonne of rare earth elements (REEs), and a single turbine can use up to 300 pounds of neodymium and praseodymium. This can amount to several REES tonnes for a sizable wind farm [45-47]. The long-term health of the rare earths market will be determined by factors such as the rate of development in demand for rare earth-based products, the availability of viable alternatives to China as a supply source, and the likelihood of increased government backing. The demand side of the supply chain, which includes domestic separation and processing capacities, has to be actively engaged more.

7.   Uses of rare earth elements in defense

Rare earth elements are used in a wide range of items: communications equipment, precision-guided weapons, night-vision goggles, and stealth technology.

A beryllium-aluminum alloy is used to make the frames of five families of fighter jets, including the popular F-35 Joint Strike Fighter (https://lifeclub.org/books/rare-keith-veronese-review-summary). Beryllium's lightness makes the alloy a superior material, perfect for optimizing aerodynamic dexterity. Within many aircrafts and drones, we will find beryllium being used to build the electrical circuits. It can also be found in radar technology and the devices used to detect bombs and other explosives. Airborne radar transmitters and electronic warfare systems require samarium-cobalt and Nd-Fe-B permanent magnets due to their unique thermal and magnetic characteristics. The superior strength of NdFeB allows for the use of smaller and lighter magnets in defense weapon systems [48,49]. Rare Earth Materials are now gain importance in flat panel television displays, rechargeable batteries, defense products, and medical devices. SmCo retains its magnetic strength at elevated temperatures and is ideal for military technologies such as precision-guided missiles, smart bombs, and aircraft (https://www.army.mil/article/227715/an-elemental-issue#:text=Rare%20earths%2C%20including%20yttrium%20and,and%20weapons%20in%20combat%20vehicles).

In defense applications, TWT's and klystrons are used in satellite communications, troposcatter communications, pulsed or continuous wave radar amplifiers, and communication links. the use of rare earth elements in a variety of defense-related applications:

1.fin actuators in missile guidance and control systems, controlling the direction of the missile;

2.disk drive motors installed in aircraft, tanks, missile systems, and command and control centers;

3.lasers for enemy mine detection, interrogators, underwater mines, and countermeasures;

4.satellite communications, radar, and sonar on submarines and surface ships; and

5.optical equipment and speakers.

Little but powerful rare-earth magnets are essential to the motors and actuators used in missile guidance and control. Yttrium and terbium are two examples of rare earths utilized in laser targeting and weapons on military vehicles (https://www.heritage.org/defense/commentary/rare-earth-elements-arent-rare-they're-vital-national-security).Rare-earth batteries have various advantages, including a higher energy density, superior discharge characteristics, and less environmental impact during disposal. Numerous electronic components found in home appliances, media players, computers, cars, phones, and even military hardware have benefited from the miniaturization made possible by high-strength rare-earth magnets. Erbium is used to strengthen the signal in fiber-optic cables, allowing them to transfer data over greater distances (Rare earth elements like yttrium are needed for equipment on military vehicles and other weaponry).

Table 1. Rare – Earth Elements applications in Aerospace Field Ref. https://www.airandspaceforces.com/article/rare-elements-of-security/ .

Rare-Earth Elements Crucial to Defense

The U.S. has identified 35 metals or minerals crucial for its industrial base. Among them are:

Name	Properties	Aerospace Uses
Gallium	Superconductivity	Computer chips, light-emitting diodes
Neodymium	Extremely powerful, durable magnets	Missile guidance systems
Samarium	High-temperature magnetism, absorbs neutrons	Nuclear reactor control rods, lasers
Praesodymium	Makes stronger, more heat-tolerant alloys, permanent magnets	Aircraft engines, satellite components
Yttrium	Alloy strengthener, glass clarifier	Microwave emitters, optical coatings, LEDs
Promethium	Low radioactivity	Long-lived batteries for missiles
Lanthanum	Glass clarifier, reacts with hydrogen	Optics and lenses, night-vision goggles, fuel cells
Europium	Phosphorescence	LEDs, plasma displays

Table 1 specifies Rare – Earth Elements applications in Aerospace Field. There is a plethora of ways in which rare earth elements can be put to use in the military industry (https://www.airandspaceforces.com/article/rare-elements-of-security/): Fin actuators are used in guidance and control systems to direct a missile. The fin surfaces of smart bombs and other precision guiding weapons are manipulated before, during, and after discharge using rare-earth actuators. A rare-earth laser interrogator is used to spy on the enemy and foil their plans. Submarine and surface ship satellite communications, radar, and sonar; and Helicopters' audible imprint can be lessened with the help of NdFeB magnets and Terfenol-D speakers, two components of stealth technology. optical gear and audio system Rare-earth magnets are used in the waveguides of microwave generators like travelling wave tubes (TWT) and klystrons, which are used in applications as diverse as •avionics, •night vision equipment, and •satellites. Electron beam focus is achieved with the aid of rare earth permanent magnets in both TWT and klystron devices.

One example of a commercial off-the-shelf product that sees heavy use in military infrastructure is the computer hard drive, which uses several rare earth minerals. The use of yttrium-iron garnets (YIG) and yttrium-gadolinium garnets (YAG) is common in phase shifters, tuners, and filters (YGG). There is no practical way to replace rare earth minerals in defence systems that rely on them, according to government and industry representatives. Fin actuators on precision-guided missiles, for instance, use rare earth magnets made of neodymium iron boron (https://www.defensemedianetwork.com/stories/rare-earths-provide-critical-weapons-support/).

Considering their extended service lives and lack of an adequate substitute, rare earth minerals will remain indispensable in many future military systems. The samarium cobalt magnets in the Aegis Spy-1 radar contribute to its long lifespan. Defense officials are quoted as saying that rare earth materials will continue to be used in the next generation of some military system components. Rare earth minerals will be required in other cases as well, namely when developing new system components [48].

As a result of their strategic importance to the operation of today's increasingly networked and high-tech military, civilian technologies are increasingly being adopted for military usage. Other uses in aeronautical technology, security systems, and lasers. As military technology advances, rare earths will find increasing utility in a wide variety of future weapons (https://www.sciencehistory.org/learn/science-matters/case-of-rare-earth-elements/manufacturer-case-study). The use of radar and sonar aids in navigation, monitoring, and accident avoidance. There is radio use in the Patriot Missile Air Defense System [48]. Gadolinium, samarium, and yttrium are three of the most crucial rare earths (for more, see: https://www.mining.com/web/rare-earths-cross-earths-new-high-tech-arms-race/). Tanks and other armoured vehicles fitted with laser cannons [49]. An object may be located and followed up to 22 miles away. In both space-based and terrestrial communication systems, rare-earth lasers are employed for line-of-sight connectivity. The use of communication lasers in geosynchronous or geostationary orbit (GEO) satellites has been limited, although there are plans to expand their use in LEO spacecraft. Generating electricity for aeroplanes' electrical systems requires the employment of samarium-cobalt permanent magnets in generators.

8. Applications of rare-earth-doped materials in diagnostic imaging and therapeutic delivery and others

(a). Applications in Bioimaging that Make Use of Rare-Earth Doping

Bioimaging seeks to cause minimal disruption to living systems and can be used to learn about an object's external three-dimensional structure. Bioimaging is the study of structures at every level of organisation, from single molecules to organ systems to complex multicellular creatures. The technique can produce images through the use of visible light, fluorescence, ultrasound, X-ray, and magnetic resonance. Fluorescence imaging is more widely utilised because of its capacity to track changes in biological tissues in real time and its role in studying the dynamic interaction between medication molecules and tumour cells. Water solubility, biocompatibility, drug-loading capability, and the ability to target distinct cancers are just a few of the benefits that can be conferred upon rare earth-doped nanoparticles by surface functionalization.

Developing a novel biocompatible, luminous bioimaging agent for application in diagnostics, imaging, and preventative medicine is essential for the advancement of biomedicine. Softer hydroxyapatite (HAp) was doped with rare earth ions ($Gd3+$, $Eu3+$) to improve its luminescent optical property.

Fan etal (50) described rare earth-doped nanoparticles have also been widely used in disease diagnosis, drug delivery, tumor therapy, and bioimaging. Among various bioimaging methods, the fluorescence imaging technology based on the rare earth-doped nanoparticles can visually display the cell activity and lesion evolution in living animals, which is a powerful tool in biological technology and has been widely applied in medical and biological fields. The rare earth-doped nanoparticles can be endowed with the water solubility, biocompatibility, drug-loading ability, and the targeting ability for different tumors by surface functionalization. This confirms its potential in the cancer diagnosis and treatment.

Rare earth doped up conversion nanoparticles (UCNPs) are a new type of luminous materials that can absorb long-wavelength near-infrared light, as described by Hong et al. (51). Fan et al. highlight the vast range of applications for nanoparticles doped with rare earth elements, which includes everything from disease diagnosis and medicine delivery to tumour therapy and bioimaging (50). The capacity to visualise cell activity and lesion progression in living animals has made fluorescence imaging based on rare earth-doped nanoparticles an invaluable tool in the medical and biological fields. Water solubility, biocompatibility, drug-loading capacity, and the ability to target different malignancies are only few of the benefits of surface functionalizing rare earth-doped nanoparticles. This demonstrates its potential for use in cancer diagnosis and treatment. emitting photons with a short wavelength in the ultraviolet-visible range. UCNPs cause minimal harm to living tissue, may penetrate deeply into tissues, have a high imaging sensitivity, and are unaffected by photobleaching.

Doping hydroxyapatite with luminous compounds for use in biomedical luminescence imaging is a novel and potentially fruitful strategy in medicine. As a result of their many desirable properties, including biodegradability, bioactivity, biocompatibility, osteoconductivity, nontoxicity, and a lack of toxicity and inflammation, as well as their ease of surface adaptation, research into luminescent materials based on hydroxyapatite is an exciting area of study. Hydroxyapatite (51), the primary inorganic component of bones, is widely acknowledged to have an important role in tissue engineering, medication and gene delivery, and other biomedical fields. Hydroxyapatite nanoparticles (HAP NPs) were emphasised by Gu et al. (52) as host materials that can be changed with different substrates and dopants. In particular, HAP NPs that have been doped with rare earth (RE) ions have garnered interest because of the unusual physicochemical and imaging features they exhibit. High brightness, high contrast, photostability, nonblinking, and narrow emission bands are just a few of the advantages that RE-doped HAP NPs show when compared to other fluorescent probes. Multiple medical fields have found utility for hydroxyapatite nanoparticles as an imaging probe, drug delivery vehicle, bone tissue engineering material, and in antibacterial research (53).

Nanoparticles that have been doped with rare earth elements have found numerous applications in bioimaging, medication delivery, tumour therapy, and illness diagnosis. Because of its capacity to clearly portray the cell activity and lesion evolution in living animals, the fluorescence imaging approach based on the rare earth-doped nanoparticles has found significant use in the medical and biological fields. Emissions in the near infrared range (700-1700 nm) have characteristics of high penetration due to their low absorbance, low photon scattering, and negligible autofluorescence interference. Water solubility, biocompatibility, drug-loading capability, and tumour targeting ability are just a few of the benefits that can be conferred on rare earth-doped nanoparticles through surface functionalization.

Fluorescence imaging in the near infrared spectrum's longer wavelength range (NIR-II, 1,000-1,700 nm) is increasingly being used by oncologists for early cancer detection. Using modified NIR-II imaging for in vitro and in vivo cancer diagnostics with rare earth nano particles (RENP). Rare-earth-doped nanoparticles show promise as an NIR-II biomedical imaging agent due to their low toxicity, great photostability, deep tissue penetration, and adaptive pharmacokinetic behavior. The use of NIR-II RENPs in bioimaging still faces challenges, despite these advancements.

The excessively huge sizes of RENPs have also been a source of concern for bioimaging professionals for quite some time. Luminescence intensity is sacrificed so that nanoparticles can be made tiny enough to penetrate biological tissues and even cells. RENPs will be used in multispectral molecular imaging and medication administration monitoring in the near future. Nanoscale materials have been used as functional components in a variety of biomedical devices. Nanoparticles, due to their ability to be functionalized with distinct features, have a wide range of potential uses in biomedicine. These include drug delivery to targeted tissues or cells [54–58] and the qualitative or quantitative detection of tumour cells (such as magnetization, fluorescence, and near-infrared absorption). Scientists' efforts are indescribable, and their quest for new information will never halt.

(b). Rare Earth-Doped Lasers

In order to function properly, nearly all contemporary technological devices require rare earth elements, sometimes known as "The Vitamins of Modern Industry" (http://metalpedia.asianmetal.com/metal/rare earth/application.shtml). Rare earth elements (REEs) are becoming increasingly important in state-of-the-art technology, driving up both demand and pricing. Due to their wide range of absorption and fluorescence transitions, rare earth ions are attractive options for use as active ions in laser materials. These aren't typically used as active ingredients in lasers, but rather as a co-dopant to accomplish things like build saturable absorbers, quell a population at a certain energy level through energy transfer processes, or act as optically passive components of laser crystals. Tb^{3+} and Dy^{3+} lasers, as well as Eu^{3+} lasers with a wavelength of 0.7 m, are all examples of lasers that emit in the visible spectrum. Some are used as phosphors, for instance in LEDs (https://www.rp-photonics.com/rare-earth-doped-laser-gain-media.html).

In addition to these unique properties, rare earths also tend to have a high quantum efficiency, a long lifetime of metastable states, and a low dependence on host materials for their emission and absorption transition wavelengths. Rare earth ions, according to their unique set of characteristics, perform admirably in a wide variety of optical applications (Gain media for rare earth doped lasers may be found at https://www.rp-photonics.com/.). The invention of fibres that magnify light by stimulated emission was a significant step forward in optical fibre technology. These not only provided the building blocks for several novel optical sources and signal processing devices, but also resulted in a huge rise in the channel capacities of fibre communication systems.

Even though rare-earth-doped laser crystals and glasses are among the most regularly utilised solid-state laser gain media, ceramic media have recently attracted a lot of attention [58,59]. This medium has been doped with trivalent rare earth ions (i.e., have a triple positive charge).

Absorption spectra [58] of rare-earth ion-doped optical fibres (Sm^{3+}, Tm^{3+}, Tb^{3+}, Pr^{3+}, and Ho^{3+}) were analysed to ensure wavelength-selective absorption in fibre lasers. When working with lasers that emit radiation of extremely high energy density (REEs), these fibres can shield the pump diode from this harmful reflected radiation. It is possible to obtain both high transmission (loss less than a few tenths of a decibel) and substantial absorption (10-20 dB) at the useful laser wavelength [60].

Rare Earth - A tribute to the late Mr. Rare Earth, Professor Karl Gschneidner   Materials Research Forum LLC
Materials Research Foundations 164 (2024) 177-210   https://doi.org/10.21741/9781644903056-4

Due of its high gain, yttrium aluminium garnet (YAG) is a fantastic material for lasers. The lasing transition in a Nd: YAG laser takes place between two energy levels of a $Nd3+$ ion. In a YAG crystal co-doped with $Er3+$ and $Ho3+$ ions, lasing is improved due to the more favourable energy transfer between the ions. Yttrium lithium fluoride is yet another typical laser host crystal. Due to their similar ionic size and outer electronic structure, materials containing Eu, Tm, or Yb can easily substitute for $Y3+$ (https://physicstoday.scitation.org/doi/10.1063/PT.3.4397).

(c) Rare Earth-Doped Fiber Lasers

The fibre laser was created by Elias Snitzer, who also showed its effectiveness in 1963. However, it wasn't until the 1990s that serious commercial applications appeared. The field of fibre lasers is one of the newest and most quickly expanding in the laser industry. Their ability to produce light at a variety of wavelengths means they have numerous applications in the manufacturing and processing sectors [61,62]. They find additional applications in the spheres of medicine and communication. The term "fibre lasers" is used to describe a wide variety of lasers and laser technologies, such as: - kW CW fibre lasers - nsec pulsed fibre lasers - ps & fs ultra-fast fibre lasers. All of these amplifiers rely on rare-earth-doped fibres [63,64]. Since the thermal load is spread out along the length of the fibre, the efficiency of a fibre laser is much higher than that of other solid-state lasers, reducing the need for expensive and inefficient cooling. Stable single mode operation is a result of the fibre design, not the laser cavity optics, which determines the fibre laser's beam quality. This revolution in power scaling from CW fibre lasers has been made possible in part by the fiber's great efficiency. In well under a decade, the output power of a "laser" made from nearly single mode fibre rocketed from the low hundreds of watts to well over ten kilowatts.

In the 1990s, photosensitive (PS) fibres were developed and fine-tuned for use in the telecommunications sector (key component in WDM systems). For the fibre laser business, these "telecom" fibres were modified to produce high power FBGs. Recent years have seen the widespread adoption of LMA versions of PS fibres such as 20/400 for application in the fabrication of high reflector (HR) and output coupler devices (OC). To minimise intra-cavity splice losses and boost laser dependability, it is essential to maintain precise control over the most important fibre characteristics [64]. Increasing the output power of Yb-doped LMA fibres without sacrificing beam quality or experiencing multimode instability (MMI) is a key area of research. The advancement of fibre technology towards higher performance lasers is continuing unabatedly. The ultimate aim of these new fibre technologies is to increase the core diameter to greater than 30 metres, which is the current upper limit for LMA fibres, without compromising beam quality. Rare earth-doped fibre lasers and amplifiers have emerged as a critical part of modern optical communication systems thanks to a number of recent technological breakthroughs.

- A summary on Therapeutic and diagnostic applications of rare earth elements in modern medicine (EC Giese, Clin Med Rep, 2018; 2(1): 1-2; doi: 10.15761/CMR.1000139).

- MRI scans with La -Lanthanum oxide nanoparticles have been shown to be successful.

- Ce – Cerium-doped lutetium orthosilicate is a scintillator most commonly employed in positron emission tomography (PET) imaging.

- Nanoparticles of pr - praseodymium oxide have been employed in radiation therapy.

- Crystals made from the element neodymium (Nd) are used in lasers for a variety of medical applications, including the elimination of hair and the treatment of skin cancer.

- Patients with malignancies that have spread into the bone can find relief from their excruciating pain with the use of the radioisotope Sm-153.

- The optical features of Eu-Europium nanoprobes make them useful in a variety of biomedical applications, including in vitro and in vivo bioimaging as well as heterogeneous and homogeneous biodetection.

- With its magnetic properties, gadolinium (Gd) is used in intravenous radio-contrast medications for MRI scans, where it improves tumor pictures.

- Tb - Tb-149 is a radioisotope used in cancer therapy.

- For the purpose of treating rheumatoid knee effusions, the radioisotope Dy-165 has been used.

- Solid-state lasers based on holmium have been utilised for non-invasive surgery on cancer and kidney stones.

- The medical and dental communities have benefited from lasers with an Er -Erbium atom at its core.

- In order to power portable X-ray machines, the radioisotope Tm-167 has been employed.

- Yb-176 is a radioactive isotope that can be used to create Lu-177, a radioisotope with interesting medical applications.

- Studies on Lu-Lutetium are being conducted to see whether or not it may be used in targeted radiotherapy to accelerate the development of new cancer medicines for diseases such as prostate cancer.

9.   Environmental impacts of rare earth Materials

Rare earth elements are crucial for eco-friendly technologies to exist (REEs). Rare earth elements are available almost everywhere. They form an integral part of the foundation upon which modern society stands. Innumerable technological advances can be traced back to their ingenuity. Some rare earth elements are highly sought after for their application in technology and consumer goods due to their exceptional magnetic, luminescent, and electrical properties. This is due to the fact that numerous essential tools rely on rare earth elements. Since these technologies are at the heart of global communication, transportation, medical, energy creation, surveillance, and even war, questions of access and applications are at the center of hot geopolitical debates. Fig.3. Lists the Rare Earth Elements availability in the World.

WHERE IN THE WORLD ARE ALL
THE RARE EARTHS?

Rare earth elements (REEs) are a group of 17 elements whose importance is critical in high technology. Their use has exploded as electronics and renewable technologies increasingly have become part of everyone's daily lives.

Rare earths are abundant in the Earth's crust but mineable concentrations are less common, making reserves potential very valuable and strategic.
*The USGS tracked the world's reserves in tons (imperial).

Fig. 3 shows the Rare Earths availability in the world. Source:
https://www.visualcapitalist.com/rare-earth-elements-where-in-the-world-are-they/

On the other hand, many of the methods used in REE mining and production are harmful to the environment. Concerns have been raised concerning the potential environmental implications of mining and producing rare earth elements (REEs) due to their rising importance in green technologies [65]. The geology of the deposit, the kind and content of the minerals, the extraction techniques, the availability of power and other resources in the area, and the presence of regulations designed to reduce negative environmental effects all play a role. Open pit and underground mining techniques, including drilling and blasting for hard rock deposits, are used to manage primary deposits. Dredge mining, manual surface collection, and mechanized dry mining utilizing standard earthmoving equipment are all taken care of by the secondary deposit managers. When extracted from the ground and processed into fine dusts, rare earth elements often coexist with other elements that are detrimental to human health, such as heavy metals, arsenic, and fluorite. Toxic trace metals like arsenic, lead, cadmium, mercury, and uranium are a focus of environmental scientists' research.

The environmental effects of products and technology can be measured through a technique called life cycle assessment (LCA). Emissions are categorised as having an acidification, eutrophication, or toxicity impact using the LCA technique, which also accounts for the greenhouse gas effect [66].

Because of the radioactive and potentially dangerous byproducts of REE processing, a great deal of hazardous waste and tailings are generated. Crushed rocks and fluids used in mills and concentrators make up tailings. There is potential for environmental harm with each and every step of the rare earth element mining process. Rare earth element access begins with mining. It then goes through a refinery before being dumped. Problems arise at every stage of the mining, processing, and disposal processes. The extraction of REEs has negative consequences, one of which being the pollution and deterioration of the water supply. The groundwater and watering holes that unique plants and animals rely on are being depleted. Farms and villages in the vicinity were contaminated by rare earth mining waste, forcing thousands of people to flee. Painful legs, diabetes, osteoporosis, and heart problems are just a few of the health difficulties that locals face [67]. The mining sector has a serious issue with its excessive water consumption. Due to the high water consumption associated with mining for rare earth elements (https://www.dw.com/en/toxic-and-radioactive-the-damage-from-mining-rare-elements/a-57148185), this practise has been criticised. Balaram recently analyzed all the issues connected to rare earth elements in detail [4].

Fig 4. Theoretical framework for evaluating risks generally (Ref.5 : Katarzyna Kapustka etal, 2019)

Depending on mine management and safety procedures, processing methods, and waste treatment systems, every type of mining can have unintended consequences (Fig.4) [68,69]. Climate, the proximity of streams, lakes, and subterranean aquifers, and the type of rock surrounding the REE deposit (and the probable presence of other metals and agents within it) are also important factors. Some of the risks to health could result from exposure to REE, while others could be caused by the overall mining activities' potential impact on the local environment. The mine's rock and dust waste could contaminate the local soil, which would have negative effects on the local fauna and flora.

Rare earth elements are found in the same rocks as radioactive elements like uranium and thorium. The radioactive waste produced during the extraction and processing of rare earth elements remains radioactive for very long periods of time. Because of the presence of radioactive isotopes in rare earth elements, the quality of the ore must be carefully controlled. Processing the ore, which can result in thorium production, is usually the main source of radiation concern. Since alpha particle emissions are produced during the bulk of thorium's decay process, a special monitoring strategy centred on such sources is warranted. Though alpha radiation is not very far-traveling, it is more dangerous to cells when ingested. Rare earth carbonates are a natural defense against the acidity that can be produced by the several acidic leaching procedures used in the refining process.

Kidney illness, heart disease, occupational lung disease, and an increased risk of developing several forms of cancer have all been cited as possible effects [70,71]. However, there is a dearth of long-term epidemiological investigations. Rarely, exposure to REE can cause nephrogenic systemic fibrosis, a condition characterised by an abnormal accumulation of connective tissue in the skin, joints, eyes, and internal organs. This condition is connected with an increased risk of heart attack. * Those who lived in an area rich in REE had abnormally low amounts of a certain protein in their blood. * IQ scores dropped dramatically in REE-exposed kids. This has * changed the way human red blood cells divide and reproduce, which is assumed to be the result of a disruption in brain neurotransmitters. It interfered with DNA replication and repair, too. Research has linked leukemia to REE pollution in the environment.

Rare earth elements Applications are shown in Fig.5.

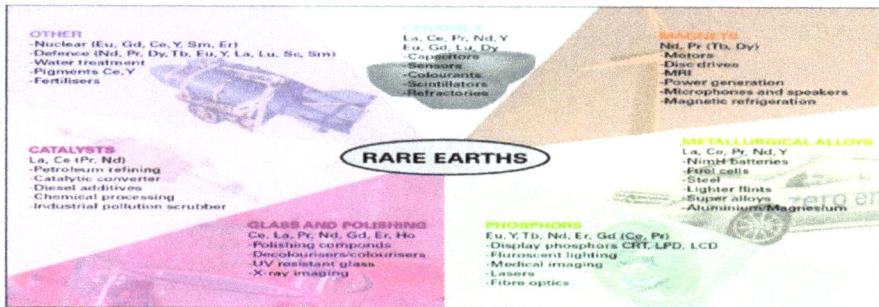

Fig.5. Rare earth elements Applications (As cited in Walters, Lusty, and Hill (2011). As stated in (Authors: Walters, A., Lusty, P., & Hill, A. Rare earth elements, a commodity profile from the British Geological Survey's online mineral profile series)

Multi-scalar social, cultural, and political processes co-evolve with the rare earth industry and its accompanying research, technology, and policy. In order to achieve sustainable development, it is essential that the environment and public health be protected from the negative effects of rare earth element mining and processing as new pollutants.

Fig.6 Rare earth element (REE) soil sources: Ramos et al. [72]

Rare earth element (REE) sources and soil distribution were studied by Ramos et al. [72]. Rare earth elements (REE) are explored in relation to their function in soils (Fig.6) and the plants they sustain. Risks to ecosystems and human health from REE in soils were also demonstrated. According to Ramos et al. [72], who talk about REE in soils and how they could move up the food chain (2016).

Bone, liver, and lungs are three major storage organs. Diseases of the heart, liver, blood, and kidneys, as well as digestive, skeletal, neurological, pulmonary, and cytogenetic effects, can all be attributed to this substance [73–75].

10.   Amazing applications of rare earths

RREs have a wide variety of applications, and as Fig.7 shows, their popularity has grown in recent years [76].

Fig. 7- The wide variety of rare earths applications. https://etn-demeter.eu/what-are-rare-earth-elements-rees/ (76).

We arbitrarily divide rare earths into two categories: There are two kinds of rare earth materials: the lighter ones are called light rare earth elements and oxides (LREEs and LREOs), while the heavier ones are called heavy rare earth elements and heavy rare earth oxides (HREEs and HREOs).

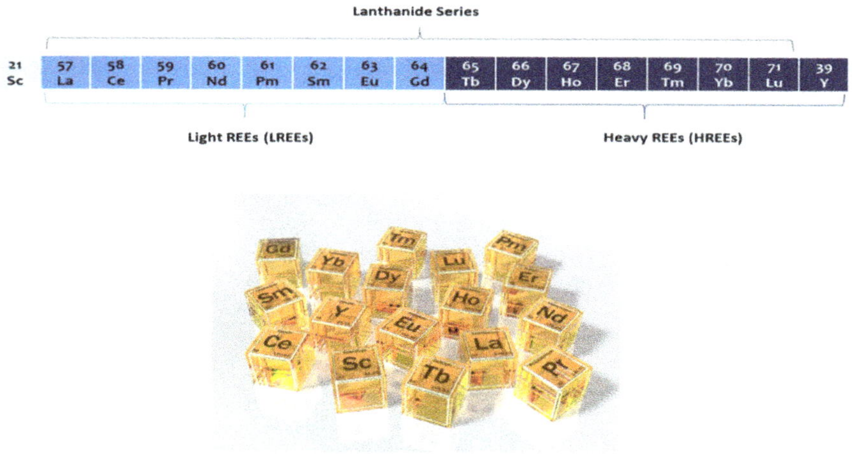

Fig. 8. Rare earth elements (REEs) are metals in the first row of the periodic table below the main body. https://www.thoughtco.com/rare-earth-elements-list-606660.

Uses for Rare Earth Elements [77–80]: (https://www.woodmac.com/news/editorial/rare-earth-elements-frequently-asked-questions/).

- The red phosphor used in color cathode ray tubes and liquid crystal displays, such those used in computers and televisions, contains europium as an essential element.
- Green phosphors for lasers and LED TVs can be made with terbium.
- In order to produce a fluid cracking catalyst that improves the efficiency with which crude oil is transformed into gasoline, the oil refining industry must rely on lanthanum.
- Fertilizers containing rare earth elements (REEs) have been employed in modern farming.
- batteries that can be charged multiple times.
- Cerium oxide is frequently used for the purpose of polishing glass. Everything from mirrors to glasses to optical lenses is manufactured from polished glass.
- Gadolinium is used in solid-state lasers and in computer memory chips.
- Car catalysts that reduce emissions.
- REEs are employed as an additive in many alloys.
- Control rods, diluents, shielding, detectors, and counters are just some of the many places you might find rare earth elements in the nuclear business.
- Rare metals lessen power transmission losses by lowering friction in power lines.

To create the powerful permanent magnets needed for high-efficiency electric motors, neodymium is required. The military relies heavily on NdFeB magnets, widely regarded as the strongest permanent magnets in the world, for a wide range of weapon systems. Since SmCo retains its magnetic strength even when heated, it is ideal for use in military equipment such guided missiles, smart bombs, and aeroplanes [80].

In order to transmit signals across great distances, fiber-optic cables use erbium-doped fibre at regular intervals to serve as laser amplifiers.

High-efficiency fluorescent lamps employ RREs. They are 70 times more energy efficient and produce far less heat than regular bulbs.

The most frequent rare earth-related components in electric vehicles are nickel-hydrogen batteries. Figure 9 depicts the wide range of technological applications for rare earth elements.

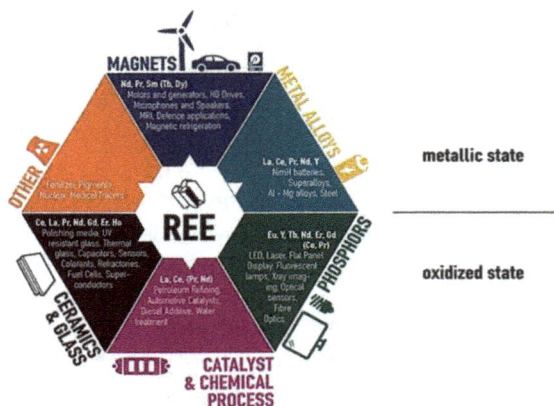

Fig.9 Rare Earth Elements usage in various technologies:
https://www.researchgate.net/figure/Rare-Earth-Elements-usage-in-various-
technologies_fig2_331438895.

Table.2 shows some more applications of the rare earths. (https://www.drishtiias.com/daily-news-analysis/rare-earth-metals-1).

End-Use Category	Description	Ai
Battery Alloys	Rare earth elements are used to produce anode materials for nickel-metal hydride ("NiMH") batteries. NiMH batteries are used in hybrid electric vehicles, consumer electronics, cordless shavers, cordless powertools, baby monitors and other applications of rechargeable batteries.	
Catalysts	Rare earth elements, such as cerium and lanthanum, are used in catalytic converters of gasoline- and diesel-powered vehicles, as well as fuel cracking catalysts and additives used by oil refiners to break down crude oil into lighter distillates, such as gasoline, diesel, kerosene and more.	
Ceramics, Pigments and Glazes	Rare earth elements are used to produce decorative ceramics, functional ceramics, structural ceramics, bio ceramics and many other types of ceramics used in everything from jet engine coatings to ceramic cutting tools, dental crowns, ceramic capacitors, ceramic tiles, and more.	
Glass Polishing Powders and Additives	Rare earth elements, such as cerium, are used to polish optical glass, hard disk drive platters, LCD display screens and gemstones, among a long list of applications. Cerium is also used as an additive in UV-filtering glass and container glass, whereas lanthanum, yttrium and gadolinium are used to produce high quality optical glass used in camera lenses, microscopes and telescopes.	
Metallurgy and Alloys	Rare earth mischmetal (a mixture of light REE metals) is used during production of some types of steel, as well as ductile iron making. Rare earth elements are also used to produce a variety of different alloys, such as ferro-cerium, ferro-holmium, ferro-gadolinium, ferro-dysprosium and a growing list of others.	
Permanent Magnets	Rare earth elements are used to produce high-strength permanent magnets that have enabled the production of ubiquitous gadgets and electronics, such as mobile phones and laptops, as well as power dense energy-efficient electric motors and generators used in electric vehicles, wind power generators, energy efficient appliances and hundreds of other applications.	
Phosphors	Rare earth elements are used in phosphors for energy efficient lamps, display screens and avionics, and are added to fiat currency in some nations as an anti-counterfeit measure.	
Other	Aside from the above described end uses and categories, rare earth elements are used in a long list of other end uses and applications, including many in defense, medicine, agriculture, high-tech and chemical industries.	

Source: Adamas Intelligence

For further information about rare earth elements refer : https://www.global-reia.org/about-rare-earth/: https://www.drishtiias.com/daily-news-analysis/rare-earth-metals-1.

In order to polish glass and add color and specific optical qualities, rare earth element (REE) raw materials are used extensively in the glass industry. Up to half of the lenses used in digital cameras and mobile phone cameras are made of lanthanum. Disk drives, both those for optical media like CDs and DVDs, and hard drives, rely on rare-earth magnets. When a disc drive spindle is powered by a rare-earth magnet, its rotational motion becomes extremely stable.

How rare earth elements and tech metals are used?

Many of these tech metals and rare earth elements are essential to the functioning of the modern technological marvels we have at our fingertips (https://www.cummins.com/news/2021/04/19/what-are-tech-metals-and-rare-earth-elements-and-how-are-they-used) : Following is a synopsis of the various RRE components and the tasks they serve :

Cerium is used in light bulbs, TVs and ovens.

- Dysprosium is mixed within alloys used in wind turbines, electric vehicles and nuclear reactors.
- Erbium is used in lasers and fiber optic cables.
- Europium is used in light bulbs, nuclear reactors and lasers.
- Gadolinium is used in magnets, nuclear reactors and magnetic resonance imaging (MRI).
- Holmium is used in magnets and nuclear reactors.
- Lanthanum is mixed within alloys that are used in batteries and hydrogen vehicles.
- Lutetium is used as a catalyst in refineries.
- Neodymium is used in magnets and lasers.
- Praseodymium is used in aircraft engines, fiber optic cables and magnets.
- Promethium is used in pacemakers and guided missiles.
- Samarium is used in microwave devices and magnets.
- Terbium is used in light bulbs, memory devices and x-rays.
- Thulium is used in lasers.
- Ytterbium is used in displays, x-ray machines and fiber-optic cables.
- Yttrium is used in radars and as an additive within alloys used in high tech devices.
- Scandium is used for fuel cells and alloys used in jet planes.
- Cobalt is used in super-alloys, jet turbines and rechargeable batteries.
- Lithium is commonly used in batteries.
- Tantalum is used in capacitors, resistors and other electronic equipment.
- Indium is often used in LCD screens.
- Gallium is used in integrated circuits, LEDs and semiconductors.
- Niobium is used in steel alloys that are a part of jet engines and rockets.
- Selenium is used in photocells and rectifiers.
- Zirconium is used in nuclear power stations.

11. Methods of rare earth element recovery

Rare earth elements are a group of seventeen chemical elements that occur together in the periodic table. The creation of state-of-the-art electronics, renewable energy sources, and automobiles all rely on ready access to rare earth elements (REEs) in key locations. Without rare earth elements, many aspects of contemporary life would be much more difficult. Rare earth metals are essential to the operation of many contemporary technology, including smartphones, tablets, laptops, televisions, hybrid vehicles, wind turbines, solar panels, and more. A term with no universally accepted definition [81], "rare earth metals" is commonly used to refer to elements required in negligibly small amounts. Rare earth elements are crucial to the operation of modern renewable energy systems and technology. (https://news.climate. The other rare earth metals are crucial to numerous cutting-edge technologies. Among the many things that fall into this category are solar panels, smartphones, computer processors, medical imaging, aircraft engines, and weapons systems. In a circular economy, rare earth element recovery from waste is a potential alternative to using easily obtained minerals. Between 2021 and 2040, rare earth element demand is projected to grow by a ratio of 3.5 to 7.5.

Recovery, disassembly, separation (preprocessing), and further processing as other metals are the four main phases of REE recycling. Due to the inadequacy of current recycling technologies in addressing the interplay between "economics," "technology," and "society," it is doubtful that Rare earth elements (REEs) will ever be recycled in large quantities. Rare earth metals (REMs) are inefficient to recover from melted garbage, and recycled rare earth oxides have low market value (REOs). Recovering REE from industrial wastes has gained a lot of attention as conventional REE material sources become depleted. The total REE levels in these secondary wastes are sometimes lower than those in REE materials [81-83], and recycling yields continue to be extremely low.

The "REE balance difficulty" is overcome when produced devices contain only the required REE, notwithstanding their low amounts in nature. Only a small amount of rare earth elements (REEs) is needed instead of isolating every lanthanide. Electronic waste recycling could be made more financially viable with the simultaneous recovery of precious metals. Fujita et al. [82] detail some of the more up-to-date techniques for REE extraction and separation, highlighting the approaches' environmental benefits such reduced reliance on harmful chemicals and lower energy costs. While recycling electronic waste is helping to alleviate the shortage of rare earth metals, it is not nearly enough to meet the ever-growing demand. Saving energy during the mining and processing processes, preserving natural resources, and decreasing pollution and greenhouse gas emissions are all additional benefits of recycling and reusing materials.

Whatever the case may be, effective separation and processing processes are crucial to the prosperity of the mining and recycling sectors. But the current situation of REE recycling is somewhat discouraging (only 2 percent of REEs are recovered by recycling processes compared with 90 percent for iron and steel). Phosphors for lamps (17%), magnets (7%), and nickel metal hydride batteries (NiMH) (5 percent). One percent. After all, we're discussing an increase of 10%. Rare earth elements can be recycled into a profitable enterprise if they are priced high enough. This means that during end-of-life recycling, the most crucial and essential REEs should be gathered first [84].

The extent to which rare earths could be improved through recycling is unknown. It's difficult to begin with an accurate assessment of the number of rare earths already present in products on the market. Rare earths can be recovered from a variety of sources, including slags from pyrometallurgical metal recycling operations, bauxite residue (red mud), phosphor gypsum, mining tailings, and industrial effluent. Rare earths can be found in all of these waste streams at concentrations low enough to be classified as a by-product, but there are many of them and they are not the primary reason for the rubbish. As a result, there are significant levels of rare earths in these waste streams, suggesting they could serve as a sustainable source of these elements. Phosphate gypsum is the source of the other rare earths, whereas red mud is a fascinating waste stream from which scandium might be recovered [85].

REE recycling has several global applications, including magnets, abrasives, phosphors, and batteries. The need to recycle rare earth elements is a relatively new problem. While recycling rare earth elements (REEs) has been widely touted, it's crucial to keep in mind that the process can expose workers to harmful chemicals. Several distinct approaches of recycling are outlined in this article:

One, Rare earth magnet, batteries and phosphors recycling

Two, reusing rare earths from battery production.

Third, recycling rare earths from light and illumination and display field.

Four, Recycling of rare earths from catalytic devices.

Due to rising demand for recycled electronics in regions like the European Union, China, and India, e-waste recycling has become increasingly common around the world [84,85]. Two primary methods exist to ensure a constant supply of rare earth elements. Coal ash, ocean sediments, and slag are all examples of secondary resources that fall under the umbrella term of "primary resources" (electronic and industrial waste). Recycling electronic waste could be a major theoretical step toward meeting the need for rare earth materials. The broad acceptance of e-waste recycling [83] is in large part due to the increasing demand for recovered electronics in places like the European Union, China, and India. There are mainly two ways to guarantee a steady supply of rare earth elements. Secondary resources like coal ash, ocean sediments, and slag are still part of the primary resource category [85].

REEs are not normally recycled in large numbers due to their poor yield and high cost; but, if mandated by legislation or if REE prices soared, recycling may be profitable. Balaram [4] has provided an in-depth analysis of the processes used for recycling and deconstructing electronic trash and the chemistry of REE extraction, elucidating the parallels between the two. New evidence reveals that recycled REEs can be a viable technological and cost-saving alternative to virgin REEs. And there's a plethora of approaches from which to pick. Due to their limited durability, rare earths require extensive physical and chemical processing before they may be recycled. It is imperative that novel recycling technologies, in vivo analytical procedures, and biomarkers be created that do not negatively damage the natural world. Due to its crucial significance in green applications like hybrid automobiles, electric car batteries, and wind turbines in the face of climate change and global warming, the rising global demand for these elements is currently generating much of attention among exploratory geochemists and technology developers.

Finding alternative materials to replace or reduce rare earths' quantities, or redesigning products to rely on them less, are also options being studied. Although rare earths have a wide variety of uses, many of them are extremely specialized, with no suitable substitutes or ones that have significantly lower performance. For instance, no material has been discovered that can produce a magnet as powerful as those made from neodymium. However, efforts are being made to build electric motors that don't require rare earths. Recently, Japan announced plans to build a motor with magnets that don't rely on rare earths, while Tesla Motors has elected to use induction motors in their mass-produced electric vehicles so as to avoid utilizing rare earths at all.

References

[1] Rajesh Kumar Jyothi, Thriveni Thenepalli, Ji Whan Ahn, Pankaj Kumar Parhi, Kyeong Woo Chung, Jin-Young Lee, 2020, Review of rare earth elements recovery from secondary resources for clean energy technologies: Grand opportunities to create wealth from waste,

Journal of Cleaner Production, Volume 267, 122048, ISSN 0959-6526.
https://doi.org/10.1016/j.jclepro.2020.122048

[2] Mubashir Mehmood. 2018, Rare Earth Elements- A Review. J Ecol & Nat Resour, 2(2): 000128. https://doi.org/10.23880/JENR-16000128

[3] Drobnjak, A., and Mastalerz, M., 2022, Rare Earth Elements-A brief overview: Indiana Geological and Water Survey, Indiana Journal of Earth Sciences, v. 4. https://doi.org/10.14434/ijes.v4i1.33628

[4] Balaram, V., 2019, Rare earth elements-A review of applications, occurrence, exploration, analysis, recycling, and environmental impact: Geoscience Frontiers 10 (4) 1285- 1303. https://doi.org/10.1016/j.gsf.2018.12.005

[5] King, H.M., 2021, REE - Rare earth elements and their uses, <https://geology.com/articles/rare-earth-elements, 14, 2021.

[6] Voncken, J.H.L., 2016, The rare earth elements-an introduction: SpringerBriefs in Earth Sciences, Springer (1st ed.) ISBN-10: 3319268074. https://doi.org/10.1007/978-3-319-26809-5

[7] Gielen, D. and M. Lyons ,2022, Critical materials for the energy transition: Rare earth elements, International Renewable Energy Agency, Abu Dhabi.

[8] Aide, M. , & Nakajima, T. , (Eds.). 2020. Rare Earth Elements and Their Minerals. IntechOpen. https://doi.org/10.5772/intechopen.77602

[9] Hobart M. King, Rare Earth Elements and their Uses, 2005-2022, https://geology.com/articles/rare-earth-elements/

[10] Gschneidner, Karl A., Jr., and Pecharsky, Vitalij K. 2022, "rare-earth element". Encyclopedia Britannica, https://www.britannica.com/science/rare-earth-element. Accessed 28 September 2022.

[11] King, H.M., 2021, REE - Rare earth elements and their uses, https://geology.com/articles/rare-earth-elements.

[12] Voncken, J.H.L., 2016, The rare earth elements-an introduction: Springer : Briefs in Earth Sciences, Springer (1sted.) ISBN-10: 3319268074, <https://link.springer.com/book/10.1007/978-3-319-26809-5>.

[13] Agnieszka Drobniak and Maria Mastalerz, 2022 ,Rare Earth Elements: A brief overview ,Indiana journal of earth Sciences, ISSN 2642-1550, Volume 4, 2022. https://doi.org/10.14434/ijes.v4i1.33628

[14] USGS, 2002. Rare Earth Elements - Critical Resources for High Technology. USGS Fact Sheet 087-02. http://pubs.usgs.gov/fs/2002/fs087-02/.

[15] Simon M Jowitt (Ed.), 2018, Criticality of the Rare Earth Elements: Current and Future Sources and Recycling, Resources,Pages: 172 , 2018, SBN 978-3-03897-017-0 (Pbk); ISBN 978-3-03897-018-7. https://doi.org/10.3390/books978-3-03897-018-7

[16] A.R. Jha, Rare, 2014, Earth Materials Properties and Applications, CRC Press, ISBN,9781466564039

[17] Subasri Arunachalam, Balakrishnan Kirubasankar, Duo Pan, Hu Liu, Chao Yan, Zhanhu Guo, Subramania Angaiah, 2020, Green Energy & Environment, Volume 5, Issue 3,2020, Pages 259-273, ISSN 2468-0257. https://doi.org/10.1016/j.gee.2020.07.021

[18] Ioana Craiciu, Mi Lei, Jake Rochman, Jonathan M. Kindem, John G. Bartholomew, Evan Miyazono, Tian Zhong, Neil Sinclair, Andrei Faraon, 2019, Nanophotonic quantum storage at telecommunications wavelength, Physical Review Applied, 12, 024062. https://doi.org/10.1103/PhysRevApplied.12.024062

[19] Mehta, B. R., Aruna, I., & Malhotra, L. K. 2011. Rare earth gadolinium nanoparticles for hydrogen induced switching, sensing and storage devices. In C. C. Thompson (Ed.), Gadolinium: Compounds, Production and Applications, pp. 341-350.

[20] X. Xian Qin, Xiaowang Liu, Wei Huang, Marco Bettinelli, and Xiaogang Liu, 2017, Lanthanide-Activated Phosphors Based on 4f-5d Optical Transitions: Theoretical and Experimental Aspects, Chemical Reviews, 117 (5),4488-4527. https://doi.org/10.1021/acs.chemrev.6b00691

[21] P.Ascenzi, M.Bettinelli,· A.Bo3· M.Botta· G.De Simone· C.Luchinat· E.Marengo· H.Mei6· S.Aime , Rare earth elements (REE) in biology and medicine, 2020 ,Rendiconti Lincei. Scienze Fisiche e Natural. https://doi.org/10.1007/s12210-020-00930-w

[22] Z. Fan Q, Cui X, Guo H, Xu Y, Zhang G, Peng B. 2020, Application of rare earth-doped nanoparticles in biological imaging and tumor treatment. J Biomater Appl.;35(2):237-263. https://doi.org/10.1177/0885328220924540

[23] A. Nethi, S.K., Bollu, V.S., P., N.A., & Patra, C.R. 2020. Rare Earth-Based Nanoparticles: Biomedical Applications, Pharmacological and Toxicological Significance. https://doi.org/10.1007/978-981-15-0391-7_1

[24] B. Qize Zhang, Stephen O'Brien, Jan Grimm, 2022, Biomedical Applications of Lanthanide Nanomaterials, for Imaging, Sensing and Therapy, Nanotheranostics; 6(2): 184-194. https://doi.org/10.7150/ntno.65530

[25] D.Teo RD, Termini J, Gray HB. 2016, Lanthanides: applications in cancer diagnosis and therapy. J Med Chem. (2016) 59:6012-24. https://doi.org/10.1021/acs.jmedchem.5b01975

[26] E. Jinyu Wang and Sheng Li, 2022, Applications of rare earth elements in cancer: Evidence mapping and scient metric analysis, Front. Med.,Sec. Nuclear Medicine, https://doi.org/10.3389/fmed.2022.946100

[27] F. Faust, Andreas. 2020,"Medical applications of rare earth compounds". Rare Earth Chemistry, edited by Rainer Pöttgen, Thomas Jüstel and Cristian A. Strassert, Berlin, Boston: De Gruyter, 2020, pp. 439-452. https://doi.org/10.1515/9783110654929-028

[28] T. Cheisson, E. J. Schelter, 2019 , Rare earth elements: Mendeleev's bane, modern marvels. Science363, 489-493. https://doi.org/10.1126/science.aau7628

[29] Gabrielle Gaustad, Eric Williams, and Alexandra Leader, 2020 ,Rare earth metals from secondary sources: Review of potential supply from waste and 3 byproducts, https://www.sciencedirect.com/science/article/pii/S0921344920305309, 2020 , Elsevier

[30] Binnemans, K., P. T. Jones, B. Blanpain, T. Van Gerven, Y. Yang, A. Walton and M. Buchert 2013. "Recycling of rare earths: a critical review." Journal of Cleaner Production 51: 1-22. https://doi.org/10.1016/j.jclepro.2012.12.037

[31] Buchert, M., A. Manhart, D. Bleher and D. Pingel ,2012. "Recycling critical raw materials from waste electronic equipment." Freiburg: Öko-Institut eV 49(0): 30-40.

[32] Costis, S., K. K. Mueller, J.-F. Blais, A. Royer-Lavallée, L. Coudert and C. M. Neculita, (2019) Review of recent work on the recovery of rare earth elements from secondary sources, INRS, Centre Eau, Terre et Environnement.

[33] Nakamura, E. and K. Sato ,2011. "Managing the scarcity of chemical elements." Nature materials 10(3): 494 158. https://doi.org/10.1038/nmat2969

[34] R. K. Jyothi, T. Thenepalli, J. W. Ahn, P. K. Parhi, K. W. Chung, J.-Y. Lee, 2020,eview of rare earth elements recovery from secondary resources for clean energy technologies: Grand opportunities to create wealth from waste. J. Clean. Prod. 267, 122048-122073 (2020). https://doi.org/10.1016/j.jclepro.2020.122048

[35] V. G. Deshmane, S. Z. Islam, R. R. Bhave, 2020,.Selective recovery of rare earth elements from a wide range of e-waste and process scalability of membrane solvent extraction. Environ. Sci. Technol. 54, 550-558 (2020). https://doi.org/10.1021/acs.est.9b05695

[36] Bing Deng, Xin Wang, Duy Xuan Luong, Robert A. Carter1, Zhe Wang, Mason B. Tomson, James M. Tour , 2022 , Rare earth elements from waste , Sci. Adv.8, 3132 . https://doi.org/10.1126/sciadv.abm3132

[37] R. K. Jyothi, T. Thenepalli, J. W. Ahn, P. K. Parhi, K. W. Chung, J.-Y. Lee, 2020 ,Review of rare earth elements recovery from secondary resources for clean energy technologies: Grand opportunities to create wealth from waste. J. Clean. Prod.267, 122048-122073. https://doi.org/10.1016/j.jclepro.2020.122048

[38] Vishwanath G. Deshmane, Syed Z. Islam, and Ramesh R. Bhave, 2020, Selective Recovery of Rare Earth Elements from a Wide Range of E-Waste and Process Scalability of Membrane Solvent Extraction, Environmental Science & Technology 2020 54 (1), 550-558. https://doi.org/10.1021/acs.est.9b05695

[39] European Commission, Joint Research Centre, Alves Dias, P., Bobba, S., Carrara, S., et al., 2020,The role of rare earth elements in wind energy and electric mobility: an analysis of future supply/demand balances, Publications Office, 2020, https://data.europa.eu/doi/10.2760/303258.

[40] W.D. Judge, Z.W. Xiao and G.J. Kipouros , 2017 ,Application of Rare Earths for Higher Efficiencies in Energy Conversion , The Minerals, Metals & Materials Society 2017,H. Kim et al. (eds.), Rare Metal Technology 2017,The Minerals, Metals & Materials Series. https://doi.org/10.1007/978-3-319-51085-9_4

[41] Talens Peiró, L., Villalba Méndez, G. 2013. Material and Energy Requirement for Rare Earth Production. JOM 65, 1327-1340 (2013). https://doi.org/10.1007/s11837-013-0719-8

[42] Gielen, D. and M. Lyons ,2022, Critical materials for the energy transition: rare earth elements, International Renewable Energy Agency, Abu Dhabi.

[43] Coey, J. 2020, 'Perspective and prospects for rare earth permanent magnets', Engineering, Vol. 6, Issue 2, pp. 119-131. https://doi.org/10.1016/j.eng.2018.11.034

[44] Deru Yan , Sunghyok Ro , Sunam O and Sehun Kim ,2020 ,On the Global Rare Earth Elements Utilization and Its Supply-Demand in the Future , IOP Conf. Series: Earth and

Environmental Science 508 (2020) 012084 IOP Publishing. https://doi.org/10.1088/1755-1315/508/1/012084

[45] Zhou, B.; Li, Z.; Chen, 2017,C. Global Potential of Rare Earth Resources and Rare Earth Demand from Clean Technologies. Minerals, 7, 203. https://doi.org/10.3390/min7110203

[46] Alonso, E.; Sherman, A.M.; Wallington, T.J.; Everson, M.P.; Field, F.R.; Roth, R.; Kirchain, R.E. ,2012, Evaluating rare earth element availability: A case with revolutionary demand from clean technologies. Environ. Sci. Technol. 2012, 46, 3406-3414. https://doi.org/10.1021/es203518d

[47] Lucas, J., Lucas, P., Le Mercier, T., Rollat, A. and Davenport, W. 2014, Rare Earths: Science, technology,production and use, Elsevier, Amsterdam. https://doi.org/10.1016/B978-0-444-62735-3.00017-6

[48] Valerie Bailey Grasso, 2013 ,Specialist in Defense Acquisition, CRS report, U.S. Department of Defense. Congressional service. Annual Industrial Capabilities Report to Congress, December.

[49] GAO-10-617, 2010, Rare Earth Materials in the Defense Supply Chain, United States Government Accountability Office Washington, DC 20548.

[50] Fan Q, Cui X, Guo H, Xu Y, Zhang G, Peng B. 2020, Application of rare earth-doped nanoparticles in biological imaging and tumor treatment. J Biomater Appl. Aug;35(2):237-263. Epub 2020 May 19. PMID: 32423319. https://doi.org/10.1177/0885328220924540

[51] Hong E, Liu L, Bai L, Xia C, Gao L, Zhang L, Wang B. 2019, Control synthesis, subtle surface modification of rare-earth-doped upconversion nanoparticles and their applications in cancer diagnosis and treatment. Mater Sci Eng C Mater Biol Appl.Dec;105:110097. Epub 2019 Aug 17. PMID: 31546381. https://doi.org/10.1016/j.msec.2019.110097

[52] Neacsu IA, Stoica AE, Vasile BS, Andronescu E. Luminescent Hydroxyapatite Doped with Rare Earth Elements for Biomedical Applications. Nanomaterials (Basel). 2019 Feb 10;9(2):239. PMID: 30744215; PMCID: PMC6409594. https://doi.org/10.3390/nano9020239

[53] Gu M, Li W, Jiang L, Li X. 2022, Recent progress of rare earth doped hydroxyapatite nanoparticles: Luminescence properties, synthesis and biomedical applications. Acta Biomater. ;148:22-43. Epub 2022 Jun 5. PMID: 35675891. https://doi.org/10.1016/j.actbio.2022.06.006

[54] Zhenfeng Yu, Christina Eich and Luis J. Cruz , 2020 ,Recent Advances in Rare-Earth-Doped Nanoparticles for NIR-II Imaging and Cancer Theranostics , Front. Chem., Sec. Nanoscience. https://doi.org/10.3389/fchem.2020.00496

[55] https://nano-magazine.com/news/2020/8/20/application-of-rare-earth-doped-nanoparticles-in-biological-imaging-and-tumor-treatment

[56] Farokhzad, O.C.; Langer, R. ,2009. Impact of nanotechnology on drug delivery. ACS Nano, 3, 16-20. https://doi.org/10.1021/nn900002m

[57] Wiglusz, R.J., 2021, Nanostructural Materials with Rare Earth Ions: Synthesis, Physicochemical Characterization, Modification and Applications. Nanomaterials , 11, 1848. https://doi.org/10.3390/nano11071848

[58] .Kochergina, T,A, Aleshkina, S.S. , khudyakov, M.M., M.V. Yashkov, D.S. Lipatov, A.N. Abramov, L.D. Iskhakova, M.M. Bubnov etal.; 2018, Use of rare-earth elements to

achieve wavelength-selective absorption in high-power fibre lasers Quantum Electronics 48 (8) 733 - 737 . https://doi.org/10.1070/QEL16740

[59] Chet R. Bhatt, Jinesh C. Christian L. Goueguel, Dustin L. McIntyre, and Jagdish P. Singh , 2017Determination of Rare Earth Elements in Geological Samples Using Laser-Induced Breakdown Spectroscopy (LIBS) ,Applied specteroscopy, Volume 72, Issue 1, https://doi.org/10.1177/0003702817734854

[60] Gaft, M., Panczer, G., Uspensky, E., & Reisfeld, R. 1999. Laser-induced time-resolved luminescence of rare-earth elements in scheelite. Mineralogical Magazine, 63(2), 199-210. https://doi.org/10.1180/002646199548439

[61] M. J. F. Digonnet, 2001 ,Rare-Earth-Doped Fiber Lasers and Amplifiers,(2001, 2nd edn., CRC Press, Boca Raton, FL. https://doi.org/10.1201/9780203904657

[62] Markus Pollnau, Chapter 296 - Rare-Earth-Doped Waveguide Amplifiers and Lasers, Editor(s): Jean-Claude G. Bünzli, Vitalij K. Pecharsky, 2017, Handbook on the Physics and Chemistry of Rare Earths, Elsevier,Volume 51,111-168, ISSN 0168-1273, ISBN 9780444638786. https://doi.org/10.1016/bs.hpcre.2017.04.001

[63] Larry D. Merkle , Rare-Earth-doped Laser Materials: Spectroscopy and Laser Properties, 2012 ,Materials Research Society symposia proceedings. Materials Research Society 1471, https://doi.org/10.1557/opl.2012.1206

[64] George Oulundsen and Bryce Samson, 2013 ,Rare Earth Doped Fibers for Use in Fiber Lasers and Amplifiers, Photonics Media Webinar.

[65] Zapp, P., Schreiber, A., Marx, J. et al. 2022, Environmental impacts of rare earth production. MRS Bulletin 47, 267-275 (2022). https://doi.org/10.1557/s43577-022-00286-6

[66] Paul, J., & Campbell, G. 2011. Investigating rare earth element mine development in epa region 8 and potential environmental impacts (908R11003). U.S. Environmental Protection Agency. Retrieved from website: http://www.epa.gov/region8/mining/ReportOnRareEarthElements.pdf

[67] Saleem H. Ali, Social and Environmental Impact of the Rare Earth Industries, Resources, 2014, 3, 123-134; doi:10.3390/resources3010123 4. US EPA. 2012. https://doi.org/10.3390/resources3010123

[68] Rare Earth Elements: A Review of Production, Processing, Recycling, and Associated Environmental Issues. www.miningwatch.ca/files/epa_ree_report_dec_2012.pdf

[69] Katarzyna Kapustka, Dorota Klimecka,Gerhard Ziegmann , 2019, The management and potential risks reduction in the processes of rare earths elements , CzOTO , volume 1, issue 1, pp. 77-84. https://doi.org/10.2478/czoto-2019-0010

[70] Julie Michelle Klinger. 2018, Rare earth elements: Development, sustainability and policy issues, The Extractive Industries and Society ,2018, https://doi.org/10.1016/j.exi.2017.

[71] Rim KT.2016, Effects of rare earth elements on the environment and human health: A literature review. Toxicology and Environmental Health Sciences volume 8, pages189-200. https://doi.org/10.1007/s13530-016-0276-y

[72] Ramos, S.J., Dinali, G.S., Oliveira, C. et al. 2016. Rare Earth Elements in the Soil Environment. Curr Pollution Rep 2, 28-50. https://doi.org/10.1007/s40726-016-0026-4

[73] Nastja Rogan Šmuc, Tadej Dolenec, Todor Serafimovski, Matej Dolenec. 2012, Geochemical characteristics of rare earth elements (REEs) in the paddy soil and rice (Oryza sativa L.) system of Kočani Field, Republic of Macedonia. Geoderma. 2012;183-184:1-11. https://doi.org/10.1016/j.geoderma.2012.03.009

[74] Li X, Chen Z, Chen Z, Zhang Y. 2013, A human health risk assessment of rare earth elements in soil and vegetables from a mining area in Fujian Province, Southeast China. Chemosphere. Oct;93(6):1240-6. Epub 2013 Jul 25. PMID: 23891580. https://doi.org/10.1016/j.chemosphere.2013.06.085

[75] Wyttenbach, A.; P. Schleppi, J. Bucher, V. Furrer, L. Tobler ,1994. The accumulation of the Rare Earth Elements and of Scandium in successive: Weedle Age classes of Norway Spruce. Biological Trace Element Research, 41:13 - 29. https://doi.org/10.1007/BF02917214

[76] https://etn-demeter.eu/what-are-rare-earth-elements-rees/.

[77] Gschneidner; 1981, Industrial Applications of Rare Earth Elements ACS Symposium Series; American Chemical Society: Washington, DC. https://doi.org/10.1021/bk-1981-0164

[78] Geology. 2012,"REE - Rare Earth Elements and their Uses." (http://geology.com/articles/rare-earth-elements/.

[79] Helmenstine, Anne Marie, Ph.D., 2020, "Rare Earth Elements List." ThoughtCo, thoughtco.com/rare-earth-elements-list-606660.

[80] T. Cheisson, E. J. Schelter, 2019, Rare earth elements: Mendeleev's bane, modern marvels. Science 363, 489-493. https://doi.org/10.1126/science.aau7628

[81] Goonan, T.G., 2011, Rare earth elements-End use and recyclability: U.S. Geological Survey Scientific Investigations Report 2011-5094, 15 https://doi.org/10.3133/sir20115094

[82] Fujita, Y., McCall, S.K. & Ginosar, D. 2022, Recycling rare earths: Perspectives and recent advances. MRS Bulletin 47, 283-288. https://doi.org/10.1557/s43577-022-00301-w

[83] R. K. Jyothi, T. Thenepalli, J. W. Ahn, P. K. Parhi, K. W. Chung, J.-Y. Lee, 2020.Review of rare earth elements recovery from secondary resources for clean energy technologies: Grand opportunities to create wealth from waste. J. Clean. Prod. 267, 122048-122073. https://doi.org/10.1016/j.jclepro.2020.122048

[84] Bing Deng, Xin Wang, Duy Xuan Luong, Robert A. Carter etal. ,2022 ; Rare earth elements from waste, Sci. Adv.8, eabm3132. https://doi.org/10.1126/sciadv.abm3132

[85] Giovanni Pagano [Ed.], 2017, Rare Earth Elements in Human and Environmental Health: At the Crossroads between Toxicity and Safety, Pan Stanford Publishing Pte. Ltd. Singapore 038988.

Rare Earth - A tribute to the late Mr. Rare Earth, Professor Karl Gschneidner   Materials Research Forum LLC
Materials Research Foundations 164 (2024) 211-229   https://doi.org/10.21741/9781644903056-5

Chapter 5

Rare-Earth Doped Materials for Optical Information Storage

C.S. Kamal[1] and D. Haranath[2*]

[1]BVM, Andhra Pradesh, India

[2]Luminescence Materials and Devices (LMD) Group, Department of Physics, National Institute of Technology Warangal, 506004, Telangana, India

haranath@nitw.ac.in

Abstract

Nowadays, in this digital era, we are constantly bombarded with information. In recent years, researchers have paid a lot of attention to new information storage formats and storage methodologies. Optical information storage offers a number of benefits that make it a viable future alternative for information storage in comparison to magnetic media and semiconductor memory. Due to the inherent $4f^n$ orbital and rich energy-level structures of rare-earth elements, research on rare-earth doped materials for optical information storage has propelled the development of state-of-the-art luminescent materials as information carriers in the 21st century by emitting polychromatic radiation covering the entire visible and near-infrared (NIR) spectral region when irradiated by NIR, visible, or ultraviolet (UV) light.

Keywords

Rare Earth Ions, Optical Storage, Optical Memory, Multiplexing, Optical Stimulated Luminescence

Contents

1.   Introduction

With the rapid advancement of IoTs, mobile tech, and AI we are beginning to reach an information explosion age [1]. At this exceptional development rate, the entire huge quantity of digital information will exceed at least 10^{11} TB by 2025, and this information explosion compels scientists and engineers to create new technologies that may significantly increase information storage capacity [2,3]. Moreover, the information must be kept indefinitely and must be readily accessible over the decades. Existing information centers built on magnetic hard-disk drives, which may have a lifespan of only two years and will consume almost one-fifth of all electricity in the globe by 2025, face an impossible barrier in meeting these strict requirements. [4-5] In order to keep up with the exponential growth of digital information produced by IoTs, mobile tech, and AI, storage modes and mediums must undergo a constant improvement. In recent years, optical and optoelectronic information storage systems have replaced magnetic information storage because to their relatively high efficiency, less energy consumption, better safety, greater capacity, and longer lifetime [5-10]. Despite the age of information explosion, conventional optical and electronic information storage (flash memory and memristive memory) still suffer from a lack of storage capacity, making it impossible to keep up with the massive amounts of information being produced daily [11,12].

Rare-earth doped materials provide new avenues for expanding the storage capacity of conventional optical and electronic information storage, opening the door to exciting new possibilities. New possibilities for light-matter interaction have emerged with the development of nanophotonics, which has made it possible to manipulate materials at the nanoscale using light [13]. Rare-earth doped materials used in optical and optoelectronic information storage can be optimized for specific purposes due to their adaptable optoelectrical features [14-17]. Further, multiplexing optical information storage offers a remarkable method to significantly increase storage space by expanding beyond a two-dimensional space into a multi-dimensional space, such as the multiplexing of wavelength, intensity, and time gate [18-22].

Rare-earth ions in all-optical information storage are indeed an emerging field with many outstanding concerns, such as how to optimally integrate a stimuli-responsive function with luminescence. Energy transfer between luminescent and photochromic units or crystal field disruption of electron transitions have been used to create rare-earth doped photochromic materials with luminescence switching behavior. In addition, rare-earth doped photochromic materials may also manipulate photochromism and luminosity by altering rare-earth ion concentrations and species. [23–31]. Recently, researchers have paid special attention to the properties of these materials and their potential use as a medium for storing visual information. [32–34] To save the information, the charge carriers must be arrested in traps before being subjected to X-rays, ultraviolet (UV), or visible light. Meanwhile, information may be read out by releasing the trapped charge carriers in response to stimulation from a high-temperature or low-energy laser. Hence, electron-trapping materials with an eye toward optical information storage, trap levels are naturally crucial.

Persistent phosphors are a kind of self-sustaining luminous material that can absorb light, store the light energy, and then release the stored energy as photon emission in response to thermal agitation at room temperature [35-38]. A material containing metastable trapped charge carriers may undergo optically stimulated luminescence when it is re-excited by an external light source and returns to its thermodynamic equilibrium state [39-42]. Erasable optical information storage has great promise because to its novel properties, such as all-optical controlled carrier storage and release. In this regard the nano-scintillators $NaLuF_4:Tb^{3+}$ $NaYF_4$ have recently been described as a high-resolution 3D X-ray extension imaging storage medium that can trap high-energy X-rays for weeks [43]. Despite these advances, ferroelectric materials doped with rare-earth ions are essential inorganic photochromic materials with excellent photoluminescence and visible light-induced color shifts. Photochromic process modulates luminescence emission intensity to readout visible light storage in rare-earth doping well. Doping rare-earth ions into ferroelectric matrix also creates transparent materials and decrease ferroelectric grain development. Small particle size reduces pores and grain dispersion, improving optical transparency. As a result, rare-earth ion doped ferroelectric materials became more popular for use in information storage processes including writing, erasing, and optical marking.

In this book chapter, we discuss about the developments in rare-earth doped materials and technology, which helps in building the next generation of optical information storage systems. First, we give an overview of optical information storage based rare-earth doped materials and what is going on in the industry right now. Then, we talk about basic aspects of rare-earth ions in solids, concept of multiplexing for optical storage information using rare-earth doped materials, followed by types of materials for Optical information storage.

2.   Basic aspects of rare-earth ions in solids

Digital information and optical information storage play critical roles in everything from global institutions to individual lives. There has been increasing focus on quantum information processing, storage, and communication in recent decades [44-50]. Quantum cryptography provides an almost unconditional layer of communication security, and these applications provide unparalleled processing powers. Getting them to work in the solid form is intriguing because it leaves room for improvement in other areas of technology.

The quantum bit (qubit) is the basic unit of information, analogous to the bit in classical computing. A qubit, or quantum bit, is a two-level quantum system with just two possible states, rather than the classical bit. Its eigenvalues, 0 and 1, are the same as the classical binary numbers 0 and 1. (see Fig.1). In contrast to the binary representations of 0 and 1 that are limited to classical bits, qubits may take on arbitrarily complex superposition states.
This superposition state can be written in the following form [44]

$$|\Psi\rangle = \cos(\theta/2)|0\rangle - \exp i\phi \, \sin(\theta/2)|1\rangle \tag{1}$$

$|\Psi\rangle$ and $\exp i\varphi \, |\Psi\rangle$ are equivalent wavefunctions and $\exp i\varphi$ a global phase factor.

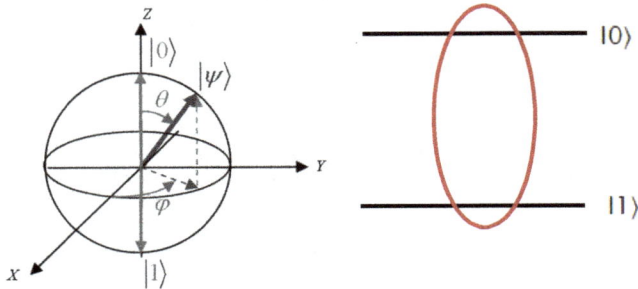

Figure. 1 A qubit in a superposition state $|\Psi>$. *Left: Bloch sphere. Right: superposition of states.*

Superposition states may be easily shown using vectors that extend out from the center of a Bloch sphere. The vector's spherical coordinates are written using the angles and from equation 1. When the vector is pointed toward the north pole or the south pole, the system transitions to the states |0> and |1>. Thus, a qubit may take on an arbitrary number of states, but only the states |0> and |1> can be recognized with any degree of accuracy. (Reproduced with permission from Ref [44])

Superposition of states provide more accurate information transfer and computation results, but only if the final states can be read out reliably. Furthermore, superposition states are very complex and easily disturbed, making isolated systems preferable and interactions with the environment are necessary for control and reading out of the superposition states. Several promising quantum systems are studied, including photons and nuclear spins in insulators and semiconductors, trapped ions, ultra-cold atoms, and superconducting circuits [51-55]. It is important to interface light to the materials in order to process and store information when light is utilized as a quantum information carrier [56]. The coherence time (T_2), is associated to the duration of the superposition states and the time available for quantum manipulation or information storage. When it comes to coherence, the choice of material makes all the difference. To date, rare-earth ion doped into dielectric host lattices have shown to be one of the most promising material classes. Using rare-earth ion transitions, it is feasible to span the whole visible range as well as the IR. These systems have lengthy optical and spin coherence states (at low temperatures). Careful compositional tweaking and the use of interactions with external electric and magnetic fields may also be used to fine-tune the degree of line broadening [44, 53, 57]. Hence, rare-earth ions are promising systems for quantum information and storage applications because their partially filled 4f shell is well shielded from the outer environment by their filled $5s^2$ and $5p^6$ shells; consequently, the ^4f_n-^4f_n transitions are only mildly perturbed by the crystalline environment. The 4f shell is so well protected that the crystal field may be seen as a little disturbance of the free ion levels. In addition, excited state lifetimes may be rather lengthy, lasting up to several microseconds. As a result, resonances of trivalent rare-earth ions may display extremely narrow inhomogeneous and homogeneous line widths (Fig.2). Line broadening processes may be classified into two types: inhomogeneous Γinh broadening and homogeneous Γh broadening. Static perturbations in the vicinity of the optical center, such as microstructural strain, point defects, and dislocations, cause the broadening of the inhomogeneous line width [58-59].

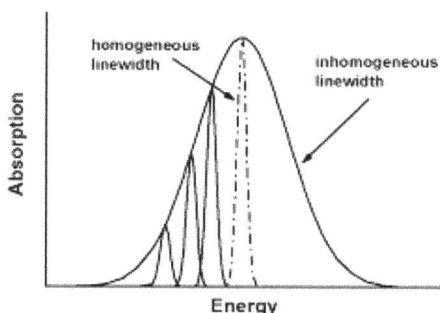

Figure 2. Example for inhomogeneous (Γinh) and homogeneous linewdiths (Γh). (Reproduced with permission from Ref [58])

Since the associated crystal field varies slightly, these disturbances induce a minor shift in the center frequencies of the optical centers. The inhomogeneous linewidth Γinh is therefore the sum of numerous homogeneously widened lines centered on their individual resonant frequencies. Typical values for inhomogeneous linewidths in rare-earth doped crystals are 0.5-

100 GHz [18]. In exceptional instances, isotopically pure elements may narrow inhomogeneous linewidths. $7LiYF_4$:$170Er^{3+}$ has 16 MHz inhomogeneous linewidths [60, 53], which is the smallest optical ensemble linewidth ever measured in a solid. However, for frequency domain storage and processing, it may be beneficial to induce disorder and increase inhomogeneous linewidths while retaining a suitable absorption coefficient and as narrow homogeneous linewidths as feasible. Therefore, in such a storage context, the figure of merit is Γinh/Γh. By contrast, dynamical events in the vicinity of the optical centers are responsible for homogenous line broadening.

3. The concept of multiplexing for optical storage information using rare-earth doped materials

The primary objective of multiplexing in information storage is to maximize the storage of information within physically constrained memory elements [61,62]. For this reason, thermochromic, magnetic, multicolor fluorescent, and optically variable color-changing inks have been employed in security printing of banknotes, identification cards, trademark tags and to determine forgery, tampering, and counterfeiting. The matrix of optical codes used in multiplexing must be able to reliably identify individual nano- or micro-sized objects at fast speed and at cheap cost. Further, one of the most widely used strategies for multiplexing is the utilization of fluorescent color codes, which have found application in many different scientific disciplines. In the past several decades, for instance, cytometry paired with fluorescent labeling has been the cutting edge of fast detection of single cells. However, three to five lasers, tens of filters, and up to 20 light detectors are needed for current multi-color flow cytometry, and the crowded spectral domain restricts it to less than 20 channels. Due to the inevitable spectrum overlap, color compensations need to be intricate and time-consuming. Therefore, developing new coding dimensions, such as mass spectra, fluorescence lifetime, and Raman spectra, is the major difficulty in multiplexing. When these new identities can be reliably decoded (detected) by a simple, fast, and low-cost technology, these additional dimensions become very valuable [63-65].

Storage capacity is a problem for optical storage technology despite significant advancements. This is due to the constraint of optical diffraction. Multiplexing the information being stored is necessary to increase storage capacity [66-71]. When exposed to light, some compounds undergo a reversible phase transition, changing from one chemical state to another and therefore altering the absorption properties of the material [72–78]. Each state may stand in for the "0" or "1" digital code, corresponding to the "off" or "on" positions, respectively. In addition, the photochromic effect was time-dependent upon light activation, which allowed for multiplexed encoding [79–81] and photochromic reaction might be used to increase information storage capacity. Hence, research into appropriate photochromic materials was stressed by many researchers to enable their use in optical storage.

In addition, linking the interactions between luminescent centers and color centers, Haiqin Sun et al. [82] created a family of reversible luminescent modulation materials based on photochromism in a variety of ferroelectric materials. In addition, their prior work has demonstrated that the optical features of rare-earth ions (such as emission bands, excitation modes) significantly impact optical information storage and recording [83], which is particularly relevant when considering ferroelectric materials. In addition, the luminescent switching behavior both before and after photochromism is clearly influenced by the excitation or emission wavelengths, demonstrating that the luminescent readout features may be successfully manipulated by excitation and emission wavelength fine-selectivity. Rare-earth ion luminescent readouts with multiple-color emissions are seen as appropriate security elements [84,85] due to their visual distinguishability. Luminescence from conventional single-doped rare-earth ions is typically monochromatic and unimodal, and a performance difference among both luminescence and photochromism may reduce the reading level. Light from the ultraviolet to the near infrared may practically be completely emitted by effective luminescent centers made from rare-earth ions. The combination of rare-earth ions should provide multicolor luminescence, as opposed to the monochromatic luminescence of traditionally single-doped rare-earth ions [86 - 88]. In $Bi_{2.5}Na_{0.5}Nb_2O_9$, which has been co-doped with Pr^{3+} and Er^{3+}, the team led by Haiqin Sun was able to create multicolor luminescence emission and tunable photo switching behavior. By adjusting the excitation wavelength and Er^{3+} ion concentration, the material's emission color shifted from red to green [89]. By altering luminescence decay durations to produce a temporal coding dimension in a broad microsecond-to-millisecond range, Yiqing Lu and his colleagues describe a novel multiplexing idea. They have found a way to precisely tune the micro- and millisecond luminescence decays and make different populations of rare-earth-doped nanocrystals with different lifetimes. This creates a new optical coding and decoding dimension, which expands the use of up-conversion materials in nanoscale photonics. Their lifetime multiplexing idea is based on the fact that lifetimes can be changed. This idea has been put into practice by co-doping Yb^{3+} ions as a sensitizer and changing the concentration of blue-emitting Tm^{3+} ions in the $NaYF_4$ nanocrystals. In this model, the lifetime can be changed by how much energy moves from the sensitizer to the emitter ion at different distances between the two (Fig. 3). We were able to change the lifetime of 40 nm $NaYF_4$:Yb,Tm nanocrystals in the blue emission band from 48 ms by using this method. One can make more than ten populations of nanocrystals with different lifetimes ranging from 25.6 ms to 662.4 ms in a single color band and decode their well-separated lifetime identities that are independent of both color and intensity.

Rare Earth - A tribute to the late Mr. Rare Earth, Professor Karl Gschneidner   Materials Research Forum LLC
Materials Research Foundations 164 (2024) 211-229   https://doi.org/10.21741/9781644903056-5

These Tm-doped "t-dots" could be used for multi - channel bioimaging, high-throughput cytometry quantification, high-density information storage, and security codes to prevent people from making counterfeits. The work of Antonio Ortu's team focused on repeaters that use time-multiplexing and on-demand read-out in time, both of which rely heavily on the extended storage periods made possible by hyperfine states. They use dynamical decoupling methods and a weak magnetic field to store six temporal modes in a $Eu^{3+}:Y_2SiO_5$ crystal for 20, 50, and 100 ms, respectively, where each temporal mode stores on average roughly one photon. By storing two time-bin qubits for 20 milliseconds, we can ensure the memory's quantum coherence. Their average memory output fidelity is F = (85 ± 2) %, with an overall number of photons per qubit of μ_{in} = 0.92 0.04. At memory read-out, they accomplish the qubit analysis with a novel composite adiabatic read-out pulse [79].

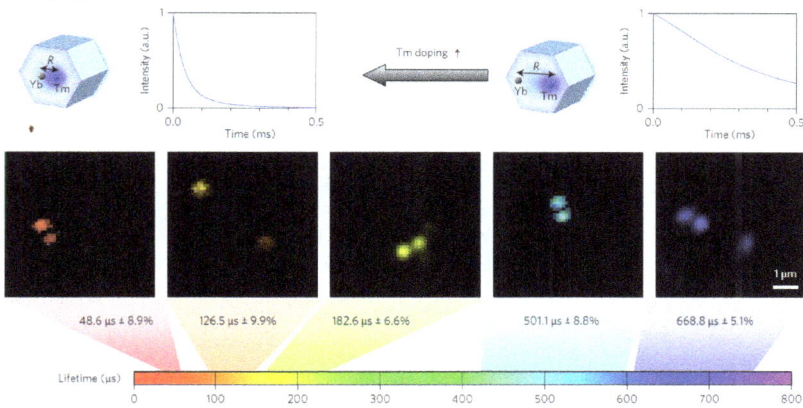

Figure 3 | Lifetime tuning scheme and time-resolved confocal images for NaYF4:Yb,Tm up conversion nanocrystals. (Reproduced with permission from Ref [79])

Zhuang et al. showed that emission intensity/wavelength multiplexing and deep trap depth engineering enabled information storage in more than one dimension. Using the optical storing information property and the isolated deep-trap, the main uses of optically stimulated luminescence materials were also found in information encryption, dynamic multimode anti-counterfeiting and stable information storing capability. Also, a study was done on $BaSi_2O_5:Eu^{2+}$, Nd^{3+} Phosphor in glass to improve its ability to store optical information by adding deep traps with a narrow energy distribution (0.16 eV) [90-94]. Therefore, such demonstration expands the potential of optical multiplexing by include the time dimension of luminous signals, which might have applications in the fields of life science, medicine, and information security.

4.   Rare doped materials for optical information storage

Due to their various fascinating luminous features, especially up-conversion fluorescence, rare-earth doped materials have garnered a lot of attention in fields as diverse as biology, optoelectronics, and photonics. Light with a shorter wavelength than the excitation light is emitted by a nonlinear optical process known as photon up-conversion or anti-Stokes-type emission. Therefore, rare-earth doped materials have play pivot role in utilizing to transform IR light into visible light. Up-conversion in rare-earth doped materials results from the accumulation of the energies of several photons absorbed by the rare-earth ions at intermediate energy levels in rapid succession. Up-conversion has excellent quantum efficiency and can be efficiently activated with relatively modest pump intensities [95], in contrast to other multiphoton absorption processes like two-photon absorption and second-harmonic generation. Excited states need to have a long lifespan for up-conversion to occur, and the energy levels should be arranged in a ladder shape with about the same sized gaps between them. Rare-earth elements like yttrium and scandium, as well as the lanthanide class, have been found to have these properties. Rare-earth ion doped materials with inorganic crystalline host matrix with small amounts of rare-earth ions furnish the light emitting centers, while the host crystalline lattice maintains their ordered, optimum orientations. For optimal luminescence, the mutual distance between dopant ions, their spatial arrangement, and their coordination number are all crucial considerations. Inorganic nanocrystals YAG, $NaYF_4$, Y_2O_3, Y_2GeO_5, $PbWO_4$, $NaPrF_4$, Sr_2SnO_4, $SrZrO_3$ $Ba_{1.6}Ca_{0.4}SiO_4$, etc. are the hosts for rare-earth ions [96-97]. As a result of the prevalence of rare-earth ion emission lines, rare-earth ions emit over the whole visible spectrum, from the blue to the red. By fine-tuning the size and content of the rare-earth ions [98], the emission color can be finely regulated, and by raising the concentration of the lanthanide dopants, light absorption may be enhanced. It's possible that radiation-free deactivation takes place at high doses, although cross-relaxation is a major roadblock. However, up to around 3% Er^{3+} is considered to be the maximum safe concentration. At this level, light absorption is inadequate, necessitating the addition of highly absorbing ions to the material to enable proper energy transfer. Yb^{3+}, which can be co-doped into the lattice at high concentrations (18-20%), is the most popular sensitizer for up-conversion based on Er^{3+}. Since the gap between Yb^{3+} and Er^{3+} is also quite small, the two ions may efficiently exchange (quasi-)resonant energy with one another. Co-doping the light sensitizer Yb^{3+} with the light emitters Er^{3+} or Tm^{3+} into the host $NaYF_4$ is an example for high-efficiency up-conversion fluorescence material (Fig.4). Near-infrared photons are absorbed by Yb^{3+} ions, which then transmit that energy, through non-radiative means, to neighboring Er^{3+} or Tm^{3+} ions [99].

In addition, by using a cation exchange approach, Yao Xie has created a heterostructure nanoparticles consisting of EuSe semiconductor and rare-earth ion incorporated up-conversion nanoparticles which exhibits exceptional downshifting and up-conversion characteristics. With the help of the cation exchange method, a transition layer rich in Eu^{3+} ions develop between the EuSe semiconductor and rare-earth ion doped up-conversion nanoparticle ($NaGdF4:Yb,Tm@$ $NaYF4:Tb$), allowing for the detection of the distinctive emission of Eu^{3+} ions to be made under both UV and NIR light irradiation. Fig. 5 demonstrates that by adjusting the kind and amount of rare-earth ion in the up-conversion nanoparticles, the nanocomposite may generate blue and white light when excited by UV light, and demonstrates color modulation upconverting emission when excited by a 980 nm laser. Time-gating technology might also be used to filter out the long-lived up-converting radiation produced by some lanthanide ions. Success in enhanced anti-counterfeiting and information storage is shown to stem from nanocomposites' multimode luminous characteristics.

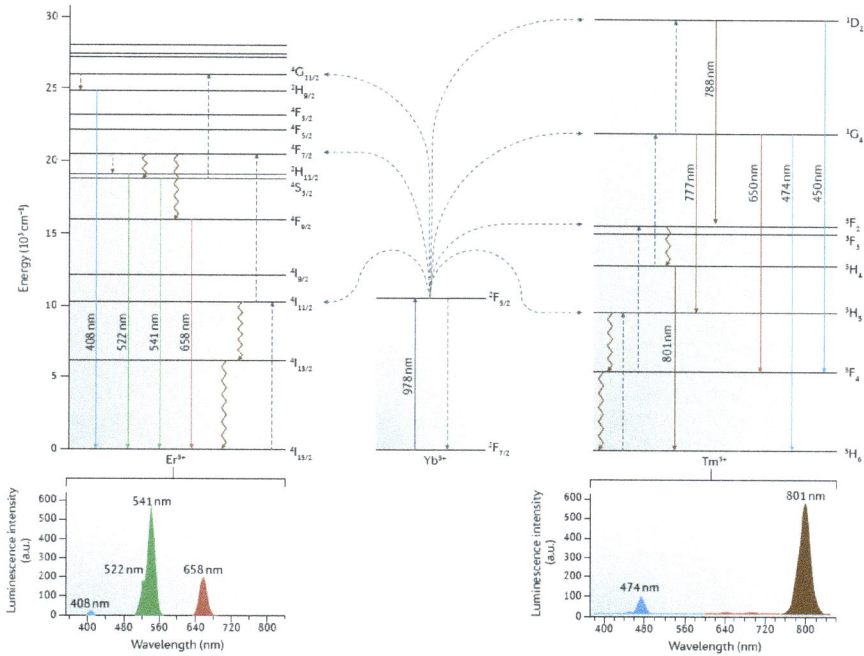

Figure 4. Energy diagram (top) and fluorescence characteristics (bottom) of co-doping rare-earth-doped nanomaterial. Yb^{3+} serves as the sensitizer, Er^{3+} or Tm^{3+} as the activator. (Reproduced with permission from Ref [99])

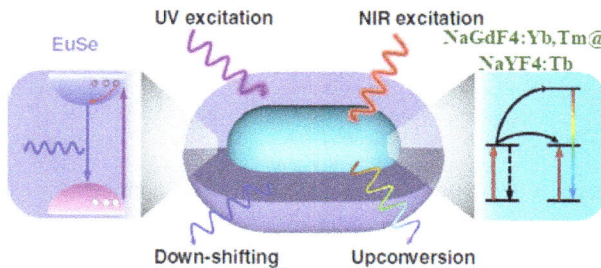

Figure. 5 Illustration of the down-shifting and up-conversion luminescence principle of heterostructure nanocomposite composed of rare-earth doped up-conversion nanoparticle and EuSe semiconductor. (Reproduced with permission from Ref [100])

Rare Earth - A tribute to the late Mr. Rare Earth, Professor Karl Gschneidner   Materials Research Forum LLC
Materials Research Foundations 164 (2024) 211-229   https://doi.org/10.21741/9781644903056-5

However, nano-crystallization greatly reduces light conversion efficiency. Thus, two techniques are devised to improve up-conversion efficiency of up-conversion NPs. First, co-dope sensitizer ions with activator ions having a similar intermediate excited state. Second, build spherical core@shell structures to regulate energy transfer to resist concentration quenching. Several applications have developed as up-conversion rare-earth doped NP luminescence intensity and optical characteristics have increased. Also, most polymer-based optical data storage was not good at withstanding high-power lasers, which made it hard to use in high-security data storage. Wang and his colleagues were able to make a new type of optical data storage medium by embedding cubic nanocrystals of Cs_3LnF_6 (Ln = Y, Yb, Lu, Sc) in glass. They did this to get high-stability optical data storage [101]. Using the screen-printing method, Cs_3YbF_6: Er nanocrystals were mixed into glass inks, which were then used to create a series of encrypted patterns with the UCL property. As shown in Figure 4d, all of the patterns changed their up-conversion emitting colors because of the photo-thermal effect caused by laser light. The Cs_3LnF_6 nanocrystals that were stabilized by glass had a noticeable photo-thermal effect were very stable over time, even when the power of the laser went from 0.5 W to 4.5 W or when they were in water for 90 days. Further, rare-earth doped materials are also good candidates for super-resolution microscopy based on up-conversion fluorescence realized using STED-like techniques. Praseodymium-doped YAG:Pr rare-earth doped materials have been used to demonstrate high-quality optical imaging with a resolution of 50 nm at a power of 29 mW [102].

Since Lindmayer's 1986 3D optical memory device, electron-trapping materials have garnered interest as optical information storage medium. By trapping charge carriers, X-rays, UV, or visible light may capture data. By releasing charge carriers, high-temperature or low-energy lasers may read data. Electron-trapping materials for optical information storage use trap levels. Most materials having stored data in shallow traps are thermally unstable at ambient temperature, which limits their use in optical information storage. [32, 103-105] To protect data, trap levels must be increased.

In 2018, Xie's team successfully demonstrated multi-dimensional macroscopic information reading and writing by increasing the trap levels of a succession of electron-trapping materials by choosing various co-dopants. [100] Through (multi-) co-doping of selected trivalent rare-earth ions, Liu et al. have also effectively customized Ba_2SiO_4 with rich trap levels to significantly enhance optical storage capabilities. [104-105]. Co-doping of selected rare-earth ions Tb^{3+} into Pr^{3+}-activated Y_2GeO_5, as described by Mingxue Dengin and colleagues, is an intriguing use of electron-trapping materials for optical information storage. They use X-ray photoelectron spectroscopy and low-temperature electronic spin resonance spectra in conjunction with first-principles simulations to determine the nature of trap levels. To actualize "0" and "1" in binary information, the emission spectrum between each pair of information points is identified using linear continuous scanning and a 515 nm femtosecond laser [106]. By (multi-)co-doping selective trivalent rare-earth ions into Eu^{2+}-activated barium orthosilicate, Dong Liu et al. created three-dimensional traps with sharply separated energy levels. Information can be successively recorded in different traps under UV/blue light illumination with the help of thermal cleaning; exponentially increasing storage capacity that can be accessed in full domain or bit-by-bit mode without crosstalk thanks to carefully designed thermal/optical stimuli. Recycle data recording has remarkably improved data persistence and excellent fatigue resistance. Evidence-based applications and theoretical understanding of charge transport dynamics and interaction with traps pave the way for high-security encryption and decryption and rewritable multi-level optical information storage

[107]. In addition, J. Du et al. found that by replacing Sn with Si, trap levels may be made both more numerous and deeper in $Sr_2SnO_4:Sm^{3+},Si^{4+}$ phosphors. Rewritable optical data storage and readout based on photon trapping and de-trapping processes are shown, and both ultraviolet light and high-energy X-ray irradiations were employed to cause reddish-orange producing persistent illumination from the as synthesized samples. One may easily encode information optically using a 365 nm light-emitting diode and decode it using heat [108].

Optical stimulated luminescence happens when an external light stimulus brings back to thermodynamic equilibrium a pre-excited material containing metastable trapped charge carriers [109-112]. There is great potential for the use of erasable optical information storage due to its novel properties, such as all-optical controllable carrier release and storage. Optical stimulated luminescence, made from rare-earth doped alkaline halides or sulfides, was the first to propose and effectively execute this idea. Still, the materials with low resistance to moisture and their dependency on X-ray irradiation for data capture severely limit their usefulness in practice [113]. Chemically stable metal oxides doped with rare-earth and transition metal ions have gained significant attention in recent years due to their excellent optical stimulated luminescence characteristics. Multimode encryption and anti-counterfeiting are two further areas where optical stimulated luminescence materials may be put to use thanks to their unique properties and the fact that they include isolated deep traps. It is difficult to regulate traps in a manner that can be controlled, and there is still a long way to go before optical stimulated luminescence materials can be used in practice. J. Zhang et al developed a novel phosphor with long-lasting luminescence and optically induced luminescence. It has been shown that intentionally co-doping with Dy^{3+} ions into the host lattice increases the trap density by 10-fold. The data is captured in traps and then read out using an optical or thermal stimulus in a controlled and reversible manner. Information storage via optical stimulated luminescence has been presented, along with the kinetics of electron mobility, through a model in its early stages of development. The as-obtained optical stimulated luminescence material was also used to fabricate a flexible film, demonstrating a proof-of-concept use for rewritable and stored optical data. Therefore, it is crucial to design new optical stimulated materials for storing the information [114].

Photochromism is when a chemical substance changes color when exposed to light and the color can be changed back to its original state by exposing the substance to light or heat at a certain wavelength. Most of the time, the physical and chemical properties of photochromic materials change when they are in different states. By controlling the two states, the material's performance can be changed. It could be used for many things, like optical storage, optical switches, smart glasses, security identification, and smart anti-counterfeiting, among other things [115-119]. Doped with rare-earth elements, inorganic photochromic materials tend to exhibit unique luminosity and photochromic behaviors. Multiple optical information storage and optical switches are possible with this material because of the overlapped absorbance of photoluminescence and photochromism. Now, the single function of inorganic photochromic materials limits usage in the fields of optical storage and anti-counterfeiting. Most of the time, a photochromic process is used to change the way light shines in a way that can be changed back. This makes the photochromic material better at storing information. Ruiting Zhang and his team made an inorganic photochromic material called $SrZrO_3:Sm^{3+}$ that changes color when exposed to light. With its red photoluminescence and ability to change color in response to light, it can store and recognize information in two ways. Doping with Sm^{3+} gives it the ability to control fluorescence, which makes it a good candidate for use as an optical switch. Lastly, a flexible film made of $SrZrO_3:Sm^{3+}$

was made to show that optical information could be written repeatedly, stored, and read [120]. Using the shift from W^{6+} to W^{5+} and Pb^{2+} to Pb^{4+} valence states, Xue Bai and coworkers reported the development of a photochromic $PbWO_4:Yb^{3+}$, Er^{3+} material. Based on the photochromism of $PbWO_4:Yb^{3+}$, Er^{3+} and using 532 nm and 808 nm laser stimulation produced optical information writing and erasing in $PbWO_4:Yb^{3+}$, Er^{3+} material. Up-conversion luminescence change due to photochromism was achieved, allowing for efficient and nondestructive optical information reading [121].

Conclusions

Optical information storage is a rapidly developing area of study that promises groundbreaking improvements in computing and almost unrestricted security mechanisms via the use of quantum cryptography. Materials with high luminescence are necessary for optical information storage. Rare-earth doped host lattices are one of the most promising types of materials investigated too far. It has been shown that trivalent rare-earth ions exhibit extended optical and spin coherence states at low temperatures, and their narrow f-f-transitions may span the visible and IR spectrum, along with wavelength at 1.5 μm. The optical and spin characteristics of new systems like ytterbium-doped crystals have been explored, and the structural possibilities for enhanced light-matter interactions and integrated optics are exciting. In addition to single crystals, rare-earth doped transparent ceramics and rare-earth doped polycrystalline (nano) powders have also made significant strides. New opportunities for the design of light-matter interfaces, long-term storage devices, and scalable computers will emerge as production techniques such as nano-structuring, nano-particles, and thin films advance. They may also be useful in other fields where tight transitions are required, such as laser stabilization, signal processing, and medical imaging.

Acknowledgment

One of the authors DH is grateful to the Department of Science and Technology (DST) of the Government of India for the financial support under the project #CRG/2021/007142.

References

[1] H. E. Lee, J. H. Park, T. J. Kim, D. Im, J. H. Shin, D. H. Kim, B. Mohammad, I.-S. Kang and K. J. Lee, Advanced Functional Materials, 2018, 28, 1801690.

[2] Zhang, Q.; Xia, Z.; Cheng, Y. B.; Gu, M. Nat. Commun. 2018, 9, 1183. https://doi.org/10.1038/s41467-018-03589-y

[3] Gu, M.; Zhang, Q.; Lamon, S. Nat. Rev. Mater. 2016, 1, 16070. https://doi.org/10.1038/natrevmats.2016.70

[4] D. Reinsel, J. Gantz and J. Rydning, IDC White Paper, 2018.

[5] W.-X. Chu, R. Wang, P.-H. Hsu and C.-C. Wang, Journal of Building Engineering, 2020, 30, 101331. https://doi.org/10.1016/j.jobe.2020.101331

[6] C. Li, M. Hu, Y. Li, H. Jiang, N. Ge, E. Montgomery, J. Zhang, W. Song, N. Dávila, C. E. Graves, Z. Li, J. P.Strachan, P. Lin, Z. Wang, M. Barnell, Q. Wu, R. S. Williams, J. J. Yang and Q. Xia, Nature Electronics, 2017, 1, 52-59. https://doi.org/10.1038/s41928-017-0002-z

[7] C. Wu, T. W. Kim, H. Y. Choi, D. B. Strukov and J. J. Yang, Nat Commun, 2017, 8, 752.

[8] P. Yao, H. Wu, B. Gao, S. B. Eryilmaz, X. Huang, W. Zhang, Q. Zhang, N. Deng, L. Shi, H. P. Wong and H. Qian, Nat Commun, 2017, 8, 15199. https://doi.org/10.1038/ncomms15199

[9] Z. Sun, E. Ambrosi, A. Bricalli and D. Ielmini, Adv Mater, 2018, 30, 1802554. https://doi.org/10.1002/adma.201802554

[10] M. Gu, Q. Zhang and S. Lamon, Nature Reviews Materials, 2016, 1, 16070.

[11] J. S. Meena, S. M. Sze, U. Chand and T. Y. Tseng, Nanoscale Res Lett, 2014, 9, 526. https://doi.org/10.1186/1556-276X-9-526

[12] Y. Park and J. S. Lee, ACS Nano, 2017, 11, 8962-8969. https://doi.org/10.1021/acsnano.7b03347

[13] M. M. Shulaker, G. Hills, R. S. Park, R. T. Howe, K. Saraswat, H. P. Wong and S. Mitra, Nature, 2017, 547, 74 https://doi.org/10.1038/nature22994

[14] T. Zhong, J. M. Kindem, J. G. Bartholomew, J. Rochman, I. Craiciu, E. Miyazono, M. Bettinelli, E. Cavalli, V. Verma, S. W. Nam, F. Marsili, M. D. Shaw, A. D. Beyer and A. Faraon, Science, 2017, 357, 1392-1395. https://doi.org/10.1126/science.aan5959

[15] J. Zhao, X. Zheng, E. P. Schartner, P. Ionescu, R. Zhang, T. L. Nguyen, D. Jin and H. Ebendorff - Heidepriem, Advanced Optical Materials, 2016, 4, 1507-1517. https://doi.org/10.1002/adom.201600296

[16] Y. Liu, Y. Lu, X. Yang, X. Zheng, S. Wen, F. Wang, X. Vidal, J. Zhao, D. Liu, Z. Zhou, C. Ma, J. Zhou, J. A. Piper, P. Xi and D. Jin, Nature, 2017, 543, 229-233. https://doi.org/10.1038/nature21366

[17] S. Wen, J. Zhou, K. Zheng, A. Bednarkiewicz, X. Liu and D. Jin, Nat Commun, 2018, 9, 2415. https://doi.org/10.1038/s41467-018-04813-5

[18] Y. Lu, J. Zhao, R. Zhang, Y. Liu, D. Liu, E.M. Goldys, X. Yang, P. Xi, A. Sunna, J. Lu, Y. Shi, R.C. Leif, Y. Huo, J. Shen, J.A. Piper, J.P. Robinson, D. Jin, Nat. Photonics 8 (2014) 32-36, https://doi.org/10.1038/nphoton.2013.322

[19] Y. Zhuang, Y. Lv, L. Wang, W. Chen, T. Zhou, T. Takeda, N. Hirosaki, R.-J. Xie, ACS Appl. Mater. Interfaces 10 (2018) 1854-1864 https://doi.org/10.1021/acsami.7b17271

[20] Y. Zhuang, L. Wang, Y. Lv, T.L. Zhou, R.J. Xie, Adv. Funct. Mater. 28 (2018) 1705769 https://doi.org/10.1002/adfm.201705769

[21] D. Liu, L. Yuan, Y. Jin, H. Wu, Y. Lv, G. Xiong, G. Ju, L. Chen, S. Yang, Y. Hu, ACS Appl. Mater. Interfaces 11 (2019) 35023-35029. https://doi.org/10.1021/acsami.9b13011

[22] W. Li, Y. Zhuang, P. Zheng, T.-L. Zhou, J. Xu, J. Ueda, S. Tanabe, L. Wang, R.-J. Xie, ACS Appl. Mater. Interfaces 10 (2018) 27150-27159, https://doi.org/10.1021/acsami.8b10713

[23] L. Xiang, X. Yujie, S. Bo, Z. Hao-Li, C. Hao, C. Huijuan, L. Weisheng, T. Yu, Angew. Chem. Int. Ed. 2017, 56, 2689.

[24] R. Jiufeng, Y. Zhengwen, H. Anjun, Z. Hailu, Q. Jianbei, S. Zhiguo, ACS Appl. Mater. Interfaces 2018, 10, 14941.

[25] Z. Qi, D. Xuelin, X. Yuxiang, Z. Binbin, L. Shan, W. Qin, L. Yonggui, Y. Yajiang, W. Hong, ACS Appl. Mater. Interfaces 2020, 12, 28539.

[26] Y. Vivian Wing-Wah, C. Alan Kwun-Wa, H. Eugene Yau-Hin, Nat. Rev. Chem. 2020, 4, 528.

[27] H. Chen, Z. Dong, W. Chen, L. Sun, X. Du, Y. Zhao, P. Chen, Z. Wu, W. Liu, Y. Zhang, Adv. Opt. Mater. 2020, 8, 1902125. https://doi.org/10.1002/adom.201902125

[28] O. Xiangyu, Q. Xian, H. Bolong, Z. Jie, W. Qinxia, H. Zhongzhu, X. Lili, B. Hongyu, Y. Zhigao, C. Xiaofeng, W. Yiming, S. Xiaorong, L. Juan, C. Qiushui, Y. Huanghao, L. Xiaogang, Nature 2021, 590, 410.

[29] Y. Zhou, S. T. Han, X. Chen, F. Wang, Y. B. Tang, V. A. Roy, Nat. Commun. 2014, 5, 4720. https://doi.org/10.1038/ncomms5720

[30] Z. Yan, S. Haiqin, J. Qiannan, G. Lili, P. Dengfeng, Z. Qiwei, H. Xihong, Adv. Opt. Mater. 2021, 9, 2001626.

[31] Y. Song, M. Lu, G. A. Mandl, Y. Xie, G. Sun, J. Chen, X. Liu, J. A. Capobianco, L. Sun, Angew. Chem. 2021, 25, 23983. https://doi.org/10.1002/ange.202109532

[32] J. Lindmayer, Solid State Technol. 1988, 31, 135. https://doi.org/10.1016/0038-1101(88)90120-7

[33] Z. Long, Y. Wen, J. Zhou, J. Qiu, H. Wu, X. Xu, X. Yu, D. Zhou, J. Yu, Q. Wang, Adv. Opt. Mater. 2019, 7, 1900006. https://doi.org/10.1002/adom.201900006

[34] L. Yuan, Y. Jin, Y. Su, H. Wu, Y. Hu, S. Yang, Laser Photonics Rev. 2020, 14, 2000123. https://doi.org/10.1002/lpor.202000123

[35] W. Zeng, Y. Wang, S. Han, W. Chen, G. Li, Y. Wang, Y. Wen, J. Mater. Chem. C 1 (2013) 3004-3011. https://doi.org/10.1039/c3tc30182f

[36] K. Kumar, A.K. Singh, S.B. Rai, Spectrochim Acta A 102 (2013) 212-218. https://doi.org/10.1016/j.saa.2012.09.054

[37] Z. Wang, Z. Song, Q. Liu, Mater. Chem. Front. 5 (2021) 333-340. https://doi.org/10.1039/D0QM00488J

[38] L. Yuan, Y. Jin, D. Zhu, Z. Mou, G. Xie, Y. Hu, ACS Sustain. Chem. Eng. 8 (2020) 6543-6550. https://doi.org/10.1021/acssuschemeng.0c01377

[39] L. Yuan, Y. Jin, Y. Su, H. Wu, Y. Hu, S. Yang, Laser Photonics Rev 14 (2020),2000123. https://doi.org/10.1002/lpor.202000123

[40] S.W.S. Mckeever, Nucl. Instrum. Meth. B 184 (2004) 29-54. https://doi.org/10.1016/S0168-583X(01)00588-2

[41] E.G. Yukihara, S.W.S. McKeever, Radiat. Prot. Dosim. 147 (2011) 619-622. https://doi.org/10.1093/rpd/ncr357

[42] I. Wieder, L.R. Sarles, Phys. Rev. Lett. 6 (1961) 95-96. https://doi.org/10.1103/PhysRevLett.6.95

[43] X. Ou, X. Qin, B. Huang, J. Zan, Q. Wu, Z. Hong, L. Xie, H. Bian, Z. Yi, X. Chen, Y. Wu, X. Song, J. Li, Q. Chen, H. Yang, X. Liu, Nature 590 (2021) 410-415. https://doi.org/10.1038/s41586-021-03251-6

[44] Ph. Goldner, A. Ferrier, O. Guillot-Noël, Vol 46 (Eds.: J.-C. G. Bünzli, V. K. Pecharsky). North Holland, Amsterdam, 2015, 1- 78. https://doi.org/10.1016/B978-0-444-63260-9.00267-4

[45] W. Tittel, T. Chanelière, R. L. Cone, S. Kröll, S. A. Moiseev, M. Sellars, Laser & Photon. Rev. 2010, 4, 244-267. https://doi.org/10.1002/lpor.200810056

[46] C. W. Thiel, T. Böttger, R. L. Cone, J. Lumin. 2011, 131, 353-361. https://doi.org/10.1016/j.jlumin.2010.12.015

[47] M. P. Hedges, J. J. Longdell, Y. Li, M. J. Sellars, Nature 2010, 465, 1052-1056. https://doi.org/10.1038/nature09081

[48] I. Usmani, M. Afzelius, H. de Riedmatten, N. Gisin, Nature Comm. 2010, 1, 1-7. https://doi.org/10.1038/ncomms1010

[49] C. Clausen, I. Usmani, F. Bussières, N. Sangouard, M. Afzelius, H. de Riedmatten, N. Gisin, Nature 2011, 469, 508-511. https://doi.org/10.1038/nature09662

[50] I. Usmani, C. Clausen, F. Bussières, N. Sangouard, M. Afzelius, N. Gisin, Nature Phot. 2012, 6, 234-237. https://doi.org/10.1038/nphoton.2012.34

[51] W. Tittel, T. Chanelière, R. L. Cone, S. Kröll, S. A. Moiseev, M. Sellars, Laser & Photon. Rev. 2010, 4, 244 https://doi.org/10.1002/lpor.200810056

[52] A. I. Lvovsky, B. C. Sanders, W. Tittel, Nature Phot. 2009, 3, 706-714. https://doi.org/10.1038/nphoton.2009.231

[53] R. M. Macfarlane, J. Lumin. 2002, 100, 1-20. https://doi.org/10.1016/S0022-2313(02)00450-7

[54] T. D. Ladd, F. Jelezko, R. Laflamme, Y. Nakamura, C. Monroe, J. L. O'Brien, Nature 2010, 464, 45-53. https://doi.org/10.1038/nature08812

[54b] R. Blatt, C. F. Roos, Nature Phys. 2012, 8, 277-284. https://doi.org/10.1038/nphys2252

[55] J. Clarke, F. K. Wilhelm, Nature 2008, 453, 1031-1042. https://doi.org/10.1038/nature07128

[56] T. E. Northup, R. Blatt, Nature Phot. 2014, 8, 356-363. https://doi.org/10.1038/nphoton.2014.53

[57] C. W. Thiel, T. Böttger, R. L. Cone, J. Lumin. 2011, 131, 353-361. https://doi.org/10.1016/j.jlumin.2010.12.015

[58] A. M. Stoneham, Proc. Phys. Soc. London 1966, 89, 909-921. https://doi.org/10.1088/0370-1328/89/4/314

[59] A. M. Stoneham, Rev. Mod. Phys. 1969, 41, 82-108. https://doi.org/10.1103/RevModPhys.41.82

[60] E. P. Chukalina, M. N. Popova, S. L. Korableva, R. Yu. Abdusabirov, Phys. Lett. A 2000, 269, 348-350. https://doi.org/10.1016/S0375-9601(00)00273-5

[61] Zijlstra, P., Chon, J. W. M. & Gu, M. Nature 459, 410-413 (2009). https://doi.org/10.1038/nature08053

[62] Jeevan, M. M. et al Nanotechnology 23, 395201 (2012). https://doi.org/10.1088/0957-4484/23/39/395201

[63] Perfetto, S. P., Chattopadhyay, P. K. & Roederer, M. Nature Rev. Immunol. 4, 648-655 (2004). https://doi.org/10.1038/nri1416

[64] Cui, H. H., Valdez, J. G., Steinkamp, J. A. & Crissman, H. A. Cytometry A 52A, 46-55 (2003). https://doi.org/10.1002/cyto.a.10022

[65] Watson, D. A. et al. Cytometry A 73A, 119-128 (2008). https://doi.org/10.1002/cyto.a.20520

[66] Betzig E, Trautman JK, Wolfe R, et al. Near-field magneto-optics and high density information storage. Appl Phys Lett, 1992, 61: 142-144 https://doi.org/10.1063/1.108198

[67] Li L, Gattass RR, Gershgoren E, et al. Science, 2009, 324: 910-913 https://doi.org/10.1126/science.1168996

[68] Li W, Zhuang Y, Zheng P, et al. ACS Appl Mater Interfaces, 2018, 10: 27150-27159 https://doi.org/10.1021/acsami.8b10713

[69] Zhuang Y, Wang L, Lv Y, et al. Adv Funct Mater, 2018, 28: 1705769

[70] Long Z, Wen Y, Zhou J, et al. Adv Opt Mater, 2019, 7: 1900006

[71] Lin S, Lin H, Huang Q, et al. Laser Photonics Rev, 2019, 13: 1970022 https://doi.org/10.1002/lpor.201900081

[72] J.C. Zhang, C. Pan, Y.F. Zhu, L.Z. Zhao, H.W. He, X.F. Liu and J.R. Qiu, Adv. Mater., 2018, 30(49), 1804644. https://doi.org/10.1002/adma.201870373

[73] J. Han, J. Sun, Y. Li, Y. Duan and T. Han, J. Mater. Chem. C, 2016, 4(39), 9287. https://doi.org/10.1039/C6TC03131E

[74] D.J. Wales, Q. Cao, K. Kastner, E. Karjalainen, G.N. Newton and V. Sans, Adv. Mater., 2018, 30(26), 1800159. https://doi.org/10.1002/adma.201800159

[75] J. Liu, H. Rijckaert, M. Zeng, K. Haustraete, B. Laforce, L. Vincze, I.V. Driessche, A.M. Kaczmarek and R. Van Deun, Adv. Funct. Mater., 2018, 28(17), 1707365. https://doi.org/10.1002/adfm.201707365

[76] K. Jiang, L. Zhang, J. Lu, C. Xu, C. Cai and H. Lin, Angew. Chem. Int. Ed., 2016, 55, 7231. https://doi.org/10.1002/anie.201602445

[77] H.Q. Sun, Y. Zhang, J. Liu, D.F. Peng, Q.W. Zhang and X.H. Hao, J. Am. Ceram. Soc., 2018, 101, 5659. https://doi.org/10.1111/jace.15885

[78] Ortu, A., Holzäpfel, A., Etesse, J. et al. npj Quantum Inf 8, 29 (2022) https://doi.org/10.1038/s41534-022-00541-3

[79] Lu, Y., Zhao, J., Zhang, R. et al. Nature Photon 8, 32-36 (2014) https://doi.org/10.1038/nphoton.2013.322

[80] Zhuang, Y.; Wang, L.; Lv, Y.; Zhou, T. L.; Xie, R. J. Adv. Funct. Mater. 2018, 28, 1705769. https://doi.org/10.1002/adfm.201705769

[81] Zhuang, Y.; Lv, Y.; Wang, L.; Chen, W.; Zhou, T. L.; Takeda, T.; Hirosaki, N.; Xie, R. J.; ACS Appl. Mater. Inter. 2018, 10, 1854-1864. https://doi.org/10.1021/acsami.7b17271

[82] H.Q. Sun, Y. Zhang, J. Liu, D.F. Peng, Q.W. Zhang and X.H. Hao, J. Am. Ceram. Soc., 2018, 101, 5659. https://doi.org/10.1111/jace.15885

[83] Q.W. Zhang, Y. Zhang, H.Q. Sun, W. Geng, X.S. Wang, X.H. Hao and S.L. An, ACS Appl. Mater. Interfaces, 2016, 8, 34581. https://doi.org/10.1021/acsami.6b11825

[84] J.C. Zhang, C. Pan, Y.F. Zhu, L.Z. Zhao, H.W. He, X.F. Liu and J.R. Qiu, Adv. Mater., 2018, [30(49), 1804644.

[85] J. Han, J. Sun, Y. Li, Y. Duan and T. Han, J. Mater. Chem. C, 2016, 4(39), 9287. https://doi.org/10.1039/C6TC03131E

[86] D.J. Wales, Q. Cao, K. Kastner, E. Karjalainen, G.N. Newton and V. Sans, Adv. Mater., 2018, [30(26), 1800159.

[87] J. Liu, H. Rijckaert, M. Zeng, K. Haustraete, B. Laforce, L. Vincze, I.V. Driessche, A.M. [Kaczmarek and R. Van Deun, Adv. Funct. Mater., 2018, 28(17), 1707365. https://doi.org/10.1002/adfm.201870112

[88] K. Jiang, L. Zhang, J. Lu, C. Xu, C. Cai and H. Lin, Angew. Chem. Int. Ed., 2016, 55, 7231. https://doi.org/10.1002/anie.201602445

[89] H. Sun, X. Li , Y. Zhu, X. Wang, Q. Zhang and X. Hao, J. Mater. Chem. C, 2019,7, 5782-5791 https://doi.org/10.1039/C9TC00834A

[90] Li, W.; Zhuang, Y.; Zheng, P.; Zhou, T. L.; Xu, J.; Ueda, J.; Tanabe, S.; Wang, L.; Xie, R. J. ACS Appl. Mater. Inter. 2018, 10, 27150-27159. https://doi.org/10.1021/acsami.8b10713

[91] Liu, Z.; Zhao, L.,Chen, W, Fan, X.; Yang, X.; Tian, S.; Yu, X.; Qiu, J.; Xu, X. J. Mater. Chem. C 2018, 6, 11137- 11143. https://doi.org/10.1039/C8TC04018D

[92] Sun, Z.; Yang, J.; Huai, L.; Wang, W, Ma, Z,Sang, J, Zhang, J, Li, H.; Ci, Z, Wang, Y. ACS Appl. Mater. Inter. 2018, 10, 21451-21457. https://doi.org/10.1021/acsami.8b08977

[93] Wang, W, Yang, J.; Zou, Z,Zhang, J, Li, H.; Wang, Y. Ceram. Int. 2018, 44, 10010-10014. https://doi.org/10.1016/j.ceramint.2018.02.224

[94] Lin, S.; Lin, S.; Huang, Q.; Cheng, Y.; Xu, J.; Wang, J.; Xiang, X.; Wang, C.; Zhang, L.; Wang, Y. A Laser & Photonics Rev. 2019, 13, 1900006. https://doi.org/10.1002/lpor.201900006

[95] Haase, M. & Schäfer, H. Angew. Chem. Int. Ed. 50, 5808-5829 (2011). https://doi.org/10.1002/anie.201005159

[96] Chang, H. et al. Nanomaterials 5, 1-25 (2014). https://doi.org/10.3390/nano5010001

[97] Wang, M., Abbineni, G., Clevenger, A., Mao, C. & Xu, S. Nanomedicine 7, 710-729 (2011). https://doi.org/10.1016/j.nano.2011.02.013

[98] Wang, F. et al. Nature 463, 1061-1065 (2010). https://doi.org/10.1038/nature08777

[99] Auzel, F. Chem. Rev. 104, 139-174 (2004). https://doi.org/10.1021/cr020357g

[100] Xie et al. Light: Science & Applications (2022) 11:150 https://doi.org/10.1038/s41377-022-00813-9

[101] S. Wang, J. Lin, Y. He, J. Chen, C. Yang, F. Huang and D. Chen, Chemical Engineering Journal, 2020, 394, 124889. https://doi.org/10.1016/j.cej.2020.124889

[102] Kolesov, R. et al. Phys. Rev. B 84, 153413 (2011). https://doi.org/10.1103/PhysRevB.84.153413

[103] Z. Long, Y. Wen, J. Zhou, J. Qiu, H. Wu, X. Xu, X. Yu, D. Zhou, J. Yu, Q. Wang, Adv. Opt. Mater. 2019, 7, 1900006. https://doi.org/10.1002/adom.201900006

[104] L. Yuan, Y. Jin, Y. Su, H. Wu, Y. Hu, S. Yang, Laser Photonics Rev. 2020, 14, 2000123. https://doi.org/10.1002/lpor.202000123

[105] Y. Li, M. Gecevicius, J. Qiu, Chem. Soc. Rev. 2016, 45, 2090. https://doi.org/10.1039/C5CS00582E

[106] Mingxue Deng, Qian Liu, Ying Zhang, Caiyan Wang, Xinjun Guo, Zhenzhen Zhou, and Xiaoke Xu Adv. Optical Mater. 2021, 2002090

[107] Dong Liu, Lifang Yuan, Yahong Jin, Haoyi Wu, Yang Lv, Guangting Xiong, Guifang Ju, Li Chen, Shihe Yang, Yihua Hu ACS Appl. Mater. Interfaces 2019, 11, 38, 35023-35029 https://doi.org/10.1021/acsami.9b13011

[108] J. Du , S. Lyu a, K. Jiang , D. Huang b, J. Li , R. Van Deun , D. Poelman , H. Lin, Materials Today Chemistry 24 (2022) 100906 https://doi.org/10.1016/j.mtchem.2022.100906

[109] L. Yuan, Y. Jin, Y. Su, H. Wu, Y. Hu, S. Yang, Laser Photonics Rev 14 (2020), 2000123. https://doi.org/10.1002/lpor.202000123

[110] S.W.S. Mckeever, Nucl. Instrum. Meth. B 184 (2004) 29-54. https://doi.org/10.1016/S0168-583X(01)00588-2

[111] E.G. Yukihara, S.W.S. McKeever Radiat. Prot. Dosim. 147 (2011) 619-622. https://doi.org/10.1093/rpd/ncr357

[112] I. Wieder, L.R. Sarles, Phys. Rev. Lett. 6 (1961) 95-96. https://doi.org/10.1103/PhysRevLett.6.95

[113] S.H. Von, Braz. J. Phys. 29 (1999) 254-268. https://doi.org/10.1590/S0103-97331999000200008

[114] Junming Zhang, Lifang Yuan, Yahong Jin, Haoyi Wu, Li Chen, Yihua Hu Journal of Luminescence 241 (2022) 118518 https://doi.org/10.1016/j.jlumin.2021.118518

[115] R. Exelby, R. Grintetr, Chem. Rev. 65 (1965) 247-260. https://doi.org/10.1021/cr60234a005

[116] G.H. Brown, Photochromism, Techniques of Chemistry (1971) 853.

[117] T. Wei, B. Jia, L. Shen, C. Zhao, L. Wu, B. Zhang, X. Tao, S. Wu, Y. Liang, J. Eur. Ceram. Soc. 40 (2020) 4153-4163. https://doi.org/10.1016/j.jeurceramsoc.2020.04.014

[118] M. Gu, X. Li, Y. Cao, , Light & Applications 3 (2014) 177.
https://doi.org/10.1038/lsa.2014.58

[119] M. Tu, H. Reinsch, S. Rodríguez-Hermida, R. Verbeke, T. Stassin, W. Egger, M.
Dickmann, B. Dieu, J. Hofkens, I. Vankelecom, Angew. Chem. Int. Ed. 58 (2019) 2423-2427.
https://doi.org/10.1002/anie.201813996

[120] Ruiting Zhang, Yahong Jin, Lifang Yuan, Chuanlong Wang, Guangting Xiong, Haoyi Wu,
Li Chen, Yihua Hu Ceramics International 48 (2022) 1836-1843
https://doi.org/10.1016/j.ceramint.2021.09.266

[121] Xue Bai, Zhengwen Yang, Yanhong Zhan, Zhen Hu, Youtao Ren, Mingjun Li, Zan Xu,
Asad Ullah, Imran Khan, Jianbei Qiu, Zhiguo Song, Bitao Liu, and Yuehui Wang ACS Appl.
Mater. Interfaces 2020, 12, 19, 21936-21943 https://doi.org/10.1021/acsami.0c05909

Rare Earth - A tribute to the late Mr. Rare Earth, Professor Karl Gschneidner   Materials Research Forum LLC
Materials Research Foundations 164 (2024) 230-256   https://doi.org/10.21741/9781644903056-6

Chapter 6

Rare Earth Element (REE) Insights for Health and Diagnostic Imaging

R. Ramakrishna Reddy[1,*] and Anthapu Pranav[2]

[1]Former UGC emeritus professor, Sri Krishnadevaraya University, Anantapur-515 001, A. P; India

[2]Biomedical Engineering, New Jersey Institute of Technology, Newark, NJ, USA

rajururreddy@gmail.com

Abstract

The medical industry is rapidly expanding its usage of rare earth elements (REE) in cutting-edge technologies including molecular imaging and radiotherapies. Because of its advantageous optical qualities, REE are included into a wide range of imaging modalities, including but not limited to X-ray, MRI, CT, ultrasound, nuclear medicine, and positron emission tomography. MRI is an imaging technique widely used in the clinic for the diagnosis of disease and visualization of injuries, which utilizes magnetic fields and electromagnetic radiation to create images of the physiology within the body or clinical analysis. It has various advantages, including quick detection time, large detection depth, no surgery need, etc. This review provides an overview of rare earths and their prospective applications in medical diagnostic imaging and other areas. Some rare earth elements may have use in biomedical imaging, cancer therapy, and image processing, all of which are briefly reviewed. The use of REE into health and medical applications is now well established. However, much of the future of diagnostic imaging analysis could depend on these paramagnetic elements.

Keywords

REE in pharmaceuticals, Biomedical imaging techniques, Rare Earth Elements in Cancer Diagnosis, Image Processing Tools

Contents

1.   Introduction

Due to numerous industrial applications, technological advancements, and residual fingerprints, the demand for REEs is rising quickly nowadays, but their availability in mineable deposits is very insufficient. The increasing use of REEs for industrial applications implicates a corresponding increase in health and environmental impacts. New developments in medical technology are expected to increase the use of surgical lasers, magnetic resonance imaging, and positron emission tomography scintillation detectors. Medical applications using REE appear to be on the rise as researchers find more and more ways to capitalize on the unique properties of these minerals, whether used singly, in combination with one or more, or with other metals.

Rare Earth Elements (REEs) biomaterials are gaining popularity due to their outstanding set of features, which includes resistance to light, magnetism, X-rays, and radioactivity. As a result of their unique properties, RE components have found applications in bioimaging, biotherapy, and biosensing materials. Due to REs' distinctive properties, new materials are being created for use in bioimaging and biosensing (biomolecules, metal ions, temperature, and pH sensing). Furthermore, the article covers how rare earths can be used in medicine. For the sake of study, diagnosis, and therapy, medical imaging techniques create an anatomically and physiologically precise picture of diverse human parts and systems. The importance of rare earth elements to reliable diagnostic imaging is well established [1-4]. This article discusses the importance of rare earth elements in science and medicine (REE). There is a clear and consistent presentation of work from many different groups. This paper presents a detailed overview of medical imaging's past, present, and probable future by analyzing the relevant literature at length and focusing on the function that rare earth elements play.

2. Importance and function of rare earths in diagnostic imaging

Besides modern technologies with their tremendous need of rare earth elements (REE), their use in medicine has gained prominence in different molecular imaging techniques and radio therapies. REE has been used in many imaging techniques such as computed tomography scans, magnetic resonance image (MRI), positron emission tomography (PET) imaging and X-rays. In medical device and equipment technology, REE are especially prized for properties that make them excellent   contrast   agents   and   the   most   powerful   magnets (https://www.metaltechnews.com/story/2020/04/29/tech-metals/rare-earth-metals-see-new-medical-uses/217.html). The REEs' supposed scarcity is greatly exaggerated. Sustainable rare earth elements (REEs) have positive implications on human health. And while they can be found everywhere across the Earth's crust, typical concentrations are quite low. They include the lanthanide elements and other metals like scandium and yttrium. Rare-earth elements (RE) are regarded as "vitamins" of the contemporary manufacturing sector. As a result of their peculiar physical and chemical properties, they have recently found widespread use in a wide variety of fields, including magnetic resonance imaging, superconductors, and lasers. Due to their high demand in cutting-edge technology, rare earth elements (REE) are finding more and more applications in the medical field, particularly in molecular imaging and radiotherapies. Rare-earth based materials can be utilized for targeted treatment by combining them with biomolecules like antibodies, tumors, peptides, or medicines [1], allowing them to cross the blood-brain barrier.

Successful rare earth element (REE) dopings of host nanostructured systems have been achieved for use in bioimaging. Materials doped with Gd^{3+}, Ho^{3+}, and Dy^{3+} ions permit excellent spatial resolution T1 -weighted, T2-weighted, and CT brain imaging because of their strong magnetic moments, large X-ray absorption coefficients, and very lengthy electron relaxation time. Multiple imaging techniques, from X-rays and MRIs to PET scans and positron emission tomography, have incorporated REE into their processes. In order to treat cancer, researchers have looked into using rare earth metals in radioimmunotherapy and photodynamic therapy [2, 3]. When compared to other materials used in biomedical settings today, RE-based Nano particles (NPs) do have some advantages. Biomedically relevant ions and molecules, including as DNA and proteins, can be detected and analysed using NPs based on RE. The luminous characteristics of the NPs are altered by the presence of the analytes, and these changes form the basis of biosensing experiments.

Bioimaging science is emerging in the domains of biomedicine and bioengineering and makes use of imaging technologies and computational tools to probe fundamental topics in the life sciences. Bioimaging allows researchers to see real-time biological processes without using needles, probes, or any other invasive tools [5].

Nanoparticles "doped" with rare earth elements have potential for use in the therapeutic contexts of tumor treatment and illness diagnosis. Multiple rare earth elements have been successfully doped into functional nanostructured systems for bioimaging [6]. Among the many potential applications for nanoparticles "doped" with rare earth elements are those in the treatment of tumours and the diagnosis of disease. Many rare earth elements have been successfully doped into suitable nanostructured systems for bioimaging applications [7,8] . It has long been understood that REE has positive effects on health. In contrast, these paramagnetic materials may offer a future solution for diagnostic imaging analysis [9].

2.1   REE in pharmaceuticals

The majority of rare earth elements (REE) are used to make catalysts like batteries, while the remaining 45 percent are utilized to make compounds, ceramics, and alloys, many of which are used to make pharmaceuticals and medical equipment. Among the many beneficial effects of lanthanum on human metabolism are decreases in LDL cholesterol, blood pressure, hunger, and the chance of developing blood clots. Reducing phosphate levels with lanthanum carbonate is a popular treatment for people with renal failure. Excessive phosphate intake has been linked to decreased calcium absorption, which can have major health effects [10].

Cancer could be treated with yttrium in at least two different ways: either by killing cancer cells on touch or by stopping blood from reaching the afflicted area. You can inject millions of radioactive Y-90 beads into the arteries giving blood to a tumor, or you can connect the element to antibodies that can precisely bind to the tumor's cells. Both yttrium-90 and yttrium-90 brachytherapy can be used to target the tumour at a specific site, which is a significant advantage. This means that just the cancer cells are destroyed, while the healthy tissues are spared. Samarium is another another rare earth element used for analgesia in cases of osteoblastic metastatic bone lesions. Additionally, it can be given to patients intravenously.

It has been recognised for some time that REE has medical and health-related uses [9,10]. The La-Lanthanum oxide nanoparticles are useful in magnetic resonance imaging (MRI). The scintillator material cerium-doped lutetium orthosilicate (CeO2) has found widespread use in positron emission tomography (PET) scans. Radon therapy has made use of praseodymium (Pr) oxide nanoparticles. Neodymium (Nd) crystals are used in a variety of laser applications, such as those used to treat skin cancer and remove unwanted hair. Patients with bone cancer have had their severe pain treated with the Sm-153 radioisotope.

Eu-optical the potential of europium-based nanoprobes for heterogeneous/homogeneous biodetection and in vitro/in vivo bioimaging is discussed. Due to their magnetic properties, gadolinium-doped MRI contrast agents and intravenous radio-contrast agents aid in the visualisation of tumours on MRI. The radioactive isotope Tb-149 has been utilised effectively in targeted therapy for the treatment of cancer. Radioactive Dy- Dy-165 has been used to treat rheumatoid arthritis knee effusions. Non-invasive medical methods using Ho -Holmium based solid-state lasers have proven effective for treating cancer and kidney stones without the need for surgery. Er -Erbium lasers have applications in dentistry and medicine. This radioactive isotope, Tm-167, can power portable X-ray generators. Using the radioactive isotope Lu-177 in medicine is a possibility.

2.2 Artificial organs

There are a lot of people that get to live healthy lives because of the metals and minerals utilised in medical equipment. An oral prosthesis such as a crown, bridge, denture, or more extensive prosthesis for the face can be securely fastened to the jawbone with the help of dental implants. Rare earth elements (REEs) have a significant impact on our daily lives because of their widespread use in the biomedical industry today. Bone is a busy organ that does more than just hold the human body together; it also controls blood calcium levels, gives muscles leverage, and shields the brain and spinal cord from injury. Recently, research has focused on the ability of rare earth elements (REEs) to accelerate bone healing. For this reason, including the appropriate rare earth elements (REEs) into bone tissue engineering (BTE) scaffolds is a practical method for enhancing bone healing outcomes in the present day [11]. The rapid growth of the electronic implant market is indicative of the industry's increasing reliance on ceramic technology. Neurostimulators are now in development as a means of pulsating nerves and easing medical symptoms, with ceramic composition playing a major role in the success of some companies' efforts.

Zirconia ceramic is substantially more robust than alumina. Due to its excellent resistance to wear, ceramic has found use in components such as the femoral ball of hip replacements. In order to keep the product from degrading too quickly, this material is occasionally combined with the metal yttrium, despite its reputation for durability. When coated with tantalum, carbon foam scaffolds can replace non-biocompatible dental or orthopaedic implants in bone regeneration. Bone fusion to a dental implant is aided by this porous substance. When a person's bones or joints start giving them trouble, they may seek out orthopaedic implants as a solution. Temporary artificial joints are available, but permanent ones are still possible. Permanent orthopaedic implants have a good chance of outliving the patient. They can go just about everywhere on the body, including the hips, knees, ankles, shoulders, elbows, and wrists [11].

The other noteworthy applications   are   also herewith given. Rare-earth elements (REEs) are used in the components of many devices used daily in our modern society, such as: the screens of smart phones, computers, and flat panel televisions; the motors of computer drives; batteries of hybrid and electric cars; and new generation light bulbs. Lanthanum-based catalysts are employed in petroleum refining. Large wind turbines use generators that contain strong permanent magnets composed   of   neodymium-iron-boron[4](https://pubs.usgs.gov/fs/2014/3078/pdf/fs2014-3078.pdf).

3. Methods

3.1 Biomedical imaging techniques

In the twenty-first century, disease will be diagnosed and treated with greater precision and less discomfort thanks to less intrusive, more technically advanced imaging and image-guided procedures. Roentgen's discovery of the x-ray has allowed for the development and spread of medical imaging. Bioimaging is the practice of studying living systems without harming them in any way. Bioimaging is a noninvasive, exterior technique used to learn about the internal structure of the studied object. The improvement of biomedical imaging has had a huge effect on healthcare. Since its inception, bioimaging has sought to simultaneously record as much data as possible about a biological activity and keep outside influences to a minimum. The historically high rate of human error in disease diagnosis and surgeries has been dramatically decreased thanks to the introduction of picture guided therapy. Bioimaging is associated with methods that covertly and reliably photograph natural cyclical phenomena [12]. This talk provides an overview of the various imaging techniques now in use and assesses how effective they are at solving pressing medical issues. An growth in the use of bioimaging for analysing chemicals, toxins, and microbiological elements is a direct result of efforts to combat climate change, which require ongoing monitoring of the environment.

X-ray computed tomography (CT), magnetic resonance imaging (MRI), positron emission tomography (PET), ionization radiography (IR), and roentgenography (RAM) are only few of the noninvasive biomedical imaging modalities that have been developed in recent decades. Molecular imaging is a subfield of medicine that arose out of the need for more contrast in medical images than could be achieved by purely anatomical distinctions. The subfield of medical imaging known as molecular imaging arose from the need to provide visual contrast beyond that offered by morphological distinctions alone. It aims to visualize molecules and biomolecular processes of medical significance within living patients without causing any harm. Common vibrational micro-spectroscopy methods include infrared (IR) and Raman [13, 14]. These two state-of-the-art methods allow for the visualization of vibrational spectra of biological components within a cell or tissue. This means that IR and Raman microscopy provide the powerful prospect of high contrast images without the need for external labelling, and instead expose the unique and individual "fingerprint" spectrum of each cell. Because of its great spatial resolution and chemical specificity, label-free vibrational (infrared (IR) and Raman) imaging has swiftly become one of the most promising molecular imaging methods available today.

3.2   Cancer diagnosis and treatment using rare earth elements

The biological properties of rare earth element ions are similar to those of calcium ions, rare earth elements are used in medicine research. In recent years, an increasing number of studies have focused on the applications of rare earth elements in cancer diagnosis and therapy. It's common knowledge at this point that REE have medical and health-related applications. But these paramagnetic ingredients could be the future of diagnostic imaging analysis. Gd is by far the most crucial rare earth element for MRI diagnoses. Intravenous radio-contrast medicines containing Gd (III) ions have been used to improve MRI images and have also been shown to increase the sensitivity and specificity of diagnostic imaging. Due to its optical qualities, REE have found use in many different types of imaging technology, from X-rays and MRIs to PET scans and CT scans.

This article (https://nanografi.com/blog/rare-earth-elements-in-cancer-diagnosis-and-therapy/)discusses the application of certain rare earth elements to the detection and treatment of cancer.

Consumers, farmers, and medical professionals are worried about the extensive use of cerium (Ce), the most abundant rare earth element on Earth.

3.2.1 Yttrium for liver cancer treatment

In terms of cancer mortality rates, hepatocellular carcinoma (HCC) ranks third. In recent years, Y90-radioembolization's effectiveness has been demonstrated through multicenter studies yielding consistent outcomes. There is evidence of a therapeutic impact in advanced HCC, especially in patients with portal vein thrombosis (PVT), and Y90-radioembolization is equally effective and more well tolerated than transarterial chemoembolization (TACE) in intermediate HCC (Hepatocellular carcinoma). Radioembolization is an effective method for treating liver cancer, and it is more than just one therapeutic method. In the case of cancers that are amenable to chemotherapy, it can be given with the latter. Further, it avoids the need for extensive medical care while having a minimal risk of problems, being well-tolerated by most patients, and requiring no more than a 24-hour hospital stay. Riaz et al. [15] provide a comprehensive description of this technique.

Radiation therapy for cancer patients now routinely includes the use of Yttrium-90 microspheres, which have been found to improve survival rates. These hermetically sealed sources release radiation that kills cancer cells. This research shows that the medication is successful in halting the growth of lesions in primary liver tumours (also known as hepatocarcinomas).

3.2.2 Vanadium applications in melanoma

Few studies have looked into vanadium's effects on skin cancer, despite the fact that melanoma is the deadliest type of skin cancer and its incidence is rising faster than that of any other malignant tumour worldwide.. New therapeutic targets and treatment alternatives for melanoma are urgently needed [16] due to the malignancy's propensity to spread to other organs and patients' recurrence due to the development of treatment resistance.

Only a small number of pre-clinical models of cancer have been undertaken with vanadium to yet, despite the fact that vanadium compounds have been the topic of extensive study, suggesting a possible therapeutic opportunity. The development of new vanadium materials and complexes, as well as the identification of the vanadium most appropriate for future development against a certain disease, are challenging endeavors that call for the coordination of specialists from a wide range of disciplines. We expect that in the next decade [17,18] basic questions about the use of metal compounds, especially vanadium, in the treatment of cancer will be answered. Possibly due to their anti-cancer properties, vanadium compounds and/or vanadium materials may be effective in the treatment of melanoma. Cancer cells are killed by the radiation from where they are kept. This research shows that the medication is successful in halting the growth of lesions in primary liver tumors (also known as hepatocarcinomas).

3.2.3 The use of neodymium in the management of skin cancer

Trivalent neodymium (Nd3 +) was the first lanthanide (a type of rare-earth element) to be used to generate laser light. Neodymium laser irradiation is an effective method with acceptable cosmetic effects [19] for treating T(1-2)N(0)M(0) facial skin carcinoma. Small movements of a neodymium magnet can increase the number of skin surface monitoring counts and promote the movement of a magnetic tracer [20]. The great majority of cancer cases in the United States are nonmelanoma skin cancers. Nd:YAG laser can be used to treat a wide variety of benign skin growths, including brown age spots (solar lentigines), freckles, naevus of Ota, naevus of Ito, lumbosacral melanocytosis, Hori naevus, and café-au-lait macules. This laser's beam may be able to penetrate deeper into tissue than others, and it speeds up the clotting of blood. Endoscopes allow Nd:YAG lasers to be put into previously inaccessible locations such as the oesophagus (food pipe) and the colon (large intestine) (colon). In addition, these light sources can be sent to a tumour via tiny, see-through tubes called flexible optical fibres, where the radiation's heat can destroy malignant cells. The depth of action of Nd:YAG lasers is significantly higher. The use of these products has the potential to cause a significant acceleration of blood clotting. The laser light is transmitted into the body via optical fibres. One laser, the Nd:YAG, has been shown to be successful in treating throat cancer. In contrast to the widespread acceptance of Nd:YAG laser therapy for rosacea, the benefits of PDL treatment are less obvious. The goal of this research was to compare the performance of PDL and Nd:YAG lasers in the treatment of rosacea.

While on long space trips, NASA uses neodymium magnets to assist astronauts maintain their strength and fitness. In orthodontic operations like molar distillation and palatal expansion [21-24], neodymium magnets can be used as a motion-generating device due to the push-and-pull forces they generate. Magnets have found their way into orthodontic operations as well. Using magnets to slowly and steadily move the buried tooth root outward over the period of 9-12 weeks is a viable treatment option for dental fractures. The root tip can then be reshaped using methods such as porcelain coating. Neodymium magnets require coatings since the material corrodes easily and loses strength with time [25].

3.2.4 MRI contrast agents using gadolinium nanoparticles

Among the newly discovered materials with favourable electromagnetic properties that are reshaping the technological landscape is gadolinium, a so-called "rare earth" element. Gadolinium is a crucial rare earth element in the detection of brain tumors. Radiology specialists frequently use gadolinium and other MRI contrast materials in the fight against malignant brain tumors. Gadolinium can be utilized to locate even the tiniest tumors due to the fact that it gives them its own color, a silvery white, when injected into the body. As gadolinium accumulates in certain structures, the T1 of such regions is shortened, resulting in enhanced signal in T1-weighted sequences. To enhance the effectiveness of magnetic resonance imaging (MRI) at low doses in the identification of malignancies, a targeted T1 contrast agent called recombinant human heavy chain (H-chain) ferritin (HFn) with gadolinium labelling (HFn-Gd) was created [26].

One of the most recent applications of gadolinium (Gd) in MRI is in contrast agents (CAs; Gd-CAs) (MRI). In view of the current status of CAs, the development of multimodal CAs that are biocompatible, non-toxic, and have a long circulation time is of the utmost importance. Gd-CAs have tremendous potential and promise to significantly advance bioimaging applications [26]. Rare and powerful, gadolinium (Gd) is used in MRI scanners, CT scanners, and neutron capture therapy for cancer. Gadolinium oxide (Gd_2O_3) nanoparticles (GNPs) are well-suited for optimising this multifunctionality as a result of the high concentration of Gd within each GNP, which allows them to serve as both diagnostic and therapeutic agents in cancer theragnosis.

3.2.5 Bone metastases treatment using samarium

Metastasis to the bone usually occurs first in people with prostate cancer. The pain from bone metastases can be managed with a number of newer drugs. A few examples are painkillers, chemotherapy, hormone therapy, surgery, biphospho-nates, internal and external beam radiation therapy, and radiopharmaceuticals. Over half of persons with prostate, breast, or lung cancer may develop bone metastases, a painful consequence of the disease. Samarium is a bright, fairly hard, silvery white metal. It is one of the lanthanide rare earth metals. With a half-life appropriate for radionuclide therapy (t1/2 = 46.3 h), Samarium-153 has gained considerable attention due to its widespread use in the clinic as a bone pain palliation agent in patients with painful bone metastases arising as complications from various cancers [27]. Samarium-153 (153Sm) is a highly interesting radionuclide within the field of targeted radionuclide therapy. Samarium is useful in treating cancer in a number of ways. Samarium-153 lexidronam, a radioactive isotope, is a highly effective treatment for bone metastases (EDTMP). Metastases to the bone are a typical consequence of cancer. After the lungs and the liver, the bone is the third most common site where metastases appear. The goals of treating bone metastases include pain relief, reduced steroid use, and mobility preservation. Samarium-153 lexidronam (153Sm-EDTMP) has been given the green light by the FDA for the treatment of painful osteoblastic bone metastases detected by a bone scan [28, 29]. The medical world has long acknowledged that systemic metabolic irradiation with samarium 153 lexidronam is a safe and effective technique to treat people with bone pain.

4.   Rare-earth doping elements play a crucial role in Bioimaging, and we need to learn more about them

Rare-earth doping can change doped nanomaterials' size, shape, and crystallographic phase, producing tunable optical responses and improved mechanical and electronic functionalities. Additionally, rare-earth doping can significantly enhance energy conversion and harvesting via tunable and scalable control over doped nanomaterials' final electrical and catalytic performance.

Use of rare earth elements in bioimaging is briefly covered in this article.

Biomedical imaging accounts for about 10% of the device business as a whole and is expanding at a rate of 5- 10% annually. Optical imaging is an essential method in biomedical research since it is noninvasive, real-time, and provides a high magnification. In recent years, fluorescence imaging and other forms of molecular imaging have received considerable attention for their potential use in the diagnosis of malignant tumors.

Bioimaging helps doctors diagnose and treat a wide variety of conditions. Gold nanoparticles, rare-earth based nanoparticles, and iron-oxide magnetic nanoparticles are just a few of the new contrast agents being investigated by the bioimaging community. Bioimaging science is a growing subdiscipline of biomedicine and bioengineering that uses cutting-edge imaging gear and software to tackle pressing problems in the life sciences. The term "bioimaging" refers to a technique that provides researchers with a real-time look at biological processes without the need for intrusive technology that must penetrate the skin or enter the body.

New applications have emerged thanks to the development of nanostructured materials that include RE elements as either primary constituents or dopants. Doping nanoparticles with rare earth elements makes them optically active, expanding their potential applications [30–32]. Various rare earth elements, such as Eu^{3+}, Gd^{3+}, La^{3+}, Ce^{3+}, Tb^{3+}, etc., have been effectively doped into a wide variety of host nanostructured systems for application in bioimaging.

Doped nanoparticles with rare earth elements have numerous uses in biomedicine, including in disease diagnostics, drug administration, tumor therapy, and even bioimaging. Although numerous bio-imaging techniques are currently in use, fluorescence imaging technology based on the rare earth-doped nanoparticles stands out as a particularly potent instrument in biological technology with wide-ranging medical and biological applications. One area where rare-earth doped laser materials shine is in quantum information storage and processing, as well as in the realm of biological imaging enabled by their high-Q 4f4f optical transitions. The luminescent, magnetic resonance, and X-ray computed tomography bioimaging applications, as well as the luminous (including down conversion, up conversion, and permanent luminescence), and magnetic properties of these rare earth-based nanoparticles, will be described.

Rare earth elements are the "treasure trove" of innovative materials (REEs). A wide range of manufacturing companies rely on them to spur innovation, from those that make consumer goods to those that make cutting-edge technologies. In recent years, there has been an increase in the usage of rare earth elements (REE) in the medical field, especially in the fields of molecular imaging and radiotherapies, due to the high demand for REE in cutting-edge technologies.

5.   Efforts in bioimaging research involving rare earth elements (REEs)

Scholarly works from the past that have made significant contributions to this area are reviewed. Since the 19th century, researchers have been curious about the 57-71 atomic number elements (rare earth elements, lanthanides). Rare earth metals are often used in medicine because of their biological resemblance to calcium ions. Anti-atherosclerotic effects of lanthanum chloride and treatment of hyperphosphatemia with lanthanum carbonate [33]. Tuberculosis was also treated with solutions containing rare earth metal ions. In the first decades of the twentieth century, rare earth elements began to be used in the medical field.

Fan et al. [6] demonstrate how rare earth-doped nanoparticles can be used for a broad variety of purposes, including illness diagnosis, drug delivery, tumour therapy, and bioimaging. The rare earth-doped nanoparticle-based fluorescence imaging approach is a powerful tool in biotechnology with widespread applications in medicine and the life sciences. Water solubility, biocompatibility, drug-loading capability, and tumor targeting specificity are just few of the desirable properties that can be imparted on rare earth-doped nanoparticles through surface functionalization. All of this points to its promising future in cancer diagnosis and treatment.

Doped nanoparticles with rare earth elements have numerous uses in biomedicine, including in disease diagnostics, drug administration, tumour therapy, and even bioimaging. There are a variety of bio-imaging methods now in use, but fluorescence imaging technology based on the rare earth-doped nanoparticles stands out as a particularly powerful instrument in biological technology with extensive medical and biological applications. Rare earth doped up conversion nanoparticles (UCNPs) were first described by Hong et al.[21] as a new type of luminescent material that can both absorb and emit light. UCNPs offer exceptional imaging sensitivity, do not suffer from photobleaching, are mostly harmless to living tissue, and may travel to vast depths within tissue. Hydroxyapatite doped with luminous chemicals is a novel and promising approach to biological luminescence imaging. Lighted materials based on hydroxyapatite are attractive because of their biodegradability, bioactivity, biocompatibility, osteoconductivity, non-toxicity, and lack of inflammation, as well as their accessibility for surface adaptation. The main inorganic component of bones, hydroxyapatite (3), has been shown to have significant potential in tissue engineering, medication and gene delivery, and other fields of biomedicine. According to a review by Gu et al.[31], hydroxyapatite nanoparticles (HAP NPs) can serve as hosts for various substrates and dopants. HAP NPs, which have been doped with rare earth (RE) ions, have garnered a lot of attention due to the unusual physics, chemistry, and imaging effects they produce. The advantages of RE-doped HAP NPs over other fluorescent probes are numerous. Researchers have found success with hydroxyapatite nanoparticles in a number of different biological applications, including imaging, drug administration, bone tissue engineering, and antibacterial studies.

Tumor-specific surgery employing CC-Nd@PEG, 12 was described by Xiao Zhang et al.[32], which combines the superior NIR-II fluorescence of RENPs with targeting of cancer cell membrane coating. Subsequent coating of NaYF4:Nd5@NaYF4 with the membrane of cancer cells has been shown to enhance tumour targeting and decrease accumulation in the liver and spleen. Potential medicinal applications for the CCNd@PEG nanoprobe are exciting.

A simple hydrothermal approach was developed to build a multimodal imaging probe using the special features of rare-earth ions. Co-doped calcium fluoride nanoparticles (CaF2:Y,Gd,Nd) are biocompatible, extremely crystalline, and exhibit a consistent shape across the particle size range. Additionally, in vitro and ex vivo studies have been conducted to determine the imaging capability of CaF2: Y,Gd,multimodal Nd, and its effective performance in NIR-II fluorescence/photoacoustic/magnetic resonance imaging has been established [33]. The authors argue that creating unique diagnosis nanoparticles will result in the creation of multifunctional nanoplatforms for illness diagnosis and therapy.

Wei et al. [35] provide a comprehensive review of numerous clinical applications involving the brain, including positron emission tomography, positron emission tomography/computed tomography, Magnetic resonance imaging, fluorescence imaging, multi-modal imaging, radiotherapy, photodynamic therapy, photothermal therapy, and many more. Researchers think that MRI, FLIM, and CT scanners could benefit from CAs made from rare-earth elements (CT).

The applications of rare-earth based materials, including brain imaging, brain diseases therapy, brain disease diagnosis and monitoring, and brain modulation through optogenetics (Fig. 1).

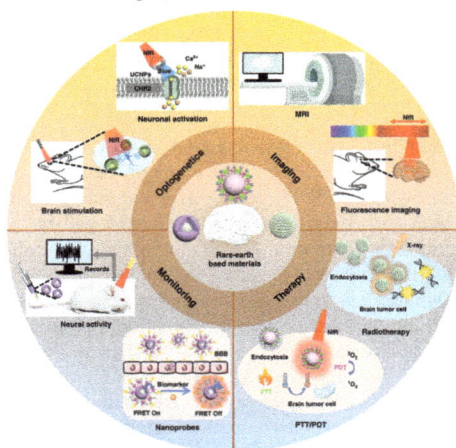

Fig.1 The applications of rare-earth based materials for brain disease diagnosis and treatments [35].

For earlier cancer cell detection, researchers have developed a technique called dual modal imaging, which combines two distinct imaging modalities. Developing a good contrast agent for dual modal imaging applications can help save lives. Though numerous attempts have been made to create contrast agents for MRI and Optical Imaging, issues of nanoparticle size, biocompatibility, and toxicity have impeded these studies. The development of MRI/optical contrast agents greatly benefits from the presence of both a high relaxivity value and robust emission.

Recent research by Bingzhu Zheng et al. [36] highlighted the growth of rare-earth doping in inorganic nanomaterials and the scope of their possible applications. We will discuss the basic criteria for rare-earth doping, such as fundamental electronic structures, lattice environments, and doping techniques, as well as the primary design principles that increase the material's electrical, optical, catalytic, and magnetic properties through rare-earth doping. We also talk about the obstacles that need to be overcome and the directions that future research should go in order to master rare-earth doping for non-traditional applications.

Due to its high spatial resolution and deep tissue penetration, NIR-II (1,000-1,700 nm) fluorescence imaging is frequently used in cancer detection and treatment. In their recent publication [37], Zhen Feng et al. address current advances in NIR-II nanoprobes, such as rare-earth-doped nanoparticles. (RENPs). The group is investigating ways to optimise RENPs for in vitro and in vivo cancer diagnostics using near-infrared (NIR)-II imaging.

In a review, Zhen Feng et al. [37] explain why RENPs are attractive for NIR-II biological imaging because to their low toxicity, strong photostability, deep tissue penetration, and variable pharmacokinetic behavior. However, researchers stressed that even with the gains made, much more work is needed to fully utilize NIR-II RENPs for bioimaging. One such factor is the diminutive size of the emission center of RENPs (the source of their fluorescence). It is well known that the emission centers of NIR-II RENPs driven by 808 nm or 980 nm lasers can be made up of any of the five rare earth elements (Nd^{3+}, Tm^{3+}, Pr^{3+}, Ho^{3+}, Er^{3+}). Existing RENPs continue to rely predominantly on Nd^{3+} and Er^{3+} as the emission centers, which poses serious challenges for the development and deployment of near-infrared probe types, as stated by Zhen Feng et al. [37].

$NaErF_4$: Ho@$NaYF_4$ nanoparticles, emitting at 1,180 nm, were fabricated by Liu et al. [38] using Er^{3+} as a sensitizer and Ho^{3+} as an emitter. Moreover, due to their disproportionately high diameters, RENPs have historically presented significant challenges for bioimaging. Size reduction of nanoparticles allows them to penetrate biological tissues and even cells, albeit at the expense of their luminescence. The luminescence intensity of commonly used core-shell structures can be increased, but nanoparticles' increased size makes it more difficult for them to enter biological tissues and lengthens the time it takes for them to be digested. In order to enhance RENPs' NIR-II bioimaging applications, it is still important to design nanoparticles of a suitable size. Recent decades have seen an uptick in research on the potential role of RENPs in multispectral molecular imaging and the monitoring of drug delivery.

Based on our current knowledge and the results of new initiatives, we can expect to see a dramatic increase in our knowledge of REE involvement in living organisms and in the increased exploitation of the peculiar properties of REE for the design of novel applications in diagnostic procedures and the establishment of powerful medical devices in the near future [39].

Applications for nanoparticles doped with rare earth elements include imaging, drug delivery, tumour therapy, and disease diagnosis. Fluorescence imaging based on rare earth-doped nanoparticles is a significant tool in biological technology [40] due to its capacity to visibly exhibit cell activity and lesion evolution in live animals.

Because of the critical need to reduce female mortality from breast cancer, Jain et al. [41] developed innovative methods for early diagnosis of the disease. Here, we present a novel approach to breast cancer diagnosis based on the use of Gd_2O_3:Eu^{3+} nanoparticles modified with folic acid (FA). Results led them to speculate that FA-conjugated Gd_2O_3:Eu^{3+} nanoparticles might one day be used in the clinic as either a standalone detection tool or to increase the sensitivity and specificity of existing technologies. The EPR result may be one reason why nanoparticles have recently attracted so much attention as a possible tool in cancer diagnosis and treatment. For the purpose of cancer imaging, nanoparticle systems compatible with computed tomography and positron emission tomography have been created (SPECT). For the detection of Folr1 breast cancer, Jain et al. [40] created a folic acid-conjugated Gd_2O_3:Eu^{3+} nanoparticles as a fluorescent probe that is safe to use in vivo. High-resolution imaging and the detection of malignancies deeper in the subcutaneous or osseous tissue may one day be possible with the use of these nanoparticles in combination with CT scans.

For screening, localisation, and therapy monitoring in cancer, state-of-the-art imaging equipment is desperately needed, as reported by Helen et al. [42]. Rare earth elements have been mostly under-utilized for in vivo imaging with the exception of gadolinium, which has found considerable use in magnetic resonance imaging. Helen et al. [42] analysed the properties of this newly discovered class of materials to determine their potential applications in cancer imaging. More recent research has also demonstrated that rare earth elements can help in the discovery of novel cancer therapies.

Bioimaging methods based on fluorescence are commonly employed to better understand these fundamental biological processes. X-ray diffraction (XRD) patterns reveal that Eu^{+3} and Sm^{+3} doped HAp NPs belong to distinguishable functional groups, as reported by Zantyea et al. [43]. Short rods of 60 nm or less in length are the typical shape that HAp NPs take after being synthesised. When stimulated at specific wavelengths, the doped HAp NPs produced lines characteristic of Eu^{+3} and Sm^{+3}.

Zantyea et al. [43] found that the as-synthesized NPs were non-toxic to HeLa cells and were readily taken up by the cells, suggesting that they would be helpful as live cell bioimaging agents. Potential possibilities for further investigation of bioimaging technologies may develop in the not-too-distant future.

The benefits of rare earth nanoparticles were discovered by Qu et al. [44]. Their NIR luminescence efficiency is great, they are non-toxic, and they are biocompatible. Their potential uses, such as in cancer diagnosis, treatment, and surgical navigation, are quite exciting. However, clinical data demonstrating the use of RE nanoparticles for surgical navigation is lacking in the published literature [43].

Using rare-earth elements, which have the potential to give extraordinary sensitivity and multiplexing in biological assays, is one manner in which chemistry may pave the way for novel approaches.

Rare Earth - A tribute to the late Mr. Rare Earth, Professor Karl Gschneidner   Materials Research Forum LLC
Materials Research Foundations 164 (2024) 230-256   https://doi.org/10.21741/9781644903056-6

6.   Techniques for diagnosing using bioimaging

Bioimaging may be a method that uses advanced bioimaging probes to visualize biological processes vivo. Bioimaging can really be interpreted in ways, particularly as imaging of organic strategies in dwelling animals and people or as imaging by way of finest imaging probes.

Biological imaging may be very noticeably rapid and objective, and produces regular effects as compared to traditional techniques. Imaging techniques, which includes optical imaging, x-ray imaging, infrared thermal energy, magnetic resonance imaging, nuclear imaging, and ultrasound are with achievement applied in medical programs.

Quite a few of us have worked in the diagnostic imaging field in the past. You should be able to jump in without any prior knowledge thanks to the foundational information provided here. Imaging modalities like MRI, CT, ultrasound, nuclear medicine, infrared sensor technology, and optical microscopy are all included in biomedical image processing, as is the process of acquiring and analysing pictures for application in medicine and biotechnology.

The field of oncology desperately needs better imaging for diagnosis, staging, and therapy outcome monitoring. By using this imaging technique, scientists are able to see, analyse, and measure molecular and cellular processes in living species like humans and other mammals. PET-CT, MRI, MRS, spectroscopy, optical imaging, and ultrasound are only few of the imaging modalities used. Researchers from many different fields, such as biology, chemistry, medical physics, pharmacology, computer science, biomedical engineering, and clinical practise, must work together to advance biomedical imaging.

Ultrasound, computed tomography (CT), and magnetic resonance imaging (MRI) have recently seen rapid technological improvements that have resulted in outstanding spatial and temporal resolution, allowing for numerous options for functional assessment of varied body systems. Radionuclide imaging (SPECT/PET) and functional magnetic resonance imaging (MRI) have greatly advanced medical science. Functional magnetic resonance (MR) encompasses a wide range of methods that reveal underlying physiological and molecular processes long before morphological changes become evident in more traditional imaging. Intraoperative MR guidance for neurosurgery improves tumour resection estimates using functional MRI and SPECT data from the same patient.

The value of these techniques, and why it is employed. Some solid justifications are as follows:

1. The first goal is to correctly identify medical problems and effectively treat and cure patients with no negative consequences.

2.Computer vision can supply 3D and 4D information that aids in human comprehension by capitalizing on texture, shape, contour, and prior knowledge in addition to contextual information from image sequence.

3.To examine the body of a patient without having to cut them open

4.We aim to aid patients in the long run without adding to the already excessive cost of medical treatment.

Here's a rundown of some of the many uses for this:

a. Diagnosis: Merging data from various imaging techniques

b. Investigating the Course of Illness: Keeping an eye on gradual shifts in dimensions, orientation, and acuity of an image

c. Radiosurgery and other surgical procedures assisted by imaging technology:

d. Integrating the patient's anatomy into the surgery plan and preoperative pictures

e. Analysis of patient data or creation of an atlas: matching a person's physical characteristics to those described in a reference atlas.

The term "medical imaging technology" is used to describe any technique that provides clinicians with a view of an area of the body that is normally out of sight. By seeing these structures, doctors are better able to diagnose diseases, plan therapies (via techniques like image-guided intervention), and keep tabs on their patients. The field of medical imaging [45] has made tremendous strides in recent years. Doctors can now use image manipulation to glean more information from the same dataset, which improves diagnostic imaging's value. The medical imaging sector is in a state of flux due to a number of factors, including rapid innovation, the creation of new business models, the arrival of AI and the IoT, and the potential disruption caused by these technologies.

All imaging methods are fundamentally the same. A detector records the amount of radiation that entered and left the patient's body during a medical imaging procedure. The wave's physical manifestation varies from one modality to the next Computed tomography (CT) utilizes x-rays, whereas magnetic resonance imaging (MRI) and single-photon emission CT (SPECT) utilize radio waves and gamma rays (SPECT).

6.1   Photography

The use of light is crucial to the photographic process. Photographic recording involves capturing images by exposing light-sensitive film or an array to radiation. Dermatology and pathology (both macroscopic and microscopic) were two areas of medicine that were rapidly impacted by photography [46,47]. Almost immediately, doctors began using photographs for patient consultation, record keeping, and teaching, and enormous image libraries were accumulated for these functions. Visible light is used to impress an image onto a light-sensitive substrate, traditionally a silver halide plate but more commonly a silicon chip. However, photons are reflected by nontransparent things such as human skin. Other types of radiation have been used to expose human interior structures, and most interestingly, these techniques are noninvasive thanks to advances in science and technology. As a result, so-called medical imaging with the use of these techniques now guides much of the medical diagnosis and therapy.

6.2   Ultrasound

Sonography, which is another name for ultrasound imaging, employs high-frequency ultrasonic vibrations to create an image of the inside of a living body. Using a high-frequency sound wave, it is possible to create an internal image of soft tissues like muscles and organs. Although it cannot resolve as finely as higher-frequency ultrasound (>10 MHz), low-frequency ultrasound (1-6 MHz) can reach deeper into tissues due to its longer wavelength. The liver and kidneys can be seen at low frequencies. Scattering or reflecting of these waves by tiny structures at high frequencies (7-

Rare Earth - A tribute to the late Mr. Rare Earth, Professor Karl Gschneidner   Materials Research Forum LLC
Materials Research Foundations 164 (2024) 230-256   https://doi.org/10.21741/9781644903056-6

18 MHz) enables imaging of superficial tissues including tendons and the newborn brain. With ultrasound imaging [48], we can see how our internal organs and blood vessels are doing in real time. Unlike X-ray imaging, ultrasound does not expose patients to potentially dangerous doses of ionizing radiation. A transducer (probe) is pushed against the patient's skin or inserted into a bodily cavity to perform an ultrasonic exam. Ultrasound waves are transmitted from the transducer through a thin layer of gel that is attached to the skin. These advancements are helpful for patients of all kinds because they improve the standard of care they receive and the accuracy of their diagnosis. For more data about state-of-the-art methods in medical imaging analysis, please keep reading.

The use of ultrasound technology has given doctors a glimpse of the workings of internal organs and blood arteries. During an ultrasonic examination, a portable transducer is utilised to make skin contact. The ultrasonic waves emitted by the transducer are reflected by the human body. Real-time images of internal organs and tissues can be generated by a computer that measures the reflected sound waves as they pass through the body. The clarity of the gathered picture depends on a number of factors, including the frequency, strength, and timing of the reflected sound wave. Ultrasound has the potential to mildly heat the tissues. This disorder causes microscopic gas bubbles to form in the body's fluids and tissues of affected persons (cavitation). The implications of this are still unclear. A significant advantage of ultrasound technology is its portability.

6.3   Magnetic resonance imaging (MRI)

At the close of the twentieth century, when worries about radiation exposure during medical imaging were at their height, magnetic resonance imaging (MRI) was developed. This imaging method takes advantage of the inherent magnetic fields of the human body to capture pictures of the insides of the body. Although MRI was previously only useful for certain diagnostic purposes, technological advancements have made this imaging method the gold standard for studying soft tissues and vascular structures. Modern MRI scanners are less claustrophobic for people since they are smaller and more open [49].

In magnetic resonance imaging (MRI), the patient is positioned on a motorised table, which is then moved into the centre of a massive, tubular scanner that produces a powerful magnetic field. Tissues are made up of positively charged atomic protons that are often dispersed at random. However, due to the extraordinarily powerful magnetic field in an MRI scanner, the protons align with it. The scanner then sends out a brief burst of radio waves to temporarily disrupt the alignment of the protons. An energy discharge occurs as the protons realign with the magnetic field (called signals). Different tissues have different maximum signal strengths. These emissions are detectable by the MRI machine. The data is sent to a computer, where it is processed to form the final image.

MRI is also used to do the following:

1.Measuring a few chemicals in the brain can distinguish between a tumour and an abscess.

2.Find any breaks in the hip or pelvis, as well as any reproductive system abnormalities.

3.Help doctors determine the severity of patients' joint injuries (such as ligament and cartilage tears in the knee).

4.Help doctors determine the source of bleeding and infection.

6.4   PICT stands for Particle Imaging Computed Tomography (PET)

Positron emission tomography (PET) has been used in clinical settings for over 15 years now. Positron emission tomography (PET) is one of the most cutting-edge non-invasive imaging techniques used to measure radioactivity in living organisms. When a positron-emitting radiopharmaceutical is administered intravenously, the next step is to scan the patient to detect and quantify accumulation patterns [50].

Each voxel's signal strength in positron emission tomography (PET) imaging is based on the radioactivity of the radionuclide-tagged radioactive tracer that was administered intravenously prior to scanning. In order to detect the photons with opposite spins that result from the indirect positron decay of PET radionuclides, the "PET scanner" makes use of a gamma photon coincidence detection system optimised for this purpose [51]. Instead of only looking at 2D images, this line of reasoning allows us to create 3D quantitative maps of radiolabeled tracers in tissue. To far, 2-[18F] fluoro-2-deoxy-D-glucose (also known as [18F] FDG) has been the most extensively used radioactive tracer for positron emission tomography (PET), although various novel tracers are currently in development that will be able to highlight a larger variety of organ and tissue metabolic activity. A comprehensive meta-analysis found that 30% of patients had their treatment plans changed as a direct result of PET.

6.5   Nuclear imaging

Radioactive compounds (radiopharmaceuticals) are used in nuclear medicine for both diagnostic purposes and therapeutic interventions. Because of their high sensitivity and need for only a tiny amount of tracer molecules to be injected, nuclear imaging techniques are widely employed for medical programmes. After a radioactive tracer material has been injected into the patient, nuclear imaging creates images by detecting radiation from various regions of the body. Both digital and film recordings of the scenes are made. Directions for getting ready for a nuclear imaging exam vary by kind [51].

Planar gamma scintigraphy, single-photon emission computed tomography (SPECT), and positron emission tomography are the three most common nuclear imaging methods used today (PET). Despite planar gamma scintigraphy's simplification of the intricate anatomical structure of the organs, the tissue distribution can still be computed as a proportion of the total dose. While radionuclides are used in both SPECT and planar gamma scintigraphy, the former can collect data in a third dimension, while the latter only allows for two. Although PET provides the most accurate pictures, its use is restricted by the short half-lives of PET radionuclides (such as 11C, 18F, and 64Cu) [52]. The versatility, adaptability, and high specificity of polymers make them an indispensable tool in nuclear imaging.

6.6   Computerized tomography (CT)

Diagnostic imaging of the skull, brain, paranasal sinuses, ventricles, and eye sockets is achieved by the use of computerized tomography (CT) or computerized axial tomography (CAT) scans [53]. Computed tomography (CT), a diagnostic imaging technique, allows for very detailed visualization of an individual's internal organs, bones, soft tissue, and blood arteries. The cross-sectional images obtained from CT scans can be seen on a monitor, printed out, or saved to a variety of digital media [54,55]. CT scans are frequently the greatest method for detecting cancers since they provide clear visual evidence of the disease and allow your doctor to make educated guesses about the tumor's size and location. CT scans allow for the rapid, painless, and accurate diagnosis and treatment of medical disorders that might otherwise require invasive surgery. In an emergency, it can identify internal bleeding and injury quickly enough to prevent death.

6.7   Radiation Therapy

Radiation therapy, also known as radiotherapy, is the therapeutic use of ionising radiation to kill cancer cells. When dealing with cancerous tumours, radiotherapy is often the first line of treatment [56]. Treatment plans often include a combination of radiotherapy, surgery, chemotherapy, and/or hormone treatment. The particular treatment goal will be determined by the nature of the tumor, its location, and its stage, as well as the patient's overall condition.

6.8   Artificial Intelligence in Healthcare

X-rays, CT scans, and MRIs are just a few examples of the complex images that medical professionals must examine to make a diagnosis. Data collection, patient response analysis, and the elimination of possible diagnoses are all facilitated by AI-based solutions. By analysing contextual textural information and learning from past experiences, Machine Learning algorithms can "improve" and "learn" to recognise patterns of sickness qualities, such as the development of breast cancers on mammograms. In addition, ML algorithms may find 'invisible' correlations in data, giving them the perfect 'tool' for better medical diagnoses.

Research on artificial intelligence (AI) and its many potential uses is among the most prolific. Recently, we have seen AI completely transform every type of medical imaging, from X-rays and ultrasound to CT scans, MRIs, fMRIs, PET scans, even SPE CT scans (SPECT). Many AI-based solutions have been created to speed up automated picture interpretation and streamline automated medical image analysis. Artificial intelligence (AI) in medical imaging to aid in illness detection and prognosis is experiencing a veritable gold rush of attention at the moment [57]. Deep learning, a branch of AI that makes use of convolutional neural networks, has excelled at clinical diagnosis, increased the efficiency of hospital workflow, and handled several administrative duties. Clinical results may be enhanced, and the value of medical data may be increased, through the use of AI. Artificial intelligence (AI) is widely employed in the pharmaceutical sector and is revolutionising every facet of health management, from population screening to epidemiology to diagnostics to follow-up to treatment planning.

Artificial intelligence holds tremendous potential to improve healthcare and life sciences. It has the ability to reveal complex data that can aid clinicians and researchers in preventing disease, speeding recovery, and saving lives. In addition, it can help them spend less time on administrative responsibilities, allowing them to devote more time to their patients or their research. Now more than ever, deep learning and machine learning are being used to improve the healthcare industry by reducing practitioners' workloads, fostering more individualized approaches to treatment, and boosting patients' overall experiences.

When medical technology can be used for disease prevention, treatment, and even a cure, it improves the lives of all people. Intel is collaborating with industry heavyweights to change health and life sciences by doing things like speeding up drug discovery and pharmaceutical development and making healthcare more accessible and affordable for everybody. This objective is greatly aided by the application of artificial intelligence (AI) in the healthcare sector, particularly in the fields of computer vision, machine learning, and deep learning [58]. Data silos have made it

Materials Research Foundations 164 (2024) 230-256   https://doi.org/10.21741/9781644903056-6

difficult for academics and healthcare systems to gain insights from large amounts of data, but AI and well-established data management infrastructure can change that.

The top three benefits of artificial intelligence in medical imaging technology are :-

1.Increases Productivity 2. Eliminates Risks and Errors 3. Analyses Comprehensively

Imaging modalities ranging from MRI and CT scans to X-rays and ultrasound are all being employed in conjunction with AI. Nonetheless, not all modalities necessitate the same algorithm. Detecting the Neurological Abnormalities

Brain traumas, blood clots, and other neurological illnesses may now be detected with a high degree of accuracy using AI in neuroimaging. When it comes to brain problems, a team of radiologists annotates and highlights the images to help train the algorithms.

Despite the limited usage of AI in renal illness at present, doctors are aware of its potential in the treatment of kidney disease. This suggests that AI could one day improve the diagnostic procedure.

In the future, radiography will rely heavily on AI to accurately diagnose a wide range of disorders, from the very mild to the fatal. Furthermore, with better or higher quality medical imaging data, the diagnosis process and prediction accuracy will be higher, resulting in more efficient and effective medical treatment and healthcare procedure.
X-ray imaging systems

Since X-rays have a frequency and intensity that allow them to pass through most materials, including the human body, they are commonly used in medical diagnostics. Shadows of internal structures are recorded by an x-ray detector (radiographic film or a digital x-ray detector) on the patient's "shadow" side. X-ray machines allow for the collection of images of bones and soft tissues, which can then be used in medical diagnostics. The X-rays used to create these photographs only expose the patient to extremely low levels of radiation. These images aid in the diagnosis and treatment planning for a wide range of medical conditions. It is normal practise for doctors to use X-rays to detect bone injuries including breaks and dislocations.

With Roentgen's discovery of X-rays in Germany [59,60], X-ray imaging in medicine has been around ever since. An X-ray tube uses a cathode emitter to generate electrons via thermal emission, which are then accelerated by a potential difference of 50-150 KV. The X-rays can only be produced if the electrons crash into the anode. Some of this energy is converted to X-rays, but it is mostly just turned into heat. When an X-ray machine is used to examine a part of the body, a flat, two-dimensional image is produced. With the help of fluoroscopy, the moving organs can be scanned. The obtained images can be viewed, saved, and shared using a wide variety of PCs. In computed tomography (CT), an image is formed with the aid of the image receptor. X-rays are used alongside a screen that has a storing phosphor device. Mammograms can tell healthy breast tissue from diseased tissue. There is little doubt that mammography is more efficient than examining bone structures using comparable amounts of energy. The used voltage is of the order of 15 to 40 kilovolts.

The X-ray source, which creates the X-rays, and the detector, which reads them out, are two crucial components of a biological X-ray imaging system. The combination of a detector with a camera allows for real-time imaging of the target's location. That which digitises analogue visual signals.

Frame grabber, an analog-to-digital (A/D) converter, and a computer for storing and processing digital images are all part of an image processing system.

In addition, numerous state-of-the-art quantitative imaging techniques have been created, such as MR/CT for perfusion imaging, MR/CT for diffusion imaging, MR/CT for functional imaging, MR/CT for elastography, MR/CT for dynamic PET imaging, and MR/CT for dynamic contrast enhancement. There is a large deal of variety in the types of imaging equipment, types of imaging techniques, types of imaging analyses, and types of imaging readers. One unfulfilled therapeutic need is the enhancement of quantitative biomedical imaging's efficacy and utility.

7. Image processing – a tool for medical diagnostics

Medical images are often corrupted by white noise, blurring and contrast defects. Consequently, important medical information may be degraded or completely masked. Advanced medical diagnostics and pathological analysis utilize information obtained from medical images. Consequently, the best techniques must be applied to capture, compress, store, retrieve and share these images.

There has been a dramatic expansion in the use of computer technology, and one of the fastest growing areas of application for these technologies is in the realm of image processing, which is notably evident in medical imaging. The pixels that make up each image are unique. A digital display device's resolution is measured in pixels, the smallest possible picture element. Image Processing is a technique that allows for the modification of visual data. In order to better understand what is meant by "Image Processing," it is helpful to break it down into its constituent parts. Magnetic resonance imaging (MRI), computed tomography (CT), positron emission tomography (PET), and ultrasound are only some of the modern medical procedures that rely on image processing. More and more areas of medicine rely substantially on medical imaging and processing tools [61,62].

Used in the medical field, it can analyse scans from a variety of imaging modalities to help diagnose a wide range of illnesses.

Different intensity levels in medical imaging can be read as markers of (1) anatomical details and/or (2) a physiological phenomenon, depending on the type of tissue being analyzed. Some of the technical details for several imaging methods are listed below: Vision field Accuracy in locating and tracking objects, in both space and time, Data Structures in the Biological and Physiological Sciences

Radiologists were limited to x-rays for diagnostic purposes until the advent of CT and MRI scanners (x-rays are a form of ionising radiation). Solid-state detectors, which transform X-rays into electrical impulses, have mostly replaced X-ray film (CCD camera). CCD cameras provide a number of benefits over traditional cameras, which require film development before being viewed. The quickness of digital communication and the simplicity of altering digital photos are two further advantages. Third, it is essential for making split-second decisions during robot-assisted surgery; fourth, it is used in digital image processing for computer-aided detection (to confirm or attract greater emphasis to problematic spots on a digital picture).

Digital subtraction angiography (DSA) is a more recent development in this area that entails taking pictures before a contrast medium is injected and then subtracting those pictures from others. The contouring phase of radiation planning necessitates thorough interpretation of medical images. Radiographic techniques are the most common medical imaging modality utilised for diagnosis, clinical research, and therapy planning. Tools for processing medical images are also crucial. Steps in Medical Image Processing are given in Fig. 2. The commonly used term "medical image processing" means the provision of digital image processing for medicine. Medical image processing faces a variety of unique difficulties. That list includes: 1) Retouching and editing photos; 2) Feature-level interest detection and segmentation that is both automated and precise Third, precise automated multimodality image registration and fusion 4) Image feature categorization, including structural labelling and classification Five) Quantitative analysis of images with an explanation of the results Sixthly, advancement of unified healthcare systems.

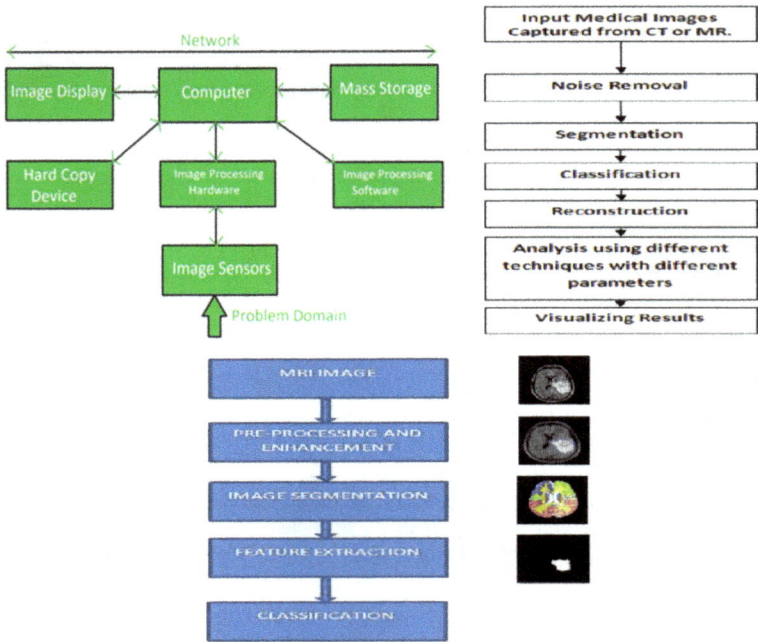

Fig.2. Steps in Medical Image Processing (Proceedings of the 65th International Research Forum Conference, Amruta Pramod Hebli and Sudha Gupta, 20 November 2016, Pune, India, ISBN: 978-93-86291-38-7).

A "process" is any action taken on data, while a "image" is any representation of that data in a visual form. Accordingly, to process an image is to work with or treat its information. Digitally captured images can be analyzed and altered by the user using a process known as "Image Processing" [63,64]. It's a method of turning an image into digital data and performing various operations on it to get meaningful insights. Images are crucial to the development of Artificial Intelligence (AI).

Medical imaging has transformed the way medical experts detect and diagnose abnormalities within the body. By producing visible images of the internal structures and visible functions of the body, medical professionals can make accurate diagnoses without conducting invasive procedures. Scientists are explored the world of medical imaging along with Artificial Intelligence (AI) in medical image processing. Medical image processing remains an exciting field of research and applications for health care, medical education, and biomedical research. In the future, integration of medical image processing into the physicians' workflow will be fostered. High-level image processing will become part of diagnosis, intervention planning, and treatment; and further standardization will be required.

The most important takeaways

1. Research in image processing has mostly focused on practical concerns such as picture creation, processing, presentation, and storage.

2. Since the memory capacities of the computers at the time were insufficient for working with such vast amounts of image data.

3. Before, image processing computation times necessitated off-line processing.

4. The automatic interpretation of images is still a highly sought-after aim at now.

Since validation is based on larger studies with higher amounts of data, improvements in segmentation, classification, and measurements of biomedical pictures are continuous and validated more accurately.

Biomedical imaging is the technique and process of imaging the inside of the human body for research or clinical analysis. It has various advantages, including quick detection time, large detection depth, no surgery need, etc. The main objective of this review was to cover the Biomedical imaging techniques, function of rare earths in diagnostic imaging and Cancer Diagnosis using Rare Eraths, Image Processing and provide a guide that the clinician can use for further research. The rare-earth based materials exhibit bright prospects in the field of integrated diagnosis and therapy. With the continual advancements of synthesis method coupled with the instrument technology, the development of rare-earth based materials for clinical brain disease treatment is highly promising.

References

[1] Wang J, Li S,2022, Applications of rare earth elements in cancer: Evidence mapping and scient metric analysis. Front Med (Lausanne).;9:946100. PMCID: PMC9399464. https://doi.org/10.3389/fmed.2022.946100

[2] Kostelnik TI, Orbig C ,2019, Radioactive main group and rare earth metals for imaging and therapy. Chemical Reviews 119: 902-956. https://doi.org/10.1021/acs.chemrev.8b00294

[3] Townley HE ,2013, Applications of the rare earth elements in cancer imaging and therapy. Current Nanoscience. https://doi.org/10.2174/15734137113099990063

[4] https://www.metaltechnews.com/story/2020/04/29/tech-metals/rare-earth-metals-see-new-medical-uses/217. https://pubs.usgs.gov/fs/2014/3078/pdf/fs2014-3078.pdf).

[5] Weissleder R, Nahrendorf M, 2016, Advancing biomedical imaging. Proc Natl Acad Sci U S A., 112:14424-8. https://doi.org/10.1073/pnas.1508524112

[6] Fan Q, Cui X, Guo H, Xu Y, Zhang G, Peng B., 2020, Application of rare earth-doped nanoparticles in biological imaging and tumor treatment. J Biomater Appl.;35(2):237-263. Epub 2020 May 19. PMID: 32423319. https://doi.org/10.1177/0885328220924540

[7] Neacsu IA, Stoica AE, Vasile BS, Andronescu E. 2019, Luminescent Hydroxyapatite Doped with Rare Earth Elements for Biomedical Applications. Nanomaterials (Basel).9(2):239. PMID: 30744215; PMCID: PMC6409594. https://doi.org/10.3390/nano9020239

[8] Giese EC, 2018, Rare earth elements: Therapeutic and diagnostic applications in modern medicine, Clinical and Medical Reports, Clin Med Rep, doi: 10.15761/CMR. 1000139, Volume 2(1): 1-2. https://doi.org/10.15761/CMR

[9] Kostova I , 2005 ,Lanthanides as Anticancer Agents. Current Medicinal Chemistry - Anti-Cancer Agents, 5: 591-602. https://doi.org/10.2174/156801105774574694

[10] https://www.metaltechnews.com/story/2020/04/29/tech-metals/rare-earth-metals-see-new-medical-uses/217.html

[11] Ascenzi, P., Bettinelli, M., Boffi, A. et al. Rare earth elements (REE) in biology and medicine. Rend. Fis. Acc. Lincei 31, 821-833 (2020). https://doi.org/10.1007/s12210-020-00930-w. https://doi.org/10.1007/s12210-020-00930-w

[12] Juanjuan Gao, Liang Feng, Baolong Chen, Biao Fu, Min Zhu, The role of rare earth elements in bone tissue engineering scaffolds - A review, Composites Part B: Engineering, Volume 235, 2022,109758,ISSN 1359-8368,https://doi.org/10.1016/j.compositesb.2022.109758. https://doi.org/10.1016/j.compositesb.2022.109758

[13] Matthäus C, Bird B, Miljković M, Chernenko T, Romeo M, Diem M. 2008; Chapter 10: Infrared and Raman microscopy in cell biology. Methods Cell Biol. 89:275-308. doi: 10.1016/S0091-679X(08)00610-9. PMID: 19118679; PMCID: PMC2830543. https://doi.org/10.1016/S0091-679X(08)00610-9

[14] Geraldes CFGC. 2020, Introduction to Infrared and Raman-Based Biomedical Molecular Imaging and Comparison with Other Modalities. Molecules. ;25(23):5547. doi: 10.3390/molecules25235547. PMID: 33256052; PMCID: PMC7731440. https://doi.org/10.3390/molecules25235547

[15] Riaz A, Kulik L, Lewandowski RJ, Ryu RK, Giakoumis Spear G, Mulcahy MF, Abecassis M, Baker T, Gates V, Nayar R, Miller FH, Sato KT, Omary RA, Salem R: 2009, Radiologic-pathologic correlation of hepatocellular carcinoma treated with internal radiation using yttrium-90 microspheres. Hepatology; 49:1185-1193. https://doi.org/10.1002/hep.22747

[16] Rastrelli, M.; Tropea, S.; Rossi, C.R.; Alaibac, M., 2014, Melanoma: Epidemiology, risk factors, pathogenesis, diagnosis and classification. Vivo, 28, 1005-1011, PMID: 25398793.

[17] Mcausland, T.M.; Vloten, J.P.V.; Santry, L.A.; Guilleman, M.M.; Rghei, A.D.; Ferreira, E.M.; Ingrao, J.C.; Arulanandam, R.; Major, P.P.; Susta, L.; et al. 2021, Combining vanadyl sulfate with Newcastle disease virus potentiates rapid innate immune-mediated regression with curative potential in murine cancer models. Mol. Ther. Oncolytics , 20, 306-324. https://doi.org/10.1016/j.omto.2021.01.009

[18] Gumerova, N.I.; Rompel, A. 2021 , Interweaving disciplines to advance chemistry: Applying polyoxometalates in biology. Inorg. Chem, 60, 6109-6114. https://doi.org/10.1021/acs.inorgchem.1c00125

[19] Moskalik K, Kozlow A, Demin E, Boiko E. 2010 , Powerful neodymium laser radiation for the treatment of facial carcinoma: 5 year follow-up data. Eur J Dermatol. 2010 Nov-Dec;20(6):738-42. doi: 10.1684/ejd..1055. Epub 2010 Nov 5. PMID: 21056940.

[20] Makita, M., Manabe, E., Kurita, T. et al. 2020, Moving a neodymium magnet promotes the migration of a magnetic tracer and increases the monitoring counts on the skin surface of sentinel lymph nodes in breast cancer. BMC Med Imaging 20, 58. https://doi.org/10.1186/s12880-020-00459-2

[21] Li Y, Wang R. 2022, Efficacy comparison of pulsed dye laser vs microsecond 1064-nm neodymium: yttrium-aluminum-garnet laser in the treatment of rosacea: A meta-analysis. Front Med (Lausanne). Published online January 20. https://doi.org/10.3389/fmed.2021.798294

[22] Noar JH, Evans RD. Rare earth magnets in orthodontics: an overview. Br J Orthod. 1999 Mar;26(1):29-37. https://doi.org/10.1093/ortho/26.1.29

[23] Yuksel C, Ankarali S, Aslan Yuksel N. 2018, The use of neodymium magnets in healthcare and their effects on health. North Clin Istanb ,5(3):268-273.

[24] Blechman, A. M. (1985). Magnetic force systems in orthodontics: clinical results of a pilot study. American Journal of Orthodontics, 87(3), 201-210. https://doi.org/10.1016/0002-9416(85)90041-7

[25] Yao Cai, Yuqing Wang, Tongwei Zhang, and Yongxin Pan, 2020, Gadolinium-Labeled Ferritin Nanoparticles as T1 Contrast Agents for Magnetic Resonance Imaging of Tumors, ACS Applied Nano Materials ,3 (9), 8771-8783. https://doi.org/10.1021/acsanm.0c01563

[26] Fatima, A.; Ahmad, M.W.;Al Saidi, A.K.A.; Choudhury, A.;Chang, Y.; Lee, G.H. 2021, Recent Advances in Gadolinium Based Contrast Agents for Bioimaging Applications. Nanomaterials, 11, 2449. https://doi.org/10.3390/nano11092449

[27] Md. Wasi Ahmad, Wenlong Xu, Sung June Kim, Jong Su Baeck, Yongmin Chang, Ji Eun Bae,Kwon Seok Chae, Ji Ae Park, Tae Jeong Kim and Gang Ho Lee, 2015, Potential dual imaging nanoparticle:Gd2O3nanoparticle . Scientific reports, 5 : 8549. https://doi.org/10.1038/srep08549

[28] Sartor O, Reid RH, Hoskin PJ, Quick DP, Ell PJ, Coleman RE, Kotler JA, Freeman LM, Olivier P, 2004, Quadramet 424Sm10/11 Study Group, Samarium-153-Lexidronam complex for treatment of painful bone metastases in hormone-refractory prostate cancer. Urology 63:940-945. https://doi.org/10.1016/j.urology.2004.01.034

[29] Das T, Banerjee S. Radiopharmaceuticals for metastatic bone pain palliation: available options in the clinical domain and their comparisons. Clin Exp Metas. 2017; 34:1-10. https://doi.org/10.1007/s10585-016-9831-9

[30] Hong E, Liu L, Bai L, Xia C, Gao L, Zhang L, Wang B. 2019, Control synthesis, subtle surface modification of rare-earth-doped upconversion nanoparticles and their applications in cancer diagnosis and treatment. Mater Sci Eng C Mater Biol Appl. https://doi.org/10.1016/j.msec.2019.110097

[31] Gu M, Li W, Jiang L, Li X. 2022, Recent progress of rare earth doped hydroxyapatite nanoparticles: Luminescence properties, synthesis and biomedical applications. Acta Biomater. 2022 Aug; 148:22-43. Epub. PMID: 35675891. https://doi.org/10.1016/j.actbio.2022.06.006

[32] Xiao Zhang, Shuqing Hea, Bingbing Dinga, Chunrong Qua etal; Cancer Cell Membrane-Coated Rare Earth Doped Nanoparticles for Tumor Surgery Navigation in NIR-II Imaging Window, https://www.sciencedirect.com/science/article/pii/S1385894719333741, Manuscript_788726923e1942a5ef920fa02b9a7bba.

[33] Kaczmarek MT, Zabiszak M, Nowak M, Jastrzab R. 2018, Lanthanides: schiff base complexes, applications in cancer diagnosis, therapy, and antibacterial activity. Coordinat Chem Rev. 370:42-5. https://doi.org/10.1016/j.ccr.2018.05.012

[34] Yu, Z.; He, Y.; Schomann, T.; Wu, K.; Hao, Y.; Suidgeest, E.; Zhang, H.; Eich, C.; Cruz, L.J. 2022, Achieving Effective Multimodal Imaging with Rare-Earth Ion-Doped CaF2 Nanoparticles. Pharmaceutics, 14, 840. https://doi.org/10.3390/pharmaceutics14040840

[35] Wei, Z., Liu, Y., Li, B. et al. 2022. Rare-earth based materials: an effective toolbox for brain imaging, therapy, monitoring and neuromodulation. Light Sci Appl 11, 175. https://doi.org/10.1038/s41377-022-00864-y

[36] Bingzhu Zheng, Jingyue Fan,Bing Chen, Xian Qin, Juan Wang , 2022 ,Rare-Earth Doping in Nanostructured Inorganic Materials ,:Chem.Rev.,122,5519−5603. https://doi.org/10.1021/acs.chemrev.1c00644

[37] Zhen Feng Yu, Christina Eich and Luis J. Cruz, 2020, Recent Advances in Rare-Earth-Doped Nanoparticles for NIR-II Imaging and Cancer Theranostics , Front. Chem.,Sec. Nanoscience. https://doi.org/10.3389/fchem.2020.00496

[38] Liu, L., Wang, S., Zhao, B., Pei, P., Fan, Y., Li, X., et al., 2018. Er3+ sensitized 1530 nm to 1180 nm second near-infrared window upconversion nanocrystals for in vivo biosensing. Angew. Chem. Int. Ed. 57, 7518-7522. https://doi.org/10.1002/anie.201802889

[39] Chistoserdova L (2016) Lanthanides: new life metals? World J Microbiol Biotechnol 32:138. https://doi.org/10.1007/s11274-016-2088-2

[40] B. Qi Fan, Xiaoxia Cui, Haitao Guo, Yantao Xu, Guangwei Zhang,Bo Peng , 2020, Application of rare earth-doped nanoparticles in biological imaging and tumor treatment ,Journal of Biomaterials Applications, Volume 35, Issue 2. https://doi.org/10.1177/0885328220924540

[41] C. Jain, A., Fournier, P.G.J., Mendoza-Lavaniegos, V. et al. 2018 ,Functionalized rare earth-doped nanoparticles for breast cancer nano diagnostic using fluorescence and CT imaging. J Nanobiotechnol 16, 26. https://doi.org/10.1186/s12951-018-0359-9

[42] D. Townley E. Helen, 2013 , Applications of the Rare Earth Elements in Cancer Imaging and Therapy, Current Nanoscience ; 9(5). https://doi.org/10.2174/15734137113099990063

[43] E. Pranjita Zantyea,, Fiona Fernandesa,, Sutapa Roy Ramananb and Meenal Kowshika, 2019 ,Rare Earth Doped Hydroxyapatite Nanoparticles for in vitro Bioimaging Applications Current Physical Chemistry, 9, 94-109 . https://doi.org/10.2174/1877946809666190828104812

[44] F. Qu Z, Shen J, Li Q, Xu F, Wang F, Zhang X, Fan C. 2020 , Near-IR emissive rare-earth nanoparticles for guided surgery. Theranostics.;10(6):2631-2644. PMID: 32194825; PMCID: PMC7052904. https://doi.org/10.7150/thno.40808

[45] Umar AA, Atabo SM. 2019 ,A review of imaging techniques in scientific research/clinical diagnosis. MOJ Anat & Physiol. ,6(5):175-183. https://doi.org/10.15406/mojap.2019.06.00269

[46] Fabian Michelangeli, 2019, Imaging the unimaginable: Medical imaging in the realm of photography, Clinics in Dermatology, Volume 37, Issue 1, Pages 38-46, ISSN 0738-081X, https://doi.org/10.1016/j.clindermatol.2018.09.008

[47] Victor I. Mikla and Victor V. Mikla, 2014, Medical Imaging Technology , ISBN , 978-0-12-417021-6 , Elsevier In. https://doi.org/10.1016/C2012-0-06086-3

[48] https://radiopaedia.org/articles/ultrasound-introduction; https://my.clevelandclinic.org/health/diagnostics/4995-ultrasound.

[49] https://www.medicalnewstoday.com/articles/146309#uses; https://www.news-medical.net/health/Magnetic-Resonance-Imaging-(MRI)-Overview.aspx; https://www.medicinenet.com/mri_scan/article.htm

[50] https://radiopaedia.org/articles/positron-emission-tomography

[51] Sibylle I. Ziegler, 2005, Positron Emission Tomography: Principles, Technology, and Recent Developments, Nuclear Physics A, Volume 752, Pages 679-687, ISSN 0375-9474, https://doi.org/10.1016/j.nuclphysa.2005.02.067

[52] https://my.clevelandclinic.org/health/diagnostics/4902-nuclear-medicine-imaging.

[53] Pat Zanzonico, 2012, Principles of Nuclear Medicine Imaging: Planar, SPECT, PET, Multi-modality, and Autoradiography Systems, Radiat Res, 177 (4): 349-364. https://doi.org/10.1667/RR2577.1

[54] Mettler FA, Wiest PW, Locken JA, Kelsey CA. 2000, CT scanning: patterns of use and dose. J Radiol Prot.; 20:353-359. https://doi.org/10.1088/0952-4746/20/4/301

[55] A.C. Kak, M. Slaney, 1988, Principles of Computerized Tomographic Imaging, IEEE Press, New York, pp. 104-107

[56] Ahmed B. Salem Salamh, Abdulrauf A. Salamah and Halil Ibrahim Akyüz, A Study of a New Technique of the CT Scan View and Disease Classification Protocol Based on Level

Challenges in Cases of Coronavirus Disease, Radiology Research and Practice, Volume 2021 | Article ID 5554408. https://doi.org/10.1155/2021/5554408

[57] Shier Nee Saw a, Kwan Hoong Ng, 2022, Current challenges of implementing artificial intelligence in medical imaging, Physica Medica, 100, 12-17. https://doi.org/10.1016/j.ejmp.2022.06.003

[58] Avanzo M, Porzio M, Lorenzon L, Milan L, Sghedoni R, Russo G, et al., 2021 Artificial intelligence applications in medical imaging: A review of the medical physics research in Italy. Physica Medica: European Journal of Medical Physics ;83: 221-41. https://doi.org/10.1016/j.ejmp.2021.04.010

[59] Berger, M., Yang, Q., Maier, A., 2018. X-ray Imaging. In: Maier, A., Steidl, S., Christlein, V., Hornegger, J. (eds) Medical Imaging Systems. Lecture Notes in Computer Science(), vol 11111. Springer, Cham. https://doi.org/10.1007/978-3-319-96520-8_7. https://doi.org/10.1007/978-3-319-96520-8_7

[60] https://www.nibib.nih.gov/science-education/science-topics/x-rays

[61] Selin veronica A,Dr. J.G.R. Sathiaseelan , 2021, Survey Of Image Processing Techniques In Medical Image Analysis And Identification , Turkish Journal of Computer and Mathematics Education Vol.12No.13, 1110-1121 .

[62] Kasban, Hany, M. A. M. El-Bendary, and D. H. Salama. 2015 "A comparative study of medical imaging techniques." International Journal of Information Science and Intelligent System 4, no. 2 : 37-5

[63] Chinmayi, P., Agilandeeswari, L., Prabukumar, M.2018., Survey of Image Processing Techniques in Medical Image Analysis: Challenges and Methodologies. In: Abraham, A., Cherukuri, A., Madureira, A., Muda, A. (eds) Proceedings of the Eighth International Conference on Soft Computing and Pattern Recognition (SoCPaR 2016). SoCPaR 2016. Advances in Intelligent Systems and Computing, vol 614. Springer, Cham. https://doi.org/10.1007/978-3-319-60618-7_45

[64] Emami, T., Janney, S.S., Chakravarty, S.2019. Elements of Medical Image Processing. In: Paul, S. (eds) Biomedical Engineering and its Applications in Healthcare. Springer, Singapore. https://doi.org/10.1007/978-981-13-3705-5_20

Rare Earth - A tribute to the late Mr. Rare Earth, Professor Karl Gschneidner   Materials Research Forum LLC
Materials Research Foundations 164 (2024) 257-278   https://doi.org/10.21741/9781644903056-7

Chapter 7

Metallic Nanoparticles in the Glasses: Advances and Current Challenges

M. Reza Dousti

Unidade Acadêmica do cabo de Santo Agostinho, Universidade Federal Rural de Pernambuco, Brasil

mohammad.rezadousti@ufrpe.br

Abstract

Glasses are fascinating materials with diverse applications. Rare earth doped glasses are well-known for their optical properties which could be used as solid-state lasers, optical sensors, scintillators, optical thermometers, optical fibers in telecommunications, nuclear waste storages, etc. Recently, by increasing the interest in the modification of materials on the nanoscale, glasses doped with metallic nanoparticles have attracted much attention. This type of doping was used historically in the coloration of the glasses, however, in the new millennium it is used also as the optical centers to enhance the radiative quantum yield of the luminescent materials. In this proposal, yet there have been confronted several challenges which need further investigations. For example, not all the cases of metal particle doping in luminescence materials yield an enhancement in emission intensities of lanthanide ions in glasses. This chapter revisits the important experimental results and discusses them as three different arguments.

Keywords

Glass, Rare Earth Ions, Metallic Nanoparticles, Optical Properties

Contents

Metallic Nanoparticles in the Glasses: Advances and Current Challenges257

1.   Introduction

Glasses are among the fascinating materials for humans. They can be used in different areas owning to their various structural, thermal, mechanical, electrical and optical properties. The term glass is usually referred to the amorphous materials, however they should be distinct as they show different thermodynamic behavior. A glass is an amorphous materials which shows glass transition temperature [1]. The history of glasses made by man has started since 4000 B.C. in Mesopotamia, western Asia. These glasses were glazed due to application of copper compounds. Colorful glasses have been prepared between fourteen and sixteen century B.C. in Egypt. In the 19th century, Europe, United States, Bohemia and many other countries developed the techniques of glass blowing. In that time, crystals became very popular. Next, fluorescent glasses, Opaline, HF-etched glasses and cobalt blue glasses were prepared [2]. The glass technology has developed gradually in different empires and territories. Recently, there has been a great increase in the number of the research and publications on the glasses. A comprehensive review paper on the statistics of published works on glasses could be find elsewhere [3]. The glassy systems - optically activated with trivalent rare earth (RE) ions (and sometimes with Eu^{2+} divalent or some transition metals) - are well-known for their potential optical properties. RE-doped glasses are usually nominated to be candidates as solid state lasers, optical amplifiers, sensors, upconverters etc [4,5]. The Stokes, anti-Stokes (upconversion) and energy transfer (ET) among the RE ions has been widely studied [6,7], although it still needs further attention. Besides, it is of utmost importance to improve the luminescence efficiency of RE ions doped glasses, as they suffer from various drawbacks such as low RE solubility in some compositions, concentration quenching, and more seriously, low absorption and emission cross-sections of embedded RE ions. In order to overcome such drawbacks, several approaches have been proposed, which has been recently listed by Zheng et al. [8], e.g.,

1) introduction of noble metallic nanoparticles (NPs) and modification of local field by surface plasmon resonance (SPR),

2) Enhanced luminescence via ET from semiconducting NPs to RE ions,

3) preparing core@shell species for efficient ET process and reducing the quenching centers,

4) modification of surrounding network and crystal structure, and

5) introduction of sensitizers to alter the local symmetry of RE ions.

Among them, SPR modified luminescence of RE ions is known to be a very challenging case, since it does not always result in enhanced luminescence, and therefore, a unique class of usually transparent glasses doped with rare earth (RE) ions and containing metallic nanoparticles (NPs) has been emerged as an interesting field of study. Such phenomenon involving the interaction of light with matters in the sub-wavelength regions is been also called as "plasmonics", which deals with special modification of structural, electrical, chemical, thermal and optical properties of matter due to special energy levels of nanoscale particles. In particular, it is of utmost importance to understand the exact nature of the metal species in the sample, since they can inter-interact with other species and exciting light in different ways. The size of the metallic particles is a determinant parameter which could drastically change the optical and electrical properties of the matter under study. The bulk metals are normally good conductors of electricity and largely reflect the light, thanks to their electronic structure consisting of free electrons. NPs, however, possess so-called SPR band and show intense color. For the particles smaller than 1 nm, the band structure becomes discontinuous and consist of several discrete energy states, which more similarity to the molecules. Since, these species are sometimes named as molecule-like particle, or nanocrystals (NCs). NCs can interact with light and result in absorption and emissions between their electronic energy levels, although do not show the SPR band [9]. Moreover, the presence of other species such as ion, atoms, and aggregates, dimers etc. is possible together with metal particles. Usually a subsequent heat-treatment strategy could led to reduce such species to neutral metal atoms from which the bigger particles could form. Indeed, it is very important to understand the optical properties of each species to understand the photophysical behavior of the RE-doped glasses containing metal particles/species.

The aim of the present chapter is to revisit some aspects of the luminescence quench and enhancement in the RE doped glasses containing metal species or main generally NPs. A selection of the results will be discussed and compared in order to clarify the stage of the art and to facilitate the understanding of the nature of phenomenon.

2. Preparation and characterization

There are several physical and chemical techniques to synthesize the NPs. One of the important physical techniques is to evaporate the metal from a source and deposit the gas on a substrate [10]. This method is more efficient since controlling the shape and size of the NPs are more facile. One of the important chemical techniques is to reduce the metal ions or metal salts with a suitable reducing agent. This method is known as most suitable technique to produce the spherical NPs [11–14] . For example, Xu et al. [15] embedded the silver NPs into the sodium-silicate glass through an ion exchange method and subsequent heat-treatments. There are also many other approaches such as electrochemistry techniques which are beyond the scope of this study [16].

The formation of NPs in the glass commonly carries out by thermodynamic reduction of NPs in the presence of an oxidant agent. Som and Karmakar showed the feasible reduction process of silver [26] and gold [27] in antimony glass, as follows

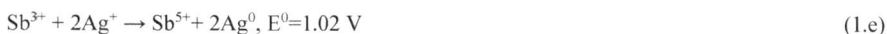

$$Sb^{5+}/Sb^{3+}, E^0=0.649 \text{ V} \tag{1.a}$$

$$Au^{3+}/Au^0, E^0=1.498 \text{ V} \tag{1.b}$$

$$Ag^+/Au^0, E^0=0.7996 \text{ V} \tag{1.c}$$

$$3Sb^{3+} + 2Au^{3+} \rightarrow 3Sb^{5+} + 2Au^0, E^0=1.05 \text{ V} \tag{1.d}$$

$$Sb^{3+} + 2Ag^+ \rightarrow Sb^{5+} + 2Ag^0, E^0=1.02 \text{ V} \tag{1.e}$$

They suggested that the reduction of gold is faster than the silver ions. This idea led to preparation of gold$_{core}$@silver$_{shell}$ nanoparticles [28]. Wu et al. [29] investigated the reduction of silver ions to neutral NPs in a bismuth-borate glass system, and the possible reduction is proposed as

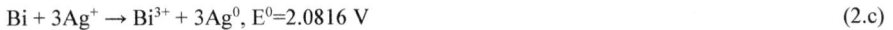

$$Bi^{3+}/Bi^0, E^0=0.3172 \text{ V} \tag{2.a}$$

$$Ag^+/Ag^0, E^0=0.7996 \text{ V} \tag{2.b}$$

$$Bi + 3Ag^+ \rightarrow Bi^{3+} + 3Ag^0, E^0=2.0816 \text{ V} \tag{2.c}$$

Dousti et al. [30] illustrated the reduction of silver ions to silver NPs in Er^{3+}-doped zinc tellurite glass. The reduction of Ag^+ particles to Ag^0 NPs can be discussed by the reduction potential (E^0) of redox system elements. The E^0 values of each component in this system are:

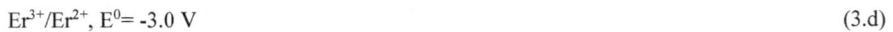

$$Te^{6+}/Te^{4+}, E^0= 1.02 \text{ V} \tag{3.a}$$

$$Ag^+/Ag^0, E^0=0.7996 \text{ V} \tag{3.b}$$

$$Er^{3+}/Er^0, E^0=-2.331 \text{ V} \tag{3.c}$$

$$Er^{3+}/Er^{2+}, E^0= -3.0 \text{ V} \tag{3.d}$$

Probable reduction process and their total potentials are as following:

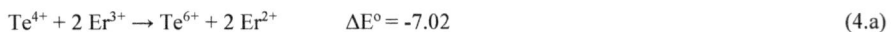

$$Te^{4+} + 2 Er^{3+} \rightarrow Te^{6+} + 2 Er^{2+} \qquad \Delta E^0 = -7.02 \tag{4.a}$$

$$3 Te^{4+} + 2 Er^{3+} \rightarrow 3 Te^{6+} + 2 Er^0 \qquad \Delta E^0 = -7.722 \tag{4.b}$$

$$Te^{4+} + 2 Ag^+ \rightarrow Te^{6+} + 2 Ag^0 \qquad \Delta E^0 = +0.5792 \tag{4.c}$$

Therefore, from the thermodynamic point of view, only the last redox reaction (34.c) is feasible. The reduction of Ag^+ ions to Ag^0 neutral particles and growth of silver NPs are also reported by addition of other reducing agents such as SnO [31].

Synthesis of the NPs includes two processes [17]: (i) formation of small seeds (nucleation) and (ii) growth process. The capping material is an important factor to determine the shape and size of NPs during the growth. If the capping material is very weak, the growth will continue to produce

Rare Earth - A tribute to the late Mr. Rare Earth, Professor Karl Gschneidner
Materials Research Foundations 164 (2024) 257-278

Materials Research Forum LLC
https://doi.org/10.21741/9781644903056-7

big crystallized particles. Contrary, if it is too strong, it may reduce or prevent the growth. Therefore, nanomaterial, capping material or reduction agent and medium are important factors to increase the efficiency of synthesize. Moreover, the concentration of agent is another factor to determine the concentration of initial seeds of metallic particles.

The clustering of the NPs can be discussed by following suggested models [18]:

1) Coagulation process; in which the pairwise particles will be destroyed through the collisions and larger particles will be formed. In a simple model [19], by considering the identical probability of collision (K_D) for different particles, one can describe the coagulation process by

$$\frac{\partial Z_n}{\partial t} = -\frac{1}{2} K_D Z^2 \tag{5}$$

where, Z_n is the concentration of particles of kind n. The solution for given time evaluation equation is

$$Z(t) = \frac{Z_0}{1+0.5K_D Z_0} \tag{6}$$

And the required time for coagulation is

$$t_c = \frac{2}{K_D Z_0} \tag{7}$$

For the coagulation of two particles with radius R_1 and R_2 and respective diffusion velocity of $u(R_1)$ and $u(R_2)$, the probability constant is:

$$K_D = \pi(R_1 + R_2)^2 [u(R_2) - u(R_1)]. \tag{8}$$

2) Ostwald ripening process; in which the large particle will be grown through feeding by smaller particles. Therefore, the smaller particles start to vanish [20]. The matter in its thermodynamic balance consists of surrounding gaseous cloud. The bigger the particle is, the lower number of particles in cloud exists. The diffusion from cloud to matter and vice versa is consistent with growth and disaggregation of particles, respectively.

3) Coalescence of particles; in this process, the particles grow due to a strong chemical or physical bonding which is consistent with coagulation in first step and the consequent Ostwald ripening effect. The critical size of particle to grow or dissolve is [21]

$$r^* = \frac{2V\gamma}{3k_B T \ln(S_r)} \tag{9}$$

where V, γ and S are the molecular volume, surface free energy per unit surface area and saturation ratio. Therefore, for a given $S>1$ and temperature (T), the particle with radius $r<r^*$ starts to grow, while the particles with $r>r^*$ will dissolve.

Figure 1 shows the possible mechanism for the formation and growth of the silver nanoparticles, as an example.

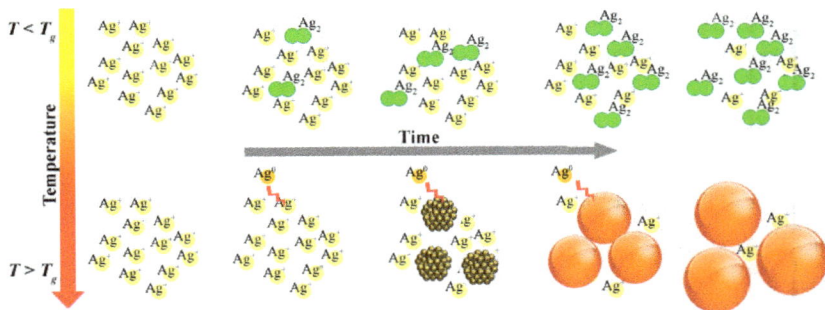

Figure 1. A possible mechanism for the growth and formation of metal NPs (case of silver as a representative).

Metal particles display fascinating physical and non-linear properties when the size lies in the quantum regime (2-20 nm). For example, some first-principle calculations based on density functional theory (DFT) to measure the density of state (DOS) and band structures for different sizes of silver NPs have been performed. For the size regime ~ 2 nm, quantum size luminescence effect is proposed with the probable energy absorption or release during the electron transfer mechanism. For sizes less than 2nm, main existence of silver is in the form of clusters and absorption or release of energy is also probable while achieving a transition between the two energy levels. When NPs grow in size, due to the overlapping of the orbitals and reduction of the energy difference between orbits, quantum size luminescence effect is vanished and electrons now have free motion (shown in Figure 2). This explanation is well suited some experimental observations mainly achieved by microscopy at nono-scale.

Properties of NPs are characterized by their extremely small size which needs appropriate apparatuses to be probed. The first instrument to observe the NPs is UV-Vis absorption spectroscope since NPs show strong absorption peak at near-UV to near-infrared region. Up to now, many instruments were developed to characterize/observe the structure of NPs such as X-Ray diffraction (XRD), atomic force microscopy (AFM), scanning electron microscopy (SEM), transmission electron microscopy (TEM), X-Ray absorption spectroscopy (XAS) and its fine extended device (EXAFS), dynamic light scattering, energy dispersion X-ray (EDX), IR and Raman [22–24].

Rare Earth - A tribute to the late Mr. Rare Earth, Professor Karl Gschneidner   Materials Research Forum LLC
Materials Research Foundations 164 (2024) 257-278   https://doi.org/10.21741/9781644903056-7

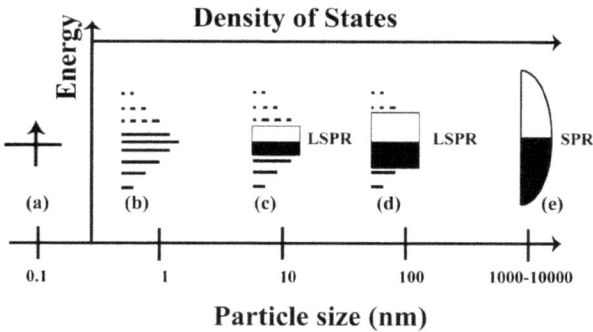

Figure 2. *Schematic energy level for silver NPs based on their size.*

XRD technique can be used to determine the crystalline phases, average inter-particle distances, and atomic structure of the nanoclusters. The size of particles, defects and strains of nanocrystals can be defined by measuring the width of diffraction lines. The broadening of the diffraction line width is directly related to reduction of nanocrystal size. However, when discussing the metallic nanoparticles in the glassy hosts, there are only few reports are available where the diffraction peals of the metal NPs are observed in the XRD profile.

AFM, STM, and chemical force microscopy are usually called in a group named: the scanning probe microscopy (SPM). It can be applied on versatile organic and inorganic materials in their gassy or liquid phases. AFM captures the image of specimen through the difference between atomic forces in short-range and long-range which are repulsive and attractive, respectively [25]. Scanning force microscopy (SFM) also is a useful tip to measure the electrostatic and magnetic interactions between the molecules. SEM is another commanding technique to observe the surface of any type of materials with a resolution of 1 nm [24]. The principle of SEM is based on the interaction of incoming electron with the sample, which will be captured by its backscattered electron. Therefore, both structural and topological properties can be defined through two captured beams.

TEM is also another powerful instrument to characterize the chemical composition and spatial structure [23]. Recently, high-resolution-TEM is developed which allow the imaging of crystals as small as 1 Å. TEM is usually used to determine the shape, size, crystallinity and inter-particle interactions in nanomaterials. An example of the TEM technique applied in the glasses doped with gold NPs is given in Figure 3(a). Figure 3(b) shows the HR-TEM of the same sample.

Rare Earth - A tribute to the late Mr. Rare Earth, Professor Karl Gschneidner   Materials Research Forum LLC
Materials Research Foundations 164 (2024) 257-278   https://doi.org/10.21741/9781644903056-7

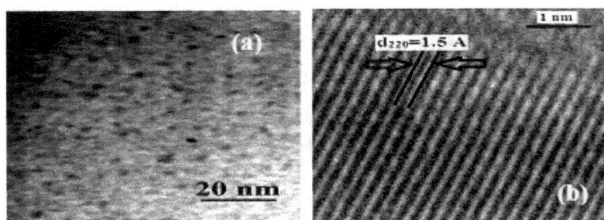

Figure 3. *TEM image of the Au NPs showing the particles with average size of 2 nm (a) and it corresponding HR-TEM (b) image in a Eu^{3+} doped tellurite glass sample. Unpublished data.*

3.   RE-doped glasses containing metallic nanoparticles

Glasses are superior candidates to embed the RE ions and metallic NPs due to their high transparency, mechanical strength, and simple preparation in any size and shape and high energy stretching vibration which extinct the interaction of ligand with metals. Therefore, metallic glasses composites containing nanoclusters or NPs are introduced as potential substrate materials (hosts) for significant local field enhancement to increase the luminescence of RE ions [26,27,32,33]. Moreover, the introduction of metallic NPs may modifies the thermal and structural properties as well as chemical durability of oxide glasses [34,35] , which are beyond the scope of this chapter. Thermal features of glasses can be also modified by incorporation of metallic NPs and may result in considerable increase in thermal diffusivity and thermal conductivity [36]. This is particularly important considering that cooling is one of the most challenging technical issues to be overcome in the areas of microelectronics and solid-state lighting. It has been reported that the thermal diffusivity of materials doped with metallic NPs depends on the size and concentration of NPs, even though the subject is still a matter of controversy in the literature. The large nonlinear absorption of silver nanoclusters and NPs are recently reported in the silicate, borate and oxyfluoride glasses, which are promising for applications in optical limiting and object's contrast enhancement due to the non-saturated and saturated nonlinear absorptions [37–39].

In the past decade, there were group of authors who contributed in the studies of effect of metallic NPs on optical properties of RE-doped glasses, glass-ceramics and thin films. The effect of asymmetric silver NPs reduced in an antimony glass system is reported by Som and Karmakar [26], where intense SPR peak were probed in UV-Vis-IR absorption spectra. They concluded that the decrease in NP-NP distances by increasing the concentration of Ag NPs enhances the localized electric field and broadens the plasmon peak by a red-shift up to 1100 nm. The formation of NPs with different shapes and size ranging between 4 and 31 nm was discussed as an Ostwald's ripening process.

3.1   The good

The introduction of metallic NPs into the glassy hosts was first reported by Malta et al. [40]. A large enhancement in the order of 5.6 for luminescence of Eu^{3+} in borate glass (emission at 612 nm and under 312 nm excitation wavelength) has been observed due to the presence of small silver particles by a concentration around 7.5wt%. The absorption peak of small silver particles in this

Rare Earth - A tribute to the late Mr. Rare Earth, Professor Karl Gschneidner   Materials Research Forum LLC
Materials Research Foundations 164 (2024) 257-278   https://doi.org/10.21741/9781644903056-7

study showed a sharp peak at 312 nm. In 1999, Hayakawa et al. [41] reported on the enhanced luminescence of Eu^{3+} ions doped in silica glass in presence of Ag NPs having sizes of about 4.3 nm (surface plasmon band was observed at 394 nm). The enhancement of $^5D_0 \rightarrow ^7F_J$ (J=0, 1, 2, 3 and 4) emissions of Eu^{3+} ions doped borosilicate glass (derived by a sol-gel method) in vicinity of polymer-protected gold NPs is also reported by Hayakawa et al. [42], where 6 times enhancement is observed under a long UV excitation light.

The luminescence of Eu^{3+}-doped lead-tellurite glass embedding gold NPs were characterized by Almeida et al. [43]. The sample containing 0.5wt% Au NPs and annealed for 41 hours (average particle diameter ~4 nm) show largest enhancements in Eu^{3+} spectra under 405 nm excitation. Since, the electric dipole transitions such as $^5D_0 \rightarrow ^7F_{4,2}(Eu^{3+})$ are sensitive to polarizability and environment around the RE, rather than those magnetic dipoles ($^5D_0 \rightarrow ^7F_{1,3}$), the presence of metallic NPs alert them progressively. They concluded that the presence of small NPs cannot result in to the energy transfer process, hence the main contribution for such intensifications are attributed to increased local field around the Eu^{3+} ions. Such enhancements are also observed in Eu^{3+}-doped GeO_2-Bi_2O_3 glasses containing Au NPs by the same research group [44], where SPR band of NPs observed at 500 nm and contributes to 1000% and 500% enhancement of $^5D_0 \rightarrow ^7F_{2,4}$ and $^5D_0 \rightarrow ^7F_{1,3}$ emissions, respectively.

The luminescence enhancements of Eu^{3+}-doped zinc-tellurite glass and lead-tellurite glass are also reported. Two-times enhancement is reported for Eu^{3+}-doped ZnO-$TeO2$ [45] and PbO-$TeO2$ [46] glasses after 12 and 9 h of heat-treatments which lead to formation of silver NPs with an average size of 14 and 10 nm, respectively. Such enhancements are attributed to intensified local field in distances between NPs and RE ions, induced by surface plasmons. In both reports, the silver NPs are grown along (200) crystallographic direction (JCPDS no. 030931), as revealed by TEM images (lattice constant of about 2-2.05 Å).

Kumar et al. reported on the enhancement of the luminescence of Eu^{3+}-doped titanosilicate glass by introduction of silver NPs [47]. The silver NPs with average particle size of 14.9 nm and TiO_2 polycrystalline are observed in SAED and XRD pattern of the glass samples. A broad absorption band in the 300-400 nm region is assigned to SPR band of NPs. The photoluminescence emissions (λ_{exc}=393 nm) and excitations (λ_{emi}=612 nm) showed enhancements after incorporation of silver NPs. The asymmetry ratio (AS, the ratio of integrated emissions bands of Eu^{3+} ions) are given as ($\int^5D_0 \rightarrow ^7F_2$ dλ)/($\int^5D_0 \rightarrow ^7F_1$ dλ), which varies by the addition of silver NPs due to modification of i) ligand filed and ii) refractive index around the RE ions. AS factor increases from 2.627 to 3.615 for singly-doped and co-doped samples, respectively.

The effect of the temperature on XRD pattern, absorption and emission spectra and Judd-Ofelt parameters in a Er^{3+}-Au NPs co-doped alumina-silicate glass is investigated by Watekar et al. [48]. A blue shift in plasmon absorption band of gold NPs is observed by increasing the annealing temperature. The radiative lifetime and integrated emission cross-section of infrared emissions of Er^{3+} ion decreased after introduction of Au NPs. Such quenches in non-resonance excitation process is attributed to ET from Au NP to Er^{3+} ion. Singh et al. [49] studied the effect of annealing time interval on the size of Ag NPs in Er^{3+}-doped tellurite glasses. They exposed the samples containing Ag NPs under the temperature below the glass transition temperature for different periods of time. They observed increased size of NPs and enhancements in up-conversion emissions in visible range (green and red lines) by increasing the annealing time.

The introduction of Au NPs in Er^{3+}-doped antimony glass raise up to preparation of dichroic nanocomposite [27], showing different colors in transmittance and reflecting surfaces. The enhancements in the order of 3.4 and 7.5 times in green (536 nm) and red (645 nm) bands through an upconversion luminescence were attributed to the enlarged local field by asymmetric Ag NPs. Diffraction peaks of Au NPs in XRD pattern was in good conformity with SEAD results for NPs with sizes about 11-30 nm. Once more, broadening of plasmon peak (612-664 nm) was observed due to presence of non-spherical metallic NPs. They conclude that optimized upconversion intensity of Er^{3+}-doped antimony glass occurs with 0.03wt% of Au. However, further increase of Au concentration results in quenched luminescence, showing the ET from Er^{3+} ion to Au NPs and reabsorption due to SPR of gold NPs [27].

Rivera et al. [50] showed that by exciting the Er^{3+} ions doped in tellurite glass containing silver NPs upon 980 nm laser, a blue shift occurs in the peaks of broadband emission ($\sim$ 1.55 µm). The modification of Stark energy levels (blue shift) was attributed to oscillator strengths of NPs which results in the ET from NP to Er^{3+} ions. The small SPR peaks were revealed in an Er^{3+}-free Ag NP-doped tellurite glass centered at 479 and 498 nm for 3 and 6 h annealed samples, respectively. Sharp peaks in XRD pattern of tellurite glass containing silver NPs ($2\theta = 44.8 \pm 0.4°$, where $d' = 2.0231 \pm 0.0169$ Å) belongs to the (hkl - 200) diffraction planes of Ag crystals (JCPDS Card File No. 4–0783.). Slight increase in FWHM and intensity of broadband emission of Er^{3+} ion ($\sim$ 1.55 µm) were observed by increasing the annealing time interval. The lifetime of 1.55 µm decreases by introduction of Ag NPs in their system comparing to Er^{3+}-single-doped tellurite glass. In another study, Rivera et al. [51] showed that the lifetime of $^4I_{13/2}$ level increases due to presence of heat-treated gold NPs. Therefore, the upconversion luminescence ($^2H_{11/2} \rightarrow {}^4I_{13/2}$, 805 nm) upon the 980 nm excitation enhances drastically ($\sim$ 75 times) due to presence of annealed Au NPs up to 7.5 h. The large enhancement is attributed to LSPR of Au NPs which is located at 800 nm, as evidenced in UV-Vis-IR absorption spectrum, which modifies the local electric field through an electric coupling by Er^{3+} ions. The XRD peaks of Au NPs are revealed at $2\theta=38.3 \pm 0.4°$ and $44.6 \pm 0.3°$, respectively corresponding to (111) and (200) diffraction plans of gold [51].

de Campos et al. [52] investigated the Er^{3+}-doped bismuth-tungsten-tellurite glasses containing silver NPs and heat-treated for 1, 24, 48 and 72 h. The maximum enhancement in upconversion luminescence were observed for sample with 24 h heat-treatments, due presence of NPs with average size of 35 nm. There was no plasmon peak reported in this study, however, they conclude that a tail on the absorption spectrum in the blue region belongs to SPR which is not clearly observable due to small amount of silver NPs.

Amjad et al. prepared the Er^{3+}-doped magnesium phosphate [53,54] and magnesium-tellurite [55] glasses embedding silver NPs. Introduction of silver NPs with average size of 37 nm enhanced the upconversion intensities of Er^{3+} ions under 797 nm excitation wavelength by a factor of 2.04 and 1.99 for 540 and 634 nm emission bands, respectively. SPR band of Ag NPs was observed in an Er^{3+}-free phosphate glass to be centered at 528 nm [53]. The enhancement is mainly attributed to enhanced local field and partly discussed in terms of ET from Ag NPs to Er^{3+} ions. HR-TEM imaging revealed the cubic closed pack structure of the silver NPs as the lattice constant were measured to be 2.13 Å which is attributed to d_{200} crystallography characterize of silver (d_{200}= 2.05Å, JCPDS No.030931). The effect of heat-treatment on Ag NPs-Er^{3+}-co-doped phosphate glass is also investigated [54]. The enhancement in visible bands up to 2.2 times was reported due to annealing the glasses up to 40 h at 300°C. Comparing to their previous report [53], the plasmon peak shows a blue-shift (observed at 488 nm) and the size of NPs are clearly smaller, about 5 nm in average. However, HR-TEM image confirms the presence of Ag NPs with d_{200}=2.17 Å. In the case of Er^{3+}-doped magnesium tellurite glasses, the addition of 0.5 mol% Ag NPs resulted in more efficient enhances in upconversion emission up to 3.33 times for the red emission after 24 h of annealing [55]. The silver NPs with average size of 12 nm were observed with SPR band located at ~ 534 nm. However, it is reported in their work that luminescence intensities in whole visible range quenched due to over-heat-treatments. The ET from Er^{3+} ions to Ag NPs and reabsorption by SPR were mentioned as main explanations for such quenches.

The effect of silver NPs on the optical and structural properties of Er^{3+}-doped zinc tellurite glass is investigated by Dousti et al. [34,56,57]. The absorption bands of Er^{3+} ions are located at 445, 488, 522, 654, 800, 976 and 1526 nm and ascribed to the electric transitions from the $^4I_{15/2}$ ground state to $^4F_{3/2}/^4F_{5/2}$, $^4F_{7/2}$, $^2H_{11/2}$, $^4F_{9/2}$, $^4I_{9/2}$, $^4I_{11/2}$ and $^4I_{13/2}$ excited states, respectively. The introduction of Ag NPs enhanced the upconversion emissions of green and red bands centered at 520, 550 and 640 nm by 4 to 6 times. The enhancement is attributed to the presence of silver NPs with average size ~ 10 nm and with SPR localized at 522 nm. By introduction of 0.5 mol% of Ag NPs with an average size of 12 nm, 3.5-fold enhancement was observed for green emission ($^2H_{11/2}{\rightarrow}^4I_{15/2}$) due to formation and growth of NPs by 8 h annealing [56]. After 2 h annealing, two peaks in UV-Vis-IR spectra were attributed to SPR, centered at 550 and 580 nm. After 8 h annealing, they observed three SPR peaks which are ascribed to different modes of oscillating particles. In the case of 1 mol% of Ag NPs in the same glassy system, enhances up to 6.5 folds were observed for upconversion emissions after 4 h annealing at temperatures above the T_g. Average size of manipulated Ag NPs were 14 nm, with two SPR bands at 560 and 594 nm, and indicative of formation of non-spherical NPs.

The influence of gold NPs on the upconversion emission and Judd-Ofelt parameters of Er^{3+}-doped tellurite glass is investigated by Awang et al. [58,59] and Sazali et al. [60]. However, the correlation of upconversion and Judd-Ofelt parameters is still not clearly understood. Refractive index, density, the quality factor, and thermal stability of this glassy system are also increased by addition of Au content. Upon heat-treatments at various temperatures, the glasses showed further enhancements of upconversion luminescence due to formation of non-spherical Au NPs [61]. The effect of the heat-treatment duration on the same glass shows improvement of green emission [62]. Surface plasmon band of silver NPs in the bismuth glass is observed at 555 nm [63]. In this glass, the infrared emission of Er^{3+} ions centered at 1554 nm experienced an enhancement in the order of 7.2 times due to local filed enhancement by SPR. Judd-Ofelt intensities parameters of these glasses are also increased by addition of Ag NPs. The enhancement of upconversion luminescence of Er^{3+} ions in zinc boro-tellurite glass is also reported [64]. Surface plasmon band is observed at 630 nm for silver NPs with average size of 4.5 nm as captured by TEM imaging for 0.1 mol% silver NPs doped tellurite glass. In another study, the luminescence of Er^{3+} ions enhanced by 4 times due to introduction of silver NPs which are embedded through an heat-treatment of borate glass , where SPR band is observed at 410 nm [65].

Nd^{3+}-doped antimony glass embedding $gold_{core}$@$silver_{shell}$ NPs are studied by different spectroscopic techniques [66]. XRD and SAED showed the formation of core-shell NPs by diffraction patterns along the (111) and (200) crystal planes. TEM imaging confirmed the presence of NPs by average size of about 22-207 nm . The plasmon peaks of NPs are observed by UV-Vis-IR absorption spectroscopy in the range of 532-675 nm, which showed a red-shift by increasing the concentration of gold$_{core}$ NPs. Five-fold intensity enhancement of upconversion emissions of Nd^{3+} ion is observed under 805 nm excitation wavelength. The enhancements of two-major bands centered at 540 nm ($^4G_{7/2} \rightarrow {}^4I_{9/2}$; green) and 649 nm ($^4G_{7/2} \rightarrow {}^4I_{13/2}$; deep-red) are attributed to local field effect induced by plasmonic core-shell metallic NPs.

Frequency upconversion emissions in Nd^{3+}-doped lead-germanate glass containing silver NPs are enhanced under 805 nm excitation wavelength [67]. The absorption band of silver NPs are not observed due to the low concentration of this specie, however, TEM showed the nanoparticle with varying size from 2 to 50 nm. Enhancement of upconversion emissions of Nd^{3+}-doped lead-tellurite glass under 800 nm excitation wavelength is also reported by Reza Dousti [68]. Sixteen-fold enhancement is attributed to large local field in vicinity of silver NPs having an average size of 18 nm. Different interactions and growth process is also described and silver crystalline peaks are observed in XRD patterns of glassy system at $2\theta=44$ degrees.

The effect of noble metallic NPs in antimony glass and glass-ceramic containing Sm^{3+} ions were reported by Som and Karmakar [66,69–71], and enhanced luminescence in samarium emissions were observed. The red upconversion luminescence (centered at 636 nm) of Sm^{3+} ions is studied in presence of core-shell bimetallic nanoparticles (Au-Ag NPs) in an antimony glass system. Two-fold enhancement is observed under the excitation at 949 nm and the surface plasmon band is observed in the range of 554-681 nm for various concentrations of metal [28]. Similar results are observed by addition of silver NPs in Sm^{3+}-doped silicate glass, where Ag ions are reduced to Ag neutral particles by antimony oxide as the oxidant agent [72].

The enhancement in the luminescence of Sm^{3+}-doped tellurite glass by introduction of silver NPs is also given in Ref. [73]. The author showed that under 406 nm excitation wavelength, the emission line at 645 nm enhances up to 130% by addition of concentration of Ag NPs up to 1mol%. The increased luminescence of the Sm^{3+} ions are attributed to localized surface plasmon resonance of silver NPs.

The effect of the silver NPs on the Sm^{3+}-doped different media are also available. For instance, Kaur et al. [74] reported on the enhanced luminescence of Sm-complex (PVA) where the Ag NPs were formed by laser irradiation at 355 nm. The SPR band is observed around 402-405 nm and emissions in the visible range are enhanced for both 355 and 400 nm excitation wavelengths. The lifetime of 595 nm emission of Sm^{3+} ions under 355 nm is increased in presence of Ag NPs. On the other hand, in another study, the upconversion emissions of Dy^{3+}-doped tellurite glasses are enhanced by heat-treated silver NPs [75]. Four-time enhancement is observed for visible emissions under 800 nm excitation wavelength. Silver NPs having an average size of 18 nm are observed and the enhancement in photoluminescence feature is described as the modification of local field due to difference between dielectric constant of medium and metallic particle.

Assumpcao et al. [76] studied the upconversion emission of Tm^{3+}-doped zinc tellurite glasses containing silver NPs. In the latter work, they investigated the infrared-to-visible and infrared-to-infrared UC process under 1050 nm excitation wavelength and the observed enhancement of luminescence is attributed to the increased local field by silver NPs after heat-treatments. Assumpcao et al. [77] also studied the frequency upconversion emissions from Tm^{3+}-Yb^{3+} co-doped germanate glasses embedding silver NPs. They showed that infrared (980 nm)-to-visible (480 nm) upconversion emission in the currents system is due to the energy transfer from Yb^{3+} ions to Tm^{3+} ions. They concluded that the absorption of two, three, and four photons result in upconversion emissions at 800, 652-477 and 542 and 455 nm, respectively. Presence of silver NPs

Rare Earth - A tribute to the late Mr. Rare Earth, Professor Karl Gschneidner   Materials Research Forum LLC
Materials Research Foundations 164 (2024) 257-278   https://doi.org/10.21741/9781644903056-7

is confirmed by TEM technique and NPs with average size of about 10 nm (isolated) and 80 nm (aggregated) are captured. Intensity of the upconversion emissions at vicinity of SPR band are enhanced up to 30%. Moreover, Assumpcao et al. [78] showed that the SPR band for this glassy system can be observed by annealing the samples at higher temperature (T>480°C) for 6 hours. As reported, the intensity of SPR absorption band and upconversion emissions increases by increasing the heat-treatment temperature up to 540°C.

The influence of silver NPs on upconversion emission of Tm^{3+}-Yb^{3+} co-doped zinc tellurite glass is also reported [79]. However, there is no SPR band observed for this glass system up to 72 hours of heat-treatment at 325°C. Upconversion emissions in this system are associated to the energy transfer from Yb^{3+} ions to Tm^{3+} ions, with only 2 and 3-photons absorption mechanism for bands at infrared and visible regions, respectively. Silver NPs enhances the upconversion emissions in order of 300% under 980 nm excitation wavelength. Kassab et al. [80] reported that the large Judd-Ofelt intensity parameters of Tm^{3+} ions (Ω_2=15.65 × 10^{-20} cm^2) in Tm^{3+}-Yb^{3+}-co-doped zinc-tellurite glass containing silver NPs can nominate them as optically stimulated quantum electronic devices and optically operated fibers.

A photoluminescence enhancement of about 1.6 times is observed in Tb^{3+}-doped silicate glass due to incorporation of silver NPs [81]. Under excitation at 488 nm, the emissions of Tb^{3+} ions at visible region (537, 578 and 612 nm) are enhanced thanks to the increased local field by metallic silver NPs. Tb^{3+}-doped silicate glass containing silver NPs are also investigated by Piasecki et al [82,83]. SPR band is observed at 480 nm and the luminescence lines are enhanced under 325 nm excitation wavlenegth. Maximum enhancement of 1.8 times is obtained for sample containing 0.5mol% of silver after 3 hours of heat-treatments.

The SPR band of silver NPs embedded Tb^{3+}-doped silicate glass is located at 420 nm and its intensity increases by increasing the concentration of silver NPs [84]. The emissions of Tb^{3+} ions in the range of 400-700 nm are enhanced for the sample containing 3mol% of $AgNO_3$ and quenched in the glass sample embedding 5mol% of silver NPs. The enhancement of about 35% in emission intensity of Tb^{3+} ions (543 nm) in a silica system is also observed as the lifetime of this fluorescence reduces. Such fluorescence enhancement and increased radiative decay rates are attributed to the increased local filed induced by silver NPs [85].

Verma et al. [86] investigated the effect of silver NPs on the fluorescence of Tb^{3+}-doped aluminosilicate glass. Silver NPs are prepared by laser ablation in distilled water, and embedded in the glass following a sol-gel technique. Surface plasmon resonance band of silver is observed at 404 nm, and their radius is estimated to be ~ 6.48 nm by measuring the FWHM of SPR band. The lifetime of the $^5D_4$ level of terbium is increased from 310 to 420 μs in the presence of silver NPs. The emission intensity of rare earth ions enhances up to 100% which is a result of the energy transfer from the excited silver NPs to Tb^{3+} ions.

Kassab et al. [33] reported that the luminescence characteristics of Pr^{3+} in lead-tellurite glass enhances due to presence of silver NPs with average size around 3.5 nm, which are formed by annealing at 350°C for 7 hours. Rai et al. also studied the influence of silver NPs on the optical properties of Pr^{3+}-doped zinc-tellurite glasses [87]. Upconversion emission of Pr^{3+} ions centered at ~482 nm and ~692 nm are observed under the excitation with a nanosecond laser operating at 590 nm. An enhancement of about 120% is achieved for the integrated intensities of those emissions after heat-treating the sample containing silver NPs up to 40 hours. Enhancement in luminescence of Pr^{3+}-doped PbO-GeO_2 glasses are also observed in presence of isolated silver

Rare Earth - A tribute to the late Mr. Rare Earth, Professor Karl Gschneidner   Materials Research Forum LLC
Materials Research Foundations 164 (2024) 257-278   https://doi.org/10.21741/9781644903056-7

particles and aggregated silver NPs with an average diameter of 2 nm and less than 100 nm, respectively [88]. They showed that the amplitude of surface plasmon band of silver NPs is centered at 464 nm in this glass and increases by increasing the heat-treatment durations. Both energy transfer mechanism from nanoparticle to ion and modified local filed around the Pr^{3+} ions are discussed as the factors of enhancement; however the exact contribution of each phenomenon was not concluded. Enhancement in emissions from Pr^{3+} ions doped zinc-tellurite glass containing silver NPs is also reported [89]. The influence of large local field on the Pr^{3+} ions is discussed as the major factor for enhancements of visible emission under both 470 nm (Stokes) and 586 (anti-Stokes) excitation wavelengths.

The enhancement in upconversion emissions of Ho^{3+}/Yb^{3+} co-doped tellurite glass is also observed [90] by incorporation of silver NPs which are formed by addition of $AgNO_3$ to the glassy host, melted at 900°C for 30 minutes and annealed above its glass transition temperature. Silver NPs with average diameter size of about 3 to 12 nm show a SPR absorption band centered at 560 nm and contribute to ~ 2.5 times enhancement in intensity of emissions at 546 and 657 nm under 980 nm excitation wavelength.

3.2   The bad

Jimenez at al. also reported on the enhanced UV-excited luminescence of Eu^{3+} ions in silver/tin-doped glass [91]. However, such enhancements are attributed to Ag^+ ions, and not Ag NPs. Moreover, the quenched PL is caused by Ag NPs, by providing "the paths for the non-radiative loss of excitation energy in europium ions through coupling with plasmon resonance modes". Eu^{3+}-doped aluminosilicate glasses containing different Ag species are reported by Li et al. [92]. The observed broadbands in UV-Vis absorption, photoluminescence excitation and emission spectra of the glasses suggest the presence of Ag ions and molecular-like Ag species. However, after 30 and 120 min of heat-treatments, the silver NPs are formed in the glasses and are discussed in terms of following redox reaction; $Eu^{2+} + Ag^+ \rightarrow Eu^{3+} + Ag^0$. The surface plasmon band of the silver NPs in this glass is observed at 440 nm by taking the difference between absorption spectra of samples with and without NPs. Although the excitation lines of Eu^{3+} ions are suppressed in the spectra of AgNPs-doped samples, the luminescence emissions in the visible region enhances under 350 nm excitation wavelength. The authors concluded that the observed enhancement can be purely associated to the energy transfer from silver aggregates to Eu^{3+} ions and not an enlarged local field by SPR of Ag NPs.

Jiao et al. [93] investigated the effect of concentration of Eu^{3+} ions on the formation and growth of silver NPs. They concluded that increasing the Eu_2O_3 content led to increase the Eu^{2+} ions and increasing the concentration of Eu^{3+} ions and Ag NPs as; $Eu^{2+} + Ag^+ \rightarrow Eu^{3+} + Ag^0$. In this regard, the enhancement and quenching of Eu^{3+} emissions are observed under 280 nm and 340 nm excitation wavelengths, respectively. These results are in good agreement with Riano et al. [94] reporting on the intense surface plasmon band of Ag NPs in presence of Eu^{3+} ions, rather than Pr^{3+} ions. However, they observed quenching of the luminescence of Eu^{3+} ions, an indicative of energy transfer from Eu^{3+} ions to Ag NPs.

The influence of Ag NPs on the luminescence decay of Dy^{3+}-doped aluminophosphate is reported by Jimenez [95]. Although the incorporation of metallic NPs in this system quenched the luminescence of Dy^{3+} ions under 450 nm excitation wavelength, the new concept of "plasmonic diluents" highlighted this work as a worthy publication to think over. The mechanism behind the plasmonic diluents is similar to lowering the effective concentrations of ions, where lower absorptions takes place in the system. On the other word, the resonance excitation of the system results in an energy transfer from ion to particle (silver NPs) which is a detrimental factor for subsequent PL processes, and results in luminescence quenching. Moreover, Jimenez showed that the increasing the volume fraction of the silver NPs by increasing the heat-treatment durations, prolongs the fast and slow decay times of Dy^{3+} ions.

Using the gold NPs in an Er^{3+}-doped SiO_2 thin film, prepared by sol-gel method, the optical absorption and emission at 1.54 μm were characterized by Fukushima et al. [96]. One hundred times enhancement emissions of sample with 1mol% Au NPs were attributed to strong filed originated from confined surface plasmon, while the excitation wavelength were selected to line in the Au plasmon band region, located at 520 nm. Lin et al. [97] fabricated the Au NPs-doped erbium optical fiber in a germane-silicate glass which showed plasmon resonance band around 498.2 nm, and the net gain of broadband emission at 1535.6 and 1551.2 nm experienced enhances under 980 nm excitation wavelength. The quenches are observed under 488 nm excitation wavelength and are attributed to absorption of the incident energy by Au NPs. They discussed the observed loss by an ET from Au NPs to lattice, and not to Er^{3+} ions.

Li et. Al [98] also investigated the effect of silver NPs on optical properties of Sm^{3+}-doped silicate glasses. Different silver species are formed in the silicate glass by an Ag^+-Na^+ ion exchange process. Although enhancement of luminescence under 270/250 nm and 355 nm is observed due to energy transfer from Ag^+ and Ag^+-Ag^+ to Sm^{3+} ions, respectively, the presence of NPs quenches the luminescence under 401 excitation wavelength. They concluded that the competitive absorptions by Sm^{3+} ions and Ag NPs (SPR ~ 420 nm) suppress the luminescence of Sm^{3+} ions.

There are not many reports on the influence of noble metallic NPs on the optical properties of Dy^{3+}-doped glasses or glass-ceramics. However, the influence of Cu NPs on the luminescence of Dy^{3+}-doped barium-phosphate glasses is studied [99]. Cu^{2+} and Cu^+ ions are reduced to Cu^0 NPs as $Cu^{2-}+Sn^{2+} \rightarrow Cu^0 + Sn^{4+}$ and $2Cu^++Sn^{2+} \rightarrow 2Cu^0+Sn^{4+}$, respectively. The glasses are annealed for 30, 60 and 120 min, but all the observed photoluminescence emissions are quenched under 350 and 450 nm excitation wavelength. The quench is attributed to non-radiative loss of excitation energy in Dy^{3+} ions with an energy transfer from ion to NP. The effect of gold is also reported on the optical properties of Dy^{3+}- and Eu^{3+}-doped silica nanoparticles [100].

Singly Tm^{3+}-doped PbO-GeO_2 oxide glasses containing silver NPs are also investigated by the latter group [101]. Upconversion emissions are enhanced in presence of silver NPs under 1050 excitation wavelength for heat-treatment up to 24 hours. Further heat-treatments (up to 72 hours) resulted in a quenching in PL spectra. Moreover, they concluded that the continuous heat-treatments stimulate the NPs to aggregate while non-continues heat-treatments with a step of 12 hours prevents the aggregation of NPs. In another study, Qi et al. [102] observed significant enhancement in emissions of Tm^{3+} ions due to participation of Ag NPs in the bismuth germanate glasses.

3.3   Challenges

Wei et al. [103] are also worked on the preparation of Ag NPs-embedded Eu^{3+}-doped oxyfluoride glasses. The enhancement in emissions of Eu^{3+} ions are attributed to presence of silver NPs, small-molecular like silver and isolated Ag^+ ions under 464 nm, 350 nm and 270 nm excitation wavelengths, respectively.

Enhancement of red-emissions of Eu^{3+} ions by Ag NPs is also reported in other medium [104] under a green light excitation which is known to be promising materials for solar cell and nano biotechnology. Moreover, white light emission is observed in Eu^{3+}-doped oxyfluoride glass containing molecular-like (ML) silver, where no proof was observed to attribute the enhancement of photoluminescence to plasmonic silver NPs [105]

In 2002, Strohhofer et al. [106] reported on the enhanced emission of Er^{3+} in a borosilicate glass by an ion-exchange process. They observed 70 and 220 times enhancements in broadband line under 488 and 360 nm excitation wavelength. Strohhofer et al. concluded that such enhancement can be attributed to the silver ions/atoms defects and an ET to Er^{3+} ions. Chiasera et al. [107] reported on the silver-sodium exchange process in soda-lime silicate glass containing Er^{3+} ions. The plasmon band of silver was observed in blue region, and its intensity increased by further heat-treatments. The silver exchange has no effect on the broadband emission of erbium; however, it increased the lifetime of this metastable state, which was in disagreement with some older reports [106,108]. The increase in lifetime of $^4I_{13/2}$ was attributed to silver-induced radiation trapping [107], while the decrements is featured by silver-induced defects in glassy hosts [106,108].

Jimenez and Sendova [109] studied the effect of silver species (Ag NPs and non-plasmonic clusters) on the luminescence intensity of Sm^{3+}-doped aluminophosphate glass as a function of holding heat-treatment time. The SPR band is not observed up to 50 min of heat-treatments, however the luminescence enhances gradually. The SPR band emerges and intensifies by further heat-treatments up to 120 min, while the luminescence intensity of Sm^{3+} ions quenches progressively. They concluded that the enhancements and quench in the luminescence are associated to the presence of non-plasmonic clusters and NPs, respectively.

The influence of gold NPs on the upconversion emission and Judd-Ofelt parameters of Er^{3+}-doped tellurite glass is investigated by Awang et al. [58,59] and Sazali et al. [60]. However, the correlation of upconversion and Judd-Ofelt parameters is still not clearly understood.

In a recent study, the effect of the heat-treatment on the upconversion luminescence of Sm^{3+}-doped borosilicate glasses containing silver NPs are examined [72]. The silver ions are reduced by Sb^{3+} ions as oxidation agents and NPs are grown gradually by increasing the time of heat-treatments up to 20 h. It is stated that the further heat-treatments result in a translucent glassy component which is not favorable for optical applications. The surface plasmon band of silver NPs in this glass is observed at ~ 436 nm which red-shifts to 450 nm by increasing the heat-treatments, indicative of a growth in the size of NPs from 8 to 14 nm.

The same doping system embedded in a tellurite glass containing silver NPs is also reported by the same group [110] where efficient mid-infrared emissions from Er^{3+} (1.55 μm) and Tm^{3+} ions (1.86 μm) are observed. Moreover, efficient energy transfer from Yb^{3+} ions to Er^{3+} and Tm^{3+} ions are concluded by time-resolved luminescence investigations. The decay lifetime of $^4F_3\rightarrow^3H_6$ (Tm^{3+}) shortened from 3.4 to 2.9 ms by addition of silver NPs and lifetime of $^4I_{13/2}\rightarrow^4I_{15/2}$ (Er^{3+}) decreased from 2.4 to 1.7 ms. $TM^{3+}/Yb^{3+}/Er^{3+}$ tri-doped oxyfluorogermanate glasses containing silver NPs are also reported [111]. The authors showed the introduction of silver NPs reduces the glass thermal stability as well as glass transition and crystallization temperatures. A broad absorption band around 400-500 nm is attributed to surface plasmon band of silver which is extended up to 800 nm for further annealing time intervals. Silver NPs are grown from 4 to 10 nm by increasing the annealing time from 34 to 51 hours. The intensity of all the observed upconversion emissions (emissions at 476, 524, 546 and 658 nm, and excitation at 980 nm) increases by increasing the annealing time up to 34 hours, while it quenches for further heat-treatments.

Perhaps, it is still to be solved a possible loss of chemical durability and optical transparency of the glasses when doped with metallic NPs, other than difficulties in the heat-treatment parameters (temperature and time), which have considerable effect in the growth rate of the particles.

4. Summary

In this chapter, recent advances and challenges in the optical properties of the rare earth doped glasses containing metallic nanoparticles are revisited. The principal methodologies to prepare and characterize the nanoparticles in the glassy hosts are described. Three classification of Good, Bad and Challenges are nominated in the cases where luminescence of rare earth ions are enhanced or quenched due to addition of metal species, and efforts to describe the phenomena behind those interactions. This chapter illustrate that there are still a lot of problems to be solved in this research line and the topic is of utmost importance for the development of new technological glasses in optics and optoelectronics.

References

[1] P.K. Gupta, J. Non. Cryst. Solids. 195 (1996) 158-164. https://doi.org/10.1016/0022-3093(95)00502-1

[2] H. Tait, ed., Five thousand years of glass, The British Museum Press, London, 1991.

[3] J.C. Mauro, and E.D. Zanotto, Int. J. Appl. Glas. Sci. 15 (2014) 1-15.

[4] A. Jha, B. Richards, G. Jose, T. Teddy-Fernandez, P. Joshi, X. Jiang, and J. Lousteau, Prog. Mater. Sci. 57 (2012) 1426-1491. https://doi.org/10.1016/j.pmatsci.2012.04.003

[5] A.B. Seddon, Z. Tang, D. Furniss, S. Sujecki, and T.M. Benson, Opt. Express. 18 (2010) 26704-26719. https://doi.org/10.1364/OE.18.026704

[6] L.G. Van Uitert, J. Chem. Phys. 44 (1966) 3514. https://doi.org/10.1063/1.1727258

[7] H. Dong, L.-D. Sun, and C.-H. Yan, Nanoscale. 5 (2013) 5703-14. https://doi.org/10.1039/c3nr34069d

[8] H. Zheng, D. Gao, Z. Fu, E. Wang, Y. Lei, Y. Tuan, and M. Cui, J. Lumin. 131 (2011) 423-428. https://doi.org/10.1016/j.jlumin.2010.09.026

[9] I. Díez, and R.H. a Ras, Nanoscale. 3 (2011) 1963-70. https://doi.org/10.1039/c1nr00006c

[10] S. Shanmukh, L. Jones, J. Driskell, Y. Zhao, R. Dluhy, and R.A. Tripp, Nano Lett. 6 (2006) 2630-2636. https://doi.org/10.1021/nl061666f

[11] A.M. Schwartzberg, C.D. Grant, T. Van Buuren, and J.Z. Zhang, J. Phys. Chem. 111 (2007) 8892-8901. https://doi.org/10.1021/jp074313t

[12] A.M. Schwartzberg, T.Y. Olson, C.E. Talley, and J.Z. Zhang, J. Phys. Chem. B. 110 (2006) 19935-19944. https://doi.org/10.1021/jp062136a

[13] L. Gou, and C.J. Murphy, Chem. Mater. 17 (2005) 3668-3672. https://doi.org/10.1021/cm050525w

Rare Earth - A tribute to the late Mr. Rare Earth, Professor Karl Gschneidner   Materials Research Forum LLC
Materials Research Foundations 164 (2024) 257-278   https://doi.org/10.21741/9781644903056-7

[14] Y. Sun, B.T. Mayers, and Y. Xia, Nano Lett. 2 (2002) 481-485.
https://doi.org/10.1021/nl025531v

[15] K. Xu, and J. Heo, J. Am. Ceram. Soc. 96 (2013) 1138-1142.
https://doi.org/10.1111/jace.12242

[16] J. V Zoval, R.M. Stiger, P.R. Biernacki, and R.M. Penner, J. Phys. Chem. 100 (1996) 837-844. https://doi.org/10.1021/jp952291h

[17] A. Henglein, Isr. J. Chem. 33 (1993) 77-88. https://doi.org/10.1002/ijch.199300013

[18] M. Quinten, Optical properties of nanoparticle systems: Mie and beyond, Wiley-VCH Verlag & Co. KGaA, Weinheim, Germany, 2011. https://doi.org/10.1002/9783527633135

[19] M. V Smoluchowski, Phys. Z. 17 (1916) 557-599.
https://doi.org/10.1163/156853216X00193

[20] W. Ostwald, Z Phys. Chem. 34 (1900) 495-503. https://doi.org/10.1515/zpch-1900-3431

[21] P. Misra, Handbook of metal physics, First, Elsevier Oxford, UK, 2009.

[22] G. Liu, S. Xu, and Y. Qian, Acc Chem. Res. 33 (2000) 457-466.
https://doi.org/10.1021/ar980081s

[23] Z.L. Wang, ed., Characterization of nanophase materials, Wiley-VCH, New York, 2000.

[24] Z.L. Wang, J. Phys. Chem. B. 104 (2000) 1153-1175. https://doi.org/10.1021/jp993593c

[25] G. Binning, C.F. Quate, and C. Gerber, Phys. Rev. Lett. 56 (1986) 930-933.
https://doi.org/10.1103/PhysRevLett.56.930

[26] T. Som, and B. Karmakar, J. Appl. Phys. 105 (2009) 013102-8.
https://doi.org/10.1063/1.3054918

[27] T. Som, and B. Karmakar, J. Opt. Soc. Am. B. 26 (2009) B21-B27.
https://doi.org/10.1364/JOSAB.26.000B21

[28] T. Som, and B. Karmakar, Nano Res. 2 (2009) 607-616. https://doi.org/10.1007/s12274-009-9061-4

[29] Y. Wu, T. Xu, X. Shen, S. Dai, Q. Nie, X. Wang, B. Song, W. Zhang, and C. Lin, Effect of Silver Nanoparticles on Spectroscopic Properties of Er 3 + -doped Bismuth Glass, in: 6th IEEE Conf. Ind. Electron. Appl., 2011: pp. 1464-1467.
https://doi.org/10.1109/ICIEA.2011.5975820

[30] M. Reza Dousti, M.R. Sahar, R.J. Amjad, S.K. Ghoshal, a. Khorramnazari, A. Dordizadeh Basirabad, and A. Samavati, Eur. Phys. J. D. 66 (2012) 237.
https://doi.org/10.1140/epjd/e2012-30089-1

[31] G.V. Rao, and H.D. Shashikala, J. Non. Cryst. Solids. 402 (2014) 204-209.
https://doi.org/10.1016/j.jnoncrysol.2014.06.007

[32] G. Speranza, L. Minati, A. Chiasera, M. Ferrari, G.C. Righini, and G. Ischia, J. Phys. Chem. C. 113 (2009) 4445-4450. https://doi.org/10.1021/jp810317q

[33] L.R. Kassab, C.B. De Araújo, R.A. Kobayashi, R.D.A. Pinto, and D.M. Silva, J. Appl. Phys. 102 (2007) 103515-4. https://doi.org/10.1063/1.2817980

[34] M. Reza Dousti, M.R. Sahar, S.K. Ghoshal, R.J. Amjad, and a. R. Samavati, J. Mol. Struct. 1035 (2013) 6-12. https://doi.org/10.1016/j.molstruc.2012.09.023

[35] M. Reza Dousti, P. Ghassemi, M.R. Sahar, and Z.A. Mahraz, Chalcogenide Lett. 11 (2014) 111-119.

[36] A.P. Silva, A.P. Carmo, V. Anjos, M.J.V. Bell, L.R.P. Kassab, and R.D.A. Pinto, Opt. Mater. (Amst). 34 (2011) 239-243. https://doi.org/10.1016/j.optmat.2011.08.018

[37] V.T. Adamiv, I.M. Bolesta, Y.V. Burak, R.V. Gamernyk, I.D. Karbovnyk, I.I. Kolych, M.G. Kovalchuk, O.O. Kushnir, M.V. Periv, and I.M. Teslyuk, Phys. B Condens. Matter. 449 (2014) 31-35. https://doi.org/10.1016/j.physb.2014.05.009

[38] H.H. Mai, V.E. Kaydashev, V.K. Tikhomirov, E. Janssens, M. V. Shestakov, M. Meledina, S. Turner, G. Van Tendeloo, V. V. Moshchalkov, and P. Lievens, J. Phys. Chem. C. (2014) 140710160814003.

[39] R. a. Ganeev, a. I. Ryasnyansky, a. L. Stepanov, and T. Usmanov, Phys. Status Solidi. 241 (2004) 935-944. https://doi.org/10.1002/pssb.200301947

[40] O.L. Malta, P.A. Santa-Cruz, G.F. De Sá, and F. Auzel, J. Lumin. 33 (1985) 261-272. https://doi.org/10.1016/0022-2313(85)90003-1

[41] T. Hayakawa, S.T. Selvan, and M. Nogami, J. Non-Cryst. Solids. 259 (1999) 16-22. https://doi.org/10.1016/S0022-3093(99)00531-1

[42] T. Hayakawa, K. Furuhashi, and M. Nogami, J. Phys. Chem. B. 108 (2004) 11301-11307. https://doi.org/10.1021/jp048247w

[43] R. De Almeida, D.M. da Silva, L.R.P. Kassab, and C.B. de Araujo, Opt. Commun. 281 (2008) 108-112. https://doi.org/10.1016/j.optcom.2007.08.072

[44] L.R.P. Kassab, D.S. da Silva, R. de Almeida, and C.B. de Araújo, Appl. Phys. Lett. 94 (2009) 101912. https://doi.org/10.1063/1.3097241

[45] R.J. Amjad, M.R. Dousti, M.R. Sahar, S.F. Shaukat, S.K. Ghoshal, E.S. Sazali, and F. Nawaz, J. Lumin. 154 (2014) 316-321. https://doi.org/10.1016/j.jlumin.2014.05.009

[46] M. Reza Dousti, M.R. Sahar, M.S. Rohani, A. Samavati, Z.A. Mahraz, R.J. Amjad, A. Awang, and R. Arifin, J. Mol. Struct. 1065-1066 (2014) 39-42. https://doi.org/10.1016/j.molstruc.2014.02.032

[47] K.V.A. Kumar, K.P. Revathy, V. Prathibha, T. Sunil, P.R. Biju, and N. V. Unnikrishnan, J. Rare Earths. 31 (2013) 441-448. https://doi.org/10.1016/S1002-0721(12)60301-9

[48] P.R. Watekar, S. Ju, and W.-T. Han, Colloids Surfaces A Physicochem. Eng. Asp. 313-314 (2008) 492-496. https://doi.org/10.1016/j.colsurfa.2007.04.178

[49] S.K. Singh, N.K. Giri, D.K. Rai, and S.B. Rai, Solid State Sci. 12 (2010) 1480-1483. https://doi.org/10.1016/j.solidstatesciences.2010.06.011

[50] V. a G. Rivera, S.P. a Osorio, Y. Ledemi, D. Manzani, Y. Messaddeq, L. a O. Nunes, and E. Marega, Opt. Express. 18 (2010) 25321-8. http://www.ncbi.nlm.nih.gov/pubmed/21164880. https://doi.org/10.1364/OE.18.025321

[51] V. a. G. a G. Rivera, Y. Ledemi, S.P. a. P. a Osorio, D. Manzani, Y. Messaddeq, L. a. O. a O. Nunes, and E. Marega, J. Non. Cryst. Solids. 358 (2012) 399-405. https://doi.org/10.1016/j.jnoncrysol.2011.10.008

[52] V.P.P. de Campos, L.R.P. Kassab, T.A.A. de Assumpção, D.S. da Silva, and C.B. de Araújo, J. Appl. Phys. 112 (2012) 063519.

[53] R.J. Amjad, M.R. Sahar, S.K. Ghoshal, M.R. Dousti, S. Riaz, and B.A. Tahir, J. Lumin. 132 (2012) 2714-2718. https://doi.org/10.1016/j.jlumin.2012.05.008

[54] R.J. Amjad, M.R. Sahar, S.K. Ghoshal, M.R. Dousti, S. Riaz, A.R. Samavati, M.N.A. Jamaludin, and S. Naseem, Chinese Phys. Lett. 30 (2013) 027301. https://doi.org/10.1088/0256-307X/30/2/027301

[55] R.J. Amjad, M.R. Sahar, S.K. Ghoshal, M.R. Dousti, S. Riaz, A.R. Samavati, R. Arifin, and S. Naseem, J. Lumin. 136 (2013) 145-149. https://doi.org/10.1016/j.jlumin.2012.11.028

[56] M. Reza Dousti, M.R. Sahar, S.K. Ghoshal, R.J. Amjad, and R. Arifin, J. Mol. Struct. 1033 (2013) 79-83. https://doi.org/10.1016/j.molstruc.2012.08.022

[57] M.R. Dousti, M.R. Sahar, S.K. Ghoshal, R.J. Amjad, and R. Arifin, J. Non. Cryst. Solids. 358 (2012) 2939-2942. https://doi.org/10.1016/j.jnoncrysol.2012.06.024

[58] A. Awang, S.K. Ghoshal, M.R. Sahar, M. Reza Dousti, R.J. Amjad, and F. Nawaz, Curr. Appl. Phys. 13 (2013) 1813-1818. https://doi.org/10.1016/j.cap.2013.06.025

[59] S.K. Ghoshal, A. Awag, M.R. Saar, R.J. Amjad, and M.R. Dousti, Chalcogenide Lett. 10 (2013) 411-420.

[60] E.S. Sazali, M.R. Sahar, S.K. Ghoshal, R. Arifin, M.S. Rohani, and a. Awang, J. Alloys Compd. 607 (2014) 85-90. https://doi.org/10.1016/j.jallcom.2014.03.175

[61] A. Awang, S.K. Ghoshal, M.R. Sahar, R. Arifin, and F. Nawaz, J. Lumin. 149 (2014) 138-143. https://doi.org/10.1016/j.jlumin.2014.01.027

[62] A. Awang, S.K. Ghoshal, M.R. Sahar, M.R. Dousti, and F. Nawaz, Adv. Mater. Res. 895 (2014) 254-259. https://doi.org/10.4028/www.scientific.net/AMR.895.254

[63] J. Qi, T. Xu, Y. Wu, X. Shen, S. Dai, and Y. Xu, Opt. Mater. (Amst). 35 (2013) 2502-2506. https://doi.org/10.1016/j.optmat.2013.07.009

[64] Z. Ashur Said Mahraz, M.R. Sahar, S.K. Ghoshal, M.R. Dousti, and R.J. Amjad, Mater. Lett. 112 (2013) 136-138. https://doi.org/10.1016/j.matlet.2013.08.131

[65] V.O. Obadina, Opt. Photonics J. 03 (2013) 45-50. https://doi.org/10.4236/opj.2013.31008

[66] T. Som, and B. Karmakar, J. Quant. Spectrosc. Radiat. Transf. 112 (2011) 2469-2479. https://doi.org/10.1016/j.jqsrt.2011.06.015

[67] D.S. da Silva, T.A.A. de Assumpção, L.R.P. Kassab, and C.B. de Araújo, J. Alloys Compd. 586 (2014) S516-S519. https://doi.org/10.1016/j.jallcom.2012.12.070

[68] M. Reza Dousti, J. Appl. Phys. 114 (2013) 113105. https://doi.org/10.1063/1.4821430

[69] T. Som, and B. Karmakar, Plasmonics. 5 (2010) 149-159. https://doi.org/10.1007/s11468-010-9129-8

[70] T. Som, and B. Karmakar, Spectrochim. Acta. A. Mol. Biomol. Spectrosc. 75 (2010) 640-646. https://doi.org/10.1016/j.saa.2009.11.032

[71] T. Som, and B. Karmakar, Appl. Surf. Sci. 255 (2009) 9447-9452. https://doi.org/10.1016/j.apsusc.2009.07.053

[72] M. Reza Dousti, Measurement. 56 (2014) 117-120. https://doi.org/10.1016/j.measurement.2014.06.024

[73] N.A. Fauzia Abdullah, M.R. Sahar, K. Hamzah, and S.K. Ghoshal, Adv. Mater. Res. 895 (2014) 260-264. https://doi.org/10.4028/www.scientific.net/AMR.895.260

[74] G. Kaur, R.K. Verma, D.K. Rai, and S.B. Rai, J. Lumin. 132 (2012) 1683-1687. https://doi.org/10.1016/j.jlumin.2012.02.014

[75] M. Reza Dousti, and S. Raheleh Hosseinian, J. Lumin. 154 (2014) 218-223. https://doi.org/10.1016/j.jlumin.2014.04.028

[76] T.A.A. de Assumpção, D.M. da Silva, M.E. Camilo, L.R.P. Kassab, A.S.L. Gomes, C.B. de Araújo, and N.U. Wetter, J. Alloys Compd. 536 (2012) S504-S506. https://doi.org/10.1016/j.jallcom.2011.12.078

[77] T. a. a. Assumpção, D.M. da Silva, L.R.P. Kassab, and C.B. de Araújo, J. Appl. Phys. 106 (2009) 063522.

[78] T. a. a. a a de Assumpção, D.M.M. Da Silva, L.R.P.R.P. Kassab, J.R.R. Martinelli, and C.B.B. De Araújo, J. Non. Cryst. Solids. 356 (2010) 2465-2467. https://doi.org/10.1016/j.jnoncrysol.2010.02.016

[79] L.R.P. Kassab, L.F. Freitas, T. a. a. Assumpção, D.M. Silva, and C.B. Araújo, Appl. Phys. B. 104 (2011) 1029-1034. https://doi.org/10.1007/s00340-011-4451-1

[80] L.P.R. Kassab, L. Ferreira Freitas, K. Ozga, M.G. Brik, and A. Wojciechowski, Opt. Laser Technol. 42 (2010) 1340-1343. https://doi.org/10.1016/j.optlastec.2010.04.016

[81] G. Bi, and L. Wang, 10 (2012) 9-11.

[82] P. Piasecki, a. Piasecki, Z. Pan, A. Ueda, R. Aga, Jr., R. Mu, and S.H. Morgan, 7757 (2010) 77572M-77572M-7.

[83] P. Piasecki, J. Nanophotonics. 4 (2010) 043522. https://doi.org/10.1117/1.3528943

[84] L. Li, Y. Yang, D. Zhou, X. Xu, and J. Qiu, J. Non. Cryst. Solids. 385 (2014) 95-99. https://doi.org/10.1016/j.jnoncrysol.2013.11.017

[85] D. Zhang, X. Hu, R. Ji, S. Zhan, J. Gao, Z. Yan, E. Liu, J. Fan, and X. Hou, J. Non. Cryst. Solids. 358 (2012) 2788-2792. https://doi.org/10.1016/j.jnoncrysol.2012.07.004

[86] R.K. Verma, K. Kumar, and S.B. Rai, Solid State Commun. 150 (2010) 1947-1950. https://doi.org/10.1016/j.ssc.2010.07.014

[87] V.K. Rai, L.D.S. Menezes, C.B. De Araújo, L.R.P. Kassab, and M. Davinson, J. Appl. Phys. 103 (2008) 093526. https://doi.org/10.1063/1.2919566

[88] L.P. Naranjo, C.B. de Araújo, O.L. Malta, P. a. S. Cruz, and L.R.P. Kassab, Appl. Phys. Lett. 87 (2005) 241914. https://doi.org/10.1063/1.2143135

[89] G. Lakshminarayana, and J. Qiu, J. Alloys Compd. 478 (2009) 630-635. https://doi.org/10.1016/j.jallcom.2008.11.146

[90] Y. Qi, Y. Zhou, L. Wu, F. Yang, S. Peng, S. Zheng, and D. Yin, J. Non. Cryst. Solids. 402 (2014) 21-27. https://doi.org/10.1016/j.jnoncrysol.2014.05.014

[91] J. a. Jiménez, S. Lysenko, and H. Liu, J. Lumin. 128 (2008) 831-833. https://doi.org/10.1016/j.jlumin.2007.11.018

[92] L. Li, Y. Yang, D. Zhou, Z. Yang, X. Xu, and J. Qiu, Opt. Mater. Express. 3 (2013) 806. https://doi.org/10.1364/OME.3.000806

[93] Q. Jiao, J. Qiu, D. Zhou, and X. Xu, Mater. Res. Bull. 51 (2014) 315-319.
https://doi.org/10.1016/j.materresbull.2013.12.044

[94] L.P. Riano, C.B. de Araujo, O.L. Malta, P. Santa Cruz, and M. a. Couto dos Santos,
Proceeding SPIE. 5622 (2004) 551-555. https://doi.org/10.1117/12.590834

[95] J. a Jiménez, Phys. Chem. Chem. Phys. 15 (2013) 17587-94.
https://doi.org/10.1039/c3cp52702f

[96] M. Fukushima, N. Managaki, M. Fujii, H. Yanagi, and S. Hayashi, J. Appl. Phys. 98 (2005)
024316. https://doi.org/10.1063/1.1990257

[97] A. Lin, S. Boo, D.S. Moon, H.J. Jeong, Y. Chung, and W.-T. Han, Opt. Express. 15 (2007)
8603-8. http://www.ncbi.nlm.nih.gov/pubmed/19547194.
https://doi.org/10.1364/OE.15.008603

[98] L. Li, Y. Yang, D. Zhou, Z. Yang, X. Xu, and J. Qiu, J. Appl. Phys. 113 (2013) 193103.
https://doi.org/10.1063/1.4807313

[99] J. a. Jiménez, and J.B. Hockenbury, J. Mater. Sci. 48 (2013) 6921-6928.
https://doi.org/10.1007/s10853-013-7497-0

[100] L. Petit, J. Griffin, N. Carlie, V. Jubera, M. García, F.E. Hernández, and K. Richardson,
Mater. Lett. 61 (2007) 2879-2882. https://doi.org/10.1016/j.matlet.2007.01.072

[101] T. a. a. Assumpção, L.R.P. Kassab, a. S.L. Gomes, C.B. Araújo, and N.U. Wetter, Appl.
Phys. B. 103 (2010) 165-169. https://doi.org/10.1007/s00340-010-4258-5

[102] J. Qi, Y. Xu, F. Huang, L. Chen, Y. Han, B. Xue, S. Zhang, T. Xu, and S. Dai, J. Am.
Ceram. Soc. 97 (2014) 1471-1474. https://doi.org/10.1111/jace.12784

[103] R. Wei, J. Li, J. Gao, and H. Guo, J. Am. Ceram. Soc. 95 (2012) 3380-3382.
https://doi.org/10.1111/j.1551-2916.2012.05459.x

[104] B. Z, V. Kumar, M. H, and C. S, Chem. Comm. 49 (2013) 9485-9487.
https://doi.org/10.1039/c3cc45267k

[105] H. Guo, X. Wang, J. Chen, and F. Li, Opt. Express. 18 (2010) 18900-5.
http://www.ncbi.nlm.nih.gov/pubmed/20940783. https://doi.org/10.1364/OE.18.018900

[106] C. Strohhöfer, and A. Polman, Appl. Phys. Lett. 81 (2002) 1414.
https://doi.org/10.1063/1.1499509

[107] A. Chiasera, M. Ferrari, M. Mattarelli, M. Montagna, S. Pelli, H. Portales, J. Zheng, and
G.C. Righini, Opt. Mater. (Amst). 27 (2005) 1743-1747.
https://doi.org/10.1016/j.optmat.2004.11.044

[108] H. Portales, M. Mattarelli, M. Montagna, A. Chiasera, M. Ferraris, A. Martucci, P.
Mazzoldi, S. Pelli, and G.C. Righini, J. Non. Cryst. Solids. 351 (2005) 1738-1742.
https://doi.org/10.1016/j.jnoncrysol.2005.04.006

[109] J. a. Jiménez, and M. Sendova, Solid State Commun. 152 (2012) 1786-1790.
https://doi.org/10.1016/j.ssc.2012.06.017

[110] G.H.H. Silva, D.P. a. P. a Holgado, V. Anjos, M.J.V.J. V Bell, L.R.P.R.P. Kassab, C.T.T.
Amâncio, and R. Moncorgè, Opt. Mater. (Amst). 2 (2014) 6-11.

[111] Y. Hu, J. Qiu, Z. Song, Z. Yang, Y. Yang, D. Zhou, Q. Jiao, and C. Ma, J. Lumin. 145
(2014) 512-517. https://doi.org/10.1016/j.jlumin.2013.08.022

Rare Earth - A tribute to the late Mr. Rare Earth, Professor Karl Gschneidner   Materials Research Forum LLC
Materials Research Foundations 164 (2024) 279-297   https://doi.org/10.21741/9781644903056-8

Chapter 8

Rare Earth Metals Doped ZnO Nanomaterials: Synthesis, Photocatalytic, and Magnetic Properties

D. Ranjith Kumar, Yuvaraj Haldorai, R.T. Rajendra Kumar*

Department of Nanoscience and Technology, Bharathiar University, Coimbatore-641046, Tamil Nadu, India

rtrkumar@buc.edu.in

Abstract

Zinc oxide (ZnO) nanomaterials have attracted increasing interest due to their unique properties. As part of this renewed interest in ZnO nanomaterials, researchers began seeking new strategies to engineer materials by doping rare earth metals on ZnO. This chapter provides a general overview of the techniques and strategies used for the synthesis of undoped and rare earth metals (Samarium, Gadolinium, Europium, Cerium, Neodymium, and Dysprosium) doped ZnO nanomaterials. In addition, optical, magnetic, and photocatalytic properties of these materials were discussed. Some key results are summarized relating to the above properties.

Keywords

Rare Earth Metals, ZnO, Doping, Photocatalytic Activity, Magnetic Properties

Contents

1.   Introduction

In the past few decades, synthesizing zinc oxide (ZnO) nanostructures have attracted great attention due to their unique properties [1]. ZnO is an inorganic semiconducting material, which is nontoxic and can be synthesized at low cost on a large and lab scale [2]. ZnO exhibits in three different forms such as hexagonal wurtzite, zinc blende, and rock salt [3] which is shown in figure 1. The hexagonal wurtzite phase is the popular structure due to its stability at room temperature and easily prepared in normal atmospheric pressure. The atomic arrangement of the wurtzite structure is comprised of four zinc ions (Zn^{2+}) occupying the corner of a tetrahedral coordinate with one oxygen ion (O_2^-) located at the center and vice versa. The growth of zinc blende structure of ZnO is challenging because the zinc blende is only stable in the cubic structure. In the rock salt structure, each Zn or O atom is surrounded by its alternative six nearest neighbors. It exists at the higher pressures (<10 GPa) and it is also not epitaxially stable [3].

Rocksalt   **Zinc blende**   **Wurtzite**

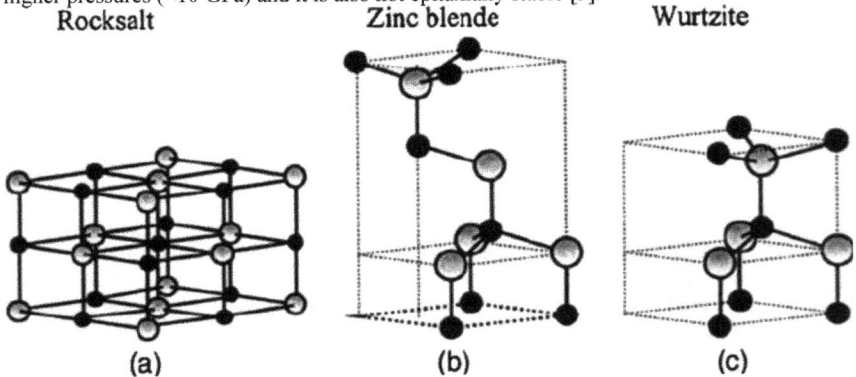

(a)   (b)   (c)

Figure 1. Stick and ball representation of ZnO crystal structures: (a) cubic rocksalt (B1), (b) cubic zinc blende (B3), and (c) hexagonal wurtzite (B4). The shaded gray and black spheres denote Zn and O atoms, respectively. Reprinted and permission from, Ü. Özgür et al., A comprehensive review of ZnO materials and devices, Journal of Applied Physics 98, 041301 (2005).

The challenges to developing these materials for many commercial applications, numerous studies are taking place by researchers to find the optimum solution for enhancing the performance properties of ZnO without changing its physiochemical properties [4]. For enhancing the properties of ZnO nanostructures, we have to introduce impurity or defect at the time of synthesis. The addition of foreign atoms or impurities to a compound by creating a defect for enhancing its physical properties is termed as doping. Here, doping is required to modify the physical properties of nanostructures.

The transition metals such as Fe [5], Ni [6], Mn [7], Co [8] and Cr [9] have been doped with ZnO to improve the magnetic and photocatalytic performances. Recently, rare-earth-doped ZnO has been a focus of numerous investigations because of their unique optical properties and promising applications [10]. Compared to transition metals-doped ZnO, doping of rare earth metals ion is interesting because of their partially filled f-orbitals. These f-orbitals carry magnetic moments and may take part in magnetic coupling as seen in the case of transition metals with partially filled d-orbitals. Further, rare earth metal ions doped ZnO are comparatively investigated less, especially concerning with regard to their magnetic properties, although such a system would be very interesting for magneto-optical applications. Thus, we can say that the properties of ZnO nanostructures depend on their crystal structure, morphology, size, and surface defects. It has been extensively proven that modifications of ZnO nanostructures such as doping of transition metals or rare-earth metals could improve their properties. The photocatalytic properties of ZnO nanoparticles were significantly increased and improved when modified with the incorporation of dopant ions. The doping of metal ions in ZnO nanostructures can lead to surface defects and enhance the fluorescence property. Moreover, ZnO-doped sample with a high concentration of oxygen defects showed excellent photocatalytic activity. The energy level or charge transfer between the energy levels helps in determining the intrinsic ferromagnetic properties of rare earth metals-doped ZnO.

Pan et al. reported that Eu-doped ZnO nanowire was prepared by chemical vapor deposition on Si substrate employing Au as a catalyst, the optical properties were changed to narrow band edge due to the presence of Eu-ions [11]. Devi et al. synthesized Sm-doped ZnO nanostructure by a simple combustion method. The charge transfer from Sm^{3+} to Sm^{2+} played an important role in PL emission process of ZnO: Sm^{3+} nanostructures [12]. Zhang et al. described that enhanced energy transfers from ZnO nanocrystal to ions which could be widely adopted in rare earth ions doped materials which is discover also provide more insights into other energy transfer problem like dye-sensitized solar cell and quantum dot solar cell [13]. Wang et al. stated that Nd-doped ZnO nanowire was prepared by vapor phase transport method which exhibited a robust ferromagnetic at room temperature [14]. He et al. prepared flower-like La_3O_2 doped ZnO nanomaterial by a simple co-precipitate method and the catalytic activity of rare earth metal oxide doped ZnO showed increased acetone response [15]. Li et al. revealed that Yd-doped ZnO thin film was prepared by a plasma enhanced physical vapor deposition and its ferromagnetism was induced by the coexistence of oxygen vacancy and Yb point defects [16]. Liang et al. reported that the Ce-doped ZnO sample exhibited improved photocatalytic performance on the decomposition of Rhodamine B dye. It is proposed that the special structural feature with a porous 3D structured and Ce modification leads to the rapid photocatalytic activity of the Ce-doped ZnO microflowers [17]. Wang et al. synthesized flower structured Ce-doped ZnO capped with luminol and utilized as an immunosensor for Aβ detection [18]. Lang et al. described that rare earth metals such as Eu, La and Sm were co-doped with ZnO by the co-precipitate method, and the optical properties were enhanced due to doping of rare earth metal ions [19].

In this chapter, the literature review examines the synthesis of undoped and rare earth metals (Sm, Gd, Eu, Ce, Nd, and Dy) doped ZnO nanomaterials using different techniques, such as the chemical vapor deposition, co-precipitation, sol-gel, etc. Owing to the extensive research activities in this field, a complete overview was not possible in this chapter. Instead, the focus is on a general overview of the techniques and strategies used to prepare rare earth metals-doped ZnO nanomaterials. Also, the optical magnetic, and photocatalytic properties of the doped ZnO were discussed. More detailed descriptions of the specific themes can be obtained from the related references.

2.   Results and discussion

ZnO is one of the most popular n-type semiconductor metal oxides, is a versatile material with a wide band gap (3.37 eV) and large exciton binding energy (60 meV) [20]. It has a variety of applications such as UV absorption, spintronics, photocatalysts, sensing and UV light-emitting devices. But, these properties of ZnO has been strongly depend on the impurities and defects [21].

It is well known that the existence of defects in a semiconductor would lead to corresponding defects energy levels in the band gap [22]. Therefore, we believe that the different types of oxygen defects such as oxygen vacancies and interstitial oxygen in ZnO nanocrystals result in changes in their photoluminescence and photocatalytic properties. In addition, magnetic properties are also affected by oxygen vacancies [23].

(i) Synthesis and characterization

Several researchers have introduced oxygen vacancies in ZnO by doping rare earth metals. There are different synthetic techniques such as chemical, physical, mechanical, co-precipitation, sol-gel, hydrothermal were used to prepare rare earth metals-doped ZnO with good crystalline nature and morphology and the data were given in Table 1.

Table 1. *Preparation of rare earth metals-doped ZnO using different techniques.*

Synthetic technique	Materials	Synthesis Condition	Morphology and applications	Ref
Vapor phase transport	ZnO, Graphite and samarium oxide	Temp-925°C, Time 45 mins	Rod-like morphology used in the photocatalytic application.	24
Chemical precipitation method	Zinc nitrate hexahydrate $(Zn(NO_3)_2.6H_2O)$, europium oxide (Eu_2O_3), ammonium bicarbonate (NH_4HCO_3)	Temp-120°C, Time-1 hrs	Eu doped ZnO nanoparticle for structural and optical properties.	25
Chemical spray technique	Zinc acetate dehydrate [Zn $(CH_3CO_2)_2,2H_2O]$, hexahydrated neodymium chloride $(NdCl_3, 6H_2O)$, terbium (III) chloride $(TbCl_3)$	Temp-350°C	Nd_3-Tb_3 co-doped ZnO thin films for electrical properties	26
Hydrothermal process	Zinc nitrate hexahydrate $(Zn(NO_3)_2.6H_2O)$, dysprosium nitrate hexahydrate $Dy(NO_3)_3.6H_2O$, sodium tungstate dihydrate $(Na_2WO_4.2H_2O)$, oxalic acid dihydrate $(C_2H_2O_4.2H_2O)$, methanol (CH_3OH) (HPLC grade)	Temp-115°C, Time-12hrs	Porous structure Dy_2 doped ZnO for photocatalytic application	27
PLD	Gd: ZnO target	(KrF laser with k-248 nm, energy-520 mJ, and substrate temperature-650°C	Gd-doped ZnO thin film for RT magnetic properties.	28

Materials Research Forum LLC
https://doi.org/10.21741/9781644903056-8

Radio frequency magnetron sputtering	ZnO and Sm_2O_3 powders for target	Chamber pressure 5.1×10^{-3} mbar and working pressure 1.2×10^{-1} mbar	Sm-doped ZnO thin film for magnetic properties	29
Electrospinning technique	PolyvinylPyrrolidone, $(Zn(Ac)_2)\cdot2H_2O$, $La(NO_3)_3\cdot6H_2O$	Voltage- 13.6 kV, Distance-23cm	La-doped ZnO nanofiber for acetone sensor	30
chemical co-precipitation method	Zinc acetate $(Zn(CH_3COO)_2.2H_2O)$, Dysprosium chloride hexahydrate $(Cl_3.Dy.6H_2O$	Temp-80°C, Time-24hrs	Dy doped ZnO nanoparticle for structural and photoluminescence properties	31
Hydrothermal method	Zinc nitrate hexahydrate $(Zn(NO_3)_2\cdot6H_2O$, AR), hexamethylenetetramine (HMT, AR), cerium nitrate hexahydrate $(Ce(NO_3)_3\cdot6H_2O$, 99.95%) and oxalic acid dehydrate $(H_2C_2O_4\cdot2H_2O$, AR)	Temp-90°C, Time-3hrs	Ce doped ZnO porous flower for photocatalytic application	32
chemical bath deposition	zinc acetate $(Zn(CH_3COO)_2.2H_2O)$, europium nitrate $(Eu(NO_3)_3.6H_2O)$, thiourea $((NH_2)2CS)$ and ammonia (25% NH_3)	Temp-80°C, Time-Overnight	Structural and optical properties of Eu doped flower-like ZnO	33
microwave assisted method	Zinc acetate, lanthanum nitrate, sodium hydroxide and high molecular weight Polyvinyl Alcohol (PVA 2000)	Input power-600W, Time-5mins	La-doped ZnO Nanorod for photocatalytic application	34
Sol-gel method	zinc acetate $(Zn-(CH_3COO)_2)$ and sodium hydroxide (NaOH), Tb $(NO_3)_3.5H_2O$	Room-Temp (under stirring), Time-24hrs	Tb3+ doped ZnO nanocrystals for photoluminescence properties	35

XRD measurements were performed to investigate the crystalline structure of synthesized ZnO. Ranjith Kumar et al. stated that the structural analysis of Sm free and Sm-doped ZnO nanorod arrays by X-ray diffraction in (figure 2). The XRD pattern showed that the high-intensity peak at $2\theta = 34.3°$ corresponds to (002) plane of ZnO grown along c-axis and has wurtzite structure [JCPDS #36-1451]. No impurity peaks were observed in the Sm-doped ZnO pattern. However, the XRD pattern of 8% Sm-doped ZnO nanorod showed additional peaks at 32.5°, 34.3° and 36.3° corresponds to (100), (002), and (101) planes respectively, which indicate the deterioration in the vertical alignment of the ZnO nanorods. As a result of Sm doping, the major peak at 34.3°

corresponds to (002) plane gets shifted to a higher 2θ value which is shown in figure 2(b). It clearly signifies that the peak intensities decreased due to the incorporation of Sm^{3+} in ZnO lattice since Sm^{3+} has a larger radius (0.0959nm) than Zn^{2+} ions (0.074 nm) [36]. Pramod Kumar et al. synthesized Er-doped ZnO with excellent structural and optical properties. The XRD pattern revealed all diffraction peaks observed for ZnO are indexed as wurtzite phase with hexagonal structure. The small amount of Er_2O_3 was incorrupted to ZnO, the (101) peak was shifted to a higher angle which is shown in figure 3 inset. The shift towards higher angle side usually occurs when a dopant with higher ionic radii is substituted in place of the host with lower ionic radii.

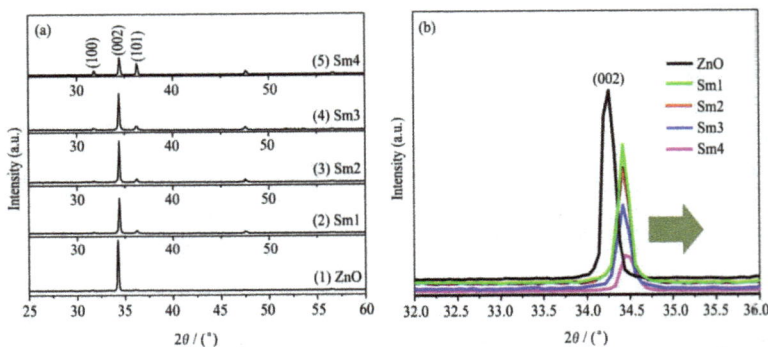

Figure 2. *(a) XRD patterns of undoped and Sm-doped ZnO nanorod arrays and (b) (002) peak of undoped and Sm-doped ZnO NR array, Reprinted and permission from, Ranjith kumar et al., Effect of samarium doping on structural, optical and magnetic properties of vertically aligned ZnO nanorod arrays, Journal of Rare Earths, 35, 10, 2017,1002.*

The analysis of the XRD peaks revealed a small shift towards higher angle side in the diffraction peaks for Er-doped ZnO (up to 3 % doping) and a shifted towards lower angle for 5 % concentration of Er. The shift in the diffraction peaks toward higher and lower angle side results in a change in the lattice parameter values. The change in lattice parameter values is because of the difference in ionic radii of Er and Zn ions. The reverse trend for higher doping concentration is attributed to the saturation limit of Er in ZnO lattice. At lower doping concentration (=3%), Er ions are going to the interstitial sites whereas, segregation/accumulation on the surface takes place for 5% Er concentration distorting the ZnO crystal structure [37].

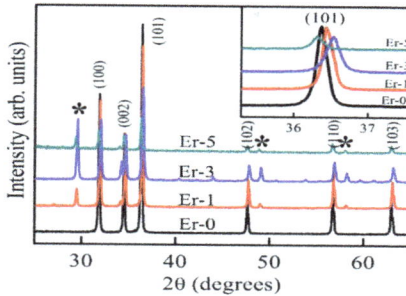

Figure 3. *XRD pattern of Er-doped ZnO (0 = x = 0.05). The inset shows the zoomed part of the most intense (101) peak and * indicates the Er2O related peak, Reprinted and permission from, Parmod Kumar et al., Understanding the origin of Ferromagnetism in Er doped ZnO System RSC Adv., 2016,6, 89242-89249.*

(ii) Optical properties

The UV-Vis spectrum is used to obtain the bandgap energy of ZnO and doped ZnO nanomaterials. Hastir et al. described that the bandgap energy of ZnO is decreased from 3.06 eV to 2.61eV with an increase in Tb^{3+} in ZnO. The narrowing of the bandgap is due to the formation of some defect level in between the conduction band and valence band due to doping [38]. Eskandarloo et al. reported that the band gap energy was decreased from 3.18 eV to 3.08 eV when increasing the Sm concentration from 0.3 mol % to 0.5 mol % in ZnO which is shown in figure 5. The results indicate that the doping induces the band gap narrowing in ZnO. This may be due to the charge transfer between the valence band and the conduction band of ZnO and f level of Sm3+ ion incorporated into the lattice of ZnO [39].

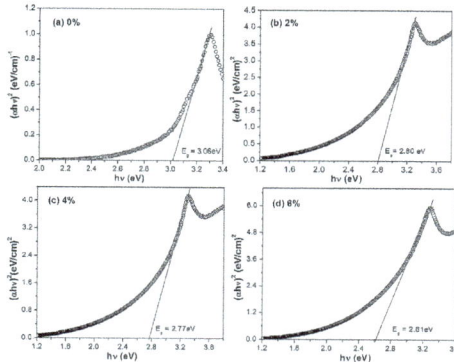

Figure. 4. *Tauc's Plot (showing the band gap of synthesized samples) of pure and Tb doped ZnO, Reprinted and permission from, Anita Hastir, Temperature dependent selective and sensitive terbium doped ZnO nanostructures, Sensors Actuators B 231 (2016) 110–119.*

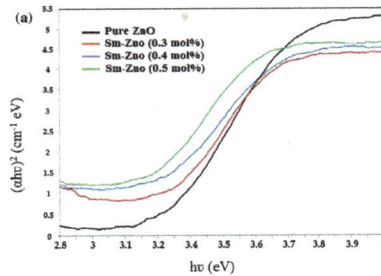

Figure 5. *UV-Vis DRS spectra of Pure and Sm-doped ZnO nanoparticles, Reprinted and permission from, Hamed Eskandarloo et al., Ultrasonic-assisted degradation of phenazopyridine with a combination of Sm-doped ZnO nanoparticles and inorganic oxidants, Ultrasonics Sonochemistry 28 (2016) 169–177.*

(iii) Photoluminance

To characterize the intrinsic defects in ZnO nanomaterial, room temperature PL measurements are performed. Hsu et al. reported that the photoluminescence properties of Pure ZnO and La-doped ZnO nanowire which is shown in figure 6. The ZnO exhibits a narrow emission peak near the band edge free exciton emission at about 390nm and a broad deep level emission at 460nm. The figure clearly shows strong peaks at about 380 nm and 378 nm with full widths at the half maxima of 14 and 12 nm, respectively. Both ZnO:La nanowire and ZnO nanowire samples undergo transitions close to the 3.3 eV ZnO band gap. Therefore, the two corresponding PL peaks were attributed to the recombination of free excitons through exciton-exciton collision. Also, PL peaks of the ZnO:La nanowires and ZnO nanowires were observed, along with strong and broad bands around 578-585 and 758-762 nm. The green and red emission band around at 2.14 eV (580 nm) and 1.63 eV (760 nm) originate from the oxygen vacancies and La impurities respectively, peak ratio of ~0.118 of ZnO:La nanowire was higher than the 762 nm/378 nm peak ratio ~0.078 of ZnO nanowire. This indicates that ZnO:La nanowire showed oxygen vacancies due to the La impurities. The oxygen vacancies and La impurities would induce the formation of new energy levels in the bandgap. Hence, the emission resulted from the radiative recombination of a photoexcited hole with an electron that occupies the oxygen vacancy and deep levels of the La impurity [40].

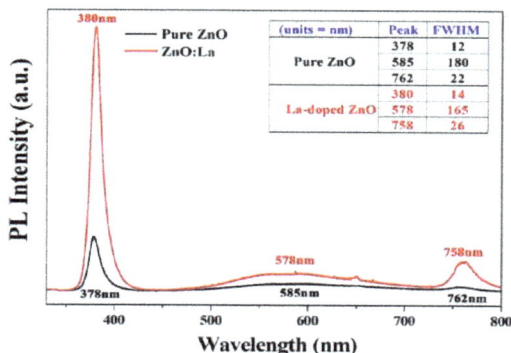

Figure 6. PL spectra of ZnO and La-doped ZnO, Reprinted and permission from, Cheng-Liang Hsu et al., Water- and Humidity-Enhanced UV Detector by Using p-Type La Doped ZnO Nanowires on Flexible Polyimide Substrate, ACS Appl. Mater. Interfaces 2013, 5, 11142-11151.

Dandan et al. stated that the PL spectra of pure and Eu doped ZnO nanowire under 325nm excitation is shown in figure 7, the pure ZnO showed a strong UV emission band at 380nm which is attributed to near band edge emission exciton recombination. The broad defect-related green emission band centered at 510nm has often been ascribed to the radioactive recombination of photogenerated holes with electron induced by oxygen vacancy. The origin of this green emission band has been the subject of a long-standing controversy and other sources such as Vzn and donor-acceptor pair recombination have also proposed [41].

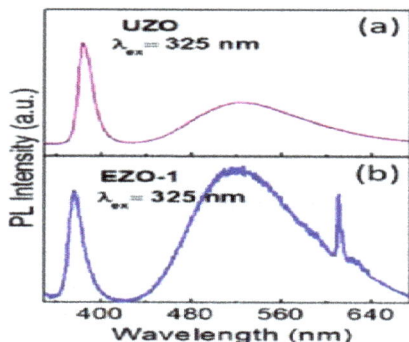

Figure 7. PL spectra of Pure ZnO and Eu-doped ZnO, Reprinted and permission from, Dandan Wang et al., Defects-Mediated Energy Transfer in Red-Light-Emitting Eu-Doped ZnO Nanowire Arrays, J. Phys. Chem. C 2011, 115, 22729–22735.

The energy transfer mechanism from the ZnO host to the Eu^{3+} ions which are illustrated in figure 8. In the bandgap excitation, the carrier relaxes to the band edge of the conduction band and valence band, where they are rapidly trapped at the defects or undergo subsequent band edge radiative emission. By means of a resonant energy transfer process, as shown in (figure 6), the trapped carriers at the oxygen vacancies could transfer their energy to the Eu^{3+} subsystem (i.e., 7F0 to 5D2). As a final step of the energetic process, Eu ions go through the radiative transition from 5D0 to 7F2 giving out of the red emission [41].

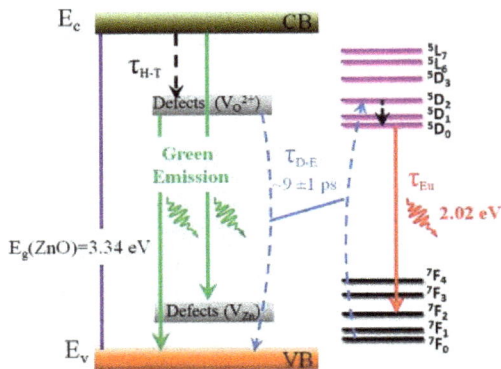

Figure 8. Schematic illustrating the proposed mechanism of energy transfer from the ZnO host to the Eu^{3+} ions. The dashed and the solid lines represent nonradiative and radiative processes, respectively. The green emission may be multi-origin in nature, and some of the possible radiative pathways are also indicated. Reprinted and permission from, Dandan Wang et al., Defects-Mediated Energy Transfer in Red-Light-Emitting Eu-Doped ZnO Nanowire Arrays, J. Phys. Chem. C 2011, 115, 22729–22735.

(iv) Magnetic property

The magnetic characterization that involves analysis of the irreversible component of magnetization. This analysis gives the activated magnetic moment upon the magnetization reversal by linking field and time changes in the irreversible component of magnetization. Subramanian et al. reported that magnetic behavior of ZnO and Gd-doped ZnO thin film which is shown in figure 9. The Vibrating sample measurement curve revealed that ZnO thin film showed a diamagnetic behavior. However, the introduction of Gd ion into ZnO exhibited ferromagnetic behavior. The ferromagnetic behavior decreased when increasing the Gd ion concentration due to antiferromagnetic alignment caused by the increased number of Gd ion occupying the adjacent position [42].

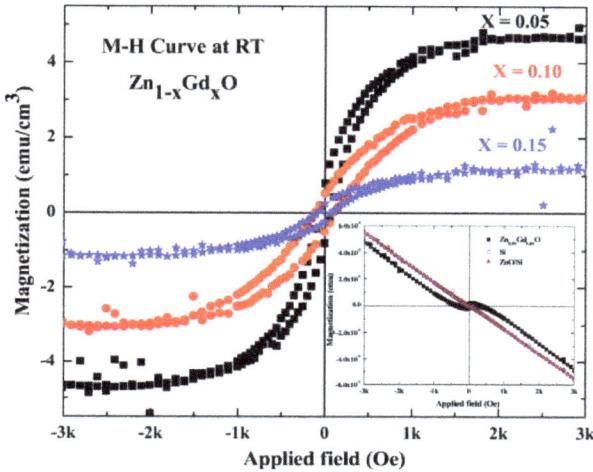

Figure 9. *Magnetic behavior of ZnO thin film and various Gd concentration and the field applied parallel to the sample plane. The inset provides magnetic data of Si, ZnO thin film and Zn0.95Gd0.05O film with the diamagnetic contribution, Reprinted and permission from, Subramanian et al., Intrinsic ferromagnetism and magnetic anisotropy in Gd-doped ZnO thin films synthesized by pulsed spray pyrolysis method, Journal of Applied Physics 108, 053904 (2010).*

Vijayaprasath et al. stated that the magnetic behavior of pure ZnO was diamagnetic and the magnetization was increased when increasing the concentration of a Nd^{3+} ion in ZnO nanoparticle which is shown in figure 10. The result showed that the diamagnetic behavior was tuned to ferromagnetic when increasing the concentration of Nd ion in ZnO nanoparticle which is due to the formation of defect like oxygen vacancy and Zn interstitial and is consistent with the bound magnetic polarons model [43]. According to the BMP model, bound electrons in defects like oxygen vacancies and zinc interstitial can couple Nd^{3+} ions and cause the ferromagnetic order of the samples.

Rare Earth - A tribute to the late Mr. Rare Earth, Professor Karl Gschneidner   Materials Research Forum LLC
Materials Research Foundations 164 (2024) 279-297   https://doi.org/10.21741/9781644903056-8

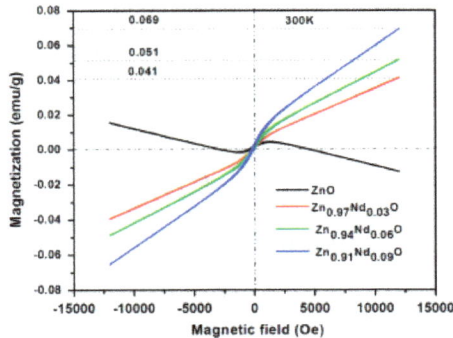

Figure 10. *Magnetic behavior of ZnO and Nd-doped ZnO nanoparticle, Reprinted and permission from, Vijayaprasath et al., Effect of Cobalt Doping on Structural, Optical, and Magnetic Properties of ZnO Nanoparticles Synthesized by Coprecipitation Method, J. Phys. Chem. C 2014, 118, 9715–9725.*

(v) Photocatalytic property

The major sources of pollution in water and air are chemicals released from industries. The textile industry is one of the major sources of the pollutants such as colored organic reagents; the dyes. The presence of such pollutants in ground and surface water are harmful to human as well as aquatic life. Some of them are carcinogenic and mutagenic as well as genotoxic and therefore, developing a technology for cleaning the contaminated water is very important. The utilization of semiconductor-based material as a photocatalyst in the detoxification of pollutants has several advantages over any other treatment method.

Khataee et al. described that a Dy-doped ZnO nanoparticle was used for the photocatalytic activity of Acid red 17 solutions under visible-light irradiation. Figure 11 shows that the degradation efficiency of pure ZnO nanoparticles was 25% and increased the degradation efficiency by concentration up to 3%. After that, efficiency decreased due to the increasing amount of Dy resulting in a higher surface barrier and narrow space charge region, leading to efficient separation of the produced electron-hole pair. Increasing the amount of Dy up to a specific value results in exceeding the space charge layer by increasing the penetration depth of the visible light into ZnO nanoparticles. This is due to the recombination of electron-hole pairs easier, causing low photocatalytic activity. In addition, the excess amount of dopant covering the surface of ZnO nanoparticles leads to a decrease in the photocatalytic activity of the photocatalyst due to an increase in the number of electron-hole recombination centers [44].

Rare Earth - A tribute to the late Mr. Rare Earth, Professor Karl Gschneidner   Materials Research Forum LLC
Materials Research Foundations 164 (2024) 279-297   https://doi.org/10.21741/9781644903056-8

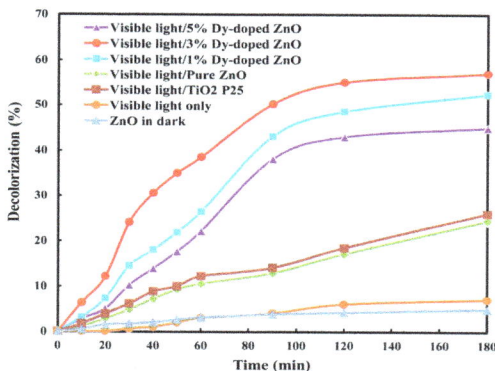

Figure 11. *Photodegradation curves of porous ZnO and Dy doped ZnO with different concentration, Reprinted and permission from, Alireza Khataee et al., Synthesis and Characterization of Dysprosium-Doped ZnO Nanoparticles for Photocatalysis of a Textile Dye under Visible Light Irradiation, Ind. Eng. Chem. Res. 2014, 53, 1924–1932.*

Ranjith Kumar et al. reported the photodegradation of methylene blue dye under visible light irradiation. Figure 12 shows that the degradation efficiency was increased from 83.02% to 97.3% for pure ZnO and up to Sm3 and decreased 95.72% for a higher concentration of Sm-doped ZnO nanorod arrays. This is due to the introduction of Sm^{3+} impurities which led to the effective carrier separation under photo-irradiation. However, the excessive space charge layer thickness could promote the recombination rate in the higher doping concentration due to the enhanced penetration depth of the light into the ZnO nanorod arrays. On increasing the recombination, it delays the carrier separation efficiency and decreases the degradation rate. The loading of impurities or Sm ions has an optimum level which offers the thickness of the space charge layer equal to light penetration depth which enhances the carrier separation rate and improves the catalytic efficiency. This kind of trend was observed in many rare earth-doped semiconductors including Sm ions. But increasing the doping concentration to 8 wt % of Sm, initiated the disturbance in alignment of the 1D arrays which affect the carrier response. The photogenerated charge transfer efficiency was influenced by the Sm doping, and it evidences the promotion of effective carrier separation which enhances the catalytic performances of Sm/ZnO photocatalysts [24].

Rare Earth - A tribute to the late Mr. Rare Earth, Professor Karl Gschneidner   Materials Research Forum LLC
Materials Research Foundations 164 (2024) 279-297   https://doi.org/10.21741/9781644903056-8

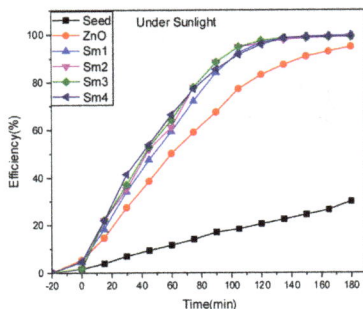

Figure 12. *Photodegradation efficiency of ZnO nanorod arrays and Sm-doped ZnO nanorod arrays with different concentration, Reprinted and permission from, Ranjith kumar et al., Effect of samarium doping on structural, optical and magnetic properties of vertically aligned ZnO nanorod arrays, Optik 154 (2018) 115–125.*

Sm doping has influences on enhancing the photocatalytic efficiency of ZnO NR arrays, as discussed below. This notable enhancement is due to the presence of Sm impurities which are responsible for electron transfer and oxygen-related defects responsible for holes transformation. Through the doping process, the oxygen defect site was enhanced which favors the effective separation of photogenerated charge carriers and prolong the life of the carriers and eventually enhance the catalytic activity [45]. The band structure in scheme 1 illustrates that much intermediate energies formed because of the deficiency which will significantly increase the light absorption in the visible range. As suggested by the previous investigations, the presence of oxygen vacancies was improvised the visible catalytic efficient. Further, Sm dopant impurities play a trap state and promote the carriers to the surface and enhance the catalytic efficiency. Naturally, the presence of Sm^{3+} ions was reacted as a trap states host lattice, which favors the reduction of Sm^{2+} ions due partially filled the f-orbitals [46]. In general, the unstable nature of Sm^{2+} which possess the 6f electron was easily trapped and promotes the formation of O_2^- which later converted to the active OH radical. This behavior evidences the effective contribution of Sm^{3+} dopant as effective carrier trap which promotes the photocatalytic efficiency. The catalytic mechanism of Sm^{3+} doped ZnO nanostructures represented by the following equations [47].

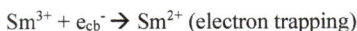

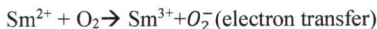

$ZnO + hv \rightarrow ZnO\ (e_{cb}^- + h_{vb}^+)$

$Sm^{3+} + e_{cb}^- \rightarrow Sm^{2+}$ (electron trapping)

$Sm^{2+} + O_2 \rightarrow Sm^{3+} + O_2^-$ (electron transfer)

$O_2^- + h^+ \rightarrow .OOH$

$.OOH + h^+ + e_{cb}^- \rightarrow H_2O_2$

$H_2O_2 + e_{cb}^- \rightarrow .OH + OH^-$

Further, separated photogenerated hole in the catalyst surface was interact with the surface bonded hydroxide species and favor to the production of active OH. radicals.

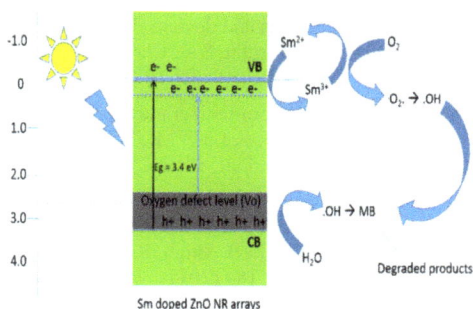

Scheme 1. *Band structure of Sm-doped ZnO NR arrays as photocatalysts under illumination with photon energy, Reprinted and permission from, Ranjith kumar et al., Effect of samarium doping on structural, optical and magnetic properties of vertically aligned ZnO nanorod arrays, Optik 154 (2018) 115–125.*

Conclusion

Rare earth metals-doped ZnO has attracted considerable interest over the last few decades providing scope for improving the optical, magnetic, and photocatalytic properties. The improvements in the functional properties of ZnO are achieved at very low rare earth metals loadings. Although significant work has been carried out on the synthesis and properties of rare earth metals-doped ZnO nanomaterials, effort is still needed to determine the interrelationships between the processing and functional properties of doped nanomaterials. The results showed that after doping, the absorption edge of ZnO shifted to the visible region as a result of a decrease in the bandgap energy due to the formation of a new energy level in the bandgap. The dopant also acts as a trapping center for the electron and holes, thereby reducing the recombination rate of charge carriers and increasing the photocatalytic efficiency and magnetic property. Understanding the relationship between the dopant and ZnO, the level of doping, and structural changes associated with doping will be very helpful for optimizing the ultimate properties of rare metals-doped ZnO. This chapter is expected to help readers with a wide range of backgrounds to understand the impact of various synthetic methods for the preparation of doped ZnO as well as dopants on the properties of ZnO.

Reference

[1] Recent advances in ZnO materials and devices, D.C. Look, Materials Science and Engineering B80 (2001) 383-387. https://doi.org/10.1016/S0921-5107(00)00604-8

[2] Fabrication and ethanol sensing characteristics of ZnO nanowire gas sensors, Q. Wan, Q. H. Li, Y. J. Chen, and T. H. Wang, Appl. Phys. Lett., Vol. 84, No. 18, 3 May 2004. https://doi.org/10.1063/1.1738932

[3] A comprehensive review of ZnO materials and devices, Ü. Özgür, Ya. I. Alivov, C. Liu, A. Teke, M. A. Reshchikov, S. Doğan, V. Avrutin, S.-J. Cho, and H. Morkoç, Journal of Applied Physics 98, 041301 (2005). https://doi.org/10.1063/1.1992666

[4] Evidence of oxygen vacancy enhanced room-temperature ferromagnetism in Co-doped ZnO, H. S. Hsu, J. C. A. Huang, Y. H. Huang, Y. F. Liao, M. Z. Lin, C. H. Lee, J. F. Lee, S. F. Chen, L. Y. Lai, and C. P. Liu, Appl. Phys. Lett. 88, 242507 (2006). https://doi.org/10.1063/1.2212277

[5] Room temperature ferromagnetism in undoped and Fe doped ZnO nanorods: Microwave-assisted synthesis, Mukta V. Limaye, Shashi B. Singh, Raja Das, Pankaj Poddar, Sulabha K. Kulkarni, Journal of Solid State Chemistry 184 (2011) 391-400. https://doi.org/10.1016/j.jssc.2010.11.008

[6] Synthesis, structural and optical properties of ZnO and Ni-doped ZnO hexagonal nanorods by Co-precipitation method, K. Raja, P.S. Ramesh b, D. Geetha, Spectrochimica Acta Part A: Molecular and Biomolecular Spectroscopy 120 (2014) 19-24. https://doi.org/10.1016/j.saa.2013.09.103

[7] Electric-Field Control of Ferromagnetism in Mn-Doped ZnO Nanowires, Li-Te Chang, Chiu-Yen Wang, Jianshi Tang, Tianxiao Nie, Wanjun Jiang, Chia-Pu Chu, Shamsul Arafin, Liang He, Manekkathodi Afsal, Lih-Juann Chen, and Kang L. Wang, Nano Lett. 2014, 14, 1823−1829. https://doi.org/10.1021/nl404464q

[8] Enhanced photocatalytic activity of Co doped ZnO nanodisks and nanorods prepared by a facile wet chemical method, Sini Kuriakose, Biswarup Satpati and Satyabrata Mohapatra, Phys. Chem. Chem. Phys., 2014, 16, 12741. https://doi.org/10.1039/c4cp01315h

[9] Cr-Doped ZnO Nanoparticles: Synthesis, Characterization, Adsorption Property, and Recyclability, Alan Meng, Jing Xing, Zhenjiang Li, and Qingdang Li, ACS Appl. Mater. Interfaces 2015, 7, 27449−27457. https://doi.org/10.1021/acsami.5b09366

[10] Rare Earth Doped Zinc Oxide Nanophosphor Powder: A Future Material for Solid State Lighting and Solar Cells, Vinod Kumar, O. M. Ntwaeaborwa, T. Soga, Viresh Dutta, and H. C. Swart, ACS Photonics 2017, 4, 2613-2637. https://doi.org/10.1021/acsphotonics.7b00777

[11] Optical and structural properties of Eu-diffused and doped ZnO nanowires, C.J. Pan, C.W. Chen, J.Y. Chen, P.J. Huang, G.C. Chi, C.Y. Chang, F. Ren, S.J. Pearton, Applied Surface Science 256 (2009) 187-190. https://doi.org/10.1016/j.apsusc.2009.07.108

[12] Photoluminescent properties of Sm3+-doped zinc oxide nanostructures, S.K. Lathika Devi n, K. Sudarsanakumar, Journal of Luminescence 130 (2010) 1221-1224. https://doi.org/10.1016/j.jlumin.2010.02.028

[13] Room temperature enhanced red emission from novel Eu3+ doped ZnO nanocrystals uniformly dispersed in nanofibers, Yongzhe Zhang, Yanxia Liu, Xiaodong Li, Qi JieWang and Erqing Xie, Nanotechnology 22 (2011) 415702. https://doi.org/10.1088/0957-4484/22/41/415702

[14] Robust Room-Temperature Ferromagnetism with Giant Anisotropy in Nd-Doped ZnO Nanowire Arrays, Dandan Wang, Qian Chen, Guozhong Xing, Jiabao Yi, Saidur Rahman Bakaul, Jun Ding,

Rare Earth - A tribute to the late Mr. Rare Earth, Professor Karl Gschneidner   Materials Research Forum LLC
Materials Research Foundations 164 (2024) 279-297   https://doi.org/10.21741/9781644903056-8

a. Jinlan Wang, and Tom Wu, Nano Lett. 2012, 12, 3994–4000.
https://doi.org/10.1021/nl301226k

[15] Enhanced acetone gas-sensing performance of La2O3-doped flowerlike ZnO

a. structure composed of nanorods, Jian-Qun He, Jing Yin, Dong Liu, Le-Xi Zhang, Feng-Shi
Cai, Li-Jian Bie, Sensors and Actuators B 182 (2013) 170- 175.
https://doi.org/10.1016/j.snb.2013.02.085

[16] Strong correlation between oxygen vacancy and ferromagnetism in Yb-doped ZnO thin
films, Fei Li, Xue-Chao Liu, Ren-Wei Zhou, Hong-Ming Chen, Shi-Yi Zhuo, and Er-Wei Shi,
J. Appl. Phys. 116, 243910 (2014). https://doi.org/10.1063/1.4905240

[17] Preparation of porous 3D Ce-doped ZnO microflowers with enhanced photocatalytic
performance, Yimai Liang, Na Guo, Linlin Li, Ruiqing Li, Guijuan Jia and Shucai Gan, RSC
Adv., 2015,5, 59887-59894. https://doi.org/10.1039/C5RA08519E

[18] Ceria Doped Zinc Oxide Nanoflowers Enhanced Luminol-Based Electrochemiluminescence
Immunosensor for Amyloid β Detection, Jing-Xi Wang, Ying Zhuo, Ying Zhou, Hai-Jun
Wang, Ruo Yuan and Ya-Qin Chai, ACS Appl. Mater. Interfaces 2016, 8, 12968–12975.
https://doi.org/10.1021/acsami.6b00021

[19] The study of structural and optical properties of (Eu, La, Sm) codoped ZnO nanoparticles
via a chemical route, Jihui Lang, Qi Zhang, Qiang Han, Yue Fang, JiayingWang, Xiuyan Li,
Yanqing Liu, Dandan Wang, Jinghai Yang, Materials Chemistry and Physics 194 (2017)
29e36. https://doi.org/10.1016/j.matchemphys.2017.03.010

[20] ZnO nanorods: synthesis, characterization and applications, Gyu-Chul Yi, Chunrui Wang
and Won Il Park, Semicond. Sci. Technol. 20 (2005) S22-S34. https://doi.org/10.1088/0268-
1242/20/4/003

[21] Synthesis and characterization of ZnO nanorods with a narrow size distribution,
Chandrakanth Reddy Chandraiahgari, Giovanni De Bellis, Paolo Ballirano, Santosh Kiran
Balijepalli, Saulius Kaciulis, Luisa Caneve, Francesca Sarto and Maria Sabrina Sarto, RSC
Adv., 2015,5, 49861-49870. https://doi.org/10.1039/C5RA02631H

[22] Controlled Defects of Zinc Oxide Nanorods for Efficient Visible Light Photocatalytic
Degradation of Phenol, Jamal Al-Sabahi, Tanujjal Bora, Mohammed Al-Abri and Joydeep
Dutta, Materials 2016, 9, 238. https://doi.org/10.3390/ma9040238

[23] Influence of Defects on the Photocatalytic Activity of ZnO, Daimei Chen, Zhihong Wang,
Tiezhen Ren, Hao Ding, Wenqing Yao, Ruilong Zong, and Yongfa Zhu, | J. Phys. Chem. C
2014, 118, 15300–15307. https://doi.org/10.1021/jp5033349

[24] Structural, optical, photocurrent and solar driven photocatalytic properties of vertically
aligned samarium doped ZnO nanorod arrays, D. Ranjith Kumar, K.S. Ranjith, R.T. Rajendra
Kumar, Optik 154 (2018) 115-125. https://doi.org/10.1016/j.ijleo.2017.10.004

[25] Effect of annealing temperature on the energy transfer in Eu-doped ZnO nanoparticles by
chemical precipitation method, Jihui Lang, Qiang Han, Xue Li, Songsong Xu, Jinghai Yang,
Lili Yang, Yongsheng Yan, Xiuyan Li, Yingrui Sui, Xiaoyan Liu, Jian Cao, Jian Wang, J
Mater Sci: Mater Electron (2013) 24:4542-4548. https://doi.org/10.1007/s10854-013-1439-0

[26] Influence of Rare Earth (Nd and Tb) Co-Doping on ZnO Thin Films Properties, Amina El Fakir, Mouaad Sekkati, Guy Schmerber, Azzam Belayachi, Zineb Edfouf, Mohammed Regragui, Fouzia Cherkaoui El Moursli, Zouheir Sekkat, Aziz Dinia, Abdelilah Slaoui, and Mohammed Abd-Lefdil, Phys. Status Solidi C 2017, 14, 1700169.

[27] Hydrothermal fabrication of natural sun light active Dy2WO6 doped ZnO and its enhanced photoelectrocatalytic activity and self-cleaning properties, Kuppulingam Thirumalai, Manohar Shanthi and Meenakshisundaram Swaminathan, RSC Adv., 2017, 7, 7509. https://doi.org/10.1039/C6RA24843H

[28] Defect-band mediated ferromagnetism in Gd-doped ZnO thin films, S. Venkatesh, J. B. Franklin, M. P. Ryan, J.-S. Lee, Hendrik Ohldag, M. A. McLachlan, N. M. Alford, and I. S. Roqan, Journal of Applied Physics 117, 013913 (2015). https://doi.org/10.1063/1.4905585

[29] Effect of electron beam rapid thermal annealing on crystallographic, structural and magnetic properties of Zn1-xSmxO thin films, Anuraj Sundararaj, Gopalakrishnan Chandrasekaran, Helen Annal Therese, Arumugam Sonachalam, Karthigeyan Annamalai, Journal of Magnetism and Magnetic Materials 378 (2015) 112-117. https://doi.org/10.1016/j.jmmm.2014.10.169

[30] Excellent acetone sensor of La-doped ZnO nanofibers with unique bead-like structures, X.L. Xu, Y. Chen, S.Y. Ma, W.Q. Li, Y.Z. Mao, Sensors and Actuators B 213 (2015) 222-233. https://doi.org/10.1016/j.snb.2015.02.073

[31] Influence of Dy dopant on structural and photoluminescence of Dy-doped ZnO nanoparticles, C. Jayachandraiah, K. Siva Kumar, G. Krishnaiah, N. Madhusudhana Rao, Journal of Alloys and Compounds 623 (2015) 248-254. https://doi.org/10.1016/j.jallcom.2014.10.067

[32] Preparation of porous 3D Ce-doped ZnO microflowers with enhanced photocatalytic performance, Yimai Liang, Na Guo, Linlin Li, Ruiqing Li, Guijuan Ji and Shucai Gan, RSC Adv., 2015,5, 59887-59894. https://doi.org/10.1039/C5RA08519E

[33] Effect of Eu3+ on the structure, morphology and optical properties of flower-like ZnO synthesized using chemical bath deposition, L.F. Koao, F.B. Dejene, R.E. Kroon, H.C. Swart, Journal of Luminescence 147 (2014) 85-89. https://doi.org/10.1016/j.jlumin.2013.10.045

[34] Highly active lanthanum doped ZnO nanorods for photodegradation of metasystox, P.V. Korake, R.S. Dhabbe, A.N. Kadam, Y.B. Gaikwad, K.M. Garadkar, Journal of Photochemistry and Photobiology B: Biology 130 (2014) 11-19. https://doi.org/10.1016/j.jphotobiol.2013.10.012

[35] Microstructural and photoluminescence properties of sol-gel derived Tb3+ doped ZnO nanocrystals, Guy L. Kabongo, Gugu H. Mhlongo, Thomas Malwela, Bakang M. Mothudi, Kenneth T. Hillie, Mokhotjwa S. Dhlamini, Journal of Alloys and Compounds 591 (2014) 156-163. https://doi.org/10.1016/j.jallcom.2013.12.075

[36] Effect of samarium doping on structural, optical and magnetic properties of vertically aligned ZnO nanorod arrays, D. Ranjith Kumar, K.S. Ranjith, L.R. Nivedita, R.T. Rajendra Kumar, Journal of Rare Earths, Vol. 35, No. 10, Oct. 2017, P. 1002. https://doi.org/10.1016/S1002-0721(17)61005-6

[37] Understanding the origin of Ferromagnetism in Er doped ZnO System, Parmod Kumar, Vikas Sharma, Ankita Sarwa, Ashish Kumar, Surbhi, Rajan Goyal, K. Sachdev, S. Annapoorni, K. Asokan and D. Kanjila, RSC Adv., 2016,6, 89242-89249. https://doi.org/10.1039/C6RA17761A

[38] Temperature dependent selective and sensitive terbium doped ZnO nanostructures, Anita Hastir, Nipin Kohli, Ravi Chand Singh, Sensors and Actuators B 231 (2016) 110-119. https://doi.org/10.1016/j.snb.2016.03.001

[39] Ultrasonic-assisted degradation of phenazopyridine with a combination of Sm-doped ZnO nanoparticles and inorganic oxidants, Hamed Eskandarloo, Alireza Badiei, Mohammad A. Behnajady, Ghodsi Mohammadi Ziarani, Ultrasonics Sonochemistry 28 (2016) 169-177. https://doi.org/10.1016/j.ultsonch.2015.07.012

[40] Water- and Humidity-Enhanced UV Detector by Using p Type La Doped ZnO Nanowires on Flexible Polyimide Substrate, Cheng-Liang Hsu, Hsieh-Heng Li and Ting-Jen Hsueh, ACS Appl. Mater. Interfaces 2013, 5, 11142−11151. https://doi.org/10.1021/am403364r

[41] Defects-Mediated Energy Transfer in Red-Light-Emitting Eu-Doped ZnO Nanowire Arrays, Dandan Wang, Guozhong Xing, Ming Gao, Lili Yang, Jinghai Yang, and Tom Wu, J. Phys. Chem. C 115, 46, 22729-22735. https://doi.org/10.1021/jp204572v

[42] Intrinsic ferromagnetism and magnetic anisotropy in Gd-doped ZnO thin films synthesized by pulsed spray pyrolysis method, M. Subramanian, P. Thakur, M. Tanemura, T. Hihara, V. Ganesan, T. Soga, K. H. Chae, R. Jayavel, and T. Jimbo, Journal of Applied Physics 108, 053904 (2010). https://doi.org/10.1063/1.3475992

[43] Effect of Cobalt Doping on Structural, Optical, and Magnetic Properties of ZnO Nanoparticles Synthesized by Coprecipitation Method, Vijayaprasath Gandhi, Ravi Ganesan, Haja Hameed Abdulrahman Syedahamed, and Mahalingam Thaiyan, J. Phys. Chem. C 2014, 118, 9715−9725. https://doi.org/10.1021/jp411848t

[44] Synthesis and Characterization of Dysprosium-Doped ZnO Nanoparticles for Photocatalysis of a Textile Dye under Visible Light Irradiation, Alireza Khataee, Reza Darvishi Cheshmeh Soltani, Younes Hanifehpour, Mahdie Safarpour, Habib Gholipour Ranjbar, and Sang Woo Joo, Ind. Eng. Chem. Res. 2014, 53, 1924−1932. https://doi.org/10.1021/ie402743u

[45] J.C. Sin, S.M. Lam, K.T. Lee, A.R. Mohamed, J. Colloid. Inter. Sci. 401, 40 (2013), J.C. Sin, https://doi.org/10.1016/j.jcis.2013.03.043

[46] S.M. Lam, K.T. Lee, A.R. Mohamed, Ceram. Inter. 39, 5833 (2013). https://doi.org/10.1016/j.ceramint.2013.01.004

[47] Facile Synthesis and Enhanced Photocatalytic Performance of Flower-like ZnO Hierarchical Microstructures, Benxia Li and Yanfen Wang, J. Phys. Chem. C 2010, 114, 890-896. https://doi.org/10.1021/jp909478q

Rare Earth - A tribute to the late Mr. Rare Earth, Professor Karl Gschneidner   Materials Research Forum LLC
Materials Research Foundations 164 (2024) 298-342   https://doi.org/10.21741/9781644903056-9

Chapter 9

The Role of Rare Earths in Fluid Catalytic Cracking (FCC) Catalyst

Aaron Akah

R&D Center, Saudi Aramco, Dhahran 31311, Saudi Arabia

aaron.akah@aramco.com

Abstract

Rare earth oxides enhance catalyst activity and prevent the loss of acid sites during the FCC unit operation, especially when feed with high metal content is used. This chapter reviews the effects of rare earth elements on the structure, activity, and stability of FCC catalysts. It also looks into the mechanism of catalyst deactivation by vanadium and how rare earths are used to combat this. The objective is to elucidate the interaction of vanadium species with the zeolite component of the FCC catalysts and to show the role of rare earth elements in countering the deleterious effects of vanadium. The recycle of spent FCC catalyst with a focus on rare earth element recovery is also outlined.

Keywords

FCC Catalyst, Rare Earth, Vanadium Trap, Zeolites

Contents

1.   Introduction

Rare earth elements (also known as the lanthanide series in the periodic table of elements) are a series of chemical elements found in the Earth's crust that are applied in many modern technologies, such as consumer electronics, computers and networks, communications, clean energy, advanced transportation, health care, environmental mitigation, national defense, and many others [1-5]. The introduction of the Welsbach incandescent lamp which made use of the oxides of zirconium, lanthanum, and yttrium during the 1880s marked the first commercial application of rare earths [1]. Since then, rare earths have found applications in various fields and their consumption has grown to over 100,000 tons annually as shown in Table 1.

Rare earths form a critical and essential part of many modern technologies as they sometimes act like technology building blocks. This is because their application in alloys and compounds can have a profound effect on the performance of complex engineered systems, some of which include automotive catalytic converters, petroleum refining catalysts, glass manufacture and polishing, ceramics, permanent magnets, metallurgical additives and alloys, and rare earth phosphors for lighting, television, computer monitors, radar and X-ray intensifying film, among a myriad of applications [1, 2, 6-8].

The demand for rare earth elements is a direct result of their applications in the end-use products, such as flat panel displays and catalysts. Catalysts represent a large market for rare earths where they provide properties desired for effective catalysis such as in FCC and in automotive catalysts [1, 9].

2.   Rare earths in FCC catalysts

Rare earth oxides have been widely investigated in catalysis as structural and electronic promoters to improve the activity, selectivity and thermal stability of catalysts [1, 2, 10-29]. Catalysis plays a major role in the consumption of rare earths with the FCC playing a major role in the catalytic application of rare earths [9]. Rare earth application in FCC emerged in the early 1960s when zeolites were introduced as cracking catalysts for oil refining. The use of rare earths can help preserve catalyst effectiveness and increase the yield of the gasoline fractions by cracking the heavier oil fractions. Rare earths, such as lanthanum, are used in FCC catalysts to refine crude oil into gasoline, distillates, lighter oil products and other fuels.

2.1   FCC technology

Fluid Catalytic Cracking (FCC) is a process for conversion of heavy oil fractions into high octane gasoline, light fuel oils and olefin-rich light gases. The process employs a catalyst in the form of microspheres, which behave as a fluid when aerated. The fluidized catalyst is continuously circulated from a reaction zone, where the cracking reactions occur, to a regeneration zone where the catalyst is reactivated as illustrated in Figure 1, based on a modified schematic by Moulijn et al [30].

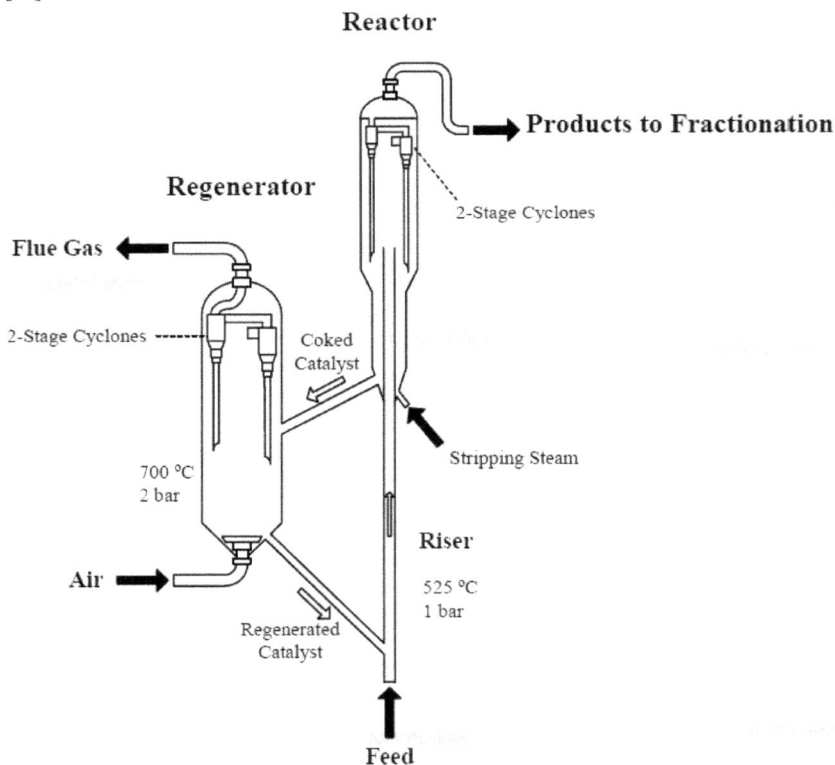

Figure 1. Schematic of a conventional FCC Unit

During the FCC process, feedstock is injected into the riser section of the FCC reactor, where it is cracked into lighter, more valuable products upon contacting hot catalyst circulated to the riser-reactor from a catalyst regenerator [31]. The catalyst and hydrocarbon vapors are carried up the riser to the disengagement section of the FCC reactor, where they are separated. Subsequently, the catalyst flows into a stripping section, where the hydrocarbon vapors entrained with the catalyst are stripped by steam injection. Following removal of occluded hydrocarbons from the spent cracking catalyst, the stripped catalyst flows through a spent catalyst standpipe and into a catalyst regenerator. As the endothermic cracking reactions take place, carbon is deposited onto the catalyst. This carbon, known as coke, reduces the activity of the catalyst and the catalyst must be regenerated .

Typically, catalyst is regenerated by introducing air into the regenerator and burning off the coke to restore catalyst activity. Coke combustion is an exothermic reaction and it is used to supply heat to the regenerated catalyst. The hot, reactivated catalyst flows through the regenerated catalyst standpipe back to the riser to complete the catalyst cycle. The coke combustion exhaust gas stream rises to the top of the regenerator and leaves the regenerator as flue gas. Flue gas generally contains nitrogen oxides (NO_x), sulfur oxides (SO_x), carbon monoxide (CO), oxygen (O_2), HCN or ammonia, nitrogen and carbon dioxide (CO_2).

In addition to providing the catalytic action, the catalyst is also the vehicle for the transfer of heat from the regeneration to the reaction zone. Catalyst performance forms an integral part of the techno-economic evaluation of the catalytic cracking process, because it affects the capital cost in terms of amount required, and the quantity and quality of the reaction products generated.

One of the strengths of the FCC process is its versatility to produce a wide variety of yield patterns by adjusting the basic operating parameters [32]. While most units have been designed mainly for gasoline production, by making appropriate adjustments and depending on the intended outcome, it is possible to operate the FCC in three modes:

Gasoline Mode

This is the most common mode of operation of the FCC unit and it is aimed at the maximum production of gasoline. This condition requires careful control of reaction severity, which must be high enough to convert a substantial portion of the feed, but not so high as to destroy the gasoline that has been produced. This balance normally is achieved by using a very active and selective catalyst and enough reaction temperature to produce the desired octane number. The catalyst-circulation rate is limited, and reaction time is confined to a very short exposure. Since this severity is carefully controlled, there is normally no need for any recycle of unconverted components.

Distillate Mode

If the reaction severity is strictly limited, the FCC unit can then be used for the production of distillates. By a change in operating conditions, a shift from the normally gasoline-oriented yield distribution to one with a more nearly equal ratio of gasoline-to-cycle oil can be accomplished. Additional distillates can be produced at the expense of gasoline by reducing the end point of the gasoline and dropping the additional material into the light-cycle-oil product. The usual limitation in this step is reached when the resulting cycle oil reaches a particular flash-point specification.

Petrochemical Mode

If reaction severity is increased, an operation producing additional light olefins and a higher octane gasoline will result. Severity may be increased by increasing reactor temperature, catalyst/oil ratio, or both. This case is sometimes described as a liquefied petroleum gas (LPG) mode or as a petrochemical FCC because of the increased quantity of light material that is produced and the

Rare Earth - A tribute to the late Mr. Rare Earth, Professor Karl Gschneidner   Materials Research Forum LLC
Materials Research Foundations 164 (2024) 298-342   https://doi.org/10.21741/9781644903056-9

increased aromatics in the gasoline product. Operating the FCC in petrochemical mode is aimed at increasing propylene yield in the product slate.

By taking into consideration the operating conditions and yields of the FCC, the propylene yield pattern can be represented in the form of a continuum varying from operating severity to process design and these can be optimized to suit the refinery specific economics [33]. The optimum process design provides refiners with the flexibility to move up or down the optimal economic range of the propylene yield curve as shown in Figure 2.

From Figure 2, it can be seen that higher propylene production comes at the expense of gasoline. For traditional refiners, maximizing gasoline yield is more important than the propylene yield, while for those interested in petrochemical applications, the target is operating at maximum propylene yield.

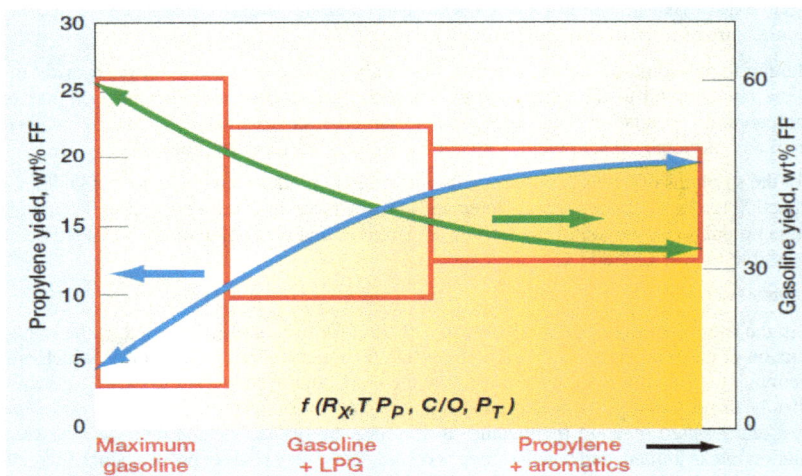

Figure 2: FCC design and operating modes [33].

Many FCC processes increase propylene by manipulating FCC reaction variables such as catalyst to oil (C/O) ratios, residence times and reaction temperatures [32]. The modifications can be put into two categories: Up Flow (Riser) and Down Flow (Downer) technologies. In the riser reactors, solid catalyst and hydrocarbon vapors flow upwards against gravity. This upward flow results in a catalyst flow that is significantly slower than the lighter hydrocarbons leading to back mixing of the catalyst and as a result there is an increase in residence time of the catalyst. This in turn can lead to undesirable secondary reactions leading to over cracking. The illustration of the flow pattern in the riser and downer is shown in Figure 3 [32].

In contrast to risers, and to overcome the issues related to back mixing, the downer reactor was developed as illustrated in Figure 3. The flow of the catalyst and the feed is in the direction of

Rare Earth - A tribute to the late Mr. Rare Earth, Professor Karl Gschneidner   Materials Research Forum LLC
Materials Research Foundations 164 (2024) 298-342   https://doi.org/10.21741/9781644903056-9

gravity and as such, back mixing is largely avoided and there is an even distribution of catalyst with an effective contact time of catalyst and feed less than that of the riser.

Figure 3: Illustration of flow in riser and downer FCC [32].

Most FCC catalysts are used in processes involving high severity operation such as high temperatures and steam, hydrothermal deactivation is an inherent problem that has to be mitigated. FCC catalysts operate between moderate to high temperatures (500–800 °C), in the presence of steam, especially during the regeneration step. These severe conditions strongly influence the performance of the catalysts because catalyst regeneration in the FCC usually takes place at temperatures as high as 800 °C and in the presence of presence of steam. Thus, thermal and hydrothermal stability of zeolites are among the most important parameters for FCC catalysts.

2.2   FCC catalyst composition

One of the factors that affect the design and operation of an FCC unit is the type of catalyst to be employed in the process. Most FCC catalysts consist of an active component (zeolite), a matrix such as amorphous silica-alumina (which also provides catalytic sites and larger pores), a binder (such as betonite clay) and filler as illustrated in Figure 4 [34]. It is composed of spherical particles, suitable for application in a fluidized circulating reactor, in which the zeolite crystals are dispersed in an active matrix of alumina or silica-alumina together with clay particles. The spherical particles contain large voids and pores necessary for allowing the mass transport of the heavy feedstock.

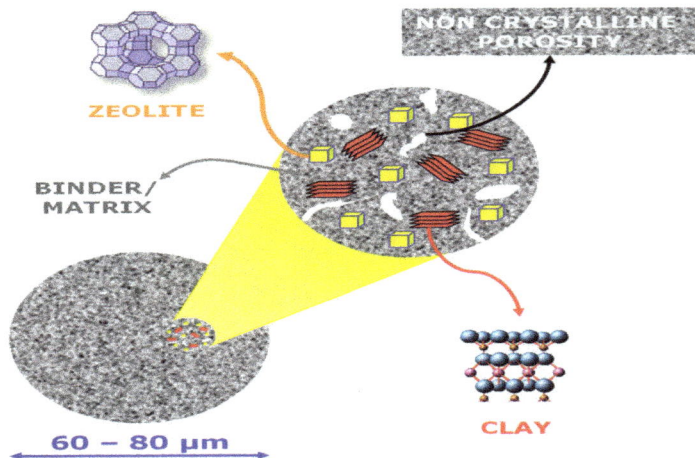

Figure 4: Schematic representation of FCC catalyst [34].

Ultra-stabilized zeolite Y (USY) is used as the main active zeolite in today's conventional FCC process. This material contains an internal porous structure in which acid sites are present, which can convert larger molecules to the desired gasoline range molecules. The matrix of an FCC catalyst serves both physical and catalytic functions [34-41]. Physical functions include providing particle integrity and attrition resistance, acting as a heat transfer medium, and providing a porous structure to allow diffusion of hydrocarbons into and out of the catalyst microspheres [18, 35-37, 39, 41, 42]. The matrix can also affect catalyst selectivity, product quality and resistance to poisons. Various alumina and silica sources are used to produce a meso- and macroporous matrix that allows access to, and pre-cracks the larger molecules in the feedstocks. In addition, these components are used to bind the system together. Additional components may comprise rare earth metals or specific metal traps for trapping Ni and V. The components are typically mixed in aqueous slurry, and then spray-dried to form more or less uniform spherical particles that can be fluidized in the regenerator.

The arrangement of an FCC catalyst particle gives rise to a hierarchical pore architecture spanning from the macro- to meso- to microporosity as shown in Figure 5 [43]. Each of these classes of pores has a defined role in the entire catalytic process. According to this scheme, the transformation of heavy molecules to valuable products (gas-oil and gasoline) occurs in the meso- and micropores [43-55].

Rare Earth - A tribute to the late Mr. Rare Earth, Professor Karl Gschneidner   Materials Research Forum LLC

Materials Research Foundations 164 (2024) 298-342   https://doi.org/10.21741/9781644903056-9

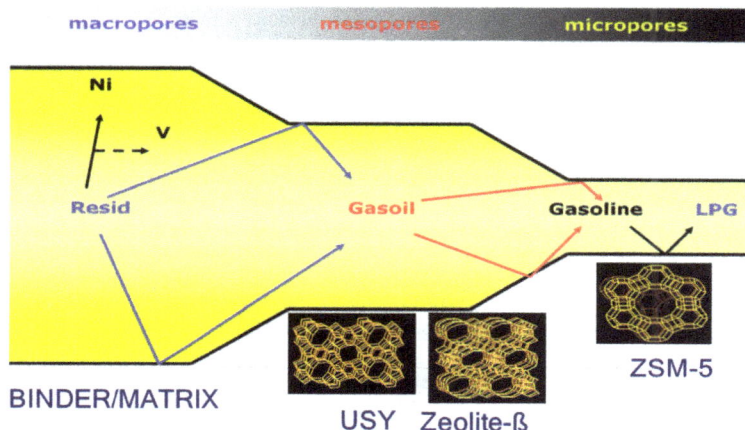

Figure 5. Schematic representation of the hierarchical pore structure in FCC catalyst [9, 43].

For the modern conventional FCC process, the desired catalyst properties are [32]:

- Good stability to high temperature and to steam. The catalysts must have the thermal stability to maintain particle and catalytic integrity under severe regenerator conditions.

- High activity to carry out conversion of the feed before any significant amount of thermal cracking sets in. Thermal cracking leads to undesirable products such as methane, ethane and some propane. On the other hand, catalytic cracking produces relatively fewer C_1 and C_2 fragments and a larger number of olefins are produced.

- Large pore sizes to crack larger molecules so that can get into smaller pores.

- Good resistance to attrition to maintain particle morphology under the severe impact and erosion forces that exist in the FCC unit.

- Low coke production so the catalyst can remain active for a longer period.

Although the FCC unit was developed purposely to help in the conversion of low value feed into more gasoline, the unit and the process have undergone several modifications, some of which are aimed at tackling the increasing demand for some of its byproducts, such as propylene. For the purpose of producing more propylene and olefins, more ZSM-5 is being used as the main active component of the catalyst in the FCC unit [56-63]. Metal contaminants usually have their biggest influence on the zeolite active components and it is through the zeolite that rare earths are usually introduced in the FCC catalysts.

The main goal of the FCC unit is to upgrade low value hydrocarbons such as residue feeds, which often contains higher levels of contaminants that can degrade catalyst activity such as nickel, vanadium, sodium, iron, and calcium [32, 64, 65]. Of all these metals, vanadium presents the most deleterious effect as it is mobile and can move from one catalyst particle to another, thereby contaminating newer active sites and aged catalysts. Vanadium is sometimes used as a determinant in the amount of fresh catalyst added to the FCC to maintain activity by measuring the amount of vanadium in the spent catalyst [23]. Vanadium also promotes dehydrogenation reactions leading to more dry gas and coke [66-70]. It also attacks the zeolite crystalline structure leading to structure collapse and a loss in surface area as the pores collapse.

2.3   FCC catalyst testing

The demand for proper catalyst testing methodologies is driven by the impact of catalysts on the performance of the FCC unit and its profitability. Therefore, the procedure for the selection and evaluation of the FCC catalysts is critical in identifying the best catalyst formulations available. There are many factors that need to be considered in the evaluation of fluid catalytic cracking (FCC) catalysts, and deactivation is considered to represent the most decisive parameter in FCC catalytic testing [71]. Due to the complexity of catalyst deactivation mechanism, the prediction of the performance of commercial catalysts is one of the most fundamental research activities. Therefore, the challenge is to simulate the deactivation of FCC catalysts in the laboratory. In order to study FCC catalysts deactivation in the laboratory, most experiments are performed using micro activity test (MAT) units. Although the MAT unit provides useful information in the study of the FCC catalysts, difficulty in matching actual FCC conditions are inherent in its design and these include [72-75]:

- The MAT unit is essentially a fixed bed reactor compared to a commercial FCC unit which is a fluidized bed unit

- In the MAT unit catalyst is contacted with the feed for a period which can be an order of magnitude longer than in the industrial unit. This has an effect on the product distribution and could explain why MAT results are usually different from real FCC data.

- The MAT unit cannot provide information about catalyst attrition since it is a fixed bed unit.

- Coke and catalyst profile can develop in the 10 cm long catalyst bed of the MAT because the catalyst particles deactivate at different rates. On the other hand, in a commercial unit all catalyst particles experience the same feed exposure leading to uniform coke concentration at the riser outlet.

- Gas residence time in the MAT unit is also significantly different than in the commercial unit, as well as temperature history, thus dry gas yield prediction can be largely incorrect.

- The MAT employs a reactant partial pressure much lower than the one of the commercial riser: 0.05 atm for MAT and 1.5 atm for the commercial riser.

It is therefore important to develop realistic deactivation techniques that can closely simulate the deactivation of commercial FCC catalyst in the laboratory.

2.4   Laboratory deactivation approaches

The ability to determine catalyst performance accurately in the laboratory and simulate its commercial performance plays a crucial role in the advancement of FCC catalyst technology are dependent upon. FCC catalysts evaluation in the laboratory helps to simulate the catalyst deactivation process in the commercial so that the performance of the evaluated catalysts can be predicted with a certain degree of accuracy. Therefore, it is vital to have realistic deactivation techniques at laboratory scale that could simulate the deactivation of catalyst in a commercial unit, under the combined action of temperature, steam, metals and thermal shocks. For this to happen, catalyst deactivation procedures have to be fine-tuned to simulate equilibrium catalyst activity, physical properties, and metal content, as well as metal activities. Wallenstein et al. [71] emphasized the usefulness applying a deactivation method capable of simulating the E-cat and to adjust the testing parameters as close as possible to the operating parameters in the FCC unit.

The deactivation of fresh FCC catalyst in the laboratory is usually accomplished by steam treatment at elevated temperatures to accelerate the hydrothermal aging which occurs in a commercial regenerator. The steamed catalysts are then tested and characterized by MAT activity, zeolite unit cell size, or surface area with the target of achieving typical equilibrium catalyst properties [76]. A wide range of steaming conditions is used in the industry. However, in order to reduce the time required to achieve the target property these conditions all deviate substantially from the commercial conditions.

The complexity involved in determination of the deactivation mechanisms means that the prediction of the commercial catalysts' performance is one of the most challenging and important research activities in the oil refining industry [77]. As a result, one of the biggest challenges in FCC research field is to first simulate how the fluid cracking catalyst is deactivated in a commercial FCC unit and then evaluate its performance in laboratory/pilot-scale testing. It should be underlined here that selecting the proper catalyst deactivation method is just as important as the pilot testing of FCC catalyst. An example of a technique that provides a close simulation of the commercial FCC operation conditions is cyclic deactivation under the combined action of steam, metals, temperature and thermal shock with the process conditions comprising of cracking, stripping and regeneration [77]. Cyclic deactivation usually consists of hydrothermal deactivation of the catalyst and metal poisoning by nickel and vanadium and is one of the best deactivation approaches to address additive deactivation.

In order to simulate catalyst deactivation in the presence of metals, FCC catalysts are impregnated with vanadium and nickel, which are the predominant metal contaminants in commercial FCC operation [71]. Vanadium is the more critical of the two metals in lab deactivations because it has two effects: vanadium destroys the zeolite and catalyzes dehydrogenation reactions [78].

Hydrothermal treatment is used to simulate the physical changes that occur in the FCC catalyst through repeated regeneration cycles. Hydrothermal treatment (steaming) destabilizes the faujasite (zeolite Y), resulting in reduced crystallinity and surface area. Further decomposition of the crystalline structure occurs in the presence of vanadium, and to a lesser extent in the presence of nickel. Vanadium is believed to form vanadic acid in a hydrothermal environment resulting in destruction of the zeolitic portion of the catalyst. Nickel's principle effect is to poison the selectivity of the FCC catalyst. Hydrogen and coke production is increased in the presence of nickel, due to the dehydrogenation activity of the metal. Vanadium also exhibits significant dehydrogenation activity, the degree of which can be influenced by the oxidation and reduction conditions prevailing throughout the deactivation process. The simulation of the metal effects that one would see commercially is part of the objective of deactivating catalysts in the laboratory.

Ihli et al [79] studied the structural changes that lead to the deactivation of FCC catalysts and from the results, they concluded that zeolite amorphization and distinct structural changes on the particle exterior as the driving forces behind catalyst deactivation. Amorphization of zeolites, in particular, close to the particle exterior, was responsible for the reduction of catalytic capacity. The congregation of the outermost particle layer into a dense amorphous silica–alumina shell further reduces the mass transport to the active sites within the composite.

2.5   FCC catalyst deactivation mechanism by metals

The processing of heavy feedstock is a major issue for refineries due to the catalyst deactivation. The deactivation taking place during catalytic cracking can be grouped into two categories: reversible deactivation (the catalytic activity can be recovered by removing contaminant species during regeneration) and irreversible deactivation (the catalytic activity is destroyed by chemical or physical means such as metals, hydrothermal and cannot be recovered) [1–3]. Therefore, the problems associated with catalyst stability, activity and selectivity are matters of priority. Reversible deactivation occurs due to coke deposition each time the catalyst passes through the reactor, and is reversed by coke burning in the unit's regenerator. Irreversible deactivation of the catalyst occurs as it ages in the unit and consists of four separate but interrelated events mechanisms [76]:

- Zeolite dealumination
- Zeolite decomposition
- Matrix surface collapse
- Contaminant effects

Zeolite dealumination, as measured by unit cell size reduction, reduces the acid site density and, hence, the inherent activity per unit of zeolite. Zeolite decomposition, measured by crystallinity or micropore surface area loss, also reduces activity. Both processes occur simultaneously in the hydrothermal atmosphere of an FCC regenerator. Matrix surface area collapse reduces activity by reducing catalytically active matrix sites, as well as by reducing porosity of the particle, which can restrict accessibility of the zeolite. Contaminants such as vanadium and sodium also contribute to deactivation in various ways.

In an effort to lower operating costs, refineries often find opportunities to purchase lower cost crudes or purchase low value feedstocks for gasoline production. Although these types of opportunity feeds can help improve profit margins, there are often severe consequences associated with them, because they may contain unknown levels of contaminants and catalyst poisons. It is therefore, important to be cognizant of the various contaminants to avoid unwanted surprises, such as unexpected catalyst deactivation and high coke deposition, which can both result in shortened cycle length and unexpected product distribution. In order to assess the impact of processing new feed streams, it is recommended to look beyond the usual bulk feed analysis which does not necessary provide enough information on the feed reactivity. Several authors have studied the impact of the addition of residual feedstocks to conventional VGO and have highlighted the challenges of co-feeding secondary refinery streams, such as using a suitable catalyst and reaction conditions to address the combined feed composition as determined by the desired product distribution [80-84].

The type and number of contaminants found in FCC feedstocks depend primarily on the source of crude oil and also on the refinery process the feed may have undergone. Ni and V contaminants are predominantly present in conventional heavy crude oils, whereas Fe contaminants are known to be present at relative high concentrations in tight and/or shale oils. These metals tend to deposit on the catalyst gradually poisoning its surface, reducing bulk accessibility, destabilizing zeolite active components, decreasing its activity and promoting dehydrogenation reactions leading coke formation [85, 86].

Meira et al [86] used x-ray nano tomography to study industrial FCC particles at differing degrees of deactivation to quantify changes in single-particle macroporosity and pore connectivity. From their study, they concluded that metals are incorporated almost exclusively in near-surface regions, severely limiting macropore accessibility as metal concentrations increase. Macropore channels act like "highways" of the pore network, therefore, blocking them prevents feedstock molecules from reaching the catalytically active domains. Consequently, metal deposition reduces conversion with time on stream because the internal pore volume, although itself unobstructed, becomes largely inaccessible.

Amongst the metals, vanadium and nickel are more deleterious because when deposited on the catalyst, they interact with the catalyst, changing the main reaction pathway and establishing parallel reactions such as dehydrogenation, leading to a shift in catalyst selectivity. The actions of these metals therefore become a problem to the petroleum refiner relying on the FCC process and hence, impacting refining cost.

Studies on vanadium and nickel have shown that these metals exhibit dissimilar behaviors on contact with the FCC catalyst. Nickel deactivates FCC catalysts by catalyzing undesirable side reactions (such as carbonization), and devaluing any residual material in which it may be isolated (such as coke). Although nickel does not contribute to zeolite destruction, it acts as a catalyst for dehydrogenation leading to a higher dehydrogenation activity than vanadium [87, 88]. This dehydrogenation results in excessive hydrogen and coke make in the FCC unit, negatively affecting unit performance usually by shifting the selectivity away from the desired cracked products. Coke arising from catalyzed dehydrogenation by contaminant metals is referred to as contaminant coke.

Reynolds [89] explains the mechanism by which Ni deactivates FCC catalysts as follows:

- In the first deactivation step, Ni deposits on the catalyst surface, probably in amorphous state, covering the surface and deactivating it locally.

- As time goes on, the surface layer continues to get thicker, forming crystalline sulfides that migrate into the porous structure. These sulfides ultimately deactivate and destroy the catalyst pores.

According to Reynolds [89], it is this two-step mechanism which allows for much higher concentrations of Ni and V to be tolerated by these catalysts without complete deactivation.

In an attempt to elucidate the mechanism of a vanadium attack on the FCC catalysts, various authors have studies and proposed different mechanisms by which this occurs [64, 90-96].

In one route and using various catalysts and characterization techniques, such as X-ray diffraction (XRD), BET and electron microprobe Wormsbecher et al. [91, 97] proposed that V_2O_5 can be transformed into volatile vanadic acid, H_3VO_4, in the regenerator by a reaction with water causing zeolite destruction due to the hydrolysis of the silica-alumina framework. The reaction of vanadium to generate vanadic acid is summarized in equations (1) and (2):

$$4V + 5O_2 \rightarrow 2V_2O_5 \tag{1}$$

$$V_2O_5 + 3H_2O \rightarrow 2\ H_3VO_4 \tag{2}$$

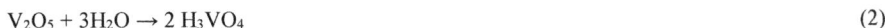

Vanadic acid is a strong acid with a pKa of 0.05 [98] and it is responsible for the mobility of vanadium and the consequent attack on the zeolite structure. According to studies, designing a catalyst regeneration system to prevent the formation of vanadic acid is not economical. Therefore, the destructive effect of vanadium on FCC catalyst can be controlled by developing suitable additives to help trap the vanadic acid. The design of suitable traps is carried out using the principle of acid-base chemistry and this helps in immobilizing the active vanadium species [99].

Small amounts of vanadium can lead to a huge loss in catalyst activity, and this has been illustrated in the study carried out by Trujillo et al. [70, 92, 100]. They used electron spin resonance techniques to study the mechanism of zeolite destruction by vanadium in the presence of steam under FCC process conditions. Based on their observations and in conjunction with the work by Wormsbecher et al. [91, 97] , they proposed the following mechanism:

- Vanadium occurs in hydrocarbon feedstock as a part of organic molecules in the form of metal porphyrins [101] and vanadium is present in the +3 or +4 oxidation state [23, 102]. Porphyrins are highly conjugated organometallic compounds which under FCC conditions will either crack or condense to coke when in contact with the catalyst surface [103]. The bare metal is then free to migrate further into the microsphere or transfer to other particles.

- In the FCC reactor, the reduced metal such as vanadium from the decomposed porphyrin deposits onto the surface of the catalyst or migrates further into the microsphere and it is transferred along with coke and to the regenerator. Such metals can promote unwanted side reactions resulting in coke and hydrogen formation and can result in destruction of zeolite and active matrix.

- Once in the regenerator the coke is burned off and vanadium is oxidized to the +4 and +5. If present at high enough concentrations, the oxidized vanadium may be present as V_2O_5.

- V_2O_5 is a liquid under regenerator conditions with a melting point around 670 °C [99, 104]. At low concentrations it can easily spread out on the high surface area solid losing its solvent properties.

- Water helps vanadium mobilization by breaking V–O–Al and V–O–Si bonds and forming hydroxylated vanadium species, which has amphoteric behavior, so it can be presented as a neutral or positively charged species avoiding electrostatic repulsion from the negatively charged framework. In this way vanadium reaches acid sites where it reacts with the zeolite leading to catalyst poisoning [100]. Also, vanadium preferentially neutralizes the strongest acid sites on the zeolite [105].

- The high temperatures and steam found in the FCC regenerator favor the formation of V_2O_5 and subsequent production of vanadic acid, which catalyzes framework hydrolysis, increasing the rate of dealumination as illustrated in equations (3) and (4):

$$V_2O_5 + 2H^+-Y \rightleftharpoons 2VO_2^+ -Y + H_2O \qquad (3)$$

$$VO_2^+ -Y +2\ H_2O \rightleftharpoons H^+-Y + H_3VO_4 \qquad (4)$$

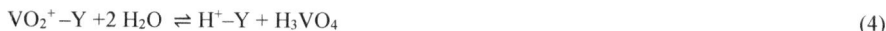

From the above mechanism and in conjunction with observations by Pine [12], it can be concluded that vanadium acts as a catalyst for zeolite framework hydrolysis because it is not consumed during zeolite destruction. Pine also advanced that rare earths do not change vanadium tolerance but that their effect is an indirect one of changing the base steam stability of the zeolite.

The mechanism proposed by Trujillo et al. [92] is also supported by Etim et al. [106] who also proposed the mechanism of mesopore formation based on accelerated dealumination by vanadic acid. They proposed that the vanadic acid migrates to the zeolite framework where it attacks and destroys strong acid sites, $Si(OH)Al$, via extraction of tetrahedral aluminum atoms in the zeolite crystal, causing it to collapse, and forms solid aluminum vanadate ($AlVO_4$). Their proposed mechanism is summarized in equations (5) to (7):

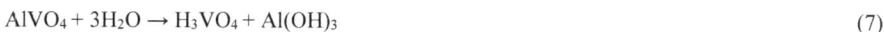

$$V_2O_5 + 3H_2O \rightarrow 2H_3VO_4 \qquad (5)$$

$$[SiO]_3Al(HO)Si + H_3VO_4 \rightarrow AlVO_4 + 4[SiOH] \qquad (6)$$

$$AlVO_4 + 3H_2O \rightarrow H_3VO_4 + Al(OH)_3 \qquad (7)$$

An alternative route to explain the destructive effect on FCC catalysts by vanadium was proposed by Xu et al. [93] in which they proposed that the key to destruction lies in the ease of formation and availability of NaOH and that the role of vanadium is to catalyze and facilitate the formation of NaOH. They suggested that it is the basic OH^- entity that attacks the framework $Si-O$ bonds and that vanadium does not engender a new destructive pathway. According to the work by Xu et al. [93] on zeolite Y, they found that at 697 °C to 827 °C in the presence of air and steam, V species reacts with cationic sodium, facilitating its release from the Y exchange site. The sodium metavanadate ($NaVO_3$) thereby formed hydrolyzes in steam to form NaOH and metavanadic acid (HVO_3), which may again react with Na^+ cations. Here, as in the V-free case, NaOH is the destructive agent, with formation of the basic OH^- entity again being the key. Xu et al [93] concluded that without sodium, V has little effect on Y zeolite stability, regardless of zeolite unit cell size. Again this study shows that vanadium acts as a catalyst for zeolite framework hydrolysis. With the aid of UV visible experiments, Xu et al. [93] explained the role of vanadium in the destruction of zeolite via the following reaction pathway:

Vanadium is present as V_2O_5, and as it is liquid at FCC regeneration condition [91, 97], it reacts in steam to give the vanadic and metavanadic acid.

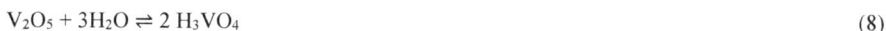

$$V_2O_5 + 3H_2O \rightleftharpoons 2\ H_3VO_4 \qquad (8)$$

$$V_2O_5 + H_2O \rightleftharpoons 2HVO_3 \qquad (9)$$

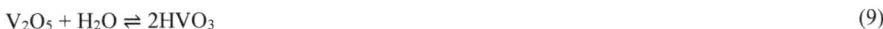

Similarly, if V is in a +5 oxidation state as an adsorbed surface $*[VO_2]^+$ species, it can also react to form an acid.

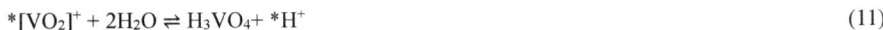

$$*[VO_2]^+ + H_2O \rightleftharpoons HVO_3 + *H^+ \qquad (10)$$

$$*[VO_2]^+ + 2H_2O \rightleftharpoons H_3VO_4 + *H^+ \qquad (11)$$

where $*$ is a surface site

Vanadic and metavanadic acid exist in equilibrium at FCC regeneration conditions.

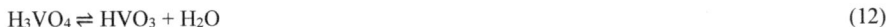

$$H_3VO_4 \rightleftharpoons HVO_3 + H_2O \qquad (12)$$

The acid reacts with the sodium cation on the Y zeolite to form sodium metavanadate.

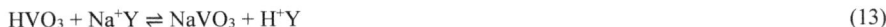

$$HVO_3 + Na^+Y \rightleftharpoons NaVO_3 + H^+Y \qquad (13)$$

And the sodium metavanadate hydrolyzes in steam to give NaOH and metavanadic acid.

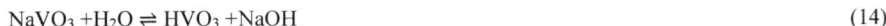

$$NaVO_3 + H_2O \rightleftharpoons HVO_3 + NaOH \qquad (14)$$

According to Xu et al. [93] the critical steps of the proposed mechanism are steps 4 and 5 containing equations 13 and 14. This route is further supported by observation by Hagiwara et al. [107].

In the case where there is no vanadium sodium removal, it would have to occur by hydrolysis of Na+Y.

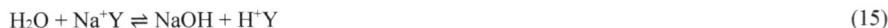

$$H_2O + Na^+Y \rightleftharpoons NaOH + H^+Y \qquad (15)$$

According to Xu et al, the uncatalyzed formation of NaOH is a difficult step and only becomes possible in the presence of vanadium.

The two different mechanisms proposed by Wormsbecher et al. [91, 97]and Trujillo et al. [70, 92, 100], and Xu et al. [93] confirm that vanadium does play the role of a catalyst in the destruction of the zeolite structure by hydrolysis. The exact mechanism to explain how this occurs is still

Rare Earth - A tribute to the late Mr. Rare Earth, Professor Karl Gschneidner   Materials Research Forum LLC
Materials Research Foundations 164 (2024) 298-342   https://doi.org/10.21741/9781644903056-9

subject to discussion. However, with the advent of more advanced analytical and characterization techniques, it is expected that the actual mechanism will be resolved.

2.6   Role of rare earths in metal passivation

From the above mechanisms, there is a better understanding of how vanadium in FCC feedstock ends up in the catalyst and subsequently deactivating the catalyst.

The common practice of maintaining the FCC unit activity is by   adjusting fresh catalyst addition, based on the level of metal contaminants in the feed [108]. Fresh catalyst additions are increased when feed metals begin to increase and the opposite applies when metal content in feed is low. Consequently, when dealing with feeds with higher metals content, adding more fresh catalyst alone may not be an effective catalyst management strategy as this will not reduce the impact of contaminant metals and the activity and stability of the catalyst will be adversely affected.

It is therefore important to have an appropriate catalyst formulation which can effectively trap metal contaminants. The metal trap technology works by capturing the volatile and mobile metal contaminants, primarily vanadium, to form a stable and catalytically inactive compound in a process known as metal passivation. The severe conditions under which the catalytic-cracking process is carried out make it particularly difficult for vanadium blocking. A suitable vanadium trap for FCC catalysts should fulfill most of the following conditions [109]:

- The substance should be stable at a temperature up to 800° C. in an oxidizing environment (regenerator) and in the presence of about 20% partial pressure of water vapor and from 60 to 2000 ppm sulfuric acid.

- That the substance be stable at a temperature of 550° C. in a strongly reducing environment (reactor) and in the presence of water vapor (stripper).

- That the substance possesses greater affinity for the vanadium than for the zeolite and for the catalyst components.

- The amount of substance required for the effective protection of the catalyst, must be low enough to avoid excessive dilution of the catalyst and in that way avoiding in consequence its loss of activity and selectivity.

- The rate of vanadium capture must be high enough to avoid damaging the catalyst.

- The substance ought to maintain its vanadium-capture power whilst it remains within the cracking unit.

- If the substance contains metallic elements, these must not be interchanged by the zeolite cations.

- The substance ought not to be damaging either for the catalyst or for the metallic structure of the unit.

- The substance ought to be able to be incorporated within the catalyst particle during its production (integral particle), and/or to be able to be prepared in the form of particles able to be fluidized having good abrasion strength, in order that they can flow together with the catalyst in the unit (dual particle).

- The substance must be cheaper than the catalyst, since it is charged to the catalyst inventory in order to decrease the fresh catalyst addition and in this manner to diminish the operation cost.

- It must be acceptable from the viewpoint of environmental preservation; the substance must not need particular handling conditions, it must not generate toxic materials during its preparation, nor can it be apt to be converted into a dangerous contaminant after being used.

- It must not possess dehydrogenating activity nor facilitate the nickel and vanadium dehydrogenating action.

In spite of a great deal of investigations carried out on the matter, no substance has been found which meets all of the requirements about an ideal trap for retaining the vanadium.

However, the use of rare earths for the preparation of vanadium tolerant FCC catalysts provides a way forward because of the following attributes [23]:

- They can process high metal feedstocks.

- They can capture and immobilize the vanadium in a nondestructive form.

- They irreversibly bind the vanadium so that it cannot migrate back to the catalyst.

- The have a high capacity to remove a considerable amount of vanadium from the catalyst.

- The show negligible interaction with other acidic species (e.g., sulfur).

- Vanadium migration to the trap is significantly faster than the migration of vanadium to the zeolite.

Metal passivation reduces the harmful effects of metals without substantial reduction in catalyst activity and without removing the metal from the unit. Nickel and vanadium, which constitute the most relevant poisons for catalytic cracking catalysts, are usually associated metal porphyrins. Prophyrins are organometallic compounds found in the higher boiling range oil fractions, and distillation concentrates the nickel and vanadium in the fractions typically sent to the FCC unit. It is during the FCC process that the metals form a deposit on the catalyst surface, damaging the zeolite structure [17, 23]. Therefore, the use of rare earths to trap vanadium and nickel will help reduce the deleterious effect of vanadium as a catalyst for zeolite framework hydrolysis.

Rare earth oxides such as La_2O_3 are basic in nature and can neutralize vanadic acid to form rare earth vanadates [23, 110-112], thereby preventing the rapid hydrolysis of the zeolite framework. The reaction of rare earths (RE) with acidic vanadium compounds forming vanadates is represented in equation (16):

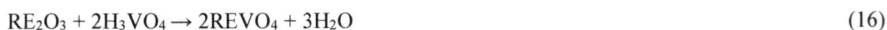

$$RE_2O_3 + 2H_3VO_4 \rightarrow 2REVO_4 + 3H_2O \hspace{2cm} (16)$$

where RE_2O_3 is representing the rare earth oxide, forming stable vanadium compounds [113].

The stabilities of the initial metal oxides and product metal vanadates are critical to the viability of this method of passivation [103]. In a study using the thermogravimetric analysis, Wang et

Rare Earth - A tribute to the late Mr. Rare Earth, Professor Karl Gschneidner   Materials Research Forum LLC
Materials Research Foundations 164 (2024) 298-342   https://doi.org/10.21741/9781644903056-9

al.[114], showed that rare earth vanadates are stable even at the highest temperatures reached in the regenerator unit. The stability and the inertness of vanadates under FCC conditions help to reduce zeolite deactivation and the formation of coke and dry gas [20, 110]. Baugis and coworkers [115] used a photoluminescence technique to study the interaction between rare earth passivators and vanadium during typical hydrothermal operating conditions of FCC units. From their study, they suggested that rare earth elements act as vanadium traps by forming rare earth vanadates, which passivate the deleterious properties of vanadium oxides. There is also research pointing to the inhibition of vanadium poisoning in the presence of nickel according to research by Yang et al. [27, 28]. They studied the interaction of vanadium and nickel in REY and USY zeolites using XRD, Fourier transform infrared spectroscopy, surface area measurement, NH_3-TPD and the n-hexane cracking reaction. They proposed that there is an interaction existing, either direct or indirect through non-framework aluminum, between vanadium and nickel. Therefore, the introduction of rare earths into the zeolite helps to reduce metal poisoning and results in the retention of the aluminum framework and an improved stable zeolite structure [21, 22, 90, 104, 116-124].

The effectiveness of the rare earth as a vanadium trap is also influenced by the method of introduction of the rare earths in the catalysts. Moreira et al. [125] studied the effectiveness of Ce-HUSY prepared by ion exchange and incipient wetness technique and from their study they found that cerium introduced by incipient wetness impregnation shows a greater vanadium tolerance than Ce–HUSY catalysts prepared by precipitation or ionic exchange. They also observed the same behavior for La-HUSY [126].

2.7   Effect of rare earths on hydrothermal stability of FCC catalysts

Most catalysts used in processes involving high severity operation, such as high temperatures and steam, face the inherent problem of hydrothermal deactivation that has to be mitigated. FCC catalysts operate between moderate to high temperatures (500 °C to 800 °C), in the presence of steam, especially during the regeneration step. These severe conditions strongly influence the performance of the catalysts especially during catalyst regeneration which usually takes place at temperatures as high as 800 °C and in the presence of steam. With such severe conditions, zeolite dealumination becomes a real problem. For instance, USY zeolites used in standard FCC catalyst usually have a framework Si/Al ratio of approximately 5 before reaction and after regeneration, the equilibrium catalyst has a Si/Al ratio close to 20, showing the extent of dealumination in the FCC regenerator [17, 127]. Therefore, thermal and hydrothermal stability of zeolites are among the most important parameters for catalyst manufacturers.

To mitigate the problem of catalyst deactivation, rare earths can be used to improve the hydrothermal stability of FCC catalysts [5, 15, 122, 127-136]. Lanthanum and cerium are the two main rare earths used in FCC catalysts [1, 2, 6-8, 137, 138]. These metals limit the extent to which zeolite dealumination occurs (thereby stabilizing the structure) under the conditions of the FCC unit [130]. A study carried out at BASF [139] and illustrated in Figure 6 shows differences in catalyst hydrothermal stability with and without rare earth. From Figure. 6, rare earth zeolite Y (REY) shows a greater thermal stability than NH_4Y. This is because rare earths provides hydrothermal stability to the zeolite by improving surface area retention, as well as inhibiting dealumination, resulting in greater preservation of acid sites [140].

MAT CONVERSION (WT. %)

90
80
70 — REY
60 — NH₄Y
50
40

MAT CONDITIONS
REACTOR TEMP.,DEF.F.   910
30   WHSV   15
C/O RATIO   5
FEED TYPE: MID-CONTINENT GAS OIL

20
1300   1350   1400   1450   1500   1550
HEAT TREAT* TEMPERATURE,° F

1150   1200   1250   1300   1350   1400
EQUIVALENT REGENERATOR TEMPERATURE,° F
*(DEG. F./4 HRS./100 % STEAM)

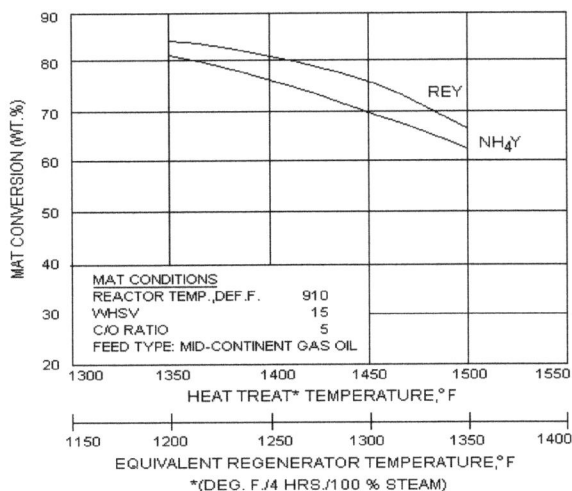

Figure 6. Effect of rare earths on the hydrothermal stability of Y zeolite [139].

Lemos et al. [15] studied the effects of Ce and La exchange on HY and compared the acidity and catalytic properties of LaHY zeolites, CeHY zeolites and HY zeolites. From their study, they concluded that the exchange of HY by La^{3+} or by Ce^{3+} improved its thermal stability, probably owing to the formation of stable oxygen complexes between rare earth cations and lattice oxygen atoms. They also found that LaHY was found to be more stable than CeHY.

Though La and Ce are commonly used in FCC catalysts, there is experimental evidence suggesting that catalysts stability improves as the ionic radius of the RE cation decreases [140]. In a study by Liu et al. [141], yttrium modified zeolite Y (YHY), showed higher stability than CeHY. They found that the stability and activity of REY zeolites increased with decreasing ionic radii of rare earth elements.

This ties in with the observation by Du et al. [142] that crystallinity retention , and consequently zeolite stability increases with decreasing ionic radius as shown in Figure 7.

Rare Earth - A tribute to the late Mr. Rare Earth, Professor Karl Gschneidner   Materials Research Forum LLC
Materials Research Foundations 164 (2024) 298-342   https://doi.org/10.21741/9781644903056-9

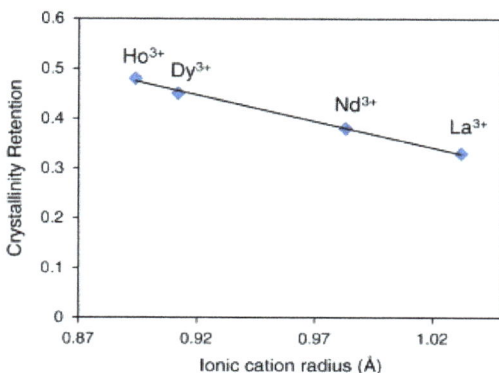

Figure 7. Crystallinity retention as a function of ionic cation radius [142].

The interaction of rare earths ion with the zeolite leads to the formation of bonds with the hydroxyl groups found in the zeolites. Using powder XRD with Rietveld refinement, Du et al. [142] studied the position, occupation and coordination of the zeolite extra framework rare earth cations and the effects these had on hydrothermal stability. To further confirm the influence of rare earth species on the hydrothermal stability of Y zeolites, the samples were studied by powder XRD after being steamed. The crystallinity retentions of the rare earth species upon steaming showed the ascending order of CeY < LaY < PrY < NdY as illustrated in Figure 8. This shows that rare earths help in stabilizing the zeolite framework and those with smaller radii provide greater stability to the framework structure.

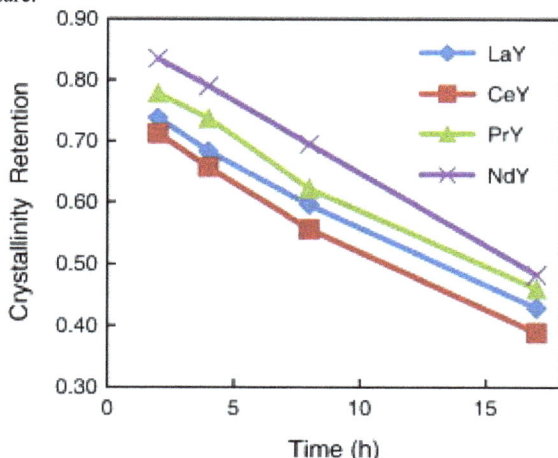

Figure 8 Crystallinity retention vs. time at 800 °C in 100% steamed catalysts [142].

Schüßler et al [18, 143] investigated the nature and location of La-species in faujasite using a combination of techniques, including DFT calculations. From their study, they concluded that the majority of the La^{3+} ions is present in the sodalite cages in multinuclear OH-bridged aggregates. The same observation was made by Nery et al [144] when they used the Rietveld method of structure refinement to determine the crystallographic positions of the rare-earth cations after the calcination step (500°C, 1 h). Their results showed that, regardless of the calcination mode, both La and Ce migrate to S2 sites that are located inside the sodalite cage, whereas Na cations migrate to sites S4. They also found that the aluminum generated by the dealumination process due to the incorporation of rare-earth cations and the calcination step is located inside the sodalite cage. Upon calcination, rare earths migrate to smaller cages and once located in these cages, they form bridges with framework oxygen atoms, stabilizing the zeolite structure [13, 17, 137, 142]. Some hydrolysis reactions take place over rare earth cations, generating Brønsted acidity, and the higher the ionic radius of the rare earths atom, the higher the degree of hydrolysis.

Du et al. [142] studied the transfer of rare earth ions into zeolite Y and from their study, in conjunction with the study by Hunger et al. [137], they proposed a mechanism to explain the transfer of rare earth ions into zeolite Y according to the following the procedure below:

- Initially, the rare earth (RE) ions usually exist as $RE(H_2O)_n^{3+}$ in aqueous solution, and those located in supercages yield $RE(OH)^{2+}$ via dehydration during calcination and hydrothermal treatments.

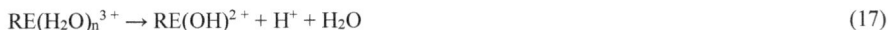

$$RE(H_2O)_n^{3+} \rightarrow RE(OH)^{2+} + H^+ + H_2O \qquad (17)$$

- The $RE(OH)^{2+}$ cation has a diameter of approximately 0.23 nm and the sodalite cage is sized 0.66 nm, which allows the migration of RE cations from the supercage to the sodalite cage. The location of RE cations in sodalite cages had earlier been observed by other authors [144-146].

$$RE(OH)^{2+}(supercage) \rightarrow RE(OH)^{2+}(sodalite\ cage) \qquad (18)$$

- Subsequently, $RE(OH)^{2+}$ cations and oxygen react to form RE oxides that then unexpectedly spread to the zeolite surface rather than move into the sodalite cages.

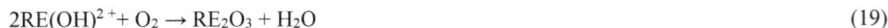

$$2RE(OH)^{2+} + O_2 \rightarrow RE_2O_3 + H_2O \qquad (19)$$

2.8   Effect of rare earths on zeolite acidity

The incorporation of rare earths influences zeolite acidity and this phenomenon has been studied by several authors [26, 147-154]. Cerqueira et al. [127] studied the influence of rare earths on zeolite acidity and from their study they suggested that rare earths influence the acidity of the zeolite through coordination bonds between rare earth cations and the framework of oxygen atoms. As a result of the coordination bonds, Brønsted acid sites are formed because of the polarization

Rare Earth - A tribute to the late Mr. Rare Earth, Professor Karl Gschneidner   Materials Research Forum LLC
Materials Research Foundations 164 (2024) 298-342   https://doi.org/10.21741/9781644903056-9

exerted by the rare earth cations, and the acid sites created by hydrolysis of the rare earth cations. Martins et al. [155] studied the effect of rare earths in zeolite beta and they found that the introduction of rare earths enhances the catalyst acidity leading to an increase in branched products. Moreover, the occurrence of hydrogen transfer reaction leads to a higher deactivation of these catalysts, by coke formation. This phenomenon was also observed on ZSM-5 by other researchers [156, 157]. Besides, a correlation between the ionic radius of the rare earths atom, the Brønsted acidity of the zeolite and the rate of hydrogen transfer reaction was observed in Fig. 7.

Deng et al. [26] also investigated the change in the amount of acid and the distribution of Lewis and Brønsted acid sites when different amount of H^+ were exchanged with rare earth metal ions in HY zeolite. From the NH_3-TPD results in Figure 9, they found that the peak intensities of strong acid sites as well as weak acid sites decreased and simultaneously the peaks of weak acid sites gradually moved to a higher temperature with the increase of La^{3+}, indicating that the introduction of La^{3+} decreases strong acid sites as well as weak acid sites, and increase the proportion of medium acid sites.

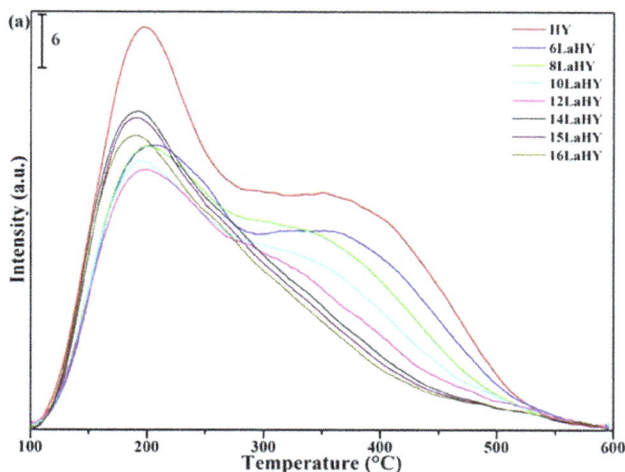

Figure 9. Effect of La loading on Y zeolite acidity [26].

Moderate rare earths (RE) ion exchange can lead to the formation of Brønsted acid sites via hydrolysis as illustrated by equations (20) and (21) [127]:

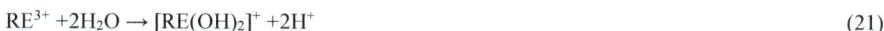

$$RE^{3+} + H_2O \rightarrow [RE(OH)]^{2+} + H^+ \qquad (20)$$

$$RE^{3+} + 2H_2O \rightarrow [RE(OH)_2]^+ + 2H^+ \qquad (21)$$

The decrease in the number of acid sites at high ion exchange can be explained by the formation of bridged hydroxyls as shown in the following equation:

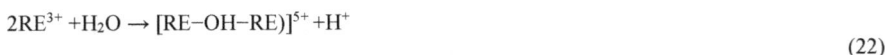

$$2RE^{3+} + H_2O \rightarrow [RE-OH-RE)]^{5+} + H^+$$

(22)

The degree of hydrolysis seems to increase with an increase in ionic radius as illustrated in Figure 10 [17, 158]. Also, the higher the ionic radius of the RE atom, the higher the Brønsted acidity of the zeolite.

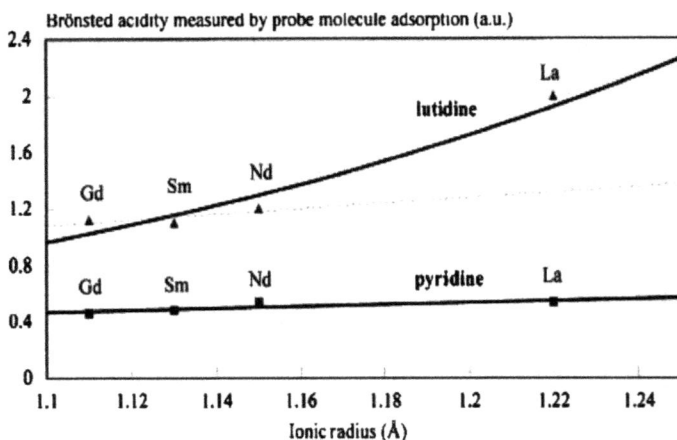

Figure 10. Influence of the ionic radius of the RE cation on the Brønsted acidity of the zeolite [158].

Since rare earth cations modify the acidic properties of the zeolite, their introduction into the zeolite might certainly have an effect on the hydrogen transfer reactions. In a study carried out by de la Puente et al. [13], looking at the influence of different rare earth ions on hydrogen transfer over Y zeolite, they found that the tendency for hydrogen transfer reactions increases with an increase in ionic radius as shown in Figure 10. They also found that Brønsted acidity increased with an increase in ionic radius of the rare earth ions. An increase in the number of acid sites will translate into an increase in activity during catalytic cracking. Upon calcination, rare earths migrate to smaller cages and once located in these cages, they form bridges with framework oxygen atoms, stabilizing the zeolite structure [13, 17, 137, 142]. Some hydrolysis reactions take place over rare earth cations, generating Brønsted acidity and the higher the ionic radius of the rare earths atom, the higher the degree of hydrolysis. Fig 8 illustrate the influence of Brønsted acidity on hydrogen transfer index.

From Figure 11, it can be seen that Brønsted acidity increases with ionic radius and as the acidity increases, so is the hydrogen transfer index (i_{HT}). This therefore, means that the higher the ionic radius of the rare earth cation, the greater the degree of hydrogen transfer reactions during catalytic cracking.

The hydrogen transfer index is defined as the paraffin/olefin ratio of C_3, linear C_4 and branched C_4 species. The relative activity of FCC catalysts for generating secondary reactions can be estimated using the hydrogen transfer index (i_{HT}) for catalysts tested under constant conditions with the same feed [16, 159-164]. Catalysts with lower i_{HT} generate fewer secondary reactions, preserving a greater quantity of the gasoline boiling range olefins, which can be subsequently cracked to lighter olefins.

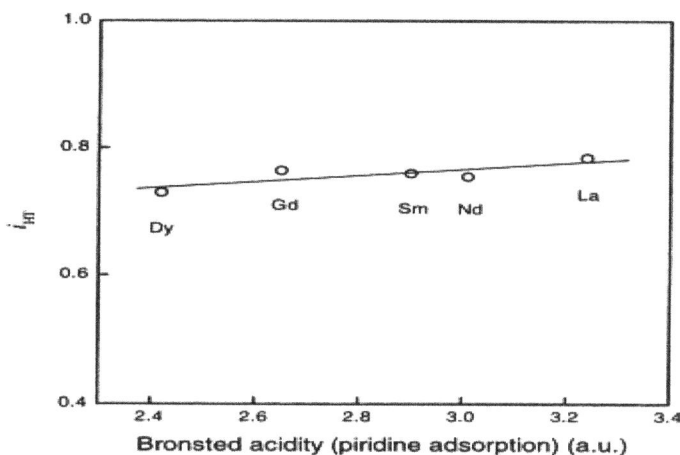

Figure 11. Hydrogen transfer index (iHT) as a function of Brønsted acidity [13, 128]

This again shows that the introduction of rare earth compounds in zeolites might not only affect the cracking activity but also the hydrogen transfer properties of the catalysts, thus affecting their product distribution as well as favoring coke formation. These results suggest that a certain degree of catalytic control can be exerted on these reactions through the selection of the rare earth elements to be loaded into commercial catalysts.

2.9   Effect of rare earths on gasoline yield and dry gas yield

The fact that rare earths inhibit the dealumination of a zeolite, means a higher concentration of acid sites will be found in a rare earth exchanged catalyst, leading to improved activity and hydrothermal stability. On average, the acid sites are weaker and in closer proximity to each other than those found in a more highly dealuminated catalyst characterized by lower unit cell size measurements.

As a result of the greater number of active sites, both the primary cracking and hydrogen transfer reactions that occur within the zeolite are enhanced. Primary cracking reactions involve the initial scission of the carbon-carbon bond to form higher valued liquid products such as gasoline. Hydrogen transfer reactions are those that occur between cracked products to terminate the cracking reactions in the gasoline range, thereby reducing the over cracking of gasoline to C_3's and C_4's. The hydrogen transfer reactions are greatly increased with the addition of rare earths to the zeolite. High acid site density leads to high intrinsic activity but poorer coke selectivity as illustrated in Figure 12.

Therefore, the incorporation of rare earths in catalytic cracking catalysts enhances gasoline yield. Figure 12 shows a plot of gasoline yield at varying conversion levels for two standard cracking catalysts with different levels of rare earths. This data suggests a strong correlation between rare earths content and gasoline yield.

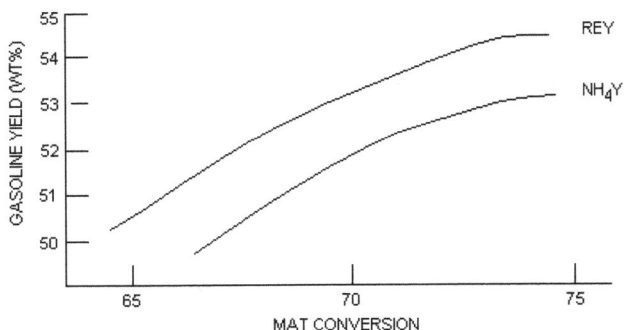

Figure 12. Effect of rare earths on gasoline yield [139].

Therefore, by restricting the loss of aluminum atoms in the zeolite, rare earth increases the activity and gasoline yield of FCC catalysts [165].

2.10   Effect of rare earths on catalyst activity

Zeolites used in cracking catalysts undergo reactions in the high temperature steam environment of the regenerator that destroy the active sites. As catalysts age in the FCC regenerator, the unit cell size drops due to dealumination of the zeolite, through the reaction of the active sites react with steam. The role of rare earths is to manipulate the active site density of the zeolite as measured by the unit cell size. The rare earth ions in the zeolite retard this deleterious reaction from occurring, thus preventing the shrinkage of the unit cell size as the catalyst ages. This active site preservation helps to maintain the activity of the catalysts. Manipulation of the active site density of the catalysts with rare earths, translates into of catalyst activity and/or selectivity profiles available to refiners. This is illustrated in Table 1 [20] which shows the effect of rare earth loading on catalysts activity at constant catalyst-to-oil ratio for the same catalysts.

Rare Earth - A tribute to the late Mr. Rare Earth, Professor Karl Gschneidner   Materials Research Forum LLC
Materials Research Foundations 164 (2024) 298-342   https://doi.org/10.21741/9781644903056-9

Table 1. Properties of laboratory deactivated Catalysts [20]

Catalyst	Wt% RE$_2$O$_3$ on Catalyst	Unit Cell Size, Å	Conversion, wt.% at constant catalyst/oil ratio
A	1.1	24.24	53
B	1.8	24.27	65
C	2.2	24.31	69
D	3.3	24.34	73
E	3.7	24.36	76
F	5.7	24.41	80

From Table 1 [20], it can be seen that the unit cell size of a series of lab deactivated catalysts increased uniformly with rare earth content. This means that the zeolite active site density in the catalyst is increasing, and hence, the activity of the catalyst is also expected to increase. This shows that the intrinsic activity of the catalyst increases with unit cell size, or rare earth content. Other studies also showed that stabilized Y zeolites with high rare earths content exhibited improved heavy oil conversion, a better coke selectivity and a higher hydrogen transfer activity [14, 17, 127].

2.11   Effect of rare earths on gasoline octane number and cetane index

The amount of rare earths, zeolite acidity as well as the degree of ultrastability of the catalyst have different effects on the ultimate octane of the quantity and quality of gasoline. A non-rare earth ultrastable catalyst will produce a higher octane gasoline than a non-ultrastable rare earths exchanged catalyst [20, 166]. The incorporation of rare earths into the zeolite inhibits the degree of unit cell size shrinkage during equilibration in the regenerator. Acid site density not only controls the intrinsic activity, but also the selectivity of the zeolite in the catalyst. High acid site density leads to high intrinsic activity but poorer coke selectivity and lower gasoline olefins (lower research octane number or RON) as illustrated in Figure 13.

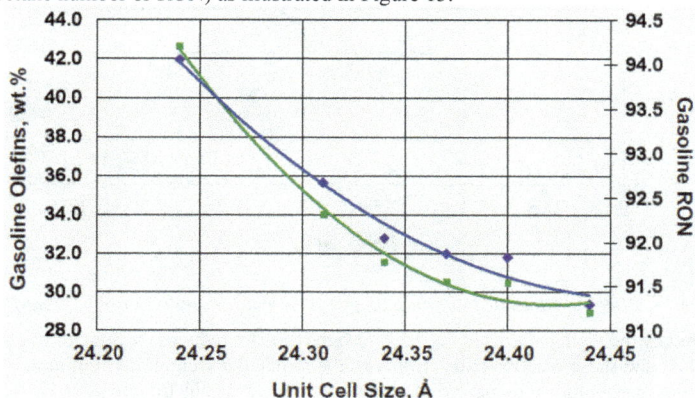

Figure 13. Effect of Unit Cell Size on Olefins yield (blue) and RON [20]

Steam in the FCC regenerator removes aluminum from the zeolite framework leading to a decrease in acidity and a reduction in unit cell size. As rare earths inhibit the extraction of aluminum from the zeolite's structure (dealumination) the equilibrium unit cell size of FCC catalysts increases. Since reducing the equilibrium unit cell size of an FCC catalyst has the effect of improving octane, adding rare earths decreases the octane [139]. Myrstad [113] also studied the effect of rare earths on octane numbers and found that they resulted in a decrease in the octane number of the gasoline produced and this was attributed to increased hydrogen transfer and decreased dehydrogenation of paraffins and naphthenes.

A high rare earth loading translates into a high hydrogen transfer index and high paraffinicity and hence, a high cetane index of the gasoline. This also means there is an increase in hydrogenation of olefins in the gasoline fraction leading to a reduced octane number.

For a rare earth exchanged cracking catalyst the hydrogen transfer reaction of interest is as follows [113, 139, 167]:

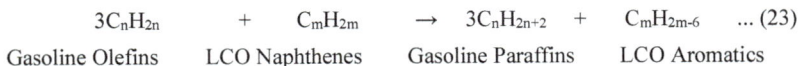

$$3C_nH_{2n} \quad + \quad C_mH_{2m} \quad \rightarrow \quad 3C_nH_{2n+2} \quad + \quad C_mH_{2m-6} \quad \dots (23)$$

Gasoline Olefins   LCO Naphthenes   Gasoline Paraffins   LCO Aromatics

The effect of rare earth loading on the gasoline yield is further illustrated by Figure 14 and Figure 15. From these figures, it can be seen that the overall gasoline yield is directly proportional to the rare earth loading while the LPG yield and RON have an inverse relationship with the increase in rare earth loading.

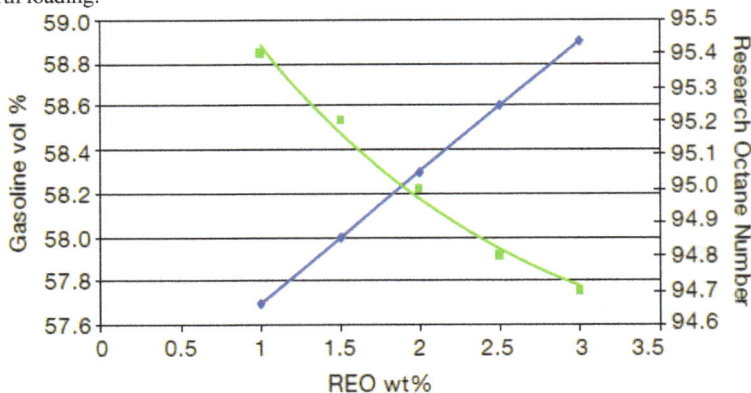

Figure 14. Effect of rare earth loading on gasoline yield (blue) and RON (green) [131]

As illustrated by the yield patterns in Fig 10, hydrogen transfer reduces the amount of olefins found in the product and these reactions also influence the molecular weight distribution of the product by terminating carbonium ions before they crack to shorter chain fragments [61]. As hydrogen transfer reactions increase relative to cracking reactions, olefin yield, light gas yield and octanes decrease, while gasoline yield increases [32].

Fig 12 illustrates the effect of rare earths on gasoline and liquefied petroleum gas (LPG) yield, and as can be seen, introducing rare earths leads to lower yield of LPG. As LPG is comprised of light olefins, this results in lower yields of light olefins too.

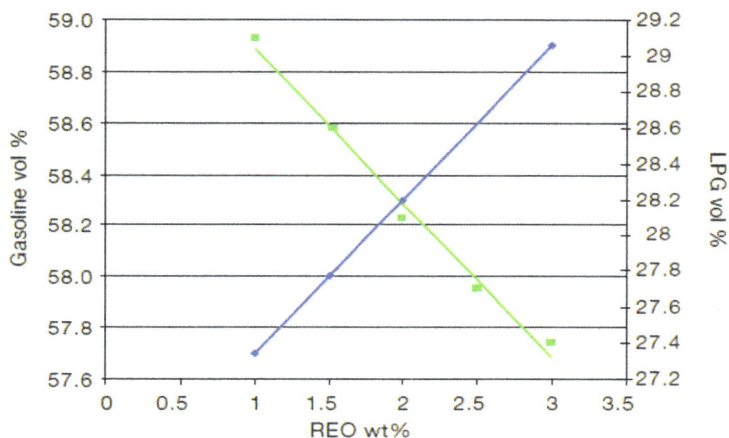

Figure 15. Effect of rare earths on gasoline (blue) and LPG yield (green) versus REO [131].

The type of zeolite and the nature of the feed may also influence the yield of light olefins, as shown by the study carried out by Xiaoning et al. [157]. They studied the effect of rare earths on the performance of HZSM-5 for catalytic cracking of butane to light olefins and their results showed that the addition of rare earth metal on the HZSM-5 catalyst greatly enhanced the selectivity to olefins, especially to propylene.

Whether a rare earth or non-rare earth catalyst is used in the refinery operation depends on the type of feed and desired products. If a refinery is interested in producing residue feed with gasoline, as the main product, a rare earth cracking catalyst will be desired. But if the focus is on higher gasoline octane, a minimum rare earth catalyst or partial rare earth catalyst will be used.

2.12 Disposal of spent fcc catalyst and recovery of rare earths from spent FCC catalysts

The oil refining industry remains one of the largest consumers of catalysts and sorbents, with applications in different processes, such as the fluid catalytic cracking (FCC) process, the hydrotreating process, the hydrocracking process, and the sorption of sulfur oxides from flue gas, among others.

During the FCC process, catalysts containing rare earth oxides are used to convert heavier oils into lighter and more valuable products. Fresh catalyst is also periodically added in order to account for reductions in performance and/or activity due to physical losses, or deactivation due to factors such as steam, temperature, time and contaminant metals contained in the feedstock [168]. After enduring numerous cycles of cracking, stripping and regeneration, the FCC zeolite catalyst become less active due to the loss of the crystal structure of the zeolite. Accordingly, the spent FCC catalyst must be periodically removed from the cracking system and replaced with fresh catalyst. If the rate of addition of fresh catalysts exceeds the physical losses of the processing unit, it becomes necessary to further withdraw spent catalyst from the unit. This is because such spent catalyst can no longer function properly in the process due to the deposition of sulfur, carbon, nickel, vanadium and other elements which inhibit or diminish the catalytic activity. The spent catalyst is referred to as equilibrium catalyst, or simply as "ECAT." Typical withdrawals from the FCC process range from a few tons per day, to as much as thirty, or more, tons per day.

In a review by Furimsky [169], the environmental, disposal and utilization aspects of spent refinery catalysts was outlined and the strategy suggested on how to handle the spent catalyst problem included: (a) minimizing spent catalyst waste generation (b) utilization to produce new catalysts and other useful materials, (c) recycling through recovery of metals and (d) treatment of spent catalysts for safe disposal.

The strategy used to dispose of this spent material vary depending on the quality of the material. For instance, material which is low in contaminant metals, and most likely high in remnant activity, is often resold and incorporated in full, or more typically, as a supplement to the new, or fresh, catalyst being added to another FCC unit. The spent catalyst may also be used during unit upset conditions, start-up of the process following a shutdown due to new unit installation, maintenance, or other planned or unplanned shutdowns. Spent catalyst that is not capable of suitable performance in another refinery is often disposed of in landfills, or by incorporating it into other industrial processes/products such into cement and road pavement. Alternatively, the spent catalyst may have other catalytic uses in other processes that require a particular property of the spent material, such as surface area/sites, or heavy molecule processing capability.

At present, most spent FCC catalyst is not considered to be hazardous waste, so the presence of the various metals contained in the catalyst are not a hindrance to normal disposal in landfills. However, environmental regulations are becoming stricter and it is possible that in the future, countries may start imposing regulations that would limit the disposal options, and/or that would add a significant economic cost to the disposal operation. But, considering the crucial importance of rare earths, their indispensable role in technologies, the high economic and environmental costs associated with their production by mining activities, spent catalysts could have a better destination. The use of spent catalyst as a source of rare earths would be an interesting application, supplementing their production from virgin ores, and also be in agreement with the recycling-recovery concept [170].

Recently, the ability to obtain rare earth metals has been greatly reduced, due at least in part to the difficulty in obtaining the rare earth materials by mining and purification techniques which would not be hazardous to the environment. Thus, many suppliers of rare earth metals have stopped mining activities due to the high costs of mining and recovering the rare earth metals without harming the surrounding environment. Accordingly, the supply of rare earth metals is being concentrated in a few countries, including China, which due to its growing economic activity, does not find it economical to readily export the rare earth metals, but instead, use such metals for domestic consumption. Therefore, to simply discard the waste zeolite catalyst, especially when such catalyst may contain significant amounts of rare earth metals, does not make economic sense. Each year, the FCC units around the world generate about 700000-900000 t/year of spent FCC catalysts [171, 172]. The typical waste catalyst can often contain rare earth oxide contents of at least 0.5 wt.%, and a zeolite content of at least about 5 wt.%.

Ferella et al [172] summarized the main steps used in the recovery of rare earths from spent FCC catalyst, as disclosed in the US Patent 2012/0156116 A1 by Gao and Owens [31] and shown in Figure16. It outlines the recovery process of rare earths in spent FCC catalysts by extraction and further precipitation of rare earth elements according to the following procedure:

The rare earth recovery start with roasting the spent FCC catalyst at 600 °C for 2 h to remove any carbon residue, followed by acid leaching. After acid leaching, the suspension was filtered to get rid of the solid, and the pH of the solution was adjusted to 5 by adding NH_4OH solution. At this stage some precipitates of Al, Fe and Ni were formed and the solution was filtered to remove these precipitates. NH_4OH solution was again added to adjust the pH of the solution to 5.7, and the mixture heated up to 70 °C. Na_2CO_3 was then added and the system kept under stirring for 15 min, leading to the precipitation of rare earth carbonates. The rare earth carbonates are recovered by filtration and drying. Rare earth oxides are obtained by roasting the carbonates at 600 °C, with a typical wt% yield of: La_2O_3 96–97%, CeO_2 1–1.2%, Al_2O_3 < 1%, SiO_2 < 1% and Na_2O as balance.

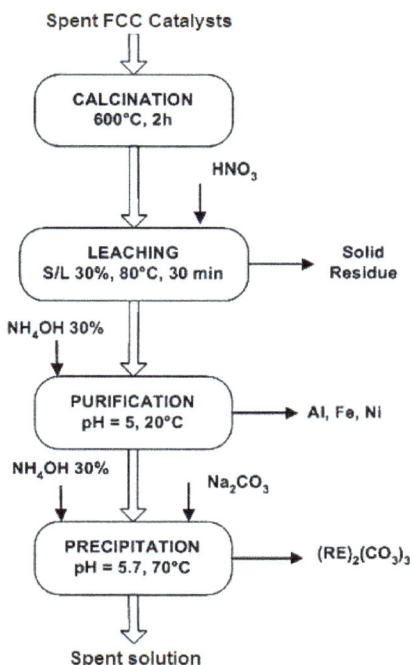

Figure 16. Schematic of recovering rare earths from spent FCC Catalysts

The recovery of rare earths from spent FCC catalysts provides an alternative way of valorizing waste catalyst meant for disposal. This is still a budding area of research and most of the work is still at laboratory scale with no proven economic evaluation. According to Ferella et al [172], the main reason is due to prices of cerium and lanthanum that are currently rather low and only a facility with very large capacity could be sustainable and viable, notwithstanding the fact that the recovery cost may be higher than the intrinsic metal value. As regards technical issues, research work is necessary to study and optimize the separation of lanthanum from cerium, and currently this can be only done by solvent extraction. Nevertheless, in the near future the extraction of rare earths could attract the attention of investors and companies since the production from primary ores are concentrated in few countries like China, Russia, South Africa, Brazil and Malaysia.

Conclusion

Rare earths are a key component of FCC catalysts, providing stability to the zeolite structures of the catalyst, as well as affecting the selectivity of the catalyst for specific reactions. The amount of rare earth influences the behavior of the zeolite active component in FCC catalysts, with regard to its response to the hydrothermal deactivation and contaminant metals. Rare earths also have a significant influence on the rate of dealumination and structural collapse of the zeolite and thus the resulting equilibrium unit cell size. This unit cell size directly impacts the catalyst activity, product distribution and octane number. The presence of rare earths influences the rates of hydrogen transfer reactions, which in turn, have an impact on catalyst deactivation by coke formation. The degree of dealumination, structural collapse and coke-on-catalyst level affect the diffusion of the feed molecules into the zeolite and the residence time of the products in the zeolite.

The continuous incorporation of rare earths in FCC catalysts is fueled by the desire for more active and hydrothermally stable catalyst with better yield performance. This is highly desirable in a situation where vanadium is predominant as a feed contaminant, because the use of rare earths can help in mitigating adverse effects of vanadium and help in stabilizing the FCC catalyst. Various studies have shown that vanadium is more mobile and migrates more readily than nickel in the catalyst particle, where it plays an important role in the effectiveness of FCC catalyst by causing structural degradation, through the interplay of hydrothermal as well as dealumination of the zeolite leading to structural collapse.

The migration of rare earth elements into smaller cages upon calcination helps in stabilizing the zeolite structures. When they are located in these cages, rare earths form bridges with framework oxygen atoms, stabilizing the zeolite structure. Some hydrolysis reactions take place over rare earth cations giving rise to Brønsted acidity. The higher the ionic radius of the rare earth cation, the higher the degree of hydrolysis. Besides, a correlation was observed between the ionic radius of the rare earth cation and the Brønsted acidity of the zeolite since an increase in the ionic radius of the rare earth cation leads an increase in Brønsted acid sites which translates into to an increase in the rate of hydrogen transfer reactions. This shows the incorporation of rare earths in commercial FCC catalysts has a knock-on effect of producing a catalyst with a tendency for high hydrogen transfer reactions, hence, a lower olefin concentration and octane number of the gasoline produced.

The demand for FCC catalysts with high activity and stability will continue to grow as rare earths will continue to make an important contribution. The use of rare earths oxides in FCC catalysts leads to enhanced catalytic activity and the prevention of loss of acid sites during FCC unit operation. It is important to realize that each FCC unit has different needs and requires different

Rare Earth - A tribute to the late Mr. Rare Earth, Professor Karl Gschneidner   Materials Research Forum LLC
Materials Research Foundations 164 (2024) 298-342   https://doi.org/10.21741/9781644903056-9

levels of rare earth loading, and that catalyst manufacturers formulate catalysts with various rare earth levels to ensure optimal performance of the unit. Feed quality, operational severity and the desired product specifications play an important role in determining the amount of rare earths in each catalyst formulation.

As the demand for FCC catalysts continue to grow, the tendency to generate more spent FCC catalysts also increases. Roughly about 700000 – 840000 t/y of spent FCC catalyst is expected to be generated, and of this amount, at least 0.5 % will be rare earth. This is an awful lot of rare earths to be wasted if all the spent catalysts were to end up in landfills. Therefore, it is important to develop a recycling strategy for spent FCC catalysts, as well as recovery techniques for rare earths.

References

[1] Swift, T.K., M.G. Moore, H.R. Rose-Glowacki, and E. Sanchez, The Economic Benefits of the North American Rare Earths Industry. 2014, Rare Earth Technology Alliance. p. 1-32.

[2] Goonan, T.G., Rare Earth Elements-End Use and Recyclability, in Scientific Investigations Report 2011-5094. 2011, U.S. Department of the Interior, U.S. Geological Survey. p. 1-22. https://doi.org/10.3133/sir20115094

[3] Curtis, N., Rare earths, we can touch them everyday. Lynas Presentation in JP Morgan Australia Corporate Access Days, 27-28 September 2010. 2010: New York.

[4] Jurd, B. and J. Nolde Lanthanum oxide product stewardship summary. https://grace.com/en-us/environment-health-and-safety/ProductStewardship/.

[5] Yung, Y. and K. Bruno, Low rare earth catalysts for FCC operations. 2012: www.digitalrefining.com/article/1000347. p. 1-10.

[6] Humphries, M., Rare Earth Elements: The Global Supply Chain, in CRS Report for Congress. 2013, Congressional Research Service, R41347. p. 1-31.

[7] Alonso, E., A.M. Sherman, T.J. Wallington, M.P. Everson, F.R. Field, R. Roth, and R.E. Kirchain, Evaluating Rare Earth Element Availability: A Case with Revolutionary Demand from Clean Technologies. Environmental Science & Technology, 2012. 46(6): p. 3406-3414. https://doi.org/10.1021/es203518d

[8] Hatch, G.P., Dynamics in the Global Market for Rare Earths. Elements, 2012. 8(5): p. 341-346. https://doi.org/10.2113/gselements.8.5.341

[9] Akah, A., Application of rare earths in fluid catalytic cracking: A review. Journal of Rare Earths, 2017. 35(10): p. 941-956. https://doi.org/10.1016/S1002-0721(17)60998-0

[10] Trovarelli, A., C. de Leitenburg, M. Boaro, and G. Dolcetti, The utilization of ceria in industrial catalysis. Catal. Today, 1999. 50(2): p. 353-367. https://doi.org/10.1016/S0920-5861(98)00515-X

[11] Occelli, M.L. and P. Ritz, The effects of Na ions on the properties of calcined rare-earths Y (CREY) zeolites. Appl. Catal. A. Gen, 1999. 183(1): p. 53-59. https://doi.org/10.1016/S0926-860X(99)00036-8

[12] Pine, L.A., Vanadium-catalyzed destruction of USY zeolites. J. Catal., 1990. 125(2): p. 514-524. https://doi.org/10.1016/0021-9517(90)90323-C

[13] de la Puente, G., E.F. Souza-Aguiar, F.M.a. Zanon Zotin, V.L. Doria Camorim, and U. Sedran, Influence of different rare earth ions on hydrogen transfer over Y zeolite. Appl. Catal. A. Gen, 2000. 197(1): p. 41-46. https://doi.org/10.1016/S0926-860X(99)00531-1

[14] Du, J., Z. Li, Y. Wang, Z. Da, J. Long, and M. He, Development of structure stabilized SSY zeolite. Stud. Surf. Sci. Catal., 2004. 154, Part C: p. 2309-2315. https://doi.org/10.1016/S0167-2991(04)80491-9

[15] Lemos, F., F. Ramo^a Ribeiro, M. Kern, G. Giannetto, and M. Guisnet, Influence of lanthanum content of LaHY catalysts on their physico-chemical and catalytic properties. Appl. Catal. A., 1988. 39: p. 227-237. https://doi.org/10.1016/S0166-9834(00)80951-3

[16] Liu, C., X. Gao, Z. Zhang, H. Zhang, S. Sun, and Y. Deng, Surface modification of zeolite Y and mechanism for reducing naphtha olefin formation in catalytic cracking reaction. Appl. Catal. A. Gen, 2004. 264(2): p. 225-228. https://doi.org/10.1016/j.gene.2011.12.006

[17] Sousa-Aguiar, E.F., F.E. Trigueiro, and F.M.Z. Zotin, The role of rare earth elements in zeolites and cracking catalysts. Catal. Today, 2013. 218-219: p. 115-122. https://doi.org/10.1016/j.cattod.2013.06.021

[18] Vogt, E.T.C. and B.M. Weckhuysen, Fluid catalytic cracking: recent developments on the grand old lady of zeolite catalysis. Chem. Soc. Rev., 2015. 44(20): p. 7342-7370. https://doi.org/10.1039/C5CS00376H

[19] Wallenstein, D., K. Schäfer, and R.H. Harding, Impact of rare earth concentration and matrix modification in FCC catalysts on their catalytic performance in a wide array of operational parameters. Appl. Catal. A. Gen, 2015. 502: p. 27-41. https://doi.org/10.1016/j.apcata.2015.05.010

[20] Wormsbecher, R., W.-C. Cheng, and D. Wallenstein, Role of the Rare Earth Elements in Fluid Catalytic Cracking. GRACE DAVISON CATALAGRAM, 2010. 108 p. 19-26.

[21] Yu, S.Q., H.P. Tian, Y.X. Zhu, Z.Y. Dai, and J. Long, Mechanism of rare earth cations on the stability and acidity of Y zeolites. Wuli Huaxue Xuebao/ Acta Physico - Chimica Sinica, 2011. 27(11): p. 2528-2534. https://doi.org/10.3866/PKU.WHXB20111101

[22] Zhan, W., Y. Guo, X. Gong, Y. Guo, Y. Wang, and G. Lu, Current status and perspectives of rare earth catalytic materials and catalysis. Chinese Journal of Catalysis, 2014. 35(8): p. 1238-1250. https://doi.org/10.1016/S1872-2067(14)60189-3

[23] Dougan, T.J., U. Alkemade, B. Lakhanpal, and L.T. Boock, New vanadium trap proven in commercial trials. Oil.Gas.J, 1994. 92(39): p. 81-91.

[24] Fei, R., L. Qianqian, and Z. Yuxia, Performance of FCC Catalyst Improved with Vanadium Trapping Components. China Petroleum Processing and Petrochemical Technology, 2014. 16(2): p. 8-11.

[25] Diddams, P., M. Evans, and R. Fletcher. Unconventional Means of Increasing Propylene Yield in Residue Operations. Paper ID : 20100362. in Petrotech-2010. 2010. New Delhi, India.

[26] Deng, C., J. Zhang, L. Dong, M. Huang, L. Bin, G. Jin, J. Gao, F. Zhang, M. Fan, L. Zhang, and Y. Gong, The effect of positioning cations on acidity and stability of the framework

structure of Y zeolite. Scientific Reports, 2016. 6: p. 23382.
https://doi.org/10.1038/srep23382

[27] Yang, S.-J., Y.-W. Chen, and L. Chiuping, The interaction of vanadium and nickel in USY zeolite. Zeolites, 1995. 15(1): p. 77-82. https://doi.org/10.1016/0144-2449(94)00010-P

[28] Yang, S.-J., Y.-W. Chen, and C. Li, Vanadium-nickel interaction in REY zeolite. Appl. Catal. A. Gen, 1994. 117(2): p. 109-123. https://doi.org/10.1016/0926-860X(94)85092-5

[29] Yang, S.-J., Y.-W. Chen, and C. Li, Metal-resistant FCC catalysts: effect of matrix. Appl. Catal. A. Gen, 1994. 115(1): p. 59-68. https://doi.org/10.1016/0378-1119(83)90167-1

[30] Moulijn, J.A., M. Makkee, and A.E. van Diepen, Chemical Process Technology. 2013: Wiley.

[31] Gao, X. and W.T. Owens, Process For Metal Recovery From Catalyst Waste, US Patent No. 20120156116A1. 2012.

[32] Akah, A. and M. Al-Ghrami, Maximizing propylene production via FCC technology. Appl Petrochem Res, 2015. 5(4): p. 377-392. https://doi.org/10.1007/s13203-015-0104-3

[33] Couch, K.A., J.P. Glavin, D.A. Wegerer, and J.A. Qafisheh FCC propylene production. Catalysis & Additives: Fluid Catalytic Cracking Propylene Maximisation, 2007. Q3, 33-43.

[34] Perego, C. and R. Millini, Porous materials in catalysis: challenges for mesoporous materials. Chem. Soc. Rev., 2013. 42(9): p. 3956-3976. https://doi.org/10.1039/C2CS35244C

[35] Silverman, L.D., W.S. Winkler, J.A. Tiethof, and A. Witoshkin, Matrix Effects in Catalytic Cracking, in NPRA Meeting. 1986, Engelhard Corporation: Los Angeles, California.

[36] Von Ballmoos, R. and C.-M.T. Hayward, Matrix vs Zeolite Contributions to the Acidity of Fluid Cracking Catalysts. Stud. Surf. Sci. Catal., 1991. 65: p. 171-183. https://doi.org/10.1016/S0167-2991(08)62905-5

[37] Gamero M, P., C. Maldonado M, J.C. Moreno M, O. Guzman M, E. Mojica M, and R. Gonzalez S, Stability of an FCC catalyst matrix for processing gas oil with resid. Stud. Surf. Sci. Catal., 1997. 111: p. 375-382. https://doi.org/10.1016/S0167-2991(97)80177-2

[38] Humphries, A. and J.R. Wilcox, Zeolite components and matrix composition determine FCC catalyst performance. Journal Name: Oil Gas J.; (United States); Journal Volume: 87:6, 1989: p. Medium: X; Size: Pages: 45-50.

[39] Al-Khattaf, S., The Influence of Alumina on the Performance of FCC Catalysts during Hydrotreated VGO Catalytic Cracking. Energy & Fuels, 2002. 17(1): p. 62-68. https://doi.org/10.1021/ef020066a

[40] Mao, R.L.V., N. Al-Yassir, and D.T.T. Nguyen, Experimental evidence for the pore continuum in hybrid catalysts used in the selective deep catalytic cracking of n-hexane and petroleum naphthas. Microporous Mesoporous Mater., 2005. 85(1-2): p. 176-182. https://doi.org/10.1016/j.micromeso.2005.05.050

[41] Yan, H.T. and R.L.V. Mao, Hybrid catalysts used in the Catalytic Steam Cracking process (CSC): Influence of the pore characteristics and the surface acidity properties of the ZSM-5

zeolite-based component on the overall catalytic performance. Appl. Catal. A. Gen, 2010 375: p. 63-69. https://doi.org/10.1016/j.apcata.2009.12.018

[42] Hargreaves, J.S.J. and A.L. Munnoch, A survey of the influence of binders in zeolite catalysis. Catalysis Science & Technology, 2013. 3 p. 1165--1171. https://doi.org/10.1039/c3cy20866d

[43] O'Connor, P. and A.P. Humphies, Accessibility of functional sites in FCC. Prepr. - Am. Chem. Soc., Div. Pet. Chem., 1993. 38: p. 598-603.

[44] Mann, R. and U.A. El-Nafaty, Probing internal structures of FCC catalyst particles: Fromparallel bundles to fractals. Stud. Surf. Sci. Catal., 1996. 100: p. 355-364. https://doi.org/10.1016/S0167-2991(96)80035-8

[45] Kuehler, C.W., R. Jonker, P. Imhof, S.J. Yanik, and P. O'Connor, Catalyst assembly technology in FCC. Part II: The influence of fresh and contaminant-affected catalyst structure on FCC performance. Stud. Surf. Sci. Catal., 2001. 134: p. 311-332. https://doi.org/10.1016/S0167-2991(01)82330-2

[46] López-Isunza, F., N. Moreno-Montiel, R. Quintana-Solórzano, J.C. Moreno-Mayorga, and F. Hernández-Beltrán, Modelling diffusion, cracking reactions and deactivation in FCC Catalysts. Stud. Surf. Sci. Catal., 2001. 133: p. 509-514. https://doi.org/10.1016/S0167-2991(01)82004-8

[47] Lu, Y., M. He, J. Song, and X. Shu, Active site accessibility of resid cracking catalysts. Stud. Surf. Sci. Catal., 2001. 134: p. 209-217. https://doi.org/10.1016/S0167-2991(01)82321-1

[48] O'Connor, P., P. Imhof, and S.J. Yanik, Catalyst assembly technology in FCC. Part I: A review of the concept, history and developments. Stud. Surf. Sci. Catal., 2001. 134: p. 299-310. https://doi.org/10.1016/S0167-2991(01)82329-6

[49] Stockwell, D.M., X. Liu, P. Nagel, P.J. Nelson, T.A. Gegan, and C.F. Keweshan, Distributed Matrix Structures-novel technology for high performance in short contact time FCC. Stud. Surf. Sci. Catal., 2004. 149: p. 257-285. https://doi.org/10.1016/S0167-2991(04)80768-7

[50] Rana, M.S., V. Sámano, J. Ancheyta, and J.A.I. Diaz, A review of recent advances on process technologies for upgrading of heavy oils and residua. Fuel, 2007. 86(9): p. 1216-1231. https://doi.org/10.1016/j.fuel.2006.08.004

[51] Sadeghbeigi, R., Fluid Catalytic Cracking Handbook: Design, Operation and Troubleshooting of FCC Facilities. 2nd ed. 2000, Austin, Texas: Gulf Publishing Company. https://doi.org/10.1016/B978-088415289-7/50009-3

[52] Rase, H.F., Handbook of Commercial Catalysts: Heterogeneous Catalysts. 2000, New York: CRC Press LLC.

[53] Andreu, P., Development of catalysts for the fluid catalytic cracking process: An example of CYTED-D program. Catal. Lett., 1993. 22(1-2): p. 135-146. https://doi.org/10.1007/BF00811774

Rare Earth - A tribute to the late Mr. Rare Earth, Professor Karl Gschneidner
Materials Research Foundations 164 (2024) 298-342

Materials Research Forum LLC
https://doi.org/10.21741/9781644903056-9

[54] Miale, J.N., N.Y. Chen, and P.B. Weisz, Catalysis by crystalline aluminosilicates: IV. Attainable catalytic cracking rate constants, and superactivity. J. Catal., 1966. 6(2): p. 278-287. https://doi.org/10.1016/0021-9517(66)90059-5

[55] Li, X., B. Shen, Q. Guo, and J. Gao, Effects of large pore zeolite additions in the catalytic pyrolysis catalyst on the light olefins production. Catal. Today, 2007. 125(3-4): p. 270-277. https://doi.org/10.1016/j.cattod.2007.03.021

[56] Zhao, X. and T.G. Roberie, ZSM-5 Additive in Fluid Catalytic Cracking. 1. Effect of Additive Level and Temperature on Light Olefins and Gasoline Olefins. Ind. Eng. Chem. Res., 1999. 38(10): p. 3847-3853. https://doi.org/10.1021/ie990179q

[57] Arandes, J.M., I. Torre, M.J. Azkoiti, J. Ereña, M. Olazar, and J. Bilbao, HZSM-5 Zeolite As Catalyst Additive for Residue Cracking under FCC Conditions. Energy & Fuels, 2009. 23(9): p. 4215-4223. https://doi.org/10.1021/ef9002632

[58] Buchanan, J.S., The chemistry of olefins production by ZSM-5 addition to catalytic cracking units. Catal. Today, 2000. 55(3): p. 207-212. https://doi.org/10.1016/S0920-5861(99)00248-5

[59] Degnan, T.F., G.K. Chitnis, and P.H. Schipper, History of ZSM-5 fluid catalytic cracking additive development at Mobil. Microporous Mesoporous Mater., 2000. 35-36(0): p. 245-252. https://doi.org/10.1016/S1387-1811(99)00225-5

[60] Abul-Hamayel, M.A., A.M. Aitani, and M.R. Saeed, Enhancement of Propylene Production in a Downer FCC Operation using a ZSM-5 Additive. Chemical Engineering & Technology, 2005. 28(8): p. 923-929. https://doi.org/10.1002/ceat.200407133

[61] Akah, A., M. Al-Ghrami, M. Saeed, and M.A.B. Siddiqui, Reactivity of naphtha fractions for light olefins production. International Journal of Industrial Chemistry, 2017. 8(2): p. 221-233. https://doi.org/10.1007/s40090-016-0106-8

[62] Arandes, J.M., I. Abajo, I. Fernández, M.J. Azkoiti, and J. Bilbao, Effect of HZSM-5 Zeolite Addition to a Fluid Catalytic Cracking Catalyst. Study in a Laboratory Reactor Operating under Industrial Conditions. Ind. Eng. Chem. Res., 2000. 39(6): p. 1917-1924. https://doi.org/10.1021/ie990335t

[63] Siddiqui, M.A.B., A.M. Aitani, M.R. Saeed, N. Al-Yassir, and S. Al-Khattaf, Enhancing propylene production from catalytic cracking of Arabian Light VGO over novel zeolites as FCC catalyst additives. Fuel, 2011. 90(2): p. 459-466. https://doi.org/10.1016/j.fuel.2010.09.041

[64] Claude, A., Holding the Key, in HydrocarbonProcessing. 2008, Hydrocarbon Engineering: www.hydrocarbonengineering.com. p. 1-7.

[65] Larocca, M., H. De Lasa, H. Farag, and S. Ng, Cracking catalysts deactivation by nickel and vanadium contaminants. Ind. Eng. Chem. Res., 1990. 29(11): p. 2181-2191. https://doi.org/10.1021/ie00107a002

[66] Schubert, P.F. and C.A. Altomare, Effects of Ni and V in Catalysts on Contaminant Coke and Hydrogen Yields. ACS Symposium Series: Fluid Catalytic Cracking, 1988. 375(375): p. 182-194. https://doi.org/10.1021/bk-1988-0375.ch011

[67] Bayraktar, O. and E.L. Kugler, Effect of pretreatment on the performance of metal-contaminated fluid catalytic cracking (FCC) catalysts. Appl. Catal. A. Gen, 2004. 260(1): p. 119-124. https://doi.org/10.1016/j.apcata.2003.10.012

[68] Etim, U.J., B. Xu, Z. Zhang, Z. Zhong, P. Bai, K. Qiao, and Z. Yan, Improved catalytic cracking performance of USY in the presence of metal contaminants by post-synthesis modification. Fuel, 2016. 178: p. 243-252. https://doi.org/10.1016/j.fuel.2016.03.060

[69] Jeon, H.J., S.K. Park, and S.I. Woo, Evaluation of vanadium traps occluded in resid fluidized catalytic cracking (RFCC) catalyst for high gasoline yield. Appl. Catal. A. Gen, 2006. 306: p. 1-7. https://doi.org/10.1016/j.apcata.2006.02.048

[70] Sandoval-Díaz, L.-E., J.-M. Martínez-Gil, and C.A. Trujillo, The combined effect of sodium and vanadium contamination upon the catalytic performance of USY zeolite in the cracking of n-butane: Evidence of path-dependent behavior in Constable-Cremer plots. J. Catal., 2012. 294: p. 89-98. https://doi.org/10.1016/j.jcat.2012.07.009

[71] Wallenstein, D., T. Roberie, and T. Bruhin, Review on the deactivation of FCC catalysts by cyclic propylene steaming. Catal. Today, 2007. 127(1-4): p. 54-69. https://doi.org/10.1016/j.cattod.2007.05.023

[72] Wallenstein, D., R.H. Harding, J. Witzler, and X. Zhao, Rational assessment of FCC catalyst performance by utilization of micro-activity testing. Appl. Catal. A. Gen, 1998. 167(1): p. 141-155. https://doi.org/10.1016/S0926-860X(97)00307-4

[73] Young, G.W., Chapter 8 Realistic Assessment of FCC Catalyst Performance in the Laboratory. Stud. Surf. Sci. Catal., 1993. 76: p. 257-292. https://doi.org/10.1016/S0167-2991(08)63831-8

[74] Boock, L.T. and X. Zhao, Recent advances in FCC catalyst evaluations : MAT vs. DCR pilot plant results. Fluid Cracking Catalysts, ed. M.L. Occelli. 1996, Washington, DC, US: American Chemical Society.

[75] Corma, A. and L. Sauvanaud, FCC testing at bench scale: New units, new processes, new feeds. Catal. Today, 2013. 218-219: p. 107-114. https://doi.org/10.1016/j.cattod.2013.03.038

[76] Catalyst deactivation model used to select laboratory procedures for fcc catalyst testing, in Catalyst Reports, Section 2: FCCU Catalyst Testing and Sampling BASF: http://www.refiningonline.com/BASFCatalystsKB/.

[77] Kostaras, K., C. Ziogou, A. Lappas, and S. Voutetakis, Design engineering, implementation and control of a flexible cyclic deactivation unit. Chemical Engineering Transactions, 2009. 18: p. 563-568.

[78] Hettinger, W.P., Catalysis challenges in fluid catalytic cracking: a 49 year personal account of past and more recent contributions and some possible new and future directions for even further improvement. Catal. Today, 1999. 53(3): p. 367-384. https://doi.org/10.1016/S0920-5861(99)00131-5

[79] Ihli, J., R.R. Jacob, M. Holler, M. Guizar-Sicairos, A. Diaz, J.C. da Silva, D. Ferreira Sanchez, F. Krumeich, D. Grolimund, M. Taddei, W.C. Cheng, Y. Shu, A. Menzel, and J.A. van Bokhoven, A three-dimensional view of structural changes caused by deactivation of

fluid catalytic cracking catalysts. Nature Communications, 2017. 8(1): p. 809.
https://doi.org/10.1038/s41467-017-00789-w

[80] Fernández, M.L., A. Lacalle, J. Bilbao, J.M. Arandes, G. de la Puente, and U. Sedran,
Recycling Hydrocarbon Cuts into FCC Units. Energy & Fuels, 2002. 16(3): p. 615-621.
https://doi.org/10.1021/ef010184i

[81] Devard, A., G. de la Puente, F. Passamonti, and U. Sedran, Processing of resid-VGO
mixtures in FCC: Laboratory approach. Applied Catalysis A: General, 2009. 353(2): p. 223-
227. https://doi.org/10.1016/j.apcata.2008.10.036

[82] Devard, A., G. de la Puente, and U. Sedran, Laboratory evaluation of the impact of the
addition of resid in FCC. Fuel Processing Technology, 2009. 90(1): p. 51-55.
https://doi.org/10.1016/j.fuproc.2008.07.009

[83] Lengyel, A., S. Magyar, and J. Hancsók, Upgrading of Delayed Coker Light Naphtha in a
Crude Oil Refinery Petroleum & Coal 2009. 51(2): p. 80-90.

[84] Torre, I., J.M. Arandes, M.J. Azkoiti, M. Olazar, and J. Bilbao, Cracking of Coker Naphtha
with Gas−Oil. Effect of HZSM-5 Zeolite Addition to the Catalyst. Energy & Fuels, 2007.
21(1): p. 11-18. https://doi.org/10.1021/ef060344w

[85] Ruiz-Martínez, J., A.M. Beale, U. Deka, M.G. O'Brien, P.D. Quinn, J.F.W. Mosselmans,
and B.M. Weckhuysen, Correlating Metal Poisoning with Zeolite Deactivation in an
Individual Catalyst Particle by Chemical and Phase-Sensitive X-ray Microscopy. Angewandte
Chemie (International Ed. in English), 2013. 52(23): p. 5983-5987.
https://doi.org/10.1002/anie.201210030

[86] Meirer, F., S. Kalirai, D. Morris, S. Soparawalla, Y. Liu, G. Mesu, J.C. Andrews, and B.M.
Weckhuysen, Life and death of a single catalytic cracking particle. Science Advances, 2015.
1(3): p. e1400199. https://doi.org/10.1126/sciadv.1400199

[87] Wallenstein, D., D. Farmer, J. Knoell, C.M. Fougret, and S. Brandt, Progress in the
deactivation of metals contaminated FCC catalysts by a novel catalyst metallation method.
Appl. Catal. A. Gen, 2013. 462-463: p. 91-99. https://doi.org/10.1016/j.apcata.2013.02.002

[88] Tangstad, E., A. Andersen, E.M. Myhrvold, and T. Myrstad, Catalytic behaviour of nickel
and iron metal contaminants of an FCC catalyst after oxidative and reductive thermal
treatments. Appl. Catal. A. Gen, 2008. 346(1-2): p. 194-199.
https://doi.org/10.1016/j.apcata.2008.05.022

[89] Reynolds, J.G., NICKEL IN PETROLEUM REFINING. Pet. Sci. Technol., 2001. 19(7-8):
p. 979-1007. https://doi.org/10.1081/LFT-100106915

[90] Roncolatto, R.E. and Y.L. Lam, Effect of vanadium on the deactivation of fcc catalysts.
Brazilian Journal of Chemical Engineering, 1998. 15(2): p. http://dx.doi.org/10.1590/S0104-
66321998000200002. https://doi.org/10.1590/S0104-66321998000200002

[91] Wormsbecher, R.F., A.W. Peters, and J.M. Maselli, Vanadium poisoning of cracking
catalysts: Mechanism of poisoning and design of vanadium tolerant catalyst system. J. Catal.,
1986. 100(1): p. 130-137. https://doi.org/10.1016/0021-9517(86)90078-3

[92] Trujillo, C.A., U.N. Uribe, P.-P. Knops-Gerrits, L.A. Oviedo A, and P.A. Jacobs, The Mechanism of Zeolite Y Destruction by Steam in the Presence of Vanadium. J. Catal., 1997. 168(1): p. 1-15. https://doi.org/10.1006/jcat.1997.1550

[93] Xu, M., X. Liu, and R.J. Madon, Pathways for Y Zeolite Destruction: The Role of Sodium and Vanadium. J. Catal., 2002. 207(2): p. 237-246. https://doi.org/10.1006/jcat.2002.3517

[94] Lerner, B. and M. Deeba, Improved Methods for Testing and Assessing Deactivation from Vanadium Interaction with Fluid Catalytic Cracking Catalyst. ACS Symposium Series: Deactivation and Testing of Hydrocarbon-Processing Catalysts, 1996. 634: p. 296-311. https://doi.org/10.1021/bk-1996-0634.ch022

[95] O'Connor, P., T. Takatsuka, and G.L. Woolery, Deactivation and Testing of Hydrocarbon-Processing Catalysts. ACS Symposium Series. Vol. 634. 1996: American Chemical Society. 468. https://doi.org/10.1021/bk-1996-0634

[96] Harding, R.H., A.W. Peters, and J.R.D. Nee, New developments in FCC catalyst technology. Appl. Catal. A. Gen, 2001. 221(1-2): p. 389-396. https://doi.org/10.1016/S0926-860X(01)00814-6

[97] Wormsbecher, R.F., W.-C. Cheng, G. Kim, and R.H. Harding, Vanadium Mobility in Fluid Catalytic Cracking. ACS Symposium Series: Deactivation and Testing of Hydrocarbon-Processing Catalysts, 1996. 634: p. 283-295. https://doi.org/10.1021/bk-1996-0634.ch021

[98] Pope, M.T., Comprehensive Coordination Chemistry, ed. G. Wilkinson. Vol. 3. 1987: Pergamon, Oxford. p1026.

[99] Huifang, P., W. Xiaofeng, T. Aijun, S. Zhihong, and Z. Gaoshan, The design of vanadium trapping system for fcc catalysts. Chin.J.Chem.Eng., 1996. V4(2): p. 120-124.

[100] Cristiano-Torres, D.V., Y. Osorio-Pérez, L.A. Palomeque-Forero, L.E. Sandoval-Díaz, and C.A. Trujillo, The action of vanadium over Y zeolite in oxidant and dry atmosphere. Appl. Catal. A. Gen, 2008. 346(1-2): p. 104-111. https://doi.org/10.1016/j.apcata.2008.05.006

[101] Sorokina, T.P., L.A. Buluchevskaya, O.V. Potapenko, and V.P. Doronin, Conversion of nickel and vanadium porphyrins under catalytic cracking conditions. Petroleum Chemistry, 2010. 50(1): p. 51-55. https://doi.org/10.1134/S096554411001007X

[102] Dechaine, G.P. and M.R. Gray, Chemistry and Association of Vanadium Compounds in Heavy Oil and Bitumen, and Implications for Their Selective Removal. Energy & Fuels, 2010. 24(5): p. 2795-2808. https://doi.org/10.1021/ef100173j

[103] Woltermann, G.M., G. Dodwell, and B. Lerner, Modern Cracking Catalyst and Residue Processing Challenges, in NPRA Annual Meeting, March 17-19, 1996. 1996, http://www.refiningonline.com/engelhardkb/npra/NPR9646.htm: San Antonio, Texas.

[104] Graaf, B.d., Y. Tang, J. Oberlin, and P. Diddams Shale crudes and FCC: A mismatch from heaven? Processing Shale Feedstocks 2014. www.digitalrefining.com/article/1000921.

[105] Osorio Pérez, Y., L.A.P. Forero, D.V.C. Torres, and C.A. Trujillo, Brønsted acid site number evaluation using isopropylamine decomposition on Y-zeolite contaminated with vanadium in a simultaneous DSC-TGA analyzer. Thermochim. Acta, 2008. 470(1-2): p. 36-39. https://doi.org/10.1016/j.tca.2008.01.016

[106] Etim, U.J., B. Xu, R. Ullah, and Z. Yan, Effect of vanadium contamination on the framework and micropore structure of ultra stable Y-zeolite. J. Colloid Interface Sci., 2016. 463: p. 188-198. https://doi.org/10.1016/j.jcis.2015.10.049

[107] Hagiwara, K., T. Ebihara, N. Urasato, S. Ozawa, and S. Nakata, Effect of vanadium on USY zeolite destruction in the presence of sodium ions and steam-studies by solid-state NMR. Appl. Catal. A. Gen, 2003. 249(2): p. 213-228. https://doi.org/10.1016/S0926-860X(03)00289-8

[108] Maholland, M.K. Improving FCC catalyst performance. Catalysis, 2006. www.digitalrefining.com/article/1000302.

[109] Trujillo, C.A., U.N. Uribe, and L.A.O. Aguiar, Vanadium traps for catalyst for catalytic cracking. 2000, Google Patents.

[110] Huai-Ping, W., W. Fang-Zhu, and W. Wen-Ru Effect of vanadium poisoning and vanadium passivation on the structure and properties of rehy zeolite and fcc catalyst. ACS Fuels 2000. 45, 623-628.

[111] Fe'ron, B.a., P. Gallezot, and M. Bourgogne, Hydrothermal aging of cracking catalysts: V. Vanadium passivation by rare-earth compounds soluble in the feedstock. J. Catal., 1992. 134(2): p. 469-478. https://doi.org/10.1016/0021-9517(92)90335-F

[112] Jeon, H.J., S.K. Park, and S.I. Woo, Evaluation of vanadium traps occluded in resid fluidized catalytic cracking (RFCC) catalyst for high gasoline yield. Appl. Catal. A. Gen, 2006. 306: p. 1-7. https://doi.org/10.1016/j.apcata.2006.02.048

[113] Myrstad, T., Effect of vanadium on octane numbers in FCC-naphtha. Appl. Catal. A. Gen, 1997. 155(1): p. 87-98. https://doi.org/10.1016/S0926-860X(96)00403-6

[114] Wang, H.-P., F.-Z. Wang, and W.-R. Wu, Effect of vanadium poisoning and vanadium passivation on the structure and properties of rehy zeolite and FCC catalyst ACS Division of Fuel Chemistry, Preprints, 2000. 45(3): p. 623-627.

[115] Baugis, G.L., H.F. Brito, W. de Oliveira, F. Rabello de Castro, and E.F. Sousa-Aguiar, The luminescent behavior of the steamed EuY zeolite incorporated with vanadium and rare earth passivators. Microporous Mesoporous Mater., 2001. 49(1-3): p. 179-187. https://doi.org/10.1016/S1387-1811(01)00416-4

[116] Biswas, J. and I.E. Maxwell, Recent process- and catalyst-related developments in fluid catalytic cracking. Appl. Catal. A., 1990. 63(1): p. 197-258. https://doi.org/10.1016/S0166-9834(00)81716-9

[117] Li, B., S. Li, N. Li, C. Liu, X. Gao, and X. Pang, Structure and acidity of REHY Zeolite in FCC catalyst. Chinese Journal of Catalysis, 2005. 26(4): p. 301-306.

[118] Corma, A., V. Fornes, J.B. Monton, and A.V. Orchilles, Structural and cracking properties of REHY zeolites. Activity, selectivity, and catalyst-decay optimization for n-heptane cracking. Industrial & Engineering Chemistry Product Research and Development, 1986. 25(2): p. 231-238. https://doi.org/10.1021/i300022a018

[119] Du, X., H. Zhang, G. Cao, L. Wang, C. Zhang, and X. Gao, Effects of La2O3, CeO2 and LaPO4 introduction on vanadium tolerance of USY zeolites. Microporous Mesoporous Mater., 2015. 206: p. 17-22. https://doi.org/10.1016/j.micromeso.2014.12.010

[120] Kugler, E.L. and D.P. Leta, Nickel and vanadium on equilibrium cracking catalysts by imaging secondary ion mass spectrometry. J. Catal., 1988. 109(2): p. 387-395. https://doi.org/10.1016/0021-9517(88)90221-7

[121] Corma, A. and J.M. López Nieto, Chapter 185 The use of rare-earth-containing zeolite catalysts, in Handbook on the Physics and Chemistry of Rare Earths. 2000, Elsevier. p. 269-313. https://doi.org/10.1016/S0168-1273(00)29008-9

[122] Baillie, C. and R. Schiller Zero and low rare earth FCC catalysts. PTQ Q4, 2011. www.digitalrefining.com/article/1000137.

[123] Du, X., H. Zhang, X. Gao, Z. He, and Z. Li, Effect of nickel and vanadium on structure and catalytic performance of FCC catalyst. Shiyou Xuebao, Shiyou JiagongActa Petrolei Sinica (Petroleum Processing Section), 2015. 31(5): p. 1063-1068.

[124] Du, X., X. Li, H. Zhang, and X. Gao, Kinetics study and analysis of zeolite Y destruction. Chinese Journal of Catalysis, 2016. 37(2): p. 316-323. https://doi.org/10.1016/S1872-2067(15)60975-5

[125] Moreira, C.R., M.H. Herbst, P.R. de la Piscina, J.-L.G. Fierro, N. Homs, and M.M. Pereira, Evidence of multi-component interaction in a V-Ce-HUSY catalyst: Is the cerium-EFAL interaction the key of vanadium trapping? Microporous Mesoporous Mater., 2008. 115(3): p. 253-260. https://doi.org/10.1016/j.micromeso.2008.01.043

[126] Moreira, C.R., N. Homs, J.L.G. Fierro, M.M. Pereira, and P. Ramírez de la Piscina, HUSY zeolite modified by lanthanum: Effect of lanthanum introduction as a vanadium trap. Microporous Mesoporous Mater., 2010. 133(1-3): p. 75-81. https://doi.org/10.1016/j.micromeso.2010.04.017

[127] Cerqueira, H.S., G. Caeiro, L. Costa, and F. Ramôa Ribeiro, Deactivation of FCC catalysts. J. Mol. Catal. A: Chem., 2008. 292(1-2): p. 1-13. https://doi.org/10.1016/j.molcata.2008.06.014

[128] Maugé, F., P. Gallezot, J.-C. Courcelle, P. Engelhard, and J. Grosmangin, Hydrothermal aging of cracking catalysts. II. Effect of steam and sodium on the structure of LaHY zeolites. Zeolites, 1986. 6(4): p. 261-266. https://doi.org/10.1016/0144-2449(86)90078-3

[129] Lemos, F., F.R. Ribeiro, M. Kern, G. Giannetto, and M. Guisnet, Influence of the cerium content of CeHY catalysts on their physicochemical and catalytic properties. Appl. Catal. A., 1987. 29(1): p. 43-54. https://doi.org/10.1016/S0166-9834(00)82605-6

[130] Topete, O., C. Baillie, and R. Schiller, Optimizing FCC Operations in a High Rare-Earth Cost World: Commercial: Update on Grace Davison's Low and Zero Rare-Earth FCC Catalysts, in GRACE DAVISON CATALAGRAM. p. 2-12.

[131] Ismail, S., Fluid Catalytic Cracking (FCC) Catalyst Optimization to Cope with High Rare Earth Oxide Price Environment, in BASF Technical Note. 2011, BASF: www.catalysts.basf.com/refining.

[132] Scherzer, J., Octane-Enhancing, Zeolitic FCC Catalysts: Scientific and Technical Aspects. Catalysis Reviews, 1989. 31(3): p. 215-354. https://doi.org/10.1080/01614948909349934

[133] Rahimi, N. and R. Karimzadeh, Catalytic cracking of hydrocarbons over modified ZSM-5 zeolites to produce light olefins: A review. Appl. Catal. A. Gen, 2011. 398: p. 1-17. https://doi.org/10.1016/j.apcata.2011.03.009

[134] Ding, F., S.H. Ng, C. Xu, and S. Yui, Reduction of light cycle oil in catalytic cracking of bitumen-derived crude HGOs through catalyst selection. Fuel Process. Technol., 2007. 88(9): p. 833-845. https://doi.org/10.1016/j.fuproc.2006.12.009

[135] Shimada, I., K. Takizawa, H. Fukunaga, N. Takahashi, and T. Takatsuka, Catalytic cracking of polycyclic aromatic hydrocarbons with hydrogen transfer reaction. Fuel, 2015. 161: p. 207-214. https://doi.org/10.1016/j.fuel.2015.08.051

[136] Gao, X., Z. Qin, B. Wang, X. Zhao, J. Li, H. Zhao, H. Liu, and B. Shen, High silica REHY zeolite with low rare earth loading as high-performance catalyst for heavy oil conversion. Appl. Catal. A. Gen, 2012. 413-414: p. 254-260. https://doi.org/10.1016/j.apcata.2011.11.015

[137] Hunger, M., G. Engelhardt, and J. Weitkamp, Solid-state 23Na, 139La, 27Al and 29Si nuclear magnetic resonance spectroscopic investigations of cation location and migration in zeolites LaNaY. Microporous Mater., 1995. 3(4): p. 497-510. https://doi.org/10.1016/0927-6513(94)00061-Y

[138] Trigueiro, F.E., D.F.J. Monteiro, F.M.Z. Zotin, and E. Falabella Sousa-Aguiar, Thermal stability of Y zeolites containing different rare earth cations. J. Alloys Compd., 2002. 344(1-2): p. 337-341. https://doi.org/10.1016/S0925-8388(02)00381-X

[139] BASF, Effects of Rare Earth Oxides in FCC Catalysts, in Catalyst Reports: Section 4- The Effects of FCC Catalyst Characteristics on FCC Yields and Product Properties. http://www.refiningonline.com/engelhardkb/. BASF Catalysts.

[140] Shu, Y., A. Travert, R. Schiller, M. Ziebarth, R. Wormsbecher, and W.-C. Cheng, Effect of Ionic Radius of Rare Earth on USY Zeolite in Fluid Catalytic Cracking: Fundamentals and Commercial Application. Top. Catal., 2015. 58(4): p. 334-342. https://doi.org/10.1007/s11244-015-0374-0

[141] Liu, X., S. Liu, and Y. Liu, A potential substitute for CeY zeolite used in fluid catalytic cracking process. Microporous Mesoporous Mater., 2016. 226: p. 162-168. https://doi.org/10.1016/j.micromeso.2015.12.046

[142] Du, X., X. Gao, H. Zhang, X. Li, and P. Liu, Effect of cation location on the hydrothermal stability of rare earth-exchanged Y zeolites. Catal. Commun., 2013. 35: p. 17-22. https://doi.org/10.1016/j.catcom.2013.02.010

[143] Schüßler, F., E.A. Pidko, R. Kolvenbach, C. Sievers, E.J.M. Hensen, R.A. van Santen, and J.A. Lercher, Nature and Location of Cationic Lanthanum Species in High Alumina Containing Faujasite Type Zeolites. J. Phys. Chem. C, 2011. 115(44): p. 21763-21776. https://doi.org/10.1021/jp205771e

[144] Nery, J.G., Y.P. Mascarenhas, T.J. Bonagamba, N.C. Mello, and E.F. Souza-Aguiar, Location of cerium and lanthanum cations in CeNaY and LaNaY after calcination. Zeolites, 1997. 18(1): p. 44-49. https://doi.org/10.1016/S0144-2449(96)00094-2

[145] Nery, J.G., M.V. Giotto, Y.P. Mascarenhas, D. Cardoso, F.M.Z. Zotin, and E.F. Sousa-Aguiar, Rietveld refinement and solid state NMR study of Nd-, Sm-, Gd-, and Dy-containing Y zeolites. Microporous Mesoporous Mater., 2000. 41(1-3): p. 281-293. https://doi.org/10.1016/S1387-1811(00)00304-8

[146] Moreira, C.R., M.M. Pereira, X. Alcobé, N. Homs, J. Llorca, J.L.G. Fierro, and P. Ramírez de la Piscina, Nature and location of cerium in Ce-loaded Y zeolites as revealed by HRTEM and spectroscopic techniques. Microporous Mesoporous Mater., 2007. 100(1-3): p. 276-286. https://doi.org/10.1016/j.micromeso.2006.11.019

[147] Scherzer, J., Chapter 5 Correlation Between Catalyst Formulation and Catalytic Properties. Stud. Surf. Sci. Catal., 1993. 76: p. 145-182. https://doi.org/10.1016/S0167-2991(08)63828-8

[148] Huang, J., Y. Jiang, V.R. Reddy Marthala, Y.S. Ooi, J. Weitkamp, and M. Hunger, Concentration and acid strength of hydroxyl groups in zeolites La,Na-X and La,Na-Y with different lanthanum exchange degrees studied by solid-state NMR spectroscopy. Microporous Mesoporous Mater., 2007. 104(1-3): p. 129-136. https://doi.org/10.1016/j.micromeso.2007.01.016

[149] Moscou, L. and M. Lakeman, Acid sites in rare-earth exchanged Y-zeolites. J. Catal., 1970. 16(2): p. 173-180. https://doi.org/10.1016/0021-9517(70)90211-3

[150] Bolton, A.P., The nature of rare-earth exchanged Y zeolites. J. Catal., 1971. 22(1): p. 9-15. https://doi.org/10.1016/0021-9517(71)90259-4

[151] Kovacheva, P., C. Bezoukhanova, and C. Dimitrov, Acidity and catalytic activity of rare-earth containing X zeolites. React. Kinet. Catal. Lett., 1978. 8(4): p. 495-499. https://doi.org/10.1007/BF02074464

[152] Noda, T., K. Suzuki, N. Katada, and M. Niwa, Combined study of IRMS-TPD measurement and DFT calculation on Brønsted acidity and catalytic cracking activity of cation-exchanged Y zeolites. J. Catal., 2008. 259(2): p. 203-210. https://doi.org/10.1016/j.jcat.2008.08.004

[153] Suzuki, K., T. Noda, N. Katada, and M. Niwa, IRMS-TPD of ammonia: Direct and individual measurement of Brønsted acidity in zeolites and its relationship with the catalytic cracking activity. J. Catal., 2007. 250(1): p. 151-160. https://doi.org/10.1016/j.jcat.2007.05.024

[154] Wang, Y., Y. Cui, Y. Suo, and W. Zhang, Influences of cerium on structure and catalytic performance of n-heptane hydroisomerization of Ni-HPW/MCM-48. Journal of Rare Earths, 2015. 33(1): p. 46-55. https://doi.org/10.1016/S1002-0721(14)60382-3

[155] Martins, A., J.M. Silva, C. Henriques, F.R. Ribeiro, and M.F. Ribeiro, Influence of rare earth elements La, Nd and Yb on the acidity of H-MCM-22 and H-Beta zeolites. Catal. Today, 2005. 107-108: p. 663-670. https://doi.org/10.1016/j.cattod.2005.07.048

[156] Lee, J., U.G. Hong, S. Hwang, M.H. Youn, and I.K. Song, Catalytic cracking of C5 raffinate to light olefins over lanthanum-containing phosphorous-modified porous ZSM-5: Effect of lanthanum content. Fuel Process. Technol., 2013. 109: p. 189-195. https://doi.org/10.1016/j.fuproc.2012.10.017

[157] Xiaoning, W., Z. Zhen, X. Chunming, D. Aijun, Z. Li, and J. Guiyuan, Effects of Light Rare Earth on Acidity and Catalytic Performance of HZSM-5 Zeolite for Catalytic Cracking of Butane to Light Olefins. Journal of Rare Earths, 2007. 25(3): p. 321-328. https://doi.org/10.1016/S1002-0721(07)60430-X

[158] Roelofsen, J.W., H. Mathies, R.L. de Groot, P.C.M. van Woerkom, and H.A. Gaur, Effect of Rare Earth Loading in Y-Zeolite on its Dealumination during Thermal Treatment. Stud. Surf. Sci. Catal., 1986. 28: p. 337-344. https://doi.org/10.1016/S0167-2991(09)60891-0

[159] Amano, T., J. Wilcox, C. Pouwels, and T. Matsuura Process and catalysis factors to maximise propylene output. Catalysts & Additives, 2012. 139, 1-11.

[160] Abbot, J. and B.W. Wojciechowski, Hydrogen transfer reactions in the catalytic cracking of paraffins. J. Catal., 1987. 107(2): p. 451-462. https://doi.org/10.1016/0021-9517(87)90309-5

[161] Corma, A., M. Faraldos, and A. Mifsud, Influence of the level of dealumination on the selective adsorption of olefins and paraffins and its implication on hydrogen transfer reactions during catalytic cracking on USY zeolites. Appl. Catal. A., 1989. 47(1): p. 125-133. https://doi.org/10.1016/S0166-9834(00)83268-6

[162] Des Rochettes, B.M., C. Marcilly, C. Gueguen, and J. Bousquet, Kinetic study of hydrogen transfer of olefins under catalytic cracking conditions. Appl. Catal. A., 1990. 58(1): p. 35-52. https://doi.org/10.1016/S0166-9834(00)82277-0

[163] Sertić-Bionda, K., V. Kuzmić, and M. Jednačak, The influence of process parameters on catalytic cracking LPG fraction yield and composition. Fuel Process. Technol., 2000. 64(1-3): p. 107-115. https://doi.org/10.1016/S0378-3820(99)00124-1

[164] Zhao, X. and R.H. Harding, ZSM-5 Additive in Fluid Catalytic Cracking. 2. Effect of Hydrogen Transfer Characteristics of the Base Cracking Catalysts and Feedstocks. Ind. Eng. Chem. Res., 1999. 38(10): p. 3854-3859. https://doi.org/10.1021/ie990180p

[165] Zhang, L., Q. Li, Y. Qin, X. Zhang, X. Gao, and L. Song, Investigation on the mechanism of adsorption and desorption behavior in cerium ions modified Y-type zeolite and improved hydrocarbons conversion. Journal of Rare Earths, 2016. 34(12): p. 1221-1227. https://doi.org/10.1016/S1002-0721(16)60157-6

[166] BASF, The Effects of FCC Catalyst Characteristics on FCC Yields and Product Properties in Catalyst Reports. Section 4: . http://www.refiningonline.com/engelhardkb/.

[167] Keyvanloo, K. and M. Sadeqzadeh. Effect of Element Modifications on Catalytic Performance of Zeolites. in Proceedings of the World Congress on Engineering and Computer Science 2008, WCECS 2008. 2008. San Francisco, USA.

[168] Vierheilig, A.A., Methods of recovering rare earth elements, US Patent No. 8216532B1 2011.

[169] Furimsky, E., Spent refinery catalysts: Environment, safety and utilization. Catal. Today, 1996. 30(4): p. 223-286. https://doi.org/10.1016/0920-5861(96)00094-6

[170] Silva, J.S.A., T.d.A. Maranhão, F.J.S.d. Oliveira, A.J. Curtius, and V.L.A. Frescura, Determination of rare earth elements in spent catalyst samples from oil refinery by dynamic

reaction cell inductively coupled plasma mass spectrometry. Journal of the Brazilian Chemical Society, 2014. 25: p. 1062-1070. https://doi.org/10.5935/0103-5053.20140079

[171] Ye, S., Y. Jing, Y. Wang, and W. Fei, Recovery of rare earths from spent FCC catalysts by solvent extraction using saponified 2-ethylhexyl phosphoric acid-2-ethylhexyl ester (EHEHPA). Journal of Rare Earths, 2017. 35(7): p. 716-722. https://doi.org/10.1016/S1002-0721(17)60968-2

[172] Ferella, F., V. Innocenzi, and F. Maggiore, Oil refining spent catalysts: A review of possible recycling technologies. Resources, Conservation and Recycling, 2016. 108: p. 10-20. https://doi.org/10.1016/j.resconrec.2016.01.010

Rare Earth - A tribute to the late Mr. Rare Earth, Professor Karl Gschneidner   Materials Research Forum LLC
Materials Research Foundations 164 (2024) 343-368   https://doi.org/10.21741/9781644903056-10

Chapter 10

Energy Transfer in Down Conversion Rare Earth Phosphors

Chaogang Lou

School of Electronic Science and Engineering, Southeast University, P.R. China

Abstract

For co-doped and tri-doped rare earth (RE) down conversion phosphors, the energy transfer between different rare earth ions plays an important role because they determine the luminous efficiency of the phosphors. In this chapter, some popular energy transfer mechanisms are presented at first, including cooperative energy transfer (CET), cross relaxation and charge transfer state (CTS). Then, the energy transfers in the co-doped and tri-doped $Y_3Al_5O_{12}$ phosphors are discussed. It is shown that the efficiency of the energy transfer between the rare earth ions varies with different ion couples, which depends on the energy levels of these ions. In the down conversion phosphors with two or more rare earth elements, the energy transfer is an important process because they have significant effects on the luminous efficiency of the phosphors [1]. In principle, the energy transfer seems a simple process in which the sensitizers (donors) in the phosphors absorb the energy of incident photons at first, then the part of the energy is emitted luminescently and the rest is transferred to neighbouring activators (acceptors). After that, the activators emit the photons whose wavelength is different from that emitted from the sensitizers. However, due to the complexity of host lattices and the abundant energy levels of rare earth ions, clarifying the mechanism of the energy transfer is a challenge. In this chapter, some proposed mechanisms of the energy transfer are presented, including cooperative energy transfer（CET), cross relaxation, charge transfer state (CTS), etc. Then, the energy transfer processes in co-doped and tri doped $Y_3Al_5O_{12}$ (YAG) phosphors are discussed.

Contents

1.   Three mechanisms of the energy transfer in rare earth phosphors

1.1   Cooperative energy transfer

The mechanism of the cooperative energy transfer is given in figure 1. After an electron of a sensitizer absorbs the energy of a photon and the sensitizer is excited from the ground state to an higher excited state, it usually relaxes to a lower excited state at first. Then part of the energy might be emitted as photons, and the rest is transferred to two neighbouring activators.

Clearly, the cooperative energy transfer requires that the difference between the excited state and the ground state of the sensitizer should be at least twice as high as that of the activator and the distance between the sensitizer and the activator should be very small. It is a one-step process in which the energy transfer to two neighbouring ions is completed simultaneously.

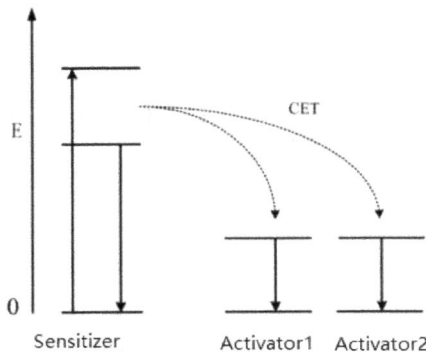

Figure 1 Schematic of the cooperative energy transfer

1.2   Cross relaxation energy transfer

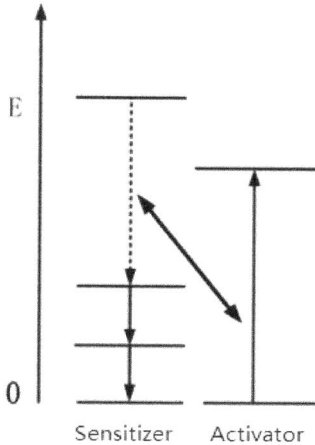

Figure 2 Schematic of the energy transfer by cross relaxation

Figure 2 shows schematically the energy transfer by cross relaxation. When the sensitizer transits from a higher excited state to a lower excited state (or the ground state), if the difference between these two states is equal to the difference between the excited state and the ground state of the neighbouring activator, the activator will be excited. So the energy transfer by cross relaxation requires that the activator should have two states whose difference is similar to that of the sensitizer. The cross relaxation is usually considered as a process of dipole-dipole interaction.

1.3   Charge transfer state (CTS)

Energy transfer by charge transfer state is another type of the energy transfer between the rare earth ions, and its mechanism is more complicated. Different from the cooperative energy transfer and the cross relaxation whose energy transfer are only relevant to the energy levels of the rare earth ions, the energy transfer by charge transfer state needs the charge transfer state which results from the bonding between sensitizers, O^{2-} ions and activators [2]. Its mechanism is schematically shown in figure 3. The charge transfer state S-O-A intersects with the excited state of the sensitizer and that of the activator. After an electron absorbs the energy of a photon and the sensitizer is excited from the ground state to its excited state, the electron may increase its energy under the assistance of phonons until it reaches the intersection of the charge transfer state and the excited state of the sensitizer. After that, the electron relaxes or rises to the intersection of the charge transfer state and the excited state of the activator and transfer the energy to the activator. In this process, the assistance of the phonons is important because they help the electron increase its energy so that it can transit from the excited state of the sensitizer to the charge transfer state and from the charge transfer state to the excited state of the activator.

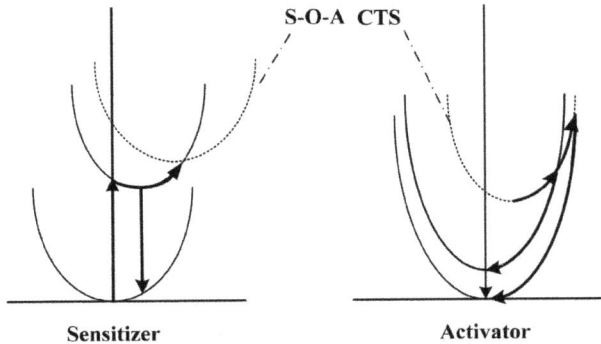

Figure 3 Schematic of the energy transfer via charge transfer state

2. Energy transfer in co-doped $Y_3Al_5O_{12}$ (YAG) phosphors

Ce^{3+}-doped YAG phosphors have been successfully used in semiconductor lighting industry for many years, so they have been investigated extensively. Here, we use this type of phosphors including co-doped and tri-doped ones as the examples to discuss the energy transfer between Ce^{3+} ions and other rare earth ions.

2.1 Energy transfer in Ce^{3+}-Yb^{3+} co-doped YAG phosphors

Before discussing the energy transfer in Ce^{3+}-Yb^{3+} co-doped YAG phosphors, let us start from singly Ce^{3+}-doped YAG phosphors. Figure 4 gives the excitation (left) and emission (right) spectra of Ce^{3+}-doped $Y_3Al_5O_{12}$ phosphors excited at 455 nm [3]. The excitation peak is around 455 nm with a weak excitation peak around 340 nm. The only emission peak is around 560 nm. This means that Ce^{3+}-doped YAG phosphors can convert ultraviolet and blue light into yellow light. The excitation process corresponds to the transition of Ce^{3+} ions from the 4f (ground) state $^2F_{5/2}$ or $^2F_{7/2}$ to a higher 5d excited state, and the emission process corresponds to the transition from a lower 5d state to $^2F_{5/2}$ or $^2F_{7/2}$ state. The difference between the excitation wavelength and the emission wavelength results from the relaxation process from the higher 5d excited state to the lower 5d state.

Rare Earth - A tribute to the late Mr. Rare Earth, Professor Karl Gschneidner   Materials Research Forum LLC
Materials Research Foundations 164 (2024) 343-368   https://doi.org/10.21741/9781644903056-10

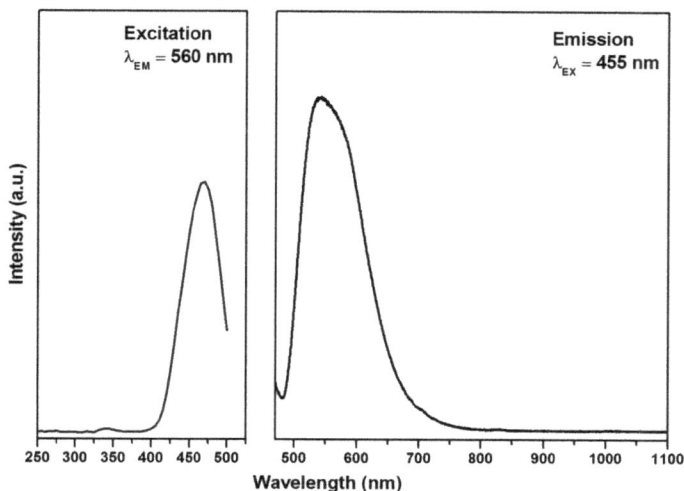

Figure 4 Typical excitation (left) and emission (right) spectra of Ce^{3+} doped $Y_3Al_5O_{12}$ phosphors excited at 455 nm

After Yb^{3+} ions are added into singly Ce^{3+}-doped YAG phosphors, the energy transfer from Ce^{3+} ions to Yb^{3+} ions occurs. Figure 5 shows schematically the energy levels of Ce^{3+} and Yb^{3+} ions. Yb^{3+} ions have two levels $^2F_{5/2}$ and $^2F_{7/2}$, corresponding the emission with the wavelength 960-1050 nm, just half the difference between 5d states and the 4f states of Ce^{3+}. Because the difference of energy levels between 4f and 5d of Ce^{3+} is about twice the difference of state $^2F_{5/2}$ and $^2F_{7/2}$ of Yb^{3+}, the energy transfer between these two ions was considered as cooperative energy transfer process. However, some works have shown that the energy transfer in Ce^{3+}-Yb^{3+} co-doped $Y_3Al_5O_{12}$ phosphors more possibly occurs through charge transfer state Ce^{3+}-O^{2-}-Yb^{3+} instead of the cooperative energy transfer process [2,4].

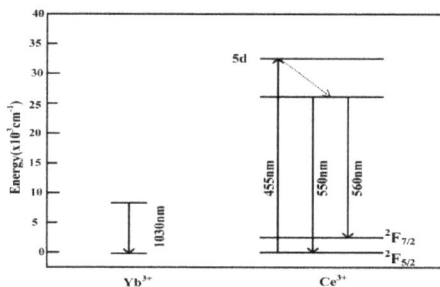

Figure 5 The energy levels of Ce^{3+} and Yb^{3+} ions.

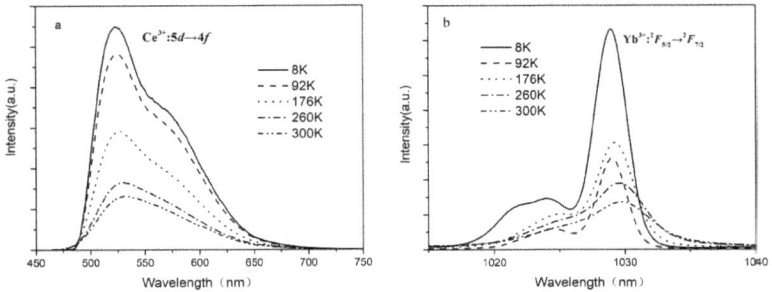

Figure 6 (a) Emission spectra of Ce^{3+} in Ce^{3+}-Yb^{3+} co-doped YAG at different temperature under 450 nm excitation; (b) Emission spectra of Yb^{3+} in Ce^{3+}-Yb^{3+} co-doped YAG at different temperature under 450 nm excitation.

The research about the influence of temperature on the luminescence of Ce^{3+}-Yb^{3+} co-doped $Y_3Al_5O_{12}$ phosphors has provided the evidence of the energy transfer via the charge transfer state [4]. Figure 6 gives the emission spectra of Ce^{3+} and Yb^{3+} in Ce^{3+}-Yb^{3+} co-doped YAG at different temperature under 450 nm excitation. As Yb^{3+} ions cannot absorb the light of 450 nm, it can be said that the near-infrared emission from Yb^{3+} is caused by the energy transfer from Ce^{3+} to Yb^{3+}.

It can be seen in figure 6 that the emission from Ce^{3+} ions decreases with increasing temperature (from 8K to 300K) monotonously, while the decrease of the emission from Yb^{3+} ions is not monotonous.

To see clearly the variation of the emission from Yb^{3+} ions, the emission intensity from Yb^{3+} in the wavelength range from 1000 nm to 1050 nm is integrated, and their changes with the temperature in Ce^{3+}-Yb^{3+} co-doped YAG are shown in figure 7. For the purpose of comparison, the emission intensity of the singly Yb^{3+}-doped YAG (YAG:Yb^{3+}) is also integrated and shown in figure 7. The emission spectra are normalized by corresponding incident intensities. Clearly, the total emission from Yb^{3+} in the singly doped YAG decreases with temperature monotonously. For the co-doped YAG, the emission from Yb^{3+} decreases with the increasing temperature until 100K. However, when the temperature is above 100K, the total near-infrared emission starts increasing. After the temperature reaches 180K, the total emission from Yb^{3+} turns to decrease again.

The degradation of emission from YAG:Yb^{3+} might be attributed to thermal quenching in the phosphors [5,6]. While for Ce^{3+}-Yb^{3+} co-doped phosphors, the reason are different. The variation of Yb^{3+} emission of the co-doped phosphors with the temperature indicates that, besides the influence of the thermal quenching, there should be another temperature-dependent process which has effects on the emission of Yb^{3+}, especially at the temperature above 100K. Different from the singly Yb^{3+}-doped YAG where Yb^{3+} is excited directly by the incident light, the energy of Yb^{3+} in the co-doped YAG comes from the energy transfer from Ce^{3+} to Yb^{3+}, so the near-infrared emission in figure 7(b) should be influenced by the energy transfer process. The variation of Yb^{3+} emission with the temperature means that the energy transfer from Ce^{3+} to Yb^{3+} in the co-doped phosphors should be related to the temperature.

Rare Earth - A tribute to the late Mr. Rare Earth, Professor Karl Gschneidner   Materials Research Forum LLC
Materials Research Foundations 164 (2024) 343-368   https://doi.org/10.21741/9781644903056-10

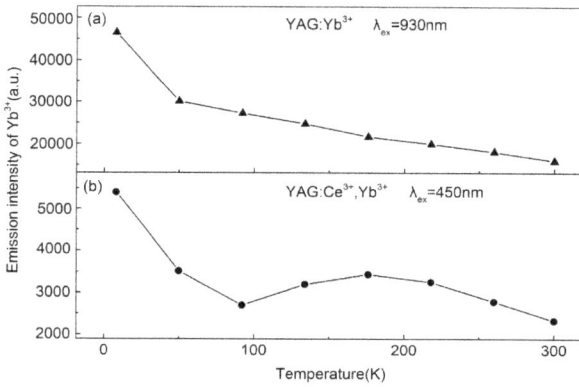

Figure 7 Emission intensity of Yb^{3+} at different temperatures (8-300K). (a) YAG:Yb^{3+} ; (b) Ce^{3+}-Yb^{3+} co-doped YAG (the emissions are normalized by the corresponding incident intensities)

On the basis of the Ce^{3+}-O^{2-}-RE^{3+} cluster compound model [7], there is a Ce-O-Yb charge transfer state in Ce^{3+}-Yb^{3+} co-doped YAG phosphors. Figure 8 shows the configuration coordinate model of energy states of Ce^{3+} and Yb^{3+}. After Ce^{3+} absorbs a photon, a $4f$ electron is excited into $5d$ state. Because the $5d$ electron has more chances to form a bond with neighboring O^{2-} than the $4f$ electron, the absorption of photons makes the bonding easier. On the other hand, the strong electronegativity of Yb^{3+} makes the ions of Yb^{3+} close to the ions of O^{2-} [8]. Thus, by the assistance of local phonons, a transient photo-induced Ce^{3+}-O^{2-}-Yb^{3+} charge transfer state is formed. In fact, there are excitation peaks around 275 nm which might be attributed to the transition from $^2F_{7/2}$ to Ce-O-Yb CTS [4].

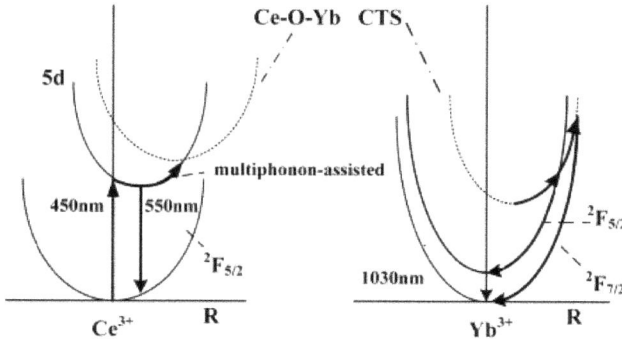

Figure 8 Configuration coordinate model of the $4f^1$ and $5d$ states of Ce^{3+}, the $4f^{13}$ state of Yb^{3+} (solid parabolas) and the Ce-O-Yb charge transfer state (dashed parabolas), illustrating the energy transfer via a Ce-O-Yb charge transfer state

After an electron at the ground state of Ce^{3+} is excited into $5d$ state by absorbing a photon of 450 nm, it has two pathways: one is the relaxation to the lowest level of 5d state followed by the transition down to the ground state of Ce^{3+} with emission; the other is the transition to Ce-O-Yb charge transfer state with the assistance of the phonons, as shown in figure 8.

For the electron transiting to Ce-O-Yb charge transfer state, it also has two pathways: one is the transition to $^2F_{5/2}$ state of Yb^{3+} with the assistance of the phonons, then relaxes to the lowest level of $^2F_{5/2}$ state, and finally transits to $^2F_{7/2}$ state with emission; the other is the transition to $^2F_{7/2}$ state of Yb^{3+} with the assistance of the phonons, then relaxes to the lowest level of $^2F_{7/2}$ state without emission, as shown in figure 8. Because the crossover of the charge transfer state and $^2F_{7/2}$ state is higher than that of the charge transfer state and $^2F_{5/2}$ state, the transition from it to $^2F_{7/2}$ state needs more phonons than that from it to $^2F_{5/2}$ state. Therefore, Yb^{3+} is an activator with high efficiency, which was usually assumed to have 100% quantum efficiency in some literature.

Now, let us to explain the curve in figure 7(b). At the temperature below 100 K, the number of the phonons is small, so the electrons are not easy to transit to the charge transfer state and $^2F_{5/2}$ state of Yb^{3+}. The thermal quenching effect makes the emission of Yb^{3+} decrease with increasing temperature. When the temperature rises to be higher than 100 K, the lattice vibration leads to the increase of the phonon number. So the charge transfer from Ce^{3+} to Yb^{3+} becomes easier via the transitions 5d state of $Ce^{3+} \rightarrow$ Ce-O-Yb CTS $\rightarrow ^2F_{5/2}$ state of Yb^{3+}, and more near-infrared photons are emitted due to the transition $^2F_{5/2} \rightarrow ^2F_{7/2}$ of Yb^{3+}. This explains why the emission of Yb^{3+} in the co-doped YAG turns to increase when the temperature rises from 100K to 180K.

If the temperature continues to rise to be higher than 180 K, more phonons are generated by the stronger lattice vibration and make it possible that the electrons at Ce-O-Yb CTS transit to $^2F_{7/2}$ state of Yb^{3+}. Then, the electrons relax to the lowest level of $^2F_{7/2}$ state without near-infrared emission. So at the temperature above 180K, the emission of Yb^{3+} decreases again.

It should be pointed out that, although the charge transfer state provide a reasonable explanation for the variation of Yb^{3+} emission of the co-doped phosphors with the temperature, we still need more evidences to clarify the energy transfer mechanism in Ce^{3+}-Yb^{3+} co-doped YAG phosphors.

2.2   Energy transfer in Ce^{3+}-Tb^{3+} co-doped YAG phosphors

Figure 9 shows the excitation and emission spectra of Ce^{3+}-Tb^{3+} co-doped $Y_3Al_5O_{12}$ phosphors excited at 275 nm. In this case, Tb^{3+} is used as a sensitizer and Ce^{3+} was used as an activator. The excitation spectrum (monitored at 520 nm) shows two main bands with peaks at 275 nm and 340 nm (the peak at 455 nm is not shown in order to make the two peaks at 275 nm and 340 nm be seen clearly). The first excitation peak at 275 nm and the second peak at 340 nm can be ascribed to the spin-allowed 4f→5d transition of Tb^{3+} and 4f→5d transition of Ce^{3+} ions, respectively [9]. Because Tb^{3+} cannot emit the light of 520 nm and Ce^{3+} cannot be excited by 275 nm, the excitation spectrum in figure 9 can be seen as the result of the energy transfer from Tb^{3+} to Ce^{3+}. The similar experiment can also be found in some literature [10].

Rare Earth - A tribute to the late Mr. Rare Earth, Professor Karl Gschneidner   Materials Research Forum LLC
Materials Research Foundations 164 (2024) 343-368   https://doi.org/10.21741/9781644903056-10

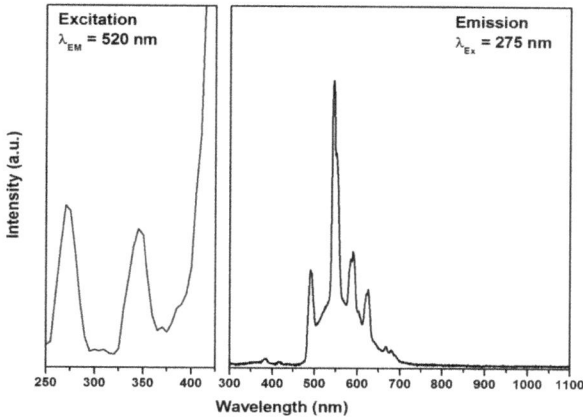

Figure 9. Typical excitation (left) and emission (right) spectra of Ce^{3+}-Tb^{3+} co-doped $Y_3Al_5O_{12}$ phosphors excited at 275 nm

Another evidence of the energy transfer from Tb^{3+} to Ce^{3+} can be also found in figure 9. Because the singly Tb^{3+}-doped phosphors has weak emissions in the range of 480-700 nm except the four peaks at 490 nm, 543 nm, 585 nm and 622 nm [3], while for Ce^{3+}-Tb^{3+} co-doped phosphors, besides these four peaks, the emissions in the range between 480 nm-700 nm is much stronger than that of singly Tb^{3+}-doped phosphors. It indicates the emission of the co-doped phosphors in the range of 480 -700 nm is partially from Ce^{3+}. Therefore, this also verifies the energy transfer from $^5D_3$ state of Tb^{3+} to the lowest-lying 5d state of Ce^{3+}.

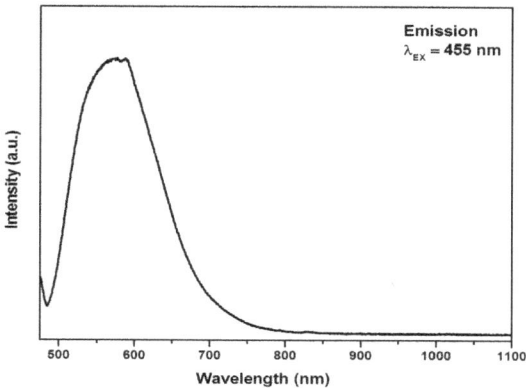

Figure 10. Typical emission spectra of Ce^{3+} - Tb^{3+} co-doped $Y_3Al_5O_{12}$ phosphors excited at 455 nm

351

Rare Earth - A tribute to the late Mr. Rare Earth, Professor Karl Gschneidner   Materials Research Forum LLC
Materials Research Foundations 164 (2024) 343-368   https://doi.org/10.21741/9781644903056-10

Figure 10 shows the emission spectra of Ce^{3+}-Tb^{3+} co-doped $Y_3Al_5O_{12}$ phosphors excited at 455 nm. In this case, Ce^{3+} is used as a sensitizer and Tb^{3+} is used as an activator. Under the excitation wavelength of 455 nm in Ce^{3+} absorption band, the emission spectrum shows a broad band from 500 nm to 700 nm which corresponds to Ce^{3+} transitions, while the emission from Tb^{3+} is not observed. The same broadband emission can also be observed under the excitation of 455 nm in singly Ce^{3+}-doped $Y_3Al_5O_{12}$ phosphors as shown in figure 4. This indicates the energy transfer from Ce^{3+} to Tb^{3+} in the phosphors is weak. So, from the spectra in Figure 9 and Figure 10, it can be concluded that there exists the efficient energy transfer from Tb^{3+} to Ce^{3+} ions, whereas the energy transfer from Ce^{3+} to Tb^{3+} ions is not obvious.

The energy transfer efficiency η_{ETE} of the rare earth YAG phosphors can be calculated by the following expression [11,12]

$$\eta_{ETE} = 1 - \frac{\tau_R}{\tau_{R0}} \tag{1}$$

where τ_{R0} is the fluorescence lifetime of the sensitizer with the absence of activators, τ_R is the fluorescence lifetime of the sensitizer with the presence of activators. The average lifetime (τ) can be calculated by using the following equation [13]

$$\tau = \frac{\int_0^\infty t\, I(t)\, dt}{\int_0^\infty I(t)\, dt} \tag{2}$$

where $I(t)$ represents the luminescence intensity at the time t.

The normalized decay curves for the Ce^{3+}-Tb^{3+} co-doped $Y_3Al_5O_{12}$ phosphors under the excitation wavelengths of 455 nm (monitored at 560 nm) and 275 nm (monitored at 543 nm) are depicted in figure 11. The fluorescence lifetime τ_{Ce} decreases slightly from 65.7 ns with the absence of Tb^{3+} to 61.4 ns with the presence of Tb^{3+} and the fluorescence lifetime of τ_{Tb} declines from 3.07 μs with the absence of Ce to 1.49 μs with the presence of Ce^{3+}. According to Equation (1), the energy transfer efficiency from Ce^{3+} to Tb^{3+} can be calculated as about 7%, and the energy transfer efficiency from Tb^{3+} to Ce^{3+} is about 51%.

In figure 11, it can also be seen that the emissions from the co-doped phosphors decay faster than the single doped phosphors whose decay curves have a single exponential nature. The reason is that the sensitizers in the co-doped phosphors can lose their energy non-radiatively through the energy transfer to the activators except for the relaxation from higher excited states to lower excited states which are the main path for the rare earth ions to lose their energy.

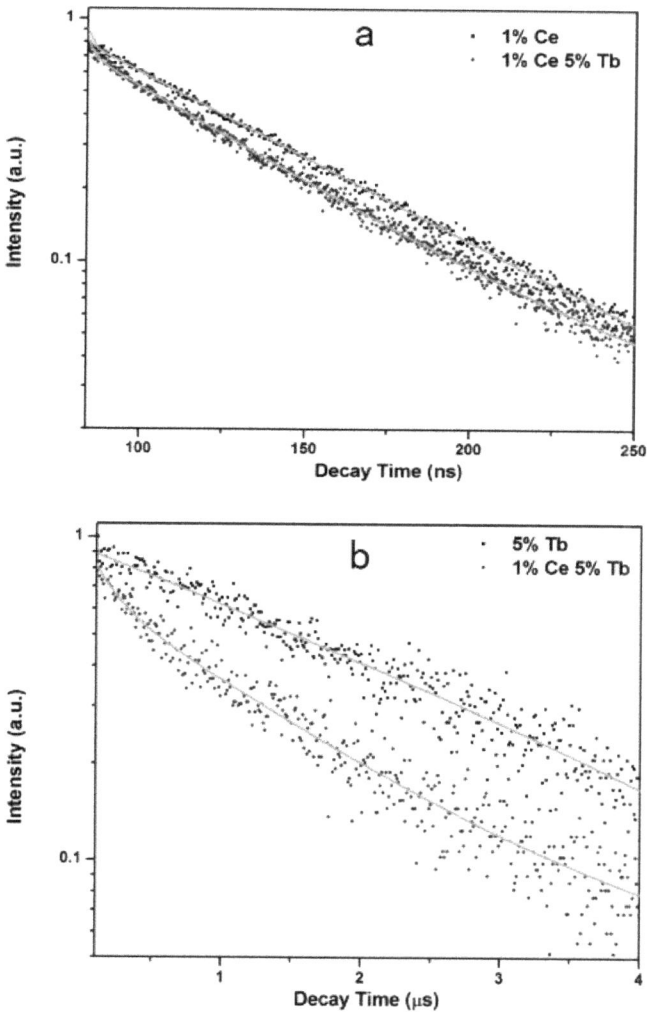

Figure 11. Decay curves of Ce^{3+}-Tb^{3+} co-doped $Y_3Al_5O_{12}$ phosphors recorded at (a) 560 nm with the excitation of 455 nm and (b) 543 nm with the excitation of 275 nm

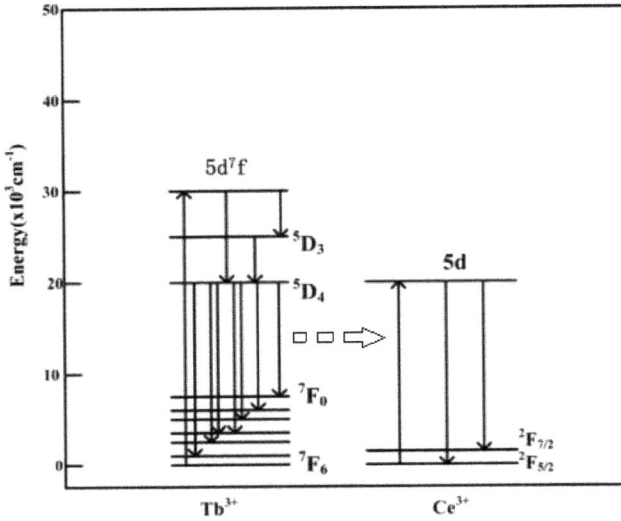

Figure 12. Schematic energy-level diagram of the energy transfer process for $Tb^{3+} \rightarrow Ce^{3+}$ in $Y_3Al_5O_{12}$ phosphors

The difference in the energy transfer efficiency between Tb^{3+} and Ce^{3+} in YAG phosphors might be explained by the nature of the transitions in Tb^{3+} and Ce^{3+} ions. The energy level diagram of Ce^{3+}-Tb^{3+} co-doped YAG phosphors is shown in Figure 12. Initially, Tb^{3+} ions are excited into 5d ^{7}f state by the excitation of 275 nm and then relaxes nonradiatively to $^5D_3$ and $^5D_4$ levels [14]. After the relaxation process, part of the energy of Tb^{3+} ions is emitted through $^5D_4 \rightarrow ^7F_J$ transition, whereas the rest is transferred into 5d states of Ce^{3+} ions. While, for the excitation wavelength of 455nm, Ce^{3+} ions are excited to higher 5d states and then relaxes to the lowest 5d state. Because the energy level of 5d states of Ce^{3+} is lower than that of $^5D_3$ and $^5D_4$ states of Tb^{3+}, the energy transfer from Ce^{3+} to Tb^{3+} is more difficult than that from Tb^{3+} to Ce^{3+} [15].

Although the difference of energy levels of the excited states between Tb^{3+} and Ce^{3-} gives a reasonable explanation, the mechanism of the energy transfer is more complicated than it. The emission spectra and the decay curves tell us that there is the energy transfer, but the detail of the transfer process needs further investigation.

2.3 Tb^{3+}-Yb^{3+} co-doped $Y_3Al_5O_{12}$

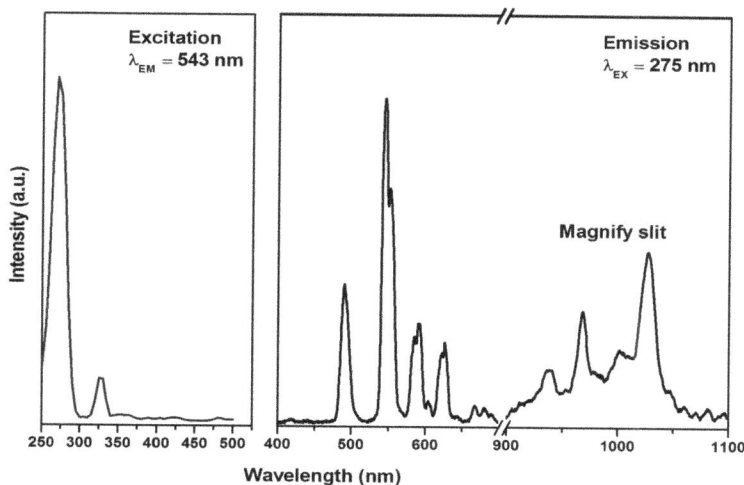

Figure 13. Typical excitation (left) and emission (right) spectra of Tb^{3+}-Yb^{3+} co-doped $Y_3Al_5O_{12}$ phosphors excited at 275 nm

In order to study the luminescence properties of Tb^{3+}-Yb^{3+} co-doped $Y_3Al_5O_{12}$ (5% Tb, 10% Yb) phosphors, the excitation and emission spectra of the phosphors are measured and shown in figure 13 (the emission from Yb^{3+} has been magnified in order to make it easier to be seen). In the excitation spectrum, the intense band at 275 nm is assigned to $^7F_6 \rightarrow {}^5D_4$ transition of Tb^{3+} which can be observed by monitoring the emissions at 543 nm and 1025 nm. Under the excitation wavelength of 275 nm, there is an occurrence of the emission peaks at 490 nm, 543 nm, 585 nm, 622 nm and 660 nm, respectively, which are attributed to the electronic transitions of $^5D_4 \rightarrow {}^7F_J$ (J = 6, 5, 4, 3, 2) of Tb^{3+} ions. Compared with the singly Tb^{3+}-doped sample [3], the emission intensity of Tb^{3+} ions is reduced after the incorporation of Yb^{3+} ions into the phosphors. Meanwhile, the clear emission centered at 1028 nm is observed, which corresponds to $^2F_{5/2} - {}^2F_{7/2}$ transitions of Yb^{3+} ions [16]. It confirms the energy transfer from $^5D_4$ state of Tb^{3+} to $^2F_{5/2}$ state of Yb^{3+}.

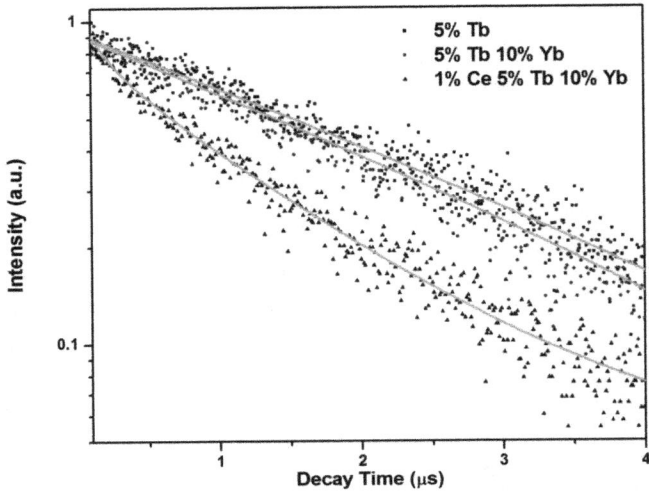

Figure 14. Decay curves of $Y_3Al_5O_{12}$ phosphors recorded at 543 nm with the excitation of 275 nm

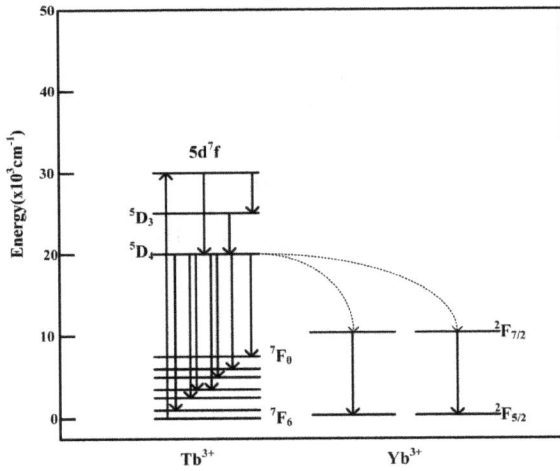

Figure 15. Schematic energy-level diagram of the ET process for $Tb^{3+} \rightarrow Yb^{3+}$ in $Y_3Al_5O_{12}$ phosphors.

Figure 14 demonstrates the normalized decay curves of Tb^{3+}-Yb^{3+} co-doped $Y_3Al_5O_{12}$ phosphors which is monitored at 543 nm under the excitation of 275 nm. The single exponential decay curve (absence of Yb^{3+}) gives a fluorescence lifetime of 3.07 μs, while the presence of Yb^{3+} reduces the lifetime to 2.76 μs. The calculated energy transfer efficiency of Tb^{3+}-Yb^{3+} co-doped YAG phosphors is 10.2 %.

Figure 15 shows a schematic energy level diagram of a proposed energy transfer mechanism from Tb^{3+} to Yb^{3+}. It can be noticed that $^5D_4$ (20500 cm^{-1}) energy level of Tb^{3+} is twice as high as the energy level of $^2F_{5/2}$ (10000 cm^{-1}) of Yb^{3+} [17]. This makes it possible that two near-infrared photons from Yb^{3+} ions are emitted after the absorption of single photon by Tb^{3+} ions. However, the confirmation of the cooperative energy transfer in Tb^{3+}-Yb^{3+} co-doped YAG phosphors needs more experimental evidences [18].

2.4 Two energy transfer mechanisms in a couple of rare earth ions

Different from the energy transfer in the phosphors mentioned above which occurs via one mechanism, the energy transfer in Pr^{3+}-Yb^{3+} co-doped $LiYF_4$ phosphors has two different mechanisms [19]. In the phosphors, the energy transfer from Pr^{3+} to Yb^{3+} is mainly through the cross relaxation with a minor energy transfer through the cooperative energy transfer. Figure 16 shows these two energy transfer mechanisms in Pr^{3+}-Yb^{3+} co-doped $LiYF_4$ phosphors. When Pr^{3+} ions transit from the state $^3P_0$ to the ground state $^3H_4$, the energy is transferred to two neighbouring Yb^{3+} ions which are excited from the ground state $^2F_{7/2}$ to $^2F_{5/2}$. Another way is to use $^1G_4$ as an intermediate state to transfer stepwise energy to Yb^{3+} ions.

The conclusion that the cross relaxation is the dominant mechanism of the energy transfer is deduced by comparing the results between modelling and experiment [19]. The simulated decay curves based on the cross relaxation are more consistent with the experiment than those based on the cooperative energy transfer, especially for higher concentration of Yb^{3+} ions. At the lower concentration of Yb^{3+} ions, the simulated curves based on both mechanisms agree well with the experiment.

In various co-doped rare earth phosphors, the energy transfer mechanism is different and depends on the rare earth ions and host lattices. It is not difficult to verify the energy transfer, but it is difficult to clarify the energy transfer mechanism directly by experiments. Combining simulation with experiment is a better way to identify the mechanism of the energy transfer between rare earth ions.

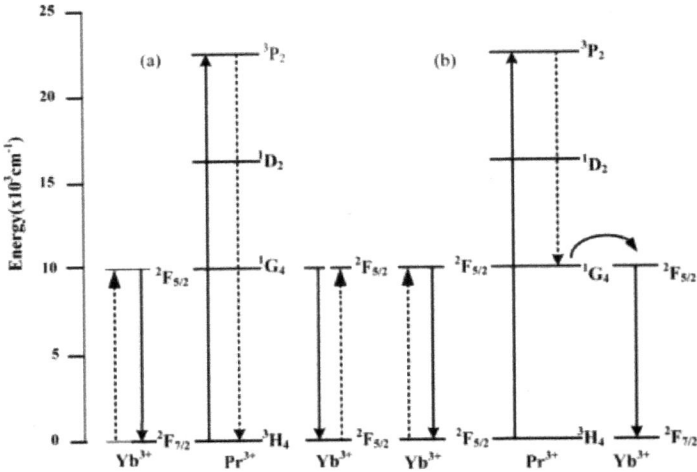

Figure 16 Two energy transfer mechanisms in Pr^{3+}-Yb^{3+} co-doped $LiYF_4$ phosphors. (a) shows the cooperative energy transfer where the energy is simultaneously transferred to two Yb^{3+} ions. (b) shows the energy transfer via cross relaxation where the energy is stepwise transferred to Yb^{3+} ions.

3. Energy transfer in tri-doped $Y_3Al_5O_{12}$ phosphors

Tri-doped rare earth phosphors include three different ions, and the energy may transfer between any couple of ions. So the energy transfer becomes more complicated than those in the co-doped phosphors. Here, the energy transfer in Ce^{3+}-Yb^{3+}-Tb^{3+} tri-doped $Y_3Al_5O_{12}$ phosphors and Bi^{3+}-Ce^{3+}-Yb^{3+} tri-doped $Y_3Al_5O_{12}$ phosphors are discussed.

3.1 Ce^{3+}-Yb^{3+}-Tb^{3+} tri-doped $Y_3Al_5O_{12}$ phosphors

Figure 17 shows the excitation (monitored at 543 nm) and emission spectra (excited at 275 nm) of Ce^{3+}-Yb^{3+}-Tb^{3+} tri-doped $Y_3Al_5O_{12}$ (1% Ce, 5% Tb, 10% Yb) phosphors [3]. It has three absorption peaks: one strong peak located at 275 nm which belongs to Tb^{3+}: $^7F_6 \rightarrow ^5D_J$ transition and two weak peaks at 340 nm and 455 nm which are ascribed to Ce^{3+}: 4f→5d transition. The emission peaks of Tb^{3+} is located at 490 nm, 543 nm, 585 nm, 622 nm and 660 nm, respectively and the emission of Yb^{3+} is presented at 1028 nm. By comparing figure 17 with figure 13 which shows the emission spectra of Tb^{3+}-Yb^{3+} co-doped phosphors, it can be noted that there is an occurrence of weak emission of Ce^{3+} in the wavelength range from 500 nm to 700 nm.

Rare Earth - A tribute to the late Mr. Rare Earth, Professor Karl Gschneidner   Materials Research Forum LLC

Materials Research Foundations 164 (2024) 343-368   https://doi.org/10.21741/9781644903056-10

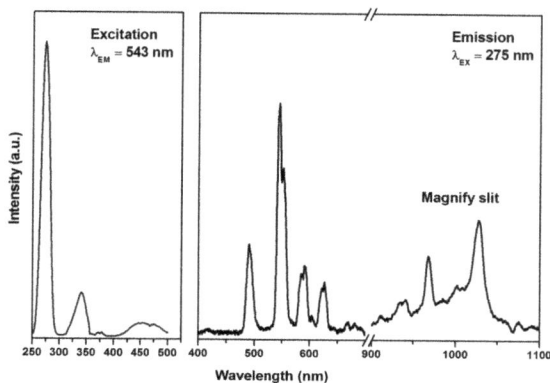

Figure 17. Typical excitation (left) and emission (right) spectra of Tb^{3+}-Ce^{3+}-Yb^{3+} tri-doped $Y_3Al_5O_{12}$ phosphors excited at 275 nm

The decay curve of Tb^{3+} in the tri-doped phosphors can be seen in figure 14. The emission intensity of Tb^{3+} in the tri-doped phosphor decreases much faster than that of Tb^{3+}-Yb^{3+} co-doped phosphors. This agrees with the results obtained from the co-doped phosphors: the energy transfer from Tb^{3+} to Ce^{3+} is more efficient than that from Tb^{3+} to Yb^{3+}.

Yb^{3+} ions in the tri-doped phosphors can get the energy through two paths: $Tb^{3+} \rightarrow Yb^{3+}$ and $Tb^{3+} \rightarrow Ce^{3+} \rightarrow Yb^{3+}$. Because it has been verified that the energy transfer from Tb^{3+} to Ce^{3+} and Ce^{3+} to Yb^{3+} are more efficient than that from Tb^{3+} to Yb^{3+}, Yb^{3+} receives the energy mainly from the path of $Tb^{3+} \rightarrow Ce^{3+} \rightarrow Yb^{3+}$ instead of $Tb^{3+} \rightarrow Yb^{3+}$.

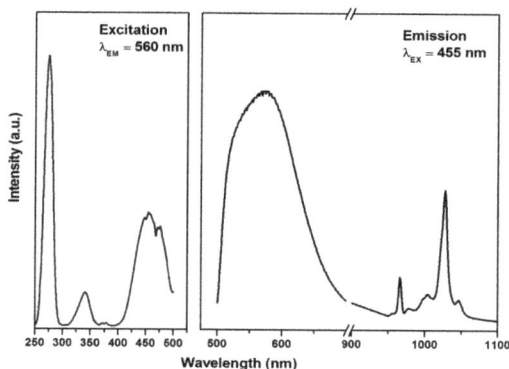

Figure 18. Typical excitation (left) and emission (right) spectra of Ce^{3+}-Tb^{3+}-Yb^{3+} tri-doped $Y_3Al_5O_{12}$ phosphors excited at 455 nm

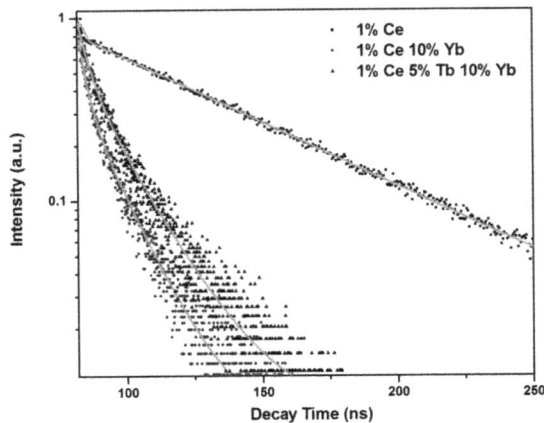

Figure 19. Decay curves of $Y_3Al_5O_{12}$ phosphors recorded at 560 nm with the excitation of 455 nm

Figure 18 exhibits the excitation (monitored at 560 nm) and emission spectra (excited at 455 nm) of Ce^{3+}-Yb^{3+}-Tb^{3+} tri-doped $Y_3Al_5O_{12}$ phosphors. In this case, Ce^{3+} is used as a sensitizer. Because of the inefficient energy transfer from Ce^{3+} to Tb^{3+}, the emission peaks from Tb^{3+} cannot be seen in figure 18 and also the transferred energy to Yb^{3+} ions is mainly from Ce^{3+} instead of Tb^{3+}. This is confirmed in figure 19 which shows that the decay curve of Ce^{3+}-Yb^{3+} co-doped YAG phosphors is nearly the same as that of the tri-doped phosphors.

According to the decay curves in figure 19, the calculated energy transfer efficiency from Ce^{3+} to other two ions was 55.4% which is lower than that the efficiency of 63.1% in Ce^{3+}-Yb^{3+} co-doped phosphors. This is an interesting result. The energy transferred to two ions Tb^{3+} and Yb^{3+} is less than to one ion Yb^{3+} in YAG lattice. The reason needs further investigation.

From the above experimental results and discussions, it can be known that the energy transfer of $Ce^{3+} \rightarrow Yb^{3+}$ and $Tb^{3+} \rightarrow Ce^{3+}$ are efficient while $Ce^{3+} \rightarrow Tb^{3+}$ and $Tb^{3+} \rightarrow Yb^{3+}$ are inefficient in YAG host materials. This leads to the efficient cascade energy transfer of $Tb^{3+} \rightarrow Ce^{3+} \rightarrow Yb^{3+}$ and the inefficient transfer of $Ce^{3+} \rightarrow Tb^{3+} \rightarrow Yb^{3+}$. The difference in energy transfer efficiency between different rare earth ions and different hosts may be helpful for designing the rare earth down-conversion materials [20,21,41-44].

3.2 Bi^{3+}-Ce^{3+}-Yb^{3+} tri-doped $Y_3Al_5O_{12}$ phosphors

Because the singly Ce^{3+} doped YAG phosphors have been proved to be stable and highly efficient, it is reasonable to improve the near-infrared emission of Ce^{3+}-Yb^{3+} co-doped YAG phosphors so that they can be used to enhance the performance of solar cells [2,22,23]. However, as mentioned above, introducing Tb^{3+} ions can not improve the near-infrared emission of Ce^{3+}-Yb^{3+} co-doped YAG phosphors. Therefore, we have to find a new way to do this. Here, Bi^{3+} ions are introduced to form Bi^{3+}-Ce^{3+}-Yb^{3+} tri-doped phosphors, and their energy transfer and the near-infrared emission are presented.

Figure 20 shows the emission spectra of Bi^{3+}-Ce^{3+}-Yb^{3+} tri-doped $Y_{3-x-y-z}Al_5O_{12}$: xBi^{3+} yCe^{3+} zYb^{3+} (x=0, 0.03; y=0.01; z=0.05, 0.1, 0.15, 0.2, 0.25) phosphors under the excitation of 455 nm. It can be seen that the emissions from Ce^{3+} and Yb^{3+} vary with Yb^{3+} concentration. When Yb^{3+} concentration is low (z=0.05, 0.1, 0.15), the near-infrared emission from Yb^{3+} ions increases with the rising Yb^{3+} concentration. It is because the energy transferred to Yb^{3+} from Ce^{3+} increases with the concentration of Yb^{3+}. This is also the reason why the emission from Ce^{3+} ions decreases with the increasing Yb^{3+} concentration. When Yb^{3+} concentration continues to rise to z=0.2 and 0.25, although the energy from Ce^{3+} to Yb^{3+} increases, Yb^{3+} emission becomes weaker due to the concentration quenching effect of Yb^{3+} [24].

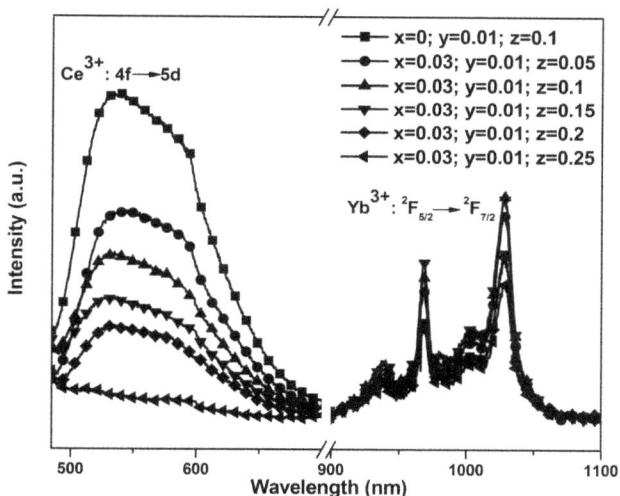

Figure 20. The emission spectra of different YAG phosphors under the excitation of 455 nm

The most desirable result in figure 20 is that, in the cases of x=0.03, y=0.01 and z=0.05, 0.10, 0.15, the intensity of near-infrared emission from Yb^{3+} ions in Bi^{3+}-Ce^{3+}-Yb^{3+} tri-doped phosphors is stronger than that in Ce^{3+}-Yb^{3+} co-doped phosphors (x=0, y=0.01 and z=0.1). Figure 21 shows the intensities of the emission at 1027 nm from the phosphors with different chemical compositions. Although the near-infrared emission of the tri-doped samples when z=0.2 and 0.25 is weaker than that of the co-doped phosphors due to the concentration quenching effect of Yb^{3+}, the stronger emission of the tri-doped ones when z=0.05, 0.10 and 0.15 indicates that the introduction of Bi^{3+} ions can improve the emission from Yb^{3+} ions.

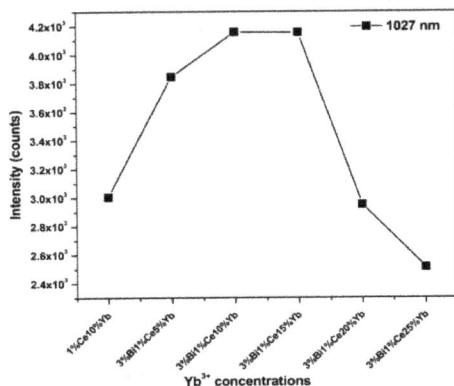

Figure 21 Emission intensities of Yb^{3+} at 1027 nm from different YAG phosphors under the excitation of 455 nm

Because Yb^{3+} ions can not absorb the photons with the wavelength of 455 nm, their emission results from the energy transfer inside the phosphors. To explain the curves in figure 20 and 21, it is necessary to clarify the energy transfers between Bi^{3+}, Ce^{3+} and Yb^{3+} ions in the tri-doped phosphors. Figure 22 shows the luminescence decay curves of Ce^{3+} (monitored at 560 nm) in singly Ce^{3+} doped ($x=0$, $y=0.01$, $z=0$), Ce^{3+}-Yb^{3+} co-doped ($x=0$, $y=0.01$, $z=0.1$) and Bi^{3+}-Ce^{3+}-Yb^{3+} tri-doped ($x=0.03$, $y=0.01$, $z=0.05$, 0.1, 0.15, 0.2) YAG phosphors under the excitation of 455 nm. Clearly, the emission from Ce^{3+} in the tri-doped phosphors decays faster than that in Ce^{3+}-Yb^{3+} co-doped phosphors. Because the tri-doped sample has Bi^{3+} ions while the co-doped sample does not, the faster decay of the emission from Ce^{3+} in the tri-doped phosphors might be attributed to the energy transfer from Ce^{3+} ions to Bi^{3+} ions.

To know the energy transfer in detail, it is a good way to calculate the efficiency of the energy transfers from Ce^{3+} to other ions. According to references [25,26] the energy transfer efficiency can be estimated by using the equation (1) , and the fluorescence lifetime can be obtained by using the equation (2) .

From the decay curves in figure 22, the fluorescence lifetimes of Ce^{3+} can be calculated as 15.3 ns in the tri-doped phosphors ($x=0.03$, $y=0.01$, $z=0.1$), 24.3 ns in Ce^{3+}-Yb^{3+} co-doped phosphors ($x=0$, $y=0.01$, $z=0.1$) and 65.7 ns in the singly Ce^{3+} doped phosphors ($x=0$, $y=0.01$, $z=0$), respectively. The energy transfer efficiency calculated from the decay curve of Ce^{3+} in the tri-doped phosphors is about 76.78%, while the energy transfer efficiency in the co-doped phosphors is 63.02%.

Rare Earth - A tribute to the late Mr. Rare Earth, Professor Karl Gschneidner   Materials Research Forum LLC
Materials Research Foundations 164 (2024) 343-368   https://doi.org/10.21741/9781644903056-10

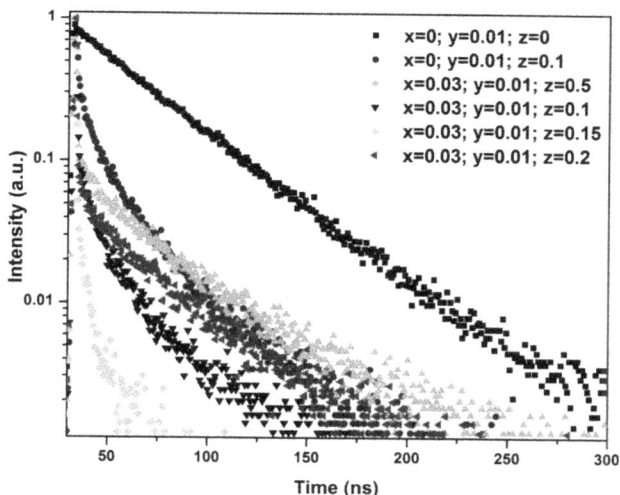

Figure 22 The decay curves of Ce³⁺ emission at 560 nm for YAG phosphors under the excitation of 455 nm

The reasonable explanation for the different energy transfer efficiency in the two kinds of phosphors is that, in the co-doped phosphors, there is only one pathway of the energy transfer: the energy is transferred from Ce^{3+} to Yb^{3+} while in the tri-doped phosphors the energy of Ce^{3+} can be transferred by two pathways: one is from Ce^{3+} to Yb^{3+} and the other is from Ce^{3+} to Bi^{3+} to Yb^{3+}.

The evidence of the energy transfer from Ce^{3+} to Bi^{3+} can be found in figure 23 which gives the decay curves of singly Ce^{3+} doped and Ce^{3+}-Bi^{3+} co-doped YAG phosphors. Under the excitation of 455 nm, the emission of Ce^{3+} in the singly doped phosphors shows nearly a single exponential decay with the fluorescence lifetime of 65.7 ns, while in Ce^{3+}-Bi^{3+} co-doped samples, the fluorescence lifetime decreases to 56.3 ns. The calculated energy transfer efficiency is 14.31%. This indicates the energy transferred from Ce^{3+} to Bi^{3+} is possible inside the phosphors although the efficiency is not as high as that from Ce^{3+} to Yb^{3+}.

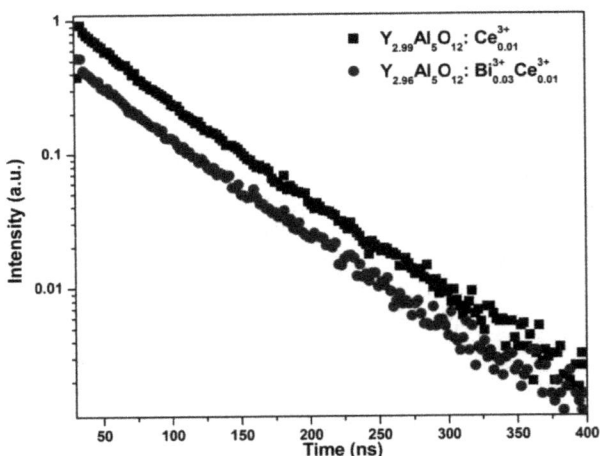

Figure 23 The Decay curves of Ce^{3+} emission at 560 nm for Ce^{3+} singly doped and Ce^{3+}-Bi^{3+} co-doped YAG phosphors under the excitation of 455 nm

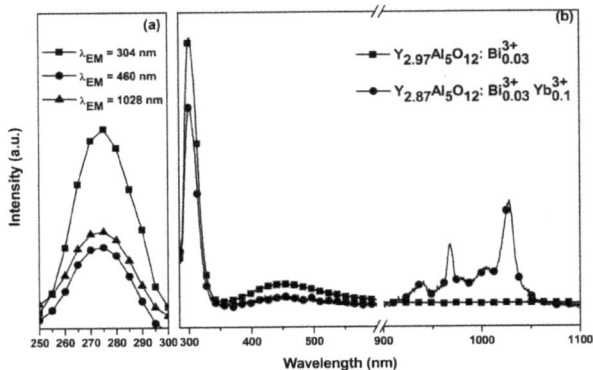

Figure 24 (a) Excitation spectra of $Y_{2.87}Al_5O_{12}$: $Bi^{3+}_{0.03}Ce^{3+}_{0.1}$ monitored at the emissions of 304 nm, 460 nm, 1028 nm and (b) Emission spectra of singly Bi^{3+} doped and Bi^{3+}-Yb^{3+} co-doped YAG phosphors under the excitation wavelength of 275 nm.

After Bi^{3+} ions obtain the energy from Ce^{3+}, the energy can be partly transferred to Yb^{3+}. The evidence of the energy transfer from Bi^{3+} to Yb^{3+} can be found in figure 24 and 25. From the figure 24 (a), it can be noted that an excitation band is centered at 275 nm owing to the transition of Bi^{3+}: $^1S_0 \rightarrow {}^1P_1$, $^3P_1$ detected by monitoring at 304 and 460 nm. Moreover, a similar excitation spectrum is also obtained for the transition of Yb^{3+}: $^2F_{5/2} \rightarrow {}^2F_{7/2}$ at the monitoring wavelength of 1028 nm.

The similarity in the shape of both excitation spectra can be considered as an evidence for an energy transfer from Bi^{3+} to Yb^{3+}. The emission spectra of the singly Bi^{3+} doped and Bi^{3+}-Yb^{3+} co-doped phosphors under the excitation of 275 nm are shown in figure 24 (b). The singly doped Bi^{3+} spectrum depicts two emission peaks centered at 304 nm and 460 nm. In the co-doped Bi^{3+}-Yb^{3+} samples, apart from the emissions of Bi^{3+} ions, a strong near-infrared emission of Yb^{3+} can also be obtained at 1028 nm which is accompanied by several weak shoulders owing to the transitions among different stark levels of ^2F_j (j = 5/2, 7/2) in Yb^{3+}. Because Yb^{3+} ions cannot absorb the photons of 275 nm, their near-infrared emission indicates the energy transfer from Bi^{3+} to Yb^{3+}.

Figure 25 is the decay curves of singly Bi^{3+} doped and Bi^{3+}-Yb^{3+} co-doped phosphors under the excitation of 275 nm. The faster decay of Bi^{3+} emission in the co-doped samples also proves the energy transfer from Bi^{3+} to Yb^{3+}. From figure 25, the fluorescence lifetime of Bi^{3+} was calculated as 768 ns in the singly doped phosphors, and decreases to 576 ns after Yb^{3+} ions are introduced. The calculated energy transfer efficiency from Bi^{3+} to Yb^{3+} is 24.95 %.

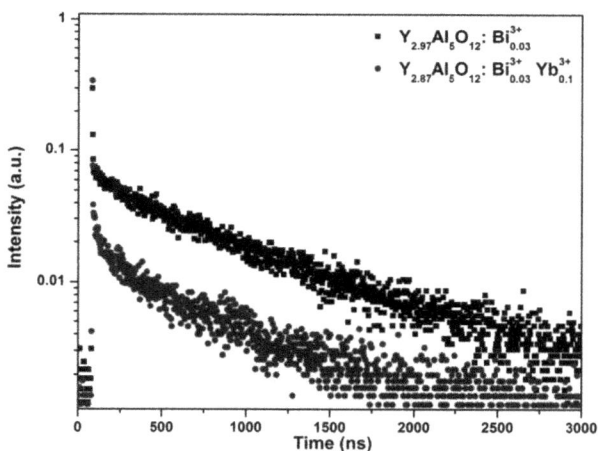

Figure 25 The Decay curves of Bi^{3+} emission at 304 nm for Bi^{3+} singly doped and Bi^{3+}-Yb^{3+} co-doped YAG phosphors under the excitation of 275 nm

In principle, the amount of the energy transferred from Ce^{3+} to Yb^{3+} in the tri-doped phosphors should be close to that in Ce^{3+}-Yb^{3+} co-doped ones because these two kinds of phosphors have the same concentrations of Ce^{3+} and Yb^{3+} [27]. This should result in the similar intensity of the near-infrared emission from Yb^{3+}. However, in the tri-doped phosphors, besides transferring energy to Yb^{3+}, Ce^{3+} can also transfer energy to Bi^{3+}. The energy transferred to Bi^{3+} is then partly transferred to Yb^{3+}, so Yb^{3+} ions obtain more energy in the tri-doped phosphors than in the co-doped phosphors. This is why the emission from Yb^{3+} in the tri-doped phosphors is stronger than that in the co-doped phosphors, as shown in figure 20.

Based on the above results, the possible energy transfer processes are analyzed and the schematic energy level diagram is shown in figure 26. In the Bi^{3+}-Ce^{3+}-Yb^{3+} tri-doped phosphors, the possible energy transfer processes takes place as four ways under the excitation of 455 nm: $Ce^{3+} \rightarrow Yb^{3+}$, $Ce^{3+} \rightarrow Bi^{3+}$, $Bi^{3+} \rightarrow Yb^{3+}$ and $Ce^{3+} \rightarrow Bi^{3+} \rightarrow Yb^{3+}$. For the first energy transfer of $Ce^{3+} \rightarrow Yb^{3+}$, the energy of 5d→4f transition of Ce^{3+} is approximately twice as high as the energy of $^2F_{5/2} \rightarrow {}^2F_{7/2}$ transition of Yb^{3+}, and the energy of Ce^{3+} transfers to Yb^{3+} by two possible ways: one is that one UV/blue photon is converted into two near-infrared photons by cooperative energy transfer [28]; other is that the energy transfers from Ce^{3+} to Yb^{3+} via a charge transfer state [2,4,29]. As mentioned above, some recent researches showed that the charge transfer state is more possible than the cooperative energy transfer.

For the second energy transfer process of $Ce^{3+} \rightarrow Bi^{3+}$, the mechanism is more complicated. The electrons of the excited 5d state of Ce^{3+} ions can transit to $^3P_2$ state of Bi^{3+} ions because these two energy levels are close to each other. On the other hand, after Bi^{3+} ions obtain the energy, part of electrons of the excited $^3P_2$ state of Bi^{3+} ions can transit back to 5d state of Ce^{3+} ions, and rest of the electrons transit to $^2F_{5/2}$ level of Yb^{3+} or the ground state $^1S_0$ of Bi^{3+}.

For the third energy transfer of $Bi^{3+} \rightarrow Yb^{3+}$, the energy level $^3P_1$ of Bi^{3+} is approximately twice as high as the energy difference between $^2F_{5/2}$ and $^2F_{7/2}$ levels of Yb^{3+}. The energy of Bi^{3+} might transfers directly to neighbouring Yb^{3+} ions by the cooperative energy transfer [30].

On the basis of the above energy transfer processes, the fourth energy transfer of $Ce^{3+} \rightarrow Bi^{3+} \rightarrow Yb^{3+}$ is easy to be understood. At the excitation of 455 nm, after Ce^{3+} ions are excited to the 5d state, part of the excited electrons transit to $^2F_{5/2}$ and $^2F_{7/2}$ states of Ce^{3+}, and part of the excited electrons transit to ^3P states of Bi^{3+} ions, then Bi^{3+} transfers energy to Yb^{3+}.

Therefore, it can be said that Bi^{3+} ions as the new pathway of the energy transfer play an important role in enhancing the near-infrared emission of Bi^{3+}-Ce^{3+}-Yb^{3+} tri-doped $Y_3Al_5O_{12}$ phosphors. This is important for broadening the application of YAG phosphors which are one of the most successful rare earth phosphors.

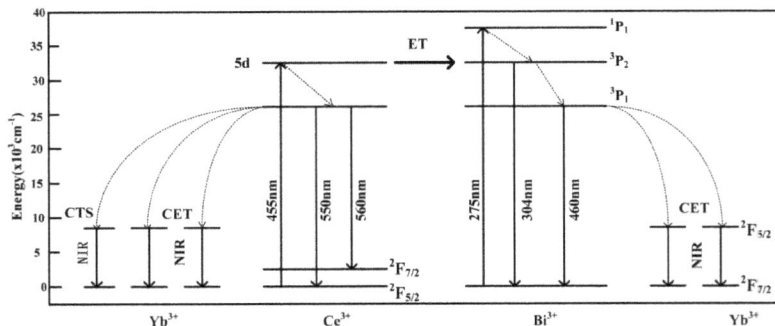

Figure 26 Schematic energy level diagram of Bi^{3+}-Ce^{3+}-Yb^{3+} tri-doped $Y_3Al_5O_{12}$ phosphors

It should be pointed out that the enhancement of the near-infrared emission from Yb^{3+} by introducing Bi^{3+} ions into Ce^{3+}-Yb^{3+} co-doped phosphors does not mean that other ions have similar effects. As mentioned above, introducing Tb^{3+} ions into Ce^{3+}-Yb^{3+} co-doped phosphors does not improve the emission of Yb^{3+} ions. Clarifying the mechanism of the energy transfer in the Bi^{3+}-Ce^{3+}-Yb^{3+} YAG phosphors is important for further improving the near-infrared emission and still needs more efforts.

In summary, this chapter presents the energy transfer in the co-doped and tri-doped phosphors, which is a significant process because it has important effects on the luminescent efficiency of the down conversion phosphors. In principle, the energy transfer between any two rare earth ions can occur, but due to the complexity inside the phosphors, the mechanism is still not clear. The combination of modelling and experiment might be a good way to clarify the energy transfer process.

References

[1] Zhang Q.Y., Huang X.Y., Progress in Materials Science, 55, 353 (2010). https://doi.org/10.1016/j.pmatsci.2009.10.001

[2] Yu D. C. , Rabouw F. T. , Boon W. Q. , Kieboom T. , Ye S. , Zhang Q.Y. and Meijerink A., Phys. Rev. B, 90 , 165126 (2014). https://doi.org/10.1103/PhysRevB.90.165126

[3] Kumar K.S., Lou C.G., Xie Y.F., Hu L., Manohari A.G., Xiao D., Ye H.Q., Tang L. and Pribat D., Journal of Rare Earths, 35,775 (2017). https://doi.org/10.1016/S1002-0721(17)60975-X

[4] Li L, Lou C.G., Sun X., Xie Y.F., Hu L. and Kumar K.S., ECS Journal of Solid State Science and Technology, 5, R146 (2016). https://doi.org/10.1149/2.0281609jss

[5] Bachmann V., Ronda C. and Meijerink A., Chem. Mater , 21, 2077 (2009). https://doi.org/10.1021/cm8030768

[6] Tian B.N., Chen B.J., Tian Y., Li X.P., Zhang J.S., Sun J.S., Zhong H.Y., Cheng L.H., Fu S.B., Zhong H., Wang Y.Z., Zhang X.Q., Xia H.P. and Hua R.N., J. Mater. Chem. C, 1, 2338 (2013). https://doi.org/10.1039/c3tc00915g

[7] Rivas-Silva J. F., Durand-Niconoff S., Schmidt T. M. and Berrondo M., Int. J. Quantum Chem, 79, 198 (2000). https://doi.org/10.1002/1097-461X(2000)79:3<198::AID-QUA5>3.0.CO;2-A

[8] Jørgensen C. K., Mol. Phys., 5, 271 (1962). https://doi.org/10.1080/00268976200100291

[9] Duan C.J., Zhang Z.J, Rösler S., Rösler S., Delsing A., Zhao J.T., Chem. Mater , 23,1851 (2011). https://doi.org/10.1021/cm103495j

[10] Jung K.Y., Lee H.W., Lumin,126, 469 (2007). https://doi.org/10.1016/j.jlumin.2006.09.009

[11] Zhang Q.Y., Yang C.H., Jiang Z.H., Ji X.H., Appl. Phys. Lett., 90,061914 (2007). https://doi.org/10.1063/1.2472195

[12] Wang Q., Ouyang S.Y., Zhang W.H, Yang B., Zhang Y.P., Xia H.P., J. Rare Earth, 33,13 (2015). https://doi.org/10.1016/S1002-0721(14)60376-8

[13] Speghini A., Bettinelli M., Riello P., Bucella S., Benedetti A., J. Mater. Res., 20, 2780 (2005). https://doi.org/10.1557/JMR.2005.0358

[14] Liu X., Wang X.J., Wang Z.K., Phys. Rev. B, 39,10633 (1989).
https://doi.org/10.1103/PhysRevB.39.10633

[15] Turos-Matysiak R, Gryk W, Grinberg M, Lin Y.S., Liu R.S., Radiat. Meas., 42, 755 (2007).
https://doi.org/10.1016/j.radmeas.2007.02.003

[16] Martin I.R., Yanes A.C., Mendez-Ramos J., Torres M.E., Rodriguez V.D., J. Appl. Phys.,
89, 2520 (2001). https://doi.org/10.1063/1.1344216

[17] Carnall W.T., Fields P.R., Rajnak K.,J. Chem. Phys., 49, 4424 (1968).
https://doi.org/10.1063/1.1669893

[18] Goget G.A., Armellini C., Chiappini A., Chiasera A., Ferrari M., Berneschi S., Brenci M.,
Pelli S., Righini G.C., Bregoli M., Maglione A., Pucker G., Speranza G., Proc. of SPIE 7725,
Photonics for solar energy systems III, (2010).

[19] Van Wijngaarden J.T., Scheidelaar S., Vlugt T.J.H., Reid M.F. and Meijerink A.,
Phys .Rev. B, 81, 155112 (2010). https://doi.org/10.1103/PhysRevB.81.155112

[20] Zhao J, Guo C F, Li T., RSC Adv, 5,28299 (2015). https://doi.org/10.1039/C5RA02728D

[21] Zhao J., Guo C.F., Li T., ECS J. Solid State Sci. Technol, 5, R3055 (2016).
https://doi.org/10.1149/2.0331602jss

[22] Li Y., Wang J., Zhou W., Zhang G., Chen Y. and Su Q., Appl. Phys. Express, 6, 082301
(2013). https://doi.org/10.7567/APEX.6.082301

[23] Kolk E.V.D., Kate O.M.T., Wiegman J.W., Biner D. and Kramer K.W., Optical Materials,
33,1024 (2011). https://doi.org/10.1016/j.optmat.2010.08.010

[24] Liao J., Lin Y., Chen Y., Luo Z., Ma E., Gong X., Tan Q. and Huang Y., J. Opt. Soc. Am.
B, 23, 2572 (2006). https://doi.org/10.1364/JOSAB.23.002572

[25] Paulose P.I., Jose G. , Thomas V., Unnikrishnan N.V. and WarrierM.K.R., J. Phys. Chem.
Solids, 64, 841 (2003). https://doi.org/10.1016/S0022-3697(02)00416-X

[26] Zhang Q.Y.,Yang C.H., Jiang Z.H. and Ji X.H., Appl. Phys. Lett, 90, 061914 (2007) .

[27] Li-Ming and Xi-Ping Jing, ECS J. Solid Sci. Technol, 1, R22 (2012).

[28] Lin H., Zhou S.M., Teng H., Li Y.K., Li W.J., Hou X.R. and Jia T.T., Appl. Phys, 107,
043107 (2010). https://doi.org/10.1063/1.3298907

[29] Zhao J, Guo C.F., Ting Li, Songa D. and Su X.Y., Phys. Chem. Chem. Phys, 17, 26330
(2015). https://doi.org/10.1039/C5CP04115E

[30] Parthasaradhi Reddy C., Naresh V., Chandra Babu B. and Buddhudu S., Adv. Mater. Phys.
Chem, 4, 165 (2014). https://doi.org/10.4236/ampc.2014.49019

Keyword Index

About the Editor

Sooraj H Nandyala, MSc., MTech., PhD., MRSC.
External Collaborator, University of Birmingham, UK.
Member of the Royal Society of Chemistry (MRSC), UK.
Healthcare Professional Member, Royal Osteoporosis Society, UK.
Editor(s) in Chief – Journal of Biomimetics, Biomaterials and Biomedical Engineering.
Emails: nandyala.sooraj@gmail.com;
sooraj.nandyala@gmail.com; nandyalash@gmail.com;
s.h.nandyala@bham.ac.uk
Websites: http://nova.sbrpc.co.uk/
Mobile: +44 -7443 080 899

Dr Sooraj Nandyala has a PhD in Physics under the guidance of Late Professor Srinivasa Buddhudu from SVU, Tirupati, India. Afterwards, he moved to Portugal for a Postdoctoral fellowship at the Institute of Biomedical Engineering, University of Porto, Portugal. Later, he joined the Institute of Science and Technology, in the same University as an Auxiliary Professor. Then, independently as a Principal Investigator (2010-2013), he worked on modified Calcium Phosphate materials for biomedical applications. He worked with different reputable international laboratories and participating in the European Project POLARIS funded by FP7 at the Biomaterials, Biodegradable and Biomimetic Materials research group in the headquarters of the European Institute of Excellence on Tissue Engineering and Regenerative Medicine, Portugal.

To date, he has published in approximately 70 journals peer-reviewed papers plus # 10 book chapters with more than 1893 citations with a h-index 29. He authored a total of # 7 books plus # 1 e-book, and # 2 books in the editing stage. So far, he supervised # 23 biomedical engineering students and one postdoc. His research has led to two keynote lecturers in Japan and Switzerland; # 8 Invited talks; #13 Oral and # 18 poster presentations at the international conferences (# 42 attended). As a teamwork, he received the best paper award - Prémio Professor Carlos Lima from the Portuguese Society for Orthopaedics and Traumatology.

Dr Sooraj has for over 5 years and also more recently has focussed his research on innovative materials that have the potential to impact both industry and medical applications. Dr Sooraj has participated in several prestigious Marie Skłodowska-Curie

EU projects in various capacities, research scientist, project manager and as a Fellow. In 2019, he was appointed by the EU Commission as an expert evaluator for the 'Marie Curie Individual Fellowship' grant applications. Currently, he is a member of the 'Research Project Evaluation' committee in the Kazakhstani Scientific Community, Republic of Kazakhstan. In the European Cost Action (E-Cost), he worked (2012-16) as a group Leader (WG3) and Core Group Member: Materials (soft, bio and nano) and Technologies for Optofluidic Devices. He has been serving as an editor in chief of the Journal of Biomimetics, Biomaterials, and Biomedical Engineering, TTP Publisher, Switzerland. Since 2021, Dr Sooraj acting as an external research collaborator in the University of Birmingham, UK.

Over the years Dr Sooraj has gained international visibility and established valuable connections with Scientists in NASA, USA, AO Research Institute, Davos, Switzerland, Horiba, UK, and Queen Mary University of London, UK. He was involved in several projects funded by the Portuguese Ministry of Science and Technology (FCT) and two bilateral joint collaborative projects between India and Portugal (2013-2015). He has also successfully completed two industrial consultancy projects with a Portuguese company – Biosckin Molecular Cell Therapies, SA, Portugal.

In 2019 as the entrepreneurial lead, Dr Sooraj successfully secured funding from Innovate UK – ICURe for the commercialisation of university research. This experience has enhanced his vision, dream, and ambition to establish a new start-up, a spin-off company related to the Bone Graft Technology (under development stage) in the UK. His main mission is to focus on biomedical antimicrobial materials to improve health and reduce the cost of antimicrobial therapies. He's standing in the field of biomaterials as an early career researcher was evidenced in April 2021 when he was admitted as a Member in the Royal Society of Chemistry (RSC) and Healthcare Professional Member, Royal Osteoporosis Society (ROS), UK.

www.ingramcontent.com/pod-product-compliance
Lightning Source LLC
Chambersburg PA
CBHW071319210326
41597CB00015B/1275